Electronic Circuits and Applications

BERNARD GROB

Technical Career Institutes, Inc.

GREGG DIVISION

McGRAW-HILL BOOK COMPANY

New York   Atlanta   Dallas   St. Louis   San Francisco
Auckland   Bogotá   Guatemala   Hamburg   Johannesburg   Lisbon
London   Madrid   Mexico   Montreal   New Delhi   Panama   Paris
San Juan   São Paulo   Singapore   Sydney   Tokyo   Toronto

Dedicated to
Francine Beth
and
Lyle Samuel

Sponsoring Editor: Gordon Rockmaker
Editing Supervisor: Patricia H. Nolan
Design Supervisor: Nancy Axelrod
Production Supervisor: Priscilla Taguer
Cover Designer: Patricia Lowy

Technical Studio: Fine Line, Inc.

Library of Congress Cataloging in Publication Data

Grob, Bernard.
    Electronic circuits and applications.

    Bibliography: p. 433
    Includes index.
    1. Electronic circuits. I. Title.
TK7867.G69   621.3815/3   81-14298
ISBN 0-07-024931-8   AACR2

1 2 3 4 5 6 7 8 9 0   DODO   8 9 8 7 6 5 4 3 2 1

ISBN 0-07-024931-8

Contents

Preface

This book is designed for courses covering basic electronic circuits and their applications. A background in algebra and a working knowledge of solid-state diodes and transistors are prerequisites for studying this text. However, where necessary to the development of new material, semiconductor theory is reviewed briefly. While not essential to the use of this text, the student's knowledge of basic vacuum tube theory will be useful in the few vacuum tube applications discussed. Although modern electronics is dominated by solid-state devices, it is unrealistic to think that technicians will not encounter vacuum tubes sometime in their work experiences. Vacuum tube applications are likely to remain for many years.

SEMICONDUCTOR DEVICES The discrete circuits discussed in this book include PNP and NPN bipolar junction transistors, field-effect transistors (FETs), two-terminal diodes including varactor, zener, and Schottky, and thyristors such as silicon controlled rectifiers, diacs, triacs, and unijunction transistors (UJTs). Integrated circuits are discussed both as building blocks for digital circuits and as single packages containing complete stages of radio and audio systems. Logic gate ICs, op amps, and other basic units are developed both in block diagram form and as a combination of discrete devices. Many circuit applications are given in terms of both the IC package and the combination of discrete devices.

COMMUNICATIONS APPLICATIONS Communications receivers and transmitters constitute the major subject area developed in this textbook. Formerly this would have meant a rather lengthy and detailed discussion of AM radio. But modern communications covers much more than that and this textbook reflects that growth. There are chapters on such traditional but essential topics as the superheterodyne AM and FM receivers, automatic volume control, and audio systems. There are also discussions of microwave systems (including applications to satellite communications), transmitters, modulation, and antennas. Modulation includes such areas as amplitude modulation, frequency modulation, pulse modulation, single side-band transmission, and multiplexing.

As noted previously, the emphasis in this book is on solid-state electronics—both discrete devices and integrated circuits. Although the preponderance of communications systems rely on analog electronics, the rapid growth of digital electronics in these systems requires that the subject be covered in some depth.

PRACTICAL APPLICATIONS There is no question that theory is an important part of any electronics program. But theory without practical applications does not serve the needs of the electronics technician or servicer. For that reason, all of the circuits discussed in this book are related to actual equipment and practical functions. Mathematically derived theory in most cases is best left to the engineer and designer. The only mathematics required to use this book is algebra though the language and techniques of trigonometry are developed in those few instances where trigonometry is required. The numerical examples presented are those technicians and servicers will likely have to solve in the course of their practical experience. These usually involve finding values of current, voltage, resistance, power, decibels, and the like.

TROUBLESHOOTING While the primary aim of this textbook is to explain how and why circuits work and how and why they function in combination to form such practical devices as radio and television receivers—there is another purpose. That second purpose is to show why the circuit or the combination of circuits are not working. What went wrong? How can we tell exactly where the trouble originated? These questions are answered by techniques that generally fall under the category of troubleshooting. We might say that all the theory and practical information gathered about circuits form the data base from which we derive our troubleshooting information. Rather than discuss circuit problems in the abstract, troubleshooting techniques and information are included together with the circuit descriptions themselves. Thus the student will learn why the circuit works as it does and what might be wrong when the circuit does not work as it should. Troubleshooting techniques are so closely related to testing and measuring that special attention is given to methods of testing diodes, transistors, and thyristors. Alignment procedures, although not strictly troubleshooting, are nevertheless so allied to problems in AM and FM receivers that they are covered in some depth. Signal injection is included since it is an important technique for localizing a defective stage in a receiver or amplifier. If one were to name the single most versatile troubleshooting instrument, the oscilloscope would probably be named first. For that reason an entire chapter has been devoted to the instrument, covering its circuits, capabilities, and precise procedures for its use. While this chapter is placed at the end of the book, it can be, and should be used throughout the course whenever oscilloscope techniques are discussed.

LEARNING AIDS The strength of a textbook lies not only on the amount of information it contains but on how it gets the information to the student and how it helps the student retain the information he or she has learned. The student aids included in this textbook are designed to do a number of things, as described below.

Introduction: Each chapter begins with a brief paragraph describing the material covered in the chapter, why the material is important, and how it relates to the overall objectives of the book. The paragraph concludes with a list of the main topics to be covered in the chapter.

Test Point Questions: After each section of text, and before a new concept is introduced, the student is given a few specific questions directly related to the section just covered. These questions are designed to produce a complete correct response from the student. All answers are given at the end of the chapter. This serves as an immediate reinforcement of the text and a confidence-builder for the material that is to follow. The questions are not framed in such a way as to trap the student or to test his or her cumulative knowledge to that point. If students find the questions easy and the answers obvious it means they are doing their work and can move on to the next section. This method of short progressive units, broken by brief test point questions, has proven to be an effective learning tool in my text, *Basic Electronics.*

Summary: At the end of each chapter the key points of the chapter are summarized in concise statements. The purpose of the summary is two-fold. First, it serves as a quick review of the material covered in the chapter, free of the details and mathematics (if any). One might say it is the framework, or skeleton of the chapter. The second purpose of the summary is to serve as a detailed outline, or preview, of the material in the chapter.

Some students find it effective to read the summary to get an overview of the chapter before delving into the heart of the material. Of course, the summary section in no way can substitute for the detailed discussions in the chapter itself.

Self-Examination Questions: Following the Summary section is a set of short answer questions, the answers for which are given in the back of the book. These questions are intended for the student's self-appraisal of his or her understanding of the material in the entire chapter as well as its relationship to material in the preceding chapters.

Essay Questions: These are questions that require fuller, more detailed answers such as comparisons, descriptions, definitions, circuit diagrams, and the like. Answers or suggestions for answers can be found within the text itself. Thus, the student must research or reread the material in the text to find the responses called for in the essay questions.

Problems: The problems in this section refer to questions requiring numerical calculations based on circuits discussed in the chapter. Answers to the odd-numbered problems are given at the back of the book.

Special Questions: The purpose of these questions is to draw the student's thinking outside the immediate material in the text and to relate their studies with experiences outside the immediate course they are taking. Thus students may be required to research certain materials in the library, to write manufacturers for data, or to describe experiences with electronic or electrical equipment they own or use. The questions are given at different levels of difficulty so that the teacher may assign the materials according to the students' needs, background, and interest.

ACKNOWLEDGMENTS The photographs of components and equipment have been provided by many manufacturers. Grateful acknowledgment is given to each source in the legend that accompanies the material. I also want to thank Charles A. Schuler for his assistance with the materials on RF circuits and Roger L. Tokheim for his assistance with the materials on digital electronics. Finally, thanks to my wife, Ruth, for her assistance in the preparation of the manuscript.

Bernard Grob

Chapter 1
Amplifier Circuits

Amplification means increasing the amplitude of a desired ac signal voltage or current such as an audio signal for sound or a video signal for a television picture. The amplifier allows a small input signal to control a larger amount of power in the output circuit. The output signal is a copy of the input signal, but it has more amplitude.

Amplifiers are necessary in most applications because the desired signal is usually too weak to be directly useful. As an example, audio output from a microphone may be as little as one millivolt, whereas the loudspeaker needs at least a few volts of audio signal. With an amplifier, however, a faint whisper can be made to fill a large room with sound.

Transistors are used as the amplifiers in most circuits. In addition, resistors, inductors, and capacitors are required to form complete amplifier circuits. They provide paths for the input and output signals. As an example, Fig. 1-1 illustrates a complete audio amplifier on a printed-circuit (PC) board. A similar amplifier may also be contained in a single integrated circuit "chip." More details of the different types of amplifier circuits are explained in the following topics:

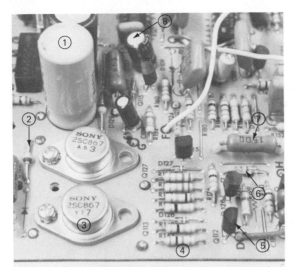

①	FILTER CAPACITOR
②	DIODE
③	POWER TRANSISTOR
④	CARBON RESISTOR
⑤	SMALL-SIGNAL TRANSISTOR
⑥	DIODE
⑦	WIRE-WOUND RESISTOR
⑧	COUPLING CAPACITOR

(a)

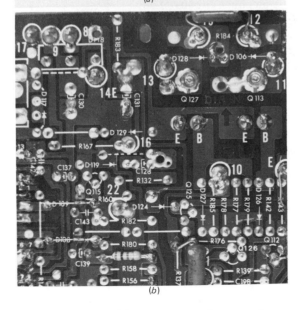

(b)

Fig. 1-1. Audio amplifier transistors and associated components on printed-circuit (PC) board. Size is 3 by 5 in [7.6 by 12.7 cm]. (a) Component side. (b) Printed wiring side.

1-1
BASIC AMPLIFIER REQUIREMENTS

A transistor or vacuum tube used as an amplifier needs dc voltages applied to its electrodes in order to conduct any current. The amplification comes from having a small ac input signal control much larger dc values in the output circuit. Furthermore, a load impedance is required in the output circuit to develop the output signal. The reason is that the current inside the transistor or tube must be made to flow in an external component. Otherwise, there is no way to take out the amplified signal. Those two basic requirements for amplifiers are illustrated in Fig. 1-2.

Types of Amplifier Load Impedance   In Fig. 1-2, Z_L can be either a resistor or a coil. A capacitor cannot be used because it would block the dc supply voltage needed for the amplifier.

A coil as the Z_L may be either a single inductor or the primary of a transformer. At audio frequencies, iron-core coils are used. At radio frequencies, the inductors can be air-core or have a ferrite core. In a tuned RF amplifier, Z_L is the high impedance of a parallel-resonant LC circuit. In general, a higher Z_L allows more amplified signal voltage in the output.

Typical DC Supply Voltages   For vacuum tubes, 90 to 280 volts (V) is generally needed for the plate and screen-grid electrodes. For transistors, the supply voltage is usually 4.5, 9, or 28 V

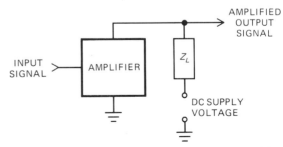

Fig. 1-2. The two basic requirements for an amplifier circuit are a dc supply voltage to make the amplifier conduct and a load impedance Z_L across which the output signal is developed.

for the collector. Higher values up to 100 V can be used with power transistors. The higher the dc supply voltage the greater the power output possible for the transistor amplifier. Integrated-circuit packages generally use 5, 12, or 20 V.

Test Point Questions 1-1
(Answers on Page 24)

Answer True or False.

a. A resistor can be used for the load in the collector circuit of a transistor amplifier.
b. The required dc supply voltage for a transistor amplifier can be provided by a 9-V battery.

1-2
TRANSISTOR CIRCUIT CONFIGURATIONS

In NPN and PNP junction transistors, the three electrodes are emitter, base, and collector. The emitter supplies charges, either electrons or holes, for current to the collector. The base controls the amount of collector current.

In the field-effect transistor (FET), the corresponding electrodes are source, gate, and drain. The source supplies charges for current to the

drain, which is controlled by the gate. The current is electron flow with an N-channel.

Either type of transistor has just three terminals. One is for input signal, and the other is for amplified output signal. The third terminal does not have any signal; it is the return connection common to the input and output circuits. Six different combinations for using the transistor in an amplifier circuit are shown in Fig. 1-3.

Common-Emitter (CE) Circuit In the circuit of Fig. 1-3a, the input signal is applied to the base and the amplified output is taken from the collector. The emitter is the common electrode. This circuit is the one generally used for transistors because the CE amplifier has the best combination of current gain and voltage gain.

The corresponding FET circuit is the common-source (CS) amplifier. The input signal is applied to the gate, and the amplified output is taken from the drain electrode (Fig. 1-3d).

Gain specifies how much the signal is amplified. As an example, when a base signal current of 2 milliamperes (mA) is increased to 60 mA of signal in the collector circuit, the gain is 60 mA/2 mA, or 30. In addition, an amplifier with voltage gain of 100 can provide a 500-millivolt (mV) output signal with 5-mV input signal.

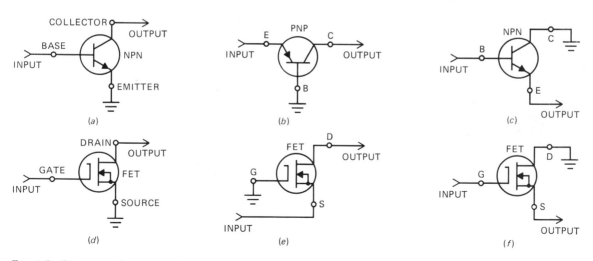

Fig. 1-3. Circuit configurations for junction transistors and field-effect transistors (FETs). (*a*) Common-emitter. (*b*) Common-base. (*c*) Common-collector or emitter-follower. (*d*) Common-source. (*e*) Common gate. (*f*) Source-follower.

Common-Base (CB) Circuit In the circuit of Fig. 1-3b, the input signal is applied to the emitter instead of to the base. The amplified output signal is still taken from the collector. The base is the common electrode.

The relatively high emitter current, compared with base current, results in a very low value of input impedance. Its input circuit, therefore, loads down the collector output circuit of the previous stage. For that reason, the CB circuit is seldom used for amplifiers.

The corresponding FET circuit is the common-gate (CG) amplifier. Input signal is applied to the source, and the amplified output is taken from the drain. The gate electrode is grounded (Fig. 1-3e).

Emitter-Follower Circuit The common-collector (CC) circuit of Fig. 1-3c has an input signal to the base, as in the CE amplifier, but the output is taken from the emitter. The collector cannot supply a signal because it is grounded. This circuit is generally called an *emitter follower*. The name means that the output signal voltage at the emitter follows the input signal at the base with the same phase but a little less amplitude. Although the voltage gain is less than 1, the emitter follower is often used for impedance matching. The stage has high input impedance at the base as the load for a preceding circuit and low output impedance at the emitter as a signal source for the next circuit.

The corresponding FET circuit is the *source follower*. An input signal is applied to the gate, and an output is taken from the source. The drain electrode is grounded (Fig. 1-3f).

Comparison of CE, CB, and CC Circuits See Table 1-1. Note that the circuit is named for the electrode that is common. That terminal may or may not be connected to chassis ground. Furthermore, the common electrode is the one that does not have the signal input or output.

Although no load impedance is shown for the amplifier circuits in Fig. 1-3, it should be noted that Z_L is in the circuit of the electrode that has the amplified output signal. In the CE and CB amplifiers, Z_L is in the collector output circuit; in the emitter follower, it is in the emitter circuit.

Test Point Questions 1-2
(Answers on Page 24)

a. Which of the amplifier circuits using junction transistors has the best gain?
b. Which FET circuit corresponds to the CE amplifier?
c. Which circuit has its output signal from the emitter?

Table 1-1
Circuits for Junction Transistors

Name	Signal Input	Signal Output	Applications
Common-emitter (CE)	Base	Collector	This is the amplifier circuit generally used because of its high current and voltage gain.
Common-base (CB)	Emitter	Collector	Little used because of low input impedance
Common-collector (CC)	Base	Emitter	Emitter-follower circuit. Often used for its high input and low output impedances.

1-3
RESISTANCE-LOADED AMPLIFIER

The circuit shown in Fig. 1-4 is often used for audio frequency (AF) amplifiers because R_L allows the same amount of gain for a wide range of audio frequencies. Also, it is economical compared with an audio transformer. The circuit of Fig. 1-4a is generally called an *RC*-coupled amplifier, but the capacitive coupling can be used with other types of output load impedance.

In Fig. 1-4a, R_L is the load in series with V^+ for the amplifier. Since R_L has the same resistance at all audio frequencies, the amplifier has uniform gain. The flat frequency response is shown by the response curve in Fig. 1-4b.

A disadvantage of using R_L is the relatively high dc voltage drop. As an example, with collector output current in a transistor amplifier the voltage drop is $I_C \times R_L$. As a result, the voltage at the collector is much less than the dc supply voltage. In Fig. 1-4a, note that V^+ is 24 V but the collector voltage would be only 10 V. The other 14 V is across R_L.

Chassis-Ground Returns The input and output circuits have a common ground return but they are shown separately in Fig. 1-4a, in order to emphasize these connections. Keep in mind the following important points:

1. V^+ is for the collector but V^- must return to the emitter. Therefore, both V^- and the emitter are connected to the common chassis return line. Without V_{CE} the transistor cannot conduct any current.
2. The amplified signal coupled by C_1 is an ac voltage from collector to chassis ground.
3. The output signal for the next stage is across R_1 from the high side to chassis ground.
4. The input circuit of the next stage must have one side connected to chassis ground in order to receive the input signal.

Low-Frequency Response At low frequencies, the output is reduced because of the reactance of

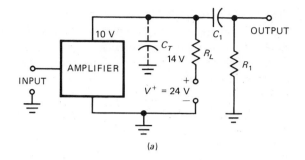

(a)

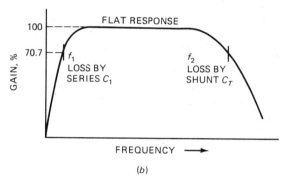

(b)

Fig. 1-4. RC coupling for amplifier output. (a) Audio amplifier circuit. R_L is the load with R_1C_1 coupling. (b) Response curve showing amplifier gain at different frequencies.

C_1. Remember that X_C increases as the frequency decreases. More of the amplifier output voltage is across C_1, therefore, and less is across R_1 in the next circuit. As a series circuit, C_1 and R_1 divide the ac signal voltage.

In Fig. 1-4b, note that the gain is down to 70.7 percent response for the frequency f_1, at the left side of the curve. At that frequency, the reactance of C_1 is exactly equal to the series resistance of R_1. Larger values for both C_1 and R_1 improve the low-frequency response.

High-Frequency Response At high frequencies, the amplifier gain decreases because of the shunting effect of the stray, distributed capacitances. The total, parallel capacitance indicated as C_T is typically 10 to 40 picofarads (pF). Even that small C can bypass high frequencies around a high value of R_L.

Actually, the amplifier load is an impedance Z_L consisting of R_L in parallel with C_T. As the frequency increases, the X_{C_T} decreases. The parallel combination then has less Z_L, and the result is less gain for the amplifier.

In Fig. 1-4b note that the gain is down to 70.7 percent response at the frequency f_2, at the right side of the curve. At that frequency, $C_T = R_L$.

Smaller C_T values improve the high-frequency response. Furthermore, smaller R_L values also extend the high-frequency gain for flat response, although the overall gain is reduced.

Test Point Questions 1-3
(Answers on Page 24)

a. In a resistance-loaded amplifier, is low-frequency response down because of C_C or C_T?
b. In a resistance-loaded amplifier, is high-frequency response down because of C_C or C_T?

1-4
RC COUPLING

The R_1C_1 coupling circuit of Fig. 1-4 is shown by itself in Fig. 1-5 so we can analyze how it blocks the dc component but passes the ac signal variations. Here the V_{in} is the output of the amplifier but is also the input to the coupling circuit. In this example, the values for V_{in} are:

10 V	Average dc level
±4 V	AC variations around 10-V axis
14 V	Maximum instantaneous value
6 V	Minimum instantaneous value

Remember that, in a coupling circuit, X_{C_1} must be very small compared with R_1. That requirement is the same as a long time constant for R_1C_1.

Since the path for both charge and discharge in the *RC* coupling circuit is the same, C_1 charges to the average dc level of 10 V. That 10-V axis for V_{in} becomes the zero axis for the ac output across R_1.

When V_{in} rises above 10 V, C_1 can take on more charge. The charging current produces positive voltage across R_1. All the changes of V_{in} between 10 and 14 V provide the positive half-cycle for V_{R_1} between 0 and 4 V.

When V_{in} drops below 10 V, C_1 discharges. The discharge current produces negative voltage across R_1. All the changes of V_{in} between 10 and 6 V provide the negative half-cycle for V_{R_1} between 0 and -4 V.

The final result is that the 10-V level is blocked as the dc voltage across C_1. It is considered blocked because the voltage across C_1 is connected to only one terminal of the next circuit. However, V_{R_1} is connected between that terminal and chassis ground. Therefore, the ac voltage across R_1 has the two connections needed to provide an input signal to the next circuit.

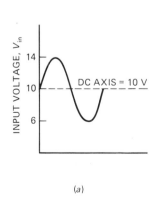

(a)

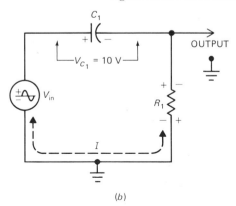

(b)

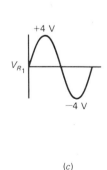

(c)

Fig. 1-5. How the *RC* coupling circuit blocks dc but passes ac signal. (*a*) Fluctuating dc input voltage. (*b*) Circuit with an average dc level of 10 V blocked as voltage across C_1. (*c*) AC component of ±4 V passed as voltage across R_1.

Test Point Questions 1-4
(Answers on Page 24)

Refer to Fig. 1-5.

a. What is the dc voltage across C_1?
b. What is the peak-to-peak (p-p) ac voltage across R_1?

1-5
IMPEDANCE COUPLING

The circuit shown in Fig. 1-6 uses an inductance, instead of R_L, as a choke for the load impedance of the amplifier. However, the R_1C_1 coupling circuit is still needed to block V^+ from the next circuit. An iron-core audio choke can be used for

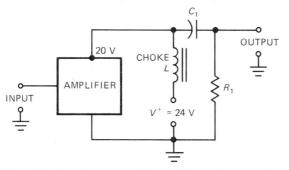

Fig. 1-6. Impedance-coupled amplifier with an audio choke L as a load impedance instead of R_L. Iron core for L shows it is an audio choke. An RF choke can be used for an RF amplifier. The R_1C_1 circuit couples the ac signal but blocks its dc component.

an AF amplifier, and an air-core choke can be used for an RF amplifier.

The advantage of the choke is low dc resistance with high ac impedance. Low resistance means a small IR drop, which allows most of V^+ to be available at the amplifier. The high ac impedance for the signal allows high gain.

However, Z_L varies with frequency. Compared with R_L, then, the impedance-coupled amplifier does not have as uniform a frequency response.

Test Point Questions 1-5
(Answers on Page 24)

Refer to Fig. 1-6.

a. Is the amplifier load R_1 or the choke L?
b. Is V^+ voltage blocked by L or by C_1?

1-6
SINGLE-TUNED AMPLIFIER

The circuit shown in Fig. 1-7a uses a parallel-resonant LC circuit to provide the required high impedance for an ac signal. This application is for an RF amplifier tuned to a specific frequency.

The dc resistance of the RF coil L is negligible. As a result, practically all the V^+ voltage is available for the amplifier. The R_1C_1 coupling circuit is used to block V^+ while coupling the RF signal. This circuit can also be considered an impedance-

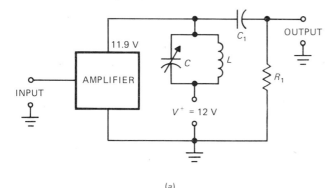

(a)

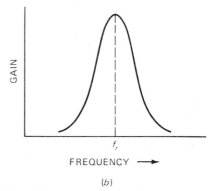

(b)

Fig. 1-7. Single-tuned amplifier. (a) Amplifier with LC resonant circuit. The R_1C_1 circuit couples the ac signal but blocks its dc component. (b) The frequency response of the amplifier is the same as the resonance curve of the tuned circuit.

coupled amplifier but its main feature is a tuned circuit for the output load.

Resonant Response   The response curve in Fig. 1-7b is the same as the resonance curve of the tuned circuit. The LC circuit is parallel-resonant, because the source is outside the tuned circuit. For parallel resonance, Z is maximum at f_r and decreases to very low values at frequencies far off resonance.

The voltage gain of the amplifier is proportional to the amplifier Z_L. Therefore, the response curve of the amplifier corresponds to the resonance curve. The tuned amplifier provides gain only for frequencies at and near the resonant frequency.

Test Point Questions 1-6
(Answers on Page 24)

Refer to Fig. 1-7.

a. What is the dc IR drop across L?
b. Is the coupling capacitor C or C_1?

1-7
TRANSFORMER COUPLING

The audio output stage shown in Fig. 1-8 has transformer coupling to drive a loudspeaker. L_P provides the primary load impedance for the amplifier. Its dc resistance is relatively low, which allows most of the V^+ supply to be used at the amplifier. The ac component of the signal current in L_P induces the desired signal voltage in L_S by transformer action. Since L_S is an isolated winding, the dc component of the primary voltage and current is blocked. No coupling capacitor is needed. Furthermore, the ac voltage in the secondary is independent of the chassis ground connections. One side of L_S may or may not be grounded.

T_1 in Fig. 1-8 is an audio output transformer, which means that it is in the last stage to drive a loudspeaker. Audio output transformers generally provide voltage step-down to match low values of 4 to 16 ohms (Ω) for the speaker impedance to the

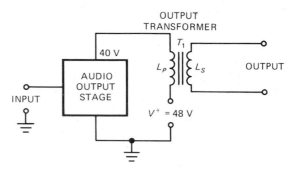

Fig. 1-8. Transformer-coupled audio amplifier. Iron core for T_1 shows it is an audio transformer. An RF transformer can be used for an RF amplifier.

higher impedance usually needed for the amplifier. The details of impedance matching with audio output transformers are explained in Sec. 4-8 in Chap. 4.

A transformer that couples signals between two amplifiers is called an *interstage transformer*. As an example, an interstage audio transformer can be used to couple signals from the collector of a transistor in the first stage of an amplifier to the base of a transistor in the output stage.

Test Point Questions 1-7
(Answers on Page 24)

Answer True or False.

a. No coupling capacitor is needed with transformer coupling.
b. In Fig. 1-8, the dc voltage drop across L_P is equal to 8 V.

1-8
DOUBLE-TUNED AMPLIFIER

The circuit shown in Fig. 1-9a is a transformer-coupled amplifier, but both the primary and secondary are tuned. L_P resonates with the capacitance C_1, and L_S resonates with C_2. The two coils have tuning slugs to align them at the same resonant frequency. A typical application for this circuit is an IF amplifier with T_1 tuned to the intermediate frequency of the receiver. The IF

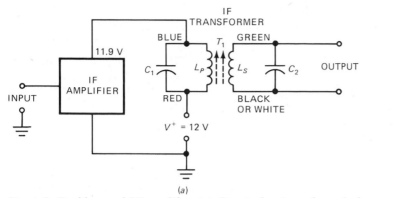
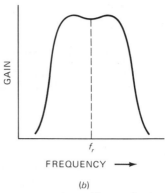

(a)   (b)

Fig. 1-9. Double-tuned RF amplifier. (a) Circuit showing color code for transformer connections. T_1 may be in a shielded metal can. (b) Frequency response of double-tuned transformer with critical coupling between L_P and L_S.

value is 455 kilohertz (kHz) in AM radios or 10.7 megahertz (MHz) in FM radios.

Resonant Response   The primary tuned circuit with L_P and C_1 is the load impedance for the amplifier. However, the secondary affects the primary by mutual coupling. The result is more bandwidth and sharper sides for better selectivity than with a single-tuned amplifier. A typical response curve for the double-tuned amplifier is shown in Fig. 1-9b.

Transformer Color Coding   The colors of the transformer leads are marked in Fig. 1-9a because they can be helpful in circuit tracing. The following colors are standard for interstage transformers:

Red—Primary return to V^+
Blue—Primary high side to collector, drain, or plate of amplifier
Green—Secondary high side to base, gate, or control grid of next stage
Black or white—Secondary return lead

The same color coding is used for AF and RF transformers. You can check continuity with an ohmmeter between red and blue for the primary winding and from green to black or white for the secondary. A meter reading of infinite ohms means that the winding is open.

Test Point Questions 1-8
(Answers on Page 24)

Refer to Fig. 1-9.

a. When T_1 is tuned to 455 kHz, what is the gain at 368 kHz?

b. What color is the terminal of T_1 that returns to V^+?

1-9
METHODS OF BIAS

In amplifiers the bias is a dc voltage or current that sets the operating point for amplifying the ac signal. The objective is to use the best part of the amplifier characteristics.

The bias is in the input circuit. For junction transistors in the CE circuit, the forward V_{BE} between base and emitter is the bias voltage. The resulting I_B is the base bias current. For an FET, the bias is between gate and source. For tubes the bias is between control grid and cathode.

Bias for NPN and PNP Junction Transistors   Forward polarity is needed for V_{BE} to turn the transistor on. The NPN type needs positive V_{BE} at the base. For a PNP transistor, the base bias is negative. Note that V_{BE} is base voltage with respect to emitter.

Table 1-2
Base-Emitter Voltages for Junction Transistors

Type	Cutoff, V	Saturation, V	Active Region, V	Middle Bias, V
Silicon	0.5 or less	0.7 or more	0.5–0.7	0.6
Germanium	0.1 or less	0.3 or more	0.1–0.3	0.2

The amount of forward V_{BE} needed is approximately 0.6 V for a silicon transistor and 0.2 V for a germanium transistor. Those values apply for any size transistor and for NPN or PNP types. The reason is that the cutoff characteristic depends only on whether the intrinsic material is silicon or germanium.

Specific values are shown by the forward characteristics of a silicon junction transistor in Fig. 1-10. With less than 0.5 V for V_{BE}, there is no forward current. A transistor is normally off; forward voltage must be applied to it to produce current. From 0.5 to 0.7 V, approximately, more V_{BE} gradually produces more current. Above 0.7 V, the current is at saturation because its increases are not proportional to the increases in V_{BE}.

The active region of 0.5 to 0.7 V shows the values of V_{BE} used for silicon transistor amplifiers. For those voltages, the base voltage controls the base current and the resultant collector output current. The values of input voltage are summarized in Table 1-2 for both silicon and germanium transistors. Note that in both cases the active region is only $\pm$ 0.1 V, or $\pm$ 100 millivolts (mV).

A typical bias for V_{BE}, therefore, is 0.6 V, for forward voltage in the middle of the active region of a silicon NPN transistor. Furthermore, with an ac signal input less than 100 mV, the V_{BE} always remains positive. The bias maintains the forward polarity for V_{BE}, although the ac signal has negative and positive half-cycles.

A specific example is shown in Fig. 1-11. The combination of 0.6 V of bias and $\pm$ 50 mV of peak ac signal results in a fluctuating dc voltage at the input. Now the ac signal varies above and below the bias axis instead of around zero. V_{BE} is always positive. Even the $-$ 50-mV peak of signal reduces V_{BE} only to 550 mV. All values of V_{BE} are in the active region between cutoff and saturation.

Types of Bias Circuits The three basic bias circuits are fixed bias, self-bias, and signal bias. Fixed bias is taken from a battery or power supply. The amount of bias voltage does not depend on the

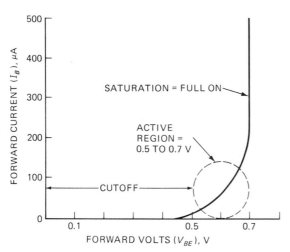

Fig. 1-10. Forward characteristics of base-emitter junction. Voltages shown are for a silicon transistor.

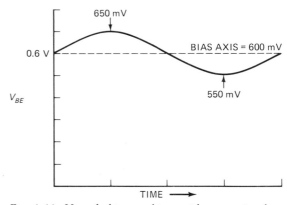

Fig. 1-11. How dc bias combines with an ac signal to offset all values of V_{BE} in the positive direction. Bias here is 0.6 V for a silicon NPN transistor with a signal input of $\pm$ 50 mV.

amount of input signal or output current in the amplifier.

In the self-bias circuit, the amplifier produces its own dc voltage from an *IR* drop across a resistor in the return circuit of the common electrode. The resistor has the output current flowing through it. For junction transistors, a resistor in the emitter circuit produces emitter bias. This bias voltage depends on the amount of emitter current, which is almost the same as I_C.

As for signal bias, the ac signal produces its own bias by rectification in the input circuit of the amplifier. For tubes, this method is called *grid-leak bias*. The amount of bias depends on the amount of input signal.

Bias Polarity   The dc bias for forward V_{BE} is positive at the base of NPN transistors and negative at the base of PNP transistors. Negative bias voltage is needed for tubes at the control grid with respect to the cathode. An N-channel FET also needs negative bias, in this case at the gate electrode with respect to the source.

Test Point Questions 1-9
(Answers on Page 24)

a. Which depends on the amount of I_C, fixed bias or emitter self-bias?
b. Which depends on the amount of ac signal input, cathode bias or grid-leak bias?

1-10
FIXED BIAS

In Fig. 1-12 a small battery or C cell is shown used for bias of -1.5 V at the control grid of the pentode amplifier tube. A pentode has five electrodes, including three grids. G_3 next to the plate is the suppressor grid; G_2 is the screen grid; and G_1 next to the cathode is the control grid. The G_1 voltage controls the amount of current from cathode to plate. The dc values for the control grid are indicated by the subscript *c*, instead of *g*, which is used for ac signals.

The positive side of the C battery is returned

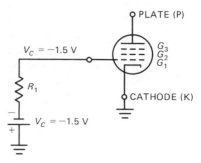

Fig. 1-12. C battery is used to provide fixed bias at the control grid of a vacuum-tube amplifier.

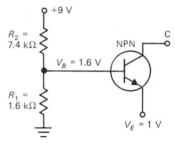

Fig. 1-13. Fixed bias is provided by an R_1R_2 divider from the 9-V supply for positive forward voltage at the base of the NPN transistor.

to the grounded cathode, so that the negative side can be connected to the control grid through R_1. There is no grid current, and there is no *IR* drop across R_1 because the grid is negative. As a result, V_C at the control grid is the same -1.5 V as the battery. This bias between grid and cathode is the dc axis for an ac signal input to the control grid.

Voltage Divider from the Power Supply   In the circuit of Fig. 1-13 the R_1R_2 divider supplies positive voltage to the base of the NPN transistor. The divider is connected across the 9 V of the V$^+$ supply.

The divided voltage across R_1, equal to 1.6 V, is the base voltage V_B to chassis ground. The emitter voltage V_E is 1 V. As a result, the potential difference for V_{BE} is $1.6 - 1 = 0.6$ V. This circuit is typical for providing forward bias for junction transistors.

Calculations for the Voltage Divider   The values of R_1 and R_2 in Fig. 1-13 are calculated for a voltage divider without any branch currents. The total R_T of R_1 in series with R_2 forms a bleeder resistance across the dc supply voltage. Its I is bleeder current. If we make the bleeder current 10 or more times the base current, approximate values for R_1 and R_2 can be calculated. In this method the effect of base current on the voltage division is ignored.

Remember that bleeder current is a steady load on a voltage source. The purpose is to swamp out the effect of any variations in current.

Assume a base current of 100 microamperes (μA). Then the bleeder current is $10 \times 100\ \mu$A $= 1000\ \mu$A, which is 1 mA.

We can find the total R_T of $R_1 + R_2$ needed for the 1-mA bleeder current.

$$R_T = \frac{9\ V}{1\ mA} = 9\ \text{kilohms (k}\Omega)$$

The total 9 V is across R_T, but R_1 must have the divided value 1.6 V. The ratio of the two voltages is

$$\frac{V_1}{V_T} = \frac{1.6\ V}{9\ V} = 0.18$$

Resistance R_1 is the same fraction of R_T, since the series voltage drops are proportional to the resistances. Then

$$R_1 = 0.18 \times 9\ k\Omega = 1.62\ k\Omega$$

We can round this off to 1.6 kΩ.

Furthermore, R_2 must be $9 - 1.6 = 7.4$ kΩ. The reason is that $R_1 + R_2$ must total 9 kΩ.

Also, V_2 is equal to $9 - 1.6\ V = 7.4\ V$. This 7.4 V across R_2 plus 1.6 V across R_1 must equal the 9-V supply voltage.

Series R from the Power Supply   The required base bias of 0.6 V can also be obtained from the supply of 9 V with a voltage-dropping resistance, R_S. It is connected between the base electrode and the 9-V supply. As shown in Fig. 1-14, the value can be calculated for an IR drop of $9 - 0.6 =$

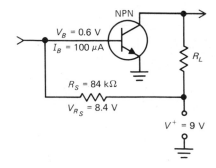

Fig. 1-14. Fixed bias is provided for the base by the series resistance R_S from the V^+ supply.

8.4 V with the required amount of base current. For an I_B of 100 μA,

$$R_S = \frac{8.4\ V}{100\ \mu A} = 0.084\ \text{megohms (M}\Omega)$$

$$R_S = 84{,}000\ \Omega$$

It should be noted, though, that the voltage-divider method provides bias voltage that is more stable because of the bleeder current.

Test Point Questions 1-10
(Answers on Page 24)

a. In Fig. 1-12, what is the value of the C-bias at the control grid?

b. In Fig. 1-13, what is the voltage across R_1?

1-11
HOW SELF-BIAS IS PRODUCED

Self-bias is probably the type of bias used most often, because it is economical and has a stabilizing effect on the dc level of output current. Three circuits are shown in Fig. 1-15 to illustrate how self-bias is applied in transistor and tube circuits. R_E in Fig. 1-15a provides emitter bias for the junction transistor; R_K in Fig. 1-15b develops cathode bias for a tube; R_S in Fig. 1-15c produces source bias for the field-effect transistor.

In all three circuits, the method is to insert a series resistor in the return circuit to the common

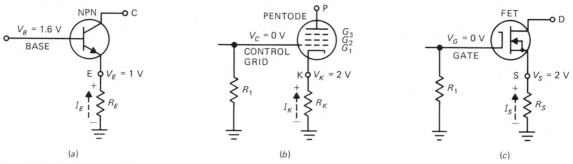

Fig. 1-15. Methods of self-bias. (a) Emitter-bias voltage equal to $I_E R_E$ for a junction transistor. (b) Cathode-bias voltage equal to $I_K R_K$ for a vacuum tube. (c) Source-bias voltage equal to $I_S R_S$ for an FET.

electrode. Then the dc level of the return current produces an IR voltage that serves as the dc bias. The effect is to offset the potential of the common electrode from chassis ground.

It should be noted that self-bias always has a reverse offset. The output current is reduced with more bias. This effect is desirable as a stabilizing bias or safety bias to prevent excessive current. In summary, then, self-bias has the advantage of stabilizing the dc level of output current because more current tends to produce more reverse bias.

Emitter Bias for Junction Transistors For the NPN transistor in Fig. 1-15a, R_E has the emitter current. The dashed arrow for I_E indicates electron flow. The direction is from the negative side of R_E at chassis ground to the emitter. Therefore, the emitter end of R_E is the positive side of V_E.

The amount of voltage shown in Fig. 1-15a is 1 V for $I_E R_E$. Then V_E is 1 V positive with respect to ground. This bias is in the reverse direction because the positive V_E is at the N-emitter. However, forward bias can be applied to the base, as shown in Fig. 1-13. The fixed bias at the base has forward polarity to turn the transistor on, and the emitter bias is used for stabilization. Its reverse offset prevents an excessive increase in collector current when the temperature rises.

Cathode Bias for Tubes For the pentode in Fig. 1-15b, R_K has the cathode current. The dashed arrow for I_K shows electron flow from the ground side of R_K to the positive end at the cathode. The

amount of voltage shown in Fig. 1-15b is 2 V for a typical value of cathode bias. Therefore, V_K is 2 V positive with respect to chassis ground.

Note that the dc voltage V_C at the control grid is zero. It is 0 V because no C battery is used. Also, there is no control-grid current, and there is no IR drop across the grid resistor R_1. Therefore, the potential difference between grid and cathode is $0 - 2 \text{ V} = -2 \text{ V}$. The control grid is at -2 V with respect to the cathode.

Source Bias for the FET The bias method of Fig. 1-15c for an FET is really the same as cathode bias, especially for the N-channel shown. The resistor R_S has the returning source current. I_S is equal to I_D, because there is no gate current. For the N-channel, I_S is electron flow. The $I_S R_S$ drop then makes the source electrode positive with respect to ground. V_S is 2 V in this example.

The V_G is 0 V because there is no gate current to produce an IR drop across R_1. With V_S at 2 V and V_G at 0 V, V_{GS} is $0 - 2 \text{ V} = -2 \text{ V}$. This bias of -2 V is the dc bias level for ac signal applied at the gate with respect to chassis ground.

Calculations for Emitter Bias In Fig. 1-16, the values of R_E and C_E are calculated to show more details for this important method of self-bias. In this example R_E is 160 Ω with I_E of 6.2 mA for the 1-V bias. C_E is 100 μF to bypass the 160-Ω R_E at 100 hertz (Hz) for an audio amplifier.

The emitter current consists of $I_C + I_B$. Assume $I_B = 0.1$ mA and $I_C = 6.1$ mA. The sum is

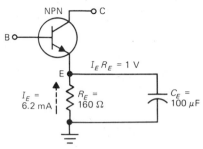

Fig. 1-16. Values of I_E and R_E for 1-V emitter bias. Capacitance C_E is 100 μF for audio frequency bypass.

6.2 mA for I_E flowing through R_E. To calculate R_E for a 1-V drop,

$$R_E = \frac{V_E}{I_E} = \frac{1\ V}{6.2\ mA} = 0.161\ k\Omega$$

$$R_E = 160\ \Omega, \text{ approximately}$$

To bypass this R_E of 160 Ω, the X_C should be $160/10 = 16\ \Omega$. At the frequency of 100 Hz, for an X_C of 16 Ω,

$$C_E = \frac{1}{2\pi f X_C} = \frac{1}{2\pi \times 100 \times 16}$$

$$= 0.0000995 \text{ farads (F)}$$

$$C_E = 100 \text{ microfarads } (\mu F), \text{ approximately}$$

It should be noted that the dc bias of V_E is present with or without the bypass C_E. The 1-V bias is $I_E R_E$ produced by the average dc value of the emitter current.

The only purpose of C_E is to remove signal variations for a steady dc voltage at 1 V. Without the bypass capacitor, V_E has signal variations above and below the 1-V axis. The result is negative feedback or degeneration of the input signal. The effect is less amplifier gain, but the distortion also is reduced.

Test Point Questions 1-11
(Answers on Page 24)

a. In Fig. 1-15a, what is the base-emitter bias V_{BE}?
b. In Fig. 1-15b, what is the grid-cathode bias V_{GK}?
c. In Fig. 1-15c, what is the gate-source bias V_{GS}?
d. In Fig. 1-16, if I_E increases to 6.5 mA, what will V_E be?

1-12
HOW SIGNAL BIAS IS PRODUCED

Signal bias is developed by rectifying the signal at the amplifier input. As a result, the amount of dc bias depends on the amount of ac signal. Without any signal, the bias is zero. The signal is needed to provide its own bias.

With tubes, the input signal is rectified in the grid-cathode circuit as shown in Fig. 1-17. This circuit is for grid-leak bias. The name derives from the fact that a small part of the charge in C_C can

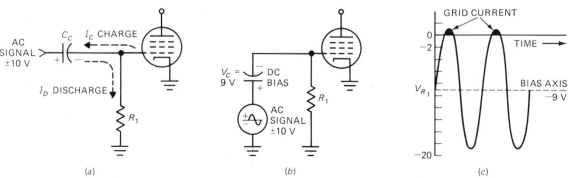

Fig. 1-17. Grid-leak bias produces dc bias voltage from an ac signal input. (a) Grid coupling circuit. (b) Equivalent circuit with dc bias and ac signal. (c) Combined waveform showing that V_R has ac signal of ± 10 V varying about the dc axis of -9 V.

leak off through R_1. With junction transistors, signal bias is produced by rectification in the base-emitter circuit.

Grid-Leak Bias   In Fig. 1-17, the grid-cathode circuit serves as a diode rectifier. C_C is charged to provide dc bias voltage proportional to the ac signal drive.

There are two requirements. First, the signal voltage must drive the control grid positive with respect to cathode. Then grid current can flow to charge C_C. This condition is satisfied automatically, when there is no other bias, because ac signals have positive half-cycles.

The second requirement is that the RC coupling circuit have a long time constant. Then C_C remains charged at a steady dc value. For sine-wave signal, low reactance for C_C compared with R_1 is the same as a long time constant.

How C_C Develops Grid-Leak Bias   The charge and discharge currents for C_C are shown in Fig. 1-17a. When the grid is driven positive by signal voltage, grid current flows. Then I_C charges C_C. Electrons flow from cathode to control grid in the tube and accumulate on the grid side of C_C, repelling electrons from the opposite side. During the time when the grid is not positive, C_C can discharge through R_1. This discharge current I_D through R_1 is the only possible path, because current cannot flow from grid to cathode in the tube.

The main factor here is that C_C charges fast from the grid current but discharges very slowly through the high resistance of R_1. After a few cycles of signal, C_C has a net charge with its grid side negative. The resultant bias V_C is a dc voltage because it has only one polarity. It can be considered a steady bias compared with the signal variations, since V_C changes little during one cycle of the ac input due to the long RC time constant.

The amount of grid-leak bias is typically 90 percent of the peak positive ac signal voltage. In Fig. 1-17, the 10-V signal peak is shown producing bias of -9 V.

The grid-leak bias action does not prevent R_1C_C from acting as a coupling circuit. For the ac signal,

C_C has symmetrical charge and discharge currents through the same resistance of R_1. The result is equal positive and negative half-cycles of signal voltage across R_1 for the grid circuit.

The combined effect of ac signal input and the dc bias is shown in Fig. 1-17b. The two voltages are effectively in series with each other for the combined voltage across R_1 as the grid-cathode voltage. In Fig. 1-17c the waveform shows the ac signal varying V_R around the dc axis of -9 V. The peak positive signal of 10 V drives the grid to $10 - 9 = 1$ V, positive. Grid current then flows, as shown at the tip of the waveform.

How the Bias Adjusts to the Signal Level   The main reason for using signal bias is that it adjusts to the level of the input signal. The effect is illustrated in Fig. 1-18 with a 10-V peak signal and a 5-V peak signal. With the 10-V signal, grid current charges C_C to -9 V for the dc bias. With the ± 5 V signal the positive peak reaches only $-9 + 5 = -4$ V. This is because grid current cannot flow with a negative bias as high as -9 V. Without grid current, then, C_C discharges through

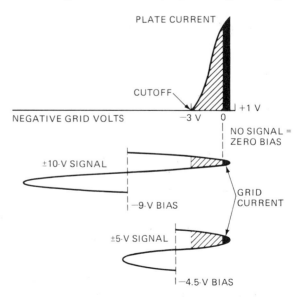

Fig. 1-18. Grid-leak bias adjusts itself to provide grid current. The amount of bias varies with the amount of signal to keep the positive peak of the signal at $+1$V for V_G.

R_1 and the bias voltage drops. When the bias is down to -4.5 V, grid current flows with the positive signal drive at 5 V. If the signal level increases again, more grid current will flow to charge C_C for more negative bias.

The bias automatically adjusts itself to the value that allows grid current for charging C_C. Therefore, the ac signal at the grid is held (or clamped) at a specific positive peak level.

Clamping Action The effect of holding the peak signal at a specific dc level, even when the amount of the ac signal changes, is called *clamping*. The grid-leak bias action is one example but the more general method of clamping is to use a diode rectifier instead of the grid-cathode circuit. Using an RC coupling circuit for the ac input signal, the diode can clamp the peak to a desired voltage level.

Signal Bias on NPN Transistor The transistor circuit in Fig. 1-19a corresponds to the grid-leak bias circuit for tubes. There are many similarities, especially with an NPN transistor.

The combined effect of the ac signal input and its dc bias is shown in Fig. 1-19b. Note that the ac signal input is clamped at a level that allows the positive peaks to produce pulses of collector current in the output. This circuit is in a pulse clipper stage. The pulse output results because the circuit does not have any fixed forward bias but instead depends on the peaks of excessive drive signals to pulse the transistor into conduction.

Test Point Questions 1-12
(Answers on Page 24)

a. Does signal bias increase or decrease with an increase in signal?
b. Is the grid signal peak in Fig. 1-18 clamped at a grid voltage of about $+1$ V or -7 V?

1-13
CLASS A, B, AND C OPERATION

Amplifier classes A, B, and C are defined by the percent of the cycle of input signal that is able to

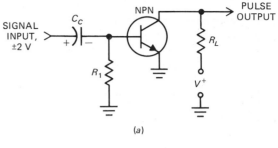

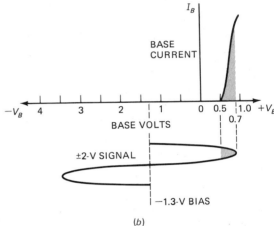

Fig. 1-19. Signal bias at the base of a silicon NPN transistor corresponding to grid-leak bias. The output pulses only for positive peaks of input signal. (*a*) Circuit. (*b*) Waveform with the base signal clamped at the positive peak.

produce output current. In other words, is any part of the input cycle cut off in the output? The amount of cutoff depends on two amplitudes: (1) the dc bias compared with the cutoff value and (2) the peak ac signal compared with the dc bias.

The input and output waveforms for class A, B, and C amplifiers are illustrated in Fig. 1-20. The output current is labeled I_O. This is plate current for tubes, collector current for junction transistors, or drain current for the FET.

Class A The output current I_O flows for the full cycle of 360° of input signal. This operation is shown by the waveform in Fig. 1-20b. An audio amplifier stage operates this way to follow the signal variations without too much distortion.

For class A operation, the dc bias allows an average I_O about one-half the maximum value.

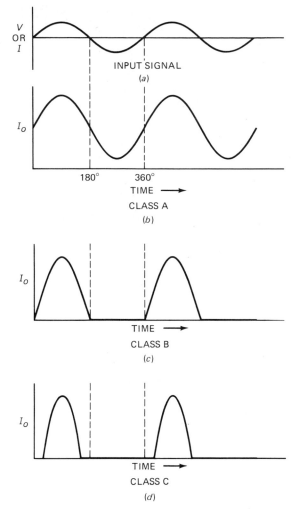

Fig. 1-20. Class of operation for amplifiers in terms of output current I_O. (a) Input signal. (b) Class A with I_O for full cycle of 360°. (c) Class B with I_O for 180°. (d) Class C with I_O for less than 180°, typically 120°.

Then the ac signal swings I_O around that middle value. I_O can vary up to its maximum value and close to zero but is never cut off.

Class B The output current I_O flows for 180°, or approximately one-half of the input cycle, as shown in Fig. 1-20b. The dc bias is at or near the cutoff value. I_O is at or close to zero, then, without any signal input. However, the positive half-cycle of signal can swing I_O up to its maximum value. The negative half-cycles of input signal are cut off

in the output, because I_O then is zero. Class B operation requires more dc bias and more ac signal drive than class A.

Class B operation with a single stage corresponds to half-wave rectification of the ac signal input. However, two stages can be used to provide opposite half-cycles of the signal in the output. A common method is the push-pull circuit described in Sec. 3-4 in Chap. 3.

Class C In class C operation the output current I_O flows for less than one-half the input cycle. Typical operation is 120° of I_O during the positive half-cycle of input, as shown in Fig. 1-20d. The result is produced by doubling the class B bias and using twice as much ac signal drive. Because of its high efficiency, class C operation is used mainly for tuned RF power amplifiers.

Characteristics of Each Class Which class of operation is used for an amplifier depends on the requirements for minimum distortion, maximum ac power output, and efficiency. The degree by which the output signal waveshape differs from the input signal waveshape is known as *distortion*. The ac power output is signal output. *Efficiency* is the ratio of ac power output to the dc power dissipated at the output electrode of the amplifier.

In class A operation distortion is lowest, but so also are ac power output and efficiency. Typical values are less than 5 to 10 percent distortion and an efficiency of 20 to 40 percent. At the opposite extreme, class C operation offers the highest efficiency, of about 80 percent, and allows the greatest ac power output but with the most distortion. Class B operation lies between A and C in distortion, power, and efficiency.

With audio amplifiers, class A must be used in a single stage for minimum distortion. Otherwise, the sound would be garbled. Also, an RF stage amplifying an amplitude-modulated signal must operate class A for minimum distortion of the modulation. In general, most small-signal amplifiers operate class A.

The reason for low efficiency in class A operation is that the middle value of I_C flows all the time, with or without ac signal input and for weak or strong signals. As a result, the dc power dissipation at the output electrode is high. Furthermore, relatively little ac signal drive can be applied without exceeding the cutoff voltage.

Class A amplifiers generally use self-bias, which is cathode bias for tubes or emitter bias and source bias for transistors. The constant value of I_O provides a steady bias in the return circuit to the common electrode. The amount of dc bias is the same with or without the ac signal.

Class B amplifiers are usually connected in pairs, each stage of which supplies opposite half-cycles of the signal input. Such a circuit is called a *push-pull amplifier*. The results approximate the low distortion of class A, but more drive can be used for more ac power output with higher efficiency. The push-pull circuit arrangement is often used for audio power output to a loudspeaker.

Class C operation is generally used for RF amplifiers with a tuned circuit in the output. Then the LC circuit can provide a full sine-wave cycle of output for each pulse of I_O. Class C amplifiers have high efficiency because the average I_O is very low compared with the peak signal amplitude. The result is relatively low dc power dissipated at the output electrode compared with the amount of ac power output. In addition, a pulse clipper circuit operates as a class C amplifier.

Class AB Class AB operation offers a compromise between the low distortion of class A and the higher power of class B. It is generally used for push-pull audio power amplifiers.

Class A$_2$ For tubes, the subscript 2 is used to indicate that grid current flows. The objective is to have more drive for the ac input signal. Plate current still flows for 360°, however, in class A$_2$ operation. The subscript 1 may be used, as in A$_1$, to indicate no grid current, but it is generally omitted. As examples, class A$_2$ or AB$_2$ operation is with grid current; class A or AB is without it. No subscripts are used for class C, even though the class C amplifier with tubes usually has grid current to provide grid-leak bias.

1-14
WIDE-BAND VIDEO AMPLIFIERS

A tuned RF stage can amplify a wide band of radio frequencies, but it cannot amplify audio frequencies. An AF amplifier can have response for a broad range of audio frequencies, but it cannot amplify radio frequencies. A *wide-band amplifier*, however, can amplify both audio and radio frequencies. A common application is amplifying the video signal that is used to reproduce the picture on television. The frequency components in the video signal can be as low as 30 Hz and extend up to 4 MHz in the RF range.

The difference between a wide-band amplifier and a broadly tuned RF amplifier, each with 4-MHz bandwidth, is illustrated in Fig. 1-21. In Fig. 1-21a, the response curve is for an RF amplifier tuned to 42 MHz. The bandwidth depends on the Q of the tuned circuit. Here the bandwidth of 4 MHz means the range of 40 to 44 MHz is amplified.

The response curve in Fig. 1-21b for the wide-band amplifier is entirely different, although the amplifier also has a bandwidth of 4 MHz. Here the bandwidth includes audio frequencies, from below 100 Hz, and radio frequencies up to 4 MHz. The flat response between those values is really an example of amplifying a wide range of frequencies, because the ratio of highest to lowest frequency is so great. Note that the horizontal axis for frequencies is marked with logarithmic spacing, in powers of 10, to compress the graph for the extremely wide frequency response.

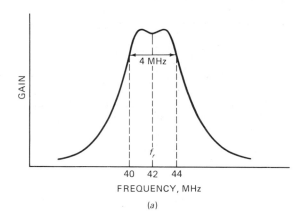

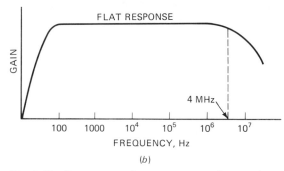

Fig. 1-21. Comparison of response curves for tuned amplifier and wideband video amplifier, each with 4-MHz bandwidth. (a) RF amplifier response at 42 MHz, ±2 MHz. (b) Video amplifier response for audio and radio frequencies up to 4 MHz.

Wide-Band Amplifier Circuit In Fig. 1-22, the basic circuit is an *RC*-coupled audio amplifier. The coupling capacitors C_1 and C_2 have values as high as possible for good low-frequency response. Also, R_L has the relatively low value of 3 kΩ for good high-frequency response. Reducing R_L reduces the overall gain but extends that gain to higher frequencies because there is less bypassing effect by the shunt capacitance C_T.

High-Frequency Compensation To improve the high-frequency response still more, the peaking coil L_1 is used in series with R_L at either end. L_1 is called a *shunt-peaking coil,* however, because it is in parallel with C_T. The coil resonates with C_T to raise the load impedance and boost the gain for high frequencies. A series-peaking coil also can be

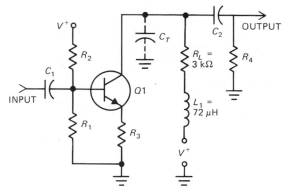

Fig. 1-22. Wideband video amplifier for 3.3-MHz response. This circuit is an *RC*-coupled audio amplifier with low R_L and RF peaking coil L_1.

used in series with the coupling path to the next stage.

Calculating the Cutoff Frequencies f_1 and f_2 More details of the basic *RC*-coupled amplifier can be analyzed in terms of its frequency response curve. As shown before in Fig. 1-4b, the gain is down to 70.7 percent response at f_1 for low frequencies because of the increasing reactance of the series coupling capacitor. Also, the gain is down to 70.7 percent response at f_2 for high frequencies because of the decreasing reactance of the shunt capacitance C_T. Those values are for the uncompensated amplifier. However, the frequency compensation is generally designed to make the amplifier have flat response between f_1 and f_2, within 10 percent, instead of being down 29.3 percent at those frequencies.

The low-frequency cutoff f_1 can be calculated as

$$f_1 = \frac{1}{2\pi RC} \tag{1-1}$$

where C is the coupling capacitor, in farads, and R is its series resistance, in ohms. The formula is derived from the fact that $R = X_C$ at f_1. As an example, assume C_2 is 10 μF and R_4 is 1 kΩ for the video amplifier of Fig. 1-22. Then

$$f_1 = \frac{1}{2\pi \times 10^3 \times 10 \times 10^{-6}} = 16 \text{ Hz}$$

With these values, the low-frequency gain is down to 70.7 percent response at 16 Hz because of the coupling capacitor.

It should be noted that the low-frequency response is perfectly flat to $10 \times f_1$ Hz. That frequency is $10 \times 16 = 160$ Hz for this example.

At the opposite end of the response curve, the high-frequency cutoff f_2 can be calculated as

$$f_2 = \frac{1}{2\pi R_L C_T} \qquad (1\text{-}2)$$

where R_L is in ohms and C_T is the total shunt capacitance, in farads. The formula is derived from the fact that $R_L = X_{C_T}$ at f_2.

As an example, consider the R_L of 3 kΩ in Fig. 1-22 with C_T of 16 pF. Then

$$f_2 = \frac{1}{2\pi \times 3 \times 10^3 \times 16 \times 10^{-12}}$$
$$= 0.0033 \times 10^9 \text{ Hz}$$
$$f_2 = 3.3 \text{ MHz}$$

With those values, the high-frequency gain is down to 70.7 percent at 3.3 MHz because the shunt C_T reduces the amplifier load impedance. However, the peaking coil L_1 boosts the high-frequency gain for uniform response up to f_2.

It should be noted that the high-frequency response is perfectly flat to $1/10 \times f_2$ Hz without compensation. In this example the frequency is $1/10 \times 3.3$ MHz $= 0.33$ MHz, or 330 kHz.

Applications for Wide-Band Amplifiers The video amplifier is very commonly used in television receivers to supply signal for the CRT to reproduce the picture. Typical frequency response is 30 Hz to 3.2 MHz. The response is limited to 3.2 MHz for minimum interference by the 3.58-MHz color signal. The amplitude is usually at least 100 V peak-to-peak (p-p) to drive the picture tube.

In oscilloscopes, the amplifiers for signal at the vertical input terminals typically have response up to 5, 10, or 50 MHz. The high-frequency response is obtained by peaking and the use of negative feedback. For the low-frequency response, oscil-loscopes usually have a switch for dc coupling of the input signal without any capacitor.

Wide-band amplifiers are generally necessary to amplify complex waveforms with strong harmonic components. The harmonic frequencies of a signal in the kilohertz range can extend up to the megahertz range. If the harmonic frequencies are not included in the amplifier response, the complex waveform will be distorted.

Test Point Questions 1-14
(Answers on Page 24)

a. Is the typical response for a wide-band video amplifier 50 to 60 MHz or 0 to 5 MHz?
b. Do peaking coils improve the response for low or high frequencies?
c. Is f_1 higher or lower with a larger coupling capacitor?
d. Is f_2 higher or lower with a smaller R_L?

1-15
DIRECT-COUPLED AMPLIFIERS

The direct-coupled amplifier circuit is similar to an RC-coupled amplifier, but, as shown in Fig. 1-23, it has no coupling capacitors. Note that the collector of Q1 is directly connected to the base of Q2. As a result of the direct coupling, both dc and ac voltages can be amplified.

DC Potentials Note in Fig. 1-23 that the base of Q2 is at the same potential, 8 V, as the collector of Q1. However, the voltages are to chassis ground. The forward bias depends on V_{BE}. Therefore, the emitter of Q2 is connected to 7.4 V to make V_{BE} equal to $8 - 7.4 = 0.6$ V. The collector of Q2 is at 16 V to make its voltage higher than the base voltage.

The required dc voltages are obtained at the $R_1 R_2 R_3$ divider from the supply. The full 28 V is for R_{L_2} to the collector of Q2. A tap at 12 V is for R_{L_1} to the collector of Q1. Finally, the lowest tap at 7.4 V is for the emitter of Q2. Note that each collector voltage V_C is lower than its supply voltage V_{CC} because of the IR drop for each R_L.

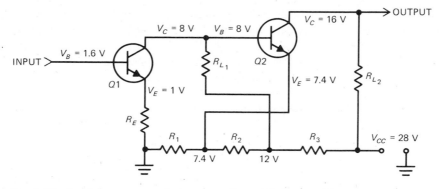

Fig. 1-23. Amplifier circuit with direct coupling. All the dc voltages for Q2 depend on the dc values in Q1. Values shown are for silicon NPN transistors.

Applications Direct-coupled circuits can be used to amplify dc control voltages. Also, audio amplifiers can be direct-coupled for good low-frequency response down to 0 Hz or direct current without any coupling capacitors. However, the dc gain depends on the stability of the dc supply voltage. In some applications, direct coupling is used, just for economy, to eliminate coupling capacitors.

Disadvantages A problem with direct coupling is that each successive stage needs progressively higher supply voltage. The full supply voltage cannot be used for all the stages. The 28-V supply of Fig. 1-22 is for the Q2 collector, but only 12 V is used for the collector of Q1.

Another problem is that any undesired change of the dc voltage in Q1 affects all the potentials for Q2. A rise of temperature in Q1 can change the Q1 bias voltage. When I_C increases in Q1, the V_C of Q1 drops to decrease V_{BE} for Q2.

Also, any change in the dc supply voltage is amplified because of the direct coupling. Finally, the power supply must have very good filtering of the 60-Hz ripple to eliminate hum, which can be amplified by the dc amplifier.

Test Point Questions 1-15
(Answers on Page 24)

Refer to Fig. 1-23.

a. What is the value of V_{BE} for Q1 and Q2?
b. What is the value of V_{CE} for Q2?

SUMMARY

1. The common-emitter (CE) circuit of junction transistors is generally used for amplifiers because it has the highest gain for both voltage and current. Input signal is applied to the base. The amplified output is taken from the collector.
2. The main types of coupling between amplifier stages are RC, impedance, and transformer coupling. In RC coupling, C_C blocks dc voltage while R has the ac signal for the next stage.
3. A resistance-loaded amplifier has uniform response over a wide range of frequencies because R_L is the load. Gain decreases at low frequencies because of the increasing reactance of C_C. Gain decreases at high frequencies because of the decreasing reactance of the stray C_T.

4. An impedance-coupled amplifier uses an RF or AF choke for Z_L. The advantage is a low dc voltage drop for the choke compared with R_L. The output signal is coupled by C_C.

5. A single-tuned amplifier uses an LC resonant circuit for Z_L. Gain is maximum at f_R.

6. A double-tuned amplifier uses an RF transformer with L_P and L_S resonant at the same frequency. The advantage is more bandwidth compared with a single-tuned amplifier.

7. The dc bias on an amplifier sets the operating point for amplifying the ac signal.

8. Typical bias voltage on junction transistors is 0.6 V for silicon and 0.2 V for germanium in class A amplifers.

9. Fixed bias comes from a battery or the dc power supply independently of the amount of signal or the amount of output current in the amplifier.

10. Cathode bias for tubes, emitter bias for a junction transistor, and source bias for the FET are methods of self-bias. The amount of bias depends on the level of average dc output current in the amplifier.

11. Grid-leak bias for tubes is a form of signal bias. The RC coupling circuit develops dc bias proportional to the amount of ac signal by rectifying the signal in the grid-cathode circuit.

12. In class A amplifiers, the output current I_O flows for the full cycle of 360° of the input signal. In class B, I_O flows for 180°. In class C, the conduction angle for I_O is less than 180°, typically 120°.

13. A wide-band video amplifier amplifies both audio and radio frequencies.

14. A direct-coupled amplifier does not have any coupling capacitors. All dc values in one stage determine the dc levels in the next.

SELF-EXAMINATION
(Answers at back of book)

Choose (a), (b), (c), or (d).

1. In a resistance-loaded, RC-coupled amplifier the dc component is blocked by (a) R_L, (b) C_C, (c) R_B, (d) the transistor.

2. The class of amplifier operation with the least distortion is (a) A, (b) B, (c) C, (d) AB_2.

3. The circuit that would be used for a 455-kHz IF amplifier is (a) resistance-loaded, (b) double-tuned transformer, (c) video amplifier, (d) class C.

4. In an RC-coupled amplifier, low-frequency response is improved with (a) lower R_L, (b) more bias, (c) less gain, (d) higher C_C.

5. An FET has a gate-source bias of -2 V. The ac input signal is ± 1.2 V. The class of operation is (a) A, (b) B, (c) C, (d) AB_2.

6. Emitter bias depends on (a) signal input, (b) I_E, (c) gain, (d) C_C.

7. In Fig. 1-23 the V_{BE} on $Q2$ is (a) 0.6 V, (b) 7.4 V, (c) 8 V, (d) 28 V.

8. For an FET, $I_S = 5$ mA and $R_S = 330\ \Omega$. The source bias is (a) 1.65 V, (b) 3.3 V, (c) 5 V, (d) 33 V.
9. The type of circuit that does not need a coupling capacitor is (a) resistance-loaded, (b) impedance-coupled, (c) single-tuned, (d) transformer-coupled.
10. The collector voltage equals the full V^+ of 28 V instead of the normal 12 V. Which of the following can be the trouble? (a) R_L is open, (b) C_C is open, (c) R_E is open, (d) C_E is shorted.

ESSAY QUESTIONS

1. Define the following types of coupling: RC, impedance, and transformer.
2. Give an advantage and disadvantage of the resistance-loaded amplifier.
3. Define the following types of bias: fixed, self, and signal.
4. Give three examples of self-bias with tubes and transistors.
5. Define the following classes of operation: A, B, C, AB, and AB_2.
6. Compare class A and class C in efficiency and distortion.
7. Draw the schematic diagram of a class A single-tuned RF amplifier for 2-MHz signal with fixed bias and emitter bias. **a.** What makes the amplifier class A? **b.** Why does the circuit amplify 2 MHz? **c.** Why is C_C needed for coupling to the next stage?
8. What is meant by clamping action?
9. Compare two characteristics of emitter bias and cathode bias.
10. Compare two characteristics of cathode bias and grid-leak bias.
11. Give two features of a wide-band video amplifier circuit.
12. Give an advantage and disadvantage of a direct-coupled amplifier.
13. Draw the schematic diagram of a resistance-loaded amplifier RC-coupled to a transformer-coupled audio output stage. No values are required. **a.** Why is class A operation used in both stages? **b.** How is the dc component blocked in both stages? **c.** What kind of bias did you use?
14. Compare two characteristics of the common-emitter amplifier circuit and the emitter follower circuit.
15. Show a triode tube and an FET as amplifiers corresponding to the common-emitter circuit.

PROBLEMS
(Answers to odd-numbered problems at back of book)

1. An NPN transistor is used with R_L for the RC-coupled amplifier in Fig. 1-4. **a.** With R_L of 2 kΩ, what is I_C? **b.** With R_1 of 1000 Ω and C_C of 5 μF, calculate f_1 for low-frequency response. **c.** With C_T of 20 pF, calculate f_2 for high-frequency response.
2. **a.** At what frequency will X_C of a 5-μF capacitor equal 1000 Ω? **b.** At what frequency will X_C of a 20-pF capacitor equal 2000 Ω?

3. Refer to the self-bias circuits in Fig. 1-15. **a.** For the 1-V emitter bias in Fig. 1-15a, what is R_E if $I_E = 300$ mA? **b.** For the 2-V cathode bias in Fig. 1-15b, what is R_K if $I_K = 4$ mA? **c.** For the 2-V source bias in Fig. 1-15c, what is R_S if $I_S = 4$ mA?
4. In the emitter-bias circuit of Fig. 1-16, what is the value of C_E to bypass R_E down to 50 Hz?
5. Refer to the R_1R_2 voltage divider for fixed bias at the base in Fig. 1-13. Calculate R_1 and R_2 for the same voltages but with 2 mA of current instead of 1 mA.
6. For the direct-coupled amplifier in Fig. 1-23, calculate: **a.** V_{BE} and V_{CE} on Q1 and **b.** V_{BE} and V_{CE} on Q2.

SPECIAL QUESTIONS

1. Give three examples of amplifiers that would operate class A.
2. Draw a circuit for a compensated video amplifier other than the one shown in Fig. 1-22.
3. Redraw the amplifiers shown in Figs. 1-4, 1-6, 1-8, and 1-9, but with an NPN silicon transistor in a CE circuit.
4. Describe briefly an application for each of the circuits in Special Question 3.

ANSWERS TO TEST POINT QUESTIONS

1-1	**a.** T		1-10	**a.** -1.5 V	
	b. T			**b.** 1.6 V	
1-2	**a.** CE		1-11	**a.** 0.6 V	
	b. CS			**b.** -2 V	
	c. Emitter follower			**c.** -2 V	
1-3	**a.** C_C			**d.** 1.04 V	
	b. C_T		1-12	**a.** Increase	
1-4	**a.** 10 V			**b.** $+1$ V	
	b. ±4 V or p-p voltage is 8 V		1-13	**a.** C	
1-5	**a.** L			**b.** A	
	b. C_1			**c.** A	
1-6	**a.** 0.1 V		1-14	**a.** 0 to 5 MHz	
	b. C_1			**b.** High	
1-7	**a.** T			**c.** Lower	
	b. T			**d.** Higher	
1-8	**a.** Practically zero		1-15	**a.** 0.6 V	
	b. Red			**b.** 8.6 V	
1-9	**a.** Emitter self-bias				
	b. Grid-leak bias				

Chapter 2
Amplifier Analysis

More details of exactly how a transistor amplifies an ac signal are analyzed here. The class A amplifier with a resistive load is used as an example. A typical circuit is examined just for dc voltages without any ac signal. The transistor is an NPN silicon small-signal amplifier. Dc electrode voltages are considered first because the amplifier must be conducting current in order to amplify the signal. Then ac signal is applied to the base electrode of the transistor. A common-emitter circuit is used and the amplified output is taken from the collector. All the ac signal waveforms of voltage and current for the input and output circuits are discussed in this chapter.

Although here the analysis is of a junction transistor, the same principles apply to any amplifier, including the FET and tubes. All operate by having a small signal in the input circuit control a larger signal in the output circuit. The specific steps in the amplification process are explained in the following topics;

2-1
DC VOLTAGES ON THE AMPLIFIER

Although it is generally used for amplifying ac signals, the transistor needs dc operating voltages. Forward bias at the base and reverse collector voltage must be applied to produce any current at all. When the transistor is conducting, its current can be varied by an ac input signal. The output variations are greater than the input signal, and the result is amplification. Both the input and output signals are actually fluctuating dc variations. The ac signal is represented by changes above and below an average dc axis.

In Fig. 2-1, transistor Q1 has reverse collector voltage applied from the dc supply of 14 V through the 2-kΩ R_L. The I_C is 3 mA; as a result, the $I_C R_L$ drop is 6 V. This voltage across R_L is calculated as

$$V_{R_L} = I_C \times R_L$$
$$= 3 \text{ mA} \times 2 \text{ k}\Omega$$
$$V_{R_L} = 6 \text{ V}$$

The collector voltage V_C is 8 V. That value is calculated as 14 V for V_{CC} minus the 6-V drop across R_L. As a formula

$$V_C = V_{CC} - V_{R_L}$$
$$= 14 - 6$$
$$V_C = 8 \text{ V}$$

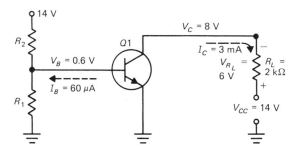

Fig. 2-1. DC operating voltages applied to a CE amplifier without an ac signal.

The voltage drop across R_L is subtracted because its polarity opposes that of the dc supply voltage.

The forward bias of 0.6 V at the base is provided by the $R_1 R_2$ divider; it also is from the dc supply of 14 V. Since the emitter of Q1 is grounded, the voltage across R_1 is V_B for base bias. The V_B is also V_{BE} because the emitter is grounded. The 0.6-V bias is needed because Q1 is silicon. The resulting base current of 60 μA is typical for a small-signal transistor.

To summarize all these dc values:

$$V_C = 8 \text{ V}$$
$$I_C = 3 \text{ mA}$$
$$V_{R_L} = 6 \text{ V}$$
$$V_B = 0.6 \text{ V}$$
$$I_B = 60 \text{ μA}$$

These are called *quiescent values*, meaning that the circuit is quiet, without any ac signal input. A quiescent value for I is also called an *idling current*. All the dc voltages and currents are just right for applying an ac signal to be amplified by Q1.

Test Point Questions 2-1
(Answers on Page 42)

Refer to Fig. 2-1.

a. What is the value of the idling current for I_C?
b. What is the value of the base bias V_{BE}?

2-2
AC SIGNAL WAVEFORMS

Figure 2-2 shows the same circuit as in Fig. 2-1 but with an ac signal input and amplified output. The quiescent dc values are the same for base bias and collector voltage. C_1 and C_2 are added for coupling the input and output signals while blocking dc voltage.

Input signal to the base is ±50 mV. The amplified output at the collector is ±3.6 V. This is 3600 mV. That value of output signal is 72 times more than the input signal.

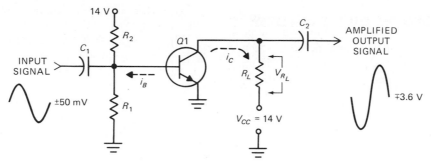

Fig. 2-2. The same CE circuit as in Fig. 2-1 but with an ac signal. The input is ±50 mV applied to base; the amplified output is ∓3.6V across R_L. Amplification shown by the waveforms in Fig. 2-3.

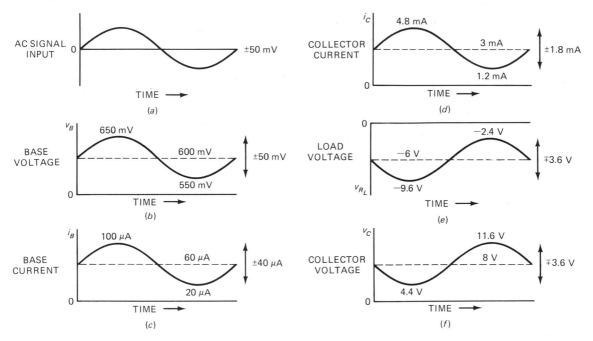

Fig. 2-3. Waveforms of the ac signals for the CE circuit of Fig. 2-2. The amplitudes are not drawn to scale. This type of figure is known as a ladder diagram.

Just how that voltage gain can be produced is illustrated by the ladder diagram of signal waveshapes in Fig. 2-3. Each current and voltage from the base input to collector output is shown in the six waveforms of Fig. 2-3a to f. The sequence of operation can be analyzed in the same six steps.

(a) AC signal voltage input. The waveform in Fig. 2-3a is an ac voltage varying around the zero axis. The peaks are 50 mV positive and negative.

The ac voltage is developed across R_1 and by the coupling capacitor C_1 as the signal from the preceding circuit. As a result, the base voltage of Q1 is varied ±50 mV by the input signal.

(b) The fluctuations in base voltage v_B. The ac input signal of ±50 mV combines with the dc bias of 600 mV across R_1 (Fig. 2-3b). Since the bias is greater than the ac signal, the base always has positive voltage for the required forward polarity.

The positive signal peak of $+50$ mV adds to the bias of 600 mV to produce 650 mV for the peak value of v_B. The negative signal peak of -50 mV decreases the bias of 600 mV to produce 550 mV for the minimum value of v_B. However, this base voltage stays positive because of the dc bias.

Note that the base signal still has the variations of ± 50 mV, but they are around the average axis of 600 mV instead of zero. The varying waveform of v_B is a fluctuating dc voltage.

(c) *The varying base current* i_B. The waveform of variations in base current (Fig. 2-3c), is produced by the variations in base voltage. Also, the bias axis of 60 μA for i_B corresponds to the bias of 600 mV for v_B.

The peak i_B of 100 μA is produced by the peak v_B of 650 mV. In the opposite direction, the minimum i_B of 20 μA corresponds to the minimum v_B of 550 mV. The i_B changes by 40 μA up or down from the average axis of 60 μA.

The changes of ± 40 μA in i_B result from the changes of ± 50 mV in v_B. Those typical values for a small-signal transistor are taken from the input characteristics shown by the graph in Fig. 2-4. The three points on the curve mark the values used in Fig. 2-3c. These changes in i_B are equal for the equal changes in v_B because the low base signal involves only a small part of the characteristic curve. The characteristic curve can be considered linear for a signal swing of approximately 50 μV or less.

It should be noted that the change of 40 μA in i_B for a change of 50 mV in v_B means that the ac input resistance is equal to 1200 Ω. The V/I ratio is calculated as

$$r_{in} = \frac{50 \times 10^{-3} \text{ V}}{40 \times 10^{-6} \text{ A}}$$

$$= 1.2 \times 10^3 \text{ } \Omega$$

$$r_{in} = 1.2 \text{ k}\Omega \text{ or } 1200 \text{ } \Omega$$

The r_{in} is essentially the resistance of the conducting base-emitter junction.

(d) *The varying collector current* i_C. The amplifying action of the transistor is in the effect of

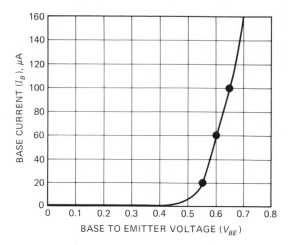

Fig. 2-4. Typical base-input characteristic curve for small-signal NPN silicon transistor. The three points on the curve indicate values used for the base current waveform in Fig. 2-3C.

changes in i_B varying the amount of collector current. When i_B increases, the i_C also increases. A decrease in i_B reduces i_C.

Furthermore, the collector current is always much larger than the base current. Relatively small changes in i_B therefore produce much larger changes in i_C. In the waveform of Fig. 2-3d, the quiescent i_C of 3 mA as the average axis corresponds to the dc current of 60 μA for i_B. A change of 40 μA in i_B produces a change of 1.8 mA in i_C.

The peak i_C is $3 + 1.8 = 4.8$ mA, which corresponds to the peak i_B of 100 μA. In the opposite direction, the minimum i_C is $3 - 1.8 = 1.2$ mA. That current corresponds to the minimum i_B of 20 μA.

As a result, the signal variations in i_C are ± 1.8 mA around the 3 mA axis. Those typical values are taken from the collector output characteristic curves for a small-signal transistor discussed in Sec. 2–3.

(e) *The varying voltage across the load* v_{R_L}. The purpose of having R_L in the output circuit is to provide an IR voltage that varies with the changes in collector current. The v_{R_L} is equal to $i_C \times R_L$.

The waveform of v_{R_L}, in Fig. 2-3e, is shown below the zero level, for negative voltage. The reason is that we are considering the voltage drop across the two ends of R_L, from the collector side at the top to the opposite side at V_{CC}. The top of R_L is more negative with electron flow of i_C into that end. Actually, the point is positive with respect to chassis ground but negative with respect to V_{CC}.

The average axis of v_{R_L} is at -6 V for the quiescent i_C of 3 mA. The calculations are 3 mA $\times$ 2 kΩ = 6 V. This value for the average axis corresponds to the base bias current of 600 μA, which produces the quiescent value of 3 mA for i_C.

For the peak i_C of 4.8 mA, the maximum negative value of v_{R_L} is -9.6 V. The calculations are 4.8 mA $\times$ 2 kΩ = 9.6 V negative at the collector side. For the minimum i_C of 1.2 mA, the minimum negative value of v_{R_L} is -2.4 V. The calculations are 1.2 mA $\times$ 2 kΩ = 2.4 V negative at the collector side.

The result is a varying voltage drop across R_L that changes with the value of i_C. Specifically, v_{R_L} varies by ∓ 3.6 V around the average axis of -6 V.

(f) *The varying collector voltage v_C.* The waveform of Fig. 2-3f corresponds to the varying voltage across R_L, but v_C is positive. The reason is that v_C at the collector is measured with respect to the emitter at chassis ground. As for calculations, v_C equals V_{CC} of 14 V minus the voltage drop across R_L. The v_{R_L} is subtracted because it is in series with the collector and its negative polarity opposes the positive supply voltage.

The average axis of 8 V for v_C results from the quiescent i_C of 6 mA. Then V_{R_L} is -6 V and v_C is $14 - 6 = 8$ V.

The minimum v_C of 4.4 V results from the maximum V_{R_L} of -9.6 V. Both values correspond to the peak i_C of 4.8 mA. The calculations for minimum v_C are $14 - 9.6 = 4.4$ V. The maximum v_C of 11.6 V results from the minimum V_{R_L} of -2.4 V. Both values correspond to the minimum i_C of 1.2 mA. The calculations for maximum v_C are $14 - 2.4 = 11.6$ V.

The result is the varying voltage values for v_C, which provide the amplified signal output. The signal variations in v_C equal ∓ 3.6 V around the average axis of 6 V.

Phase Inversion It is important to note the inverse relation between the waveforms for i_C, in Fig. 2-3d, and v_C, in Fig. 2-3f. When i_C increases, v_C decreases because of a larger voltage drop across R_L. A decrease in i_C allows more v_C, with less voltage across R_L.

Furthermore, the waveform of v_C is inverted with respect to the input signal in Fig. 2-3a. When the base voltage increases in the positive direction, v_B and i_B increase to raise i_C but the positive collector voltage decreases.

As a result, the input and output signals are 180° out of phase. The inversion of the signal is a characteristic of the common-emitter amplifier. The reason for the inversion is simply that R_L in the collector circuit reduces v_C when i_C increases. The polarity inversion applies to both NPN and PNP transistors.

Test Point Questions 2-2
(Answers on Page 42)

With reference to the waveforms in Fig. 2-3 for the amplifier in Fig. 2-2, give the p-p signal swing for:

a. Base voltage   c. Collector current
b. Base current   d. Collector voltage

2-3
COLLECTOR CHARACTERISTIC CURVES

The curves in Fig. 2-5 show the volt-ampere characteristics for the collector. I_C on the vertical axis is plotted against V_{CE} on the horizontal axis. Each curve is for a specific I_B. The different curves specify how I_C increases with increases in I_B.

The characteristic curves are provided by the manufacturer in a transistor manual or application notes. For the CE circuit, the collector curves are for different values of I_B; for a CB circuit, they would be for different values of I_E.

An Experimental Circuit for I_C Figure 2-5a is a circuit in which the transistor voltages can be

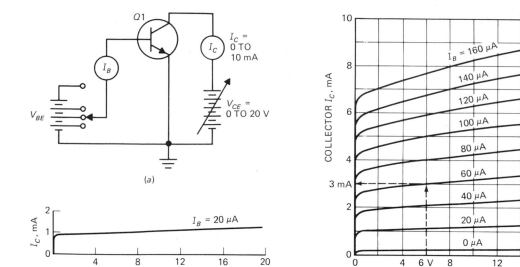

Fig. 2-5. Collector characteristic curves showing how I_C varies with V_{CE} for a specific I_B. (a) The CE circuit used to determine values experimentally. The circuit is not loaded. (b) Typical collector characteristic curve for an I_B of 20 μA. (c) Family of collector curves for different values of I_B. Arrows shown for example in text.

varied experimentally in order to determine the effect on I_C. One value of base voltage is used for a specific I_B while V_{CE} is varied to see how much I_C changes. Then the base voltage is set for another value of I_B and V_{CE} is varied again while I_C is measured. The V_{CE} and I_C values for each I_B setting are used for one curve.

The collector and base voltages are labeled V_{CE} and V_{BE} for the general case of a potential difference with respect to emitter. However, they are the same as V_C and V_B with the emitter grounded.

Note that no load resistance is used for the experimental circuit of Fig. 2-5a. Also, there is no signal input or output. The circuit is not an amplifier for signal voltage; it is only an experimental arrangement for measuring the volt-ampere characteristics of the transistor itself without any load.

One Typical Curve The results for one value of I_B at 20 μA are shown by the curve in Fig. 2-5b. I_C rises from 0 to 1 mA when V_C is increased from 0 to 1 V, approximately, but there is only a slight rise in I_C to about 1.3 mA when V_C is further increased to 20 V. The reason for the small second

increase in I_C is that the increase is limited by the value of I_B permitted by the forward voltage at the base junction. For more I_C, the transistor needs more base current. As an example of reading values from the curve, I_C is exactly 1 mA when V_{CE} is 4 V.

The Family of Curves The results for different values of base current are shown by the family of collector characteristics in Fig. 2-5c. Each curve represents the relationship of I_C and V_C for a specific I_B. Note that the curve for 20 μA of I_B is the same as in Fig. 2-5b.

There are several curves, but only one is read at a time. As an example, arrows are shown for the fourth curve up from the bottom, for I_B of 60 μA. The values there are 3 mA of collector current with 6 V at the collector.

The family of curves really shows how the values of I_C and I_B are related for the same collector voltage. If the vertical arrow in Fig. 2-5c is extended up to the next curve, for 80 μA of I_B, the I_C reading will increase to 4 mA with the same 6 V at the collector. For the opposite change, when

I_B decreases to 40 μA, the I_C is reduced to 2 mA, all with 6 V for V_{CE}.

Test Point Questions 2-3
(Answers on Page 42)

Refer to the curves in Fig. 2-5.

a. What is the base current for the collector characteristic curve in Fig. 2-5b?
b. From the characteristic curves in Fig. 2-5c, what is I_C with 8 V for V_{CE} and 140 μA of I_B?

2-4
BETA AND ALPHA CHARACTERISTICS

The beta and alpha values determine how good a junction transistor is as an amplifier. The beta, indicated by Greek letter β, specifies how the base current controls the amount of collector current. This factor is important for the CE amplifier. The alpha, α, compares I_C with I_E for the CB circuit.

Calculations for Beta Specifically, beta is the ratio of collector current to base current. The formula is

$$\beta = \frac{I_C}{I_B} \qquad \text{(2-1)}$$

As an example, when 60 μA of I_B produces 3 mA of I_C,

$$\beta = \frac{I_C}{I_B}$$

$$= \frac{3 \text{ mA}}{60 \text{ }\mu\text{A}}$$

$$= \frac{3000 \text{ }\mu\text{A}}{60 \text{ }\mu\text{A}}$$

$$\beta = 50$$

The values for I_C and I_B are taken from the family of collector curves in Fig. 2-5c. The value for β

has no unit since it is a ratio of two currents. Typical values of β are 40 to 300 for small-signal transistors and 10 to 30 for power transistors.

We can use the beta ratio in two ways to relate I_C and I_B:

$$I_C = \beta \times I_B$$

or

$$I_B = I_C \div \beta$$

For instance, when I_B is 2 mA and β is 90, the I_C is 90 × 2 = 180 mA. Or when I_C is 180 mA and β is 90, the I_B must be 180/90 = 2 mA.

Formula 2-1 gives the static or dc value of β without ac signal input. The dynamic or ac value is calculated for small changes in the currents. As an example, assume a 20-μA change in i_B changes i_C by 1 mA. Then

$$\beta_{ac} = \frac{\Delta i_C}{\Delta i_B} = \frac{1 \text{ mA}}{20 \text{ }\mu\text{A}} = \frac{1000 \text{ }\mu\text{A}}{20 \text{ }\mu\text{A}}$$

$$\beta_{ac} = 50$$

This ac beta happens to be the same value as the dc beta, but the two can be different. The ac beta characteristic of the transistor is the current-transfer ratio for the common-emitter circuit from base input to collector output. Specifically, β is the value of current gain.

Calculations for Alpha The ratio α compares collector current to emitter current. Then

$$\alpha = \frac{I_C}{I_E} \qquad \text{(2-2)}$$

As an example, for a transistor with 3 mA of I_C and 3.06 mA of I_E, the calculations are

$$\alpha = \frac{I_C}{I_E}$$

$$= \frac{3 \text{ mA}}{3.06 \text{ mA}} = 0.98, \text{ approximately}$$

The alpha must be less than 1 because I_C must be less than I_E, which includes the base current. In fact, the 3.06 mA of I_E is the sum of 3 mA of I_C and 60 μA of I_B. In this example, α has its dc or static value. The ac or dynamic α is calculated from changes in i_C and i_E.

It should be noted that α and β are related to each other because each includes the effects of I_B, I_C, and I_E. The relation is

$$\beta = \frac{\alpha}{1 - \alpha} \qquad (2\text{-}3)$$

For the example of alpha equal to 0.98, the corresponding beta can be calculated as

$$\beta = \frac{0.98}{1 - 0.98} = \frac{0.98}{0.02}$$

$$\beta = 49$$

The β is not exactly 50 because of the approximate value of 0.98 for α.

Input-Output Forward-Transfer Characteristic

The curve in Fig. 2-6 shows how I_B controls I_C. This characteristic curve is in the forward direction, from the input circuit at the base to amplified output from the collector. The curve is plot-

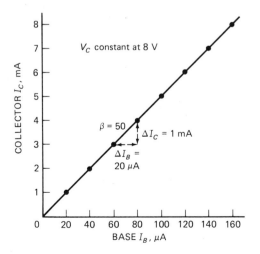

Fig. 2-6. Input-output forward transfer characteristic. Values are obtained from the collector curves in Fig. 2-5c.

ted from the family of collector characteristic curves in Fig. 2-5c. However, the collector voltage is held constant at 8 V, whereas values of I_C are taken for different values of I_B. The result, in Fig. 2-6, is a forward-transfer characteristic that shows how the input current controls the output current.

The slope of the transfer characteristic curve is the ac beta of the transistor. As shown in Fig. 2-6, the change in i_C, or Δi_C, is a vertical distance. Also, the change in i_B, or Δi_B, is a horizontal distance. The slope of the curve is the ratio of the vertical to the horizontal intervals.

The ratio $\Delta i_C / \Delta i_B$, for the slope, is the ac beta for the transistor. For this example β is equal to 50. A higher value for β would result in a sharper slope for the transfer characteristic curve.

Test Point Questions 2-4
(Answers on Page 42)

a. With reference to Fig. 2-5c, calculate β at the point of 140 μA for I_B and 7 mA for I_C with V_C at 8 V.
b. Calculate α at the same point as in a.
c. A power transistor has 5 A of I_C and a β of 25. What is the value of I_B?

2-5
LOAD-LINE GRAPHICAL ANALYSIS

The characteristic curves show voltages and currents for the transistor itself without any external load. However, a load impedance is needed to provide amplified output voltage. Those requirements are illustrated in Fig. 2-7, where the circuit of Fig. 2-2 is analyzed by means of the transistor characteristic curves. The same ac signal input of ± 40 μA and amplified output of ∓ 3.6 V is analyzed in Fig. 2-7 with the characteristic curves. The circuit is a CE amplifier operating class A.

Although the transistor curves are nonlinear, the load resistance has a linear volt-ampere characteristic. To see the effect of R_L on the collector voltage and current, the straight-line characteristic of R_L is superimposed on the collector curves. This method is called *graphical analysis*.

Constructing the Load Line As shown in Fig. 2-7, we need only the values of R_L and the V_{CC} supply because those values determine the end points of the load line. Then all values of V_C and I_C are on the load line for the specified R_L.

One point is at V_{CC} of 14 V on the horizontal axis where I_C is zero. The value is one operating point because the collector voltage equals V_{CC} when there is no I_C and there is no voltage drop across R_L. The transistor is actually cut off here.

Another point is where I_C is enough to make V_C equal to zero. In other words, the voltage drop across R_L is equal to the V_{CC} so that the voltage of the collector is zero. This end of the load line is at $I_C = V_{CC}/R_L$. For the example here, that point is at 14 V/2 kΩ = 7 mA on the vertical axis, where $V_C = 0$. The transistor is near maximum I_C for this point of operation.

The straight line drawn between 7 mA on the vertical axis and 14 V on the horizontal axis is the load line for $R_L = 2$kΩ with a 14-V supply. For any value of i_C the corresponding v_C must be on the load line. The reason is that the load line shows the effect of the voltage drop produced across R_L by i_C.

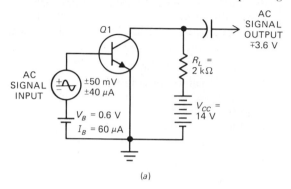

(a)

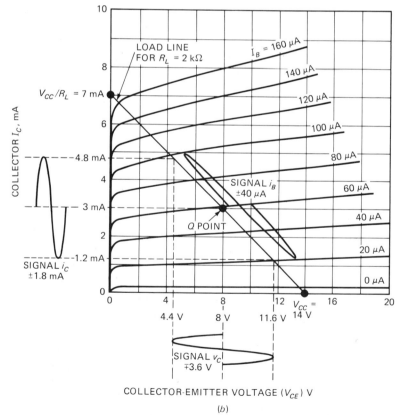

(b)

Fig. 2-7. Load-line analysis. (a) CE circuit equivalent to Fig. 2-2. (b) Construction of a load line for $R_L = 2$ kΩ and $V_{CC} = 14$ V. See text for details.

Q Point The quiescent, or Q, point specifies the dc values for the operating bias without signal. In Fig. 2-7, the base current of 60 μA is chosen for the bias because it is a middle value. That bias allows room for the signal swing without reaching cutoff or saturation of the collector current. As a result, the amplifier operates class A.

In this example the Q point is where the load line intersects the I_B curve of 60 μA for bias. From that point we can determine the resulting values for the collector quiescent current and voltage. Follow the Q point down to the horizontal axis at $V_C = 8$ V. Also, take the Q point across to the vertical axis at $I_C = 3$ mA. Those are the average dc values in the collector circuit for the base bias current of 60 μA. The values at the Q point, indicated by the subscript Q, can be summarized as follows:

$$I_{BQ} = 60 \ \mu A$$
$$I_{CQ} = 3 \ mA$$
$$V_{CQ} = 8 \ V$$

AC Signal Swing When there is signal input to the base, the ac drive varies the base current up to the peak of 100 μA and down to the minimum of 20 μA. In Fig. 2-7, those values of i_B are two curves up and down from the Q point at 60 μA.

The signal values for base current can be summarized as follows:

Peak	100 μA
Average	60 μA
Minimum	20 μA
P-P	80 μA
Swing	±40 μA

These signal swings of base current are shown next to the load line. Only the intercepts exactly on the load line are read. At the peak for i_B of 120 μA, the maximum i_C is 4.8 μA. For the minimum i_B of 20 μA the minimum i_C is 1.2 mA. Those values are read by projecting the points of intersection to the left on the vertical axis for collector current.

The signal values for collector current can be summarized as follows:

Peak	4.8 mA
Average	3 mA
Minimum	1.2 mA
P-P	3.6 mA
Swing	±1.8 mA

In the same way, we can read the minimum and peak values of v_C by projecting the points of intersection down to the horizontal axis for collector voltage. The values are 4.4 and 11.6 V. Note that the minimum v_C corresponds to the maximum i_C because of the voltage drop across R_L.

The signal values for collector voltage are:

Minimum	4.4 V
Average	8 V
Peak	11.6 V
P-P	7.2 V
Swing	∓3.6 V

These values for the ac signal show that the base current variations of ±40 μA swing the collector current ±1.8 mA. Then the collector voltage swings ∓3.6 V because of the varying voltage drop across the 2-kΩ R_L. The collector voltage decreases when the collector current increases.

Furthermore, intermediate values can be determined. For instance, an i_B of 80 μA results in 4 mA for i_C. At that point on the load line, the v_C is 6 V.

It should be noted that the load line in Fig. 2-7 is called a *static or dc load line*. It is used for the ac signal variations, but the value of R_L does not include any shunting effect of the signal by the input impedance of the next stage. The *dynamic or ac load line* is drawn with an equivalent load equal to R_L in parallel with the shunt impedance. The method of construction is the same, but the dynamic load line has a sharper slope for less R and more I because the resistance of the parallel combination must be less than R_L.

Test Point Questions 2-5
(Answers on Page 42)

Refer to Fig. 2-7 for the following values:

a. Peak base current.

b. Peak collector current.
c. Minimum collector voltage.
d. i_C and v_C when i_B is 80 μA.

2-6
LETTER SYMBOLS FOR TRANSISTORS

Because of the combination of an ac component on a dc axis, it is important to distinguish between the different voltages and currents in a transistor amplifier. In general, there are letter symbols for three kinds of values:

1. Average dc values
2. Instantaneous values of the fluctuating dc waveforms
3. Values for the ac signal variations alone

All these are summarized in Table 2-1, which shows how the various letters are used to indicate the different voltages or currents.

The capital letters V and I and their subscripts are used for average dc values. The subscript also is a capital letter. An example is V_C for average dc collector voltage.

Double subscripts that are repeated, as in V_{CC}, indicate the supply voltage that does not change. Also, V_{EE} is used to denote the dc supply voltage for the emitter.

The small letters v and i are used for instantaneous values that vary with the fluctuating dc waveform. As an example, v_C is an instantaneous value of the varying dc collector voltage.

A small letter in the subscript indicates the ac component.

The rms value, or effective value, of the ac component is a capital letter. However, its subscript is a small letter. As an example, V_c is the rms value of the ac component of collector voltage.

At the bottom of Table 2-1, additional letter symbols for transistors are listed. For example, I_{CBO} denotes reverse leakage current. The letter O shows which electrode is open when leakage current between the other two electrodes is measured. Therefore, I_{CBO} is leakage current between collector and base with the emitter open.

In the symbol h_{fe}, the h stands for *hybrid parameters,* which are combinations of voltage and current ratios in the forward and reverse directions. In the subscripts, f indicates a forward characteristic from the base input to collector output. The e indicates the common-emitter circuit. The symbol h_{fe} is used often, therefore, because its forward current-transfer ratio is the same as the small-signal or ac beta of the transistor in the CE circuit. Also, h_{ie} is the input resistance of the base-emitter junction.

Table 2-1
Letter Symbols for Transistors

Symbol	Definition	Notes
V_{CC}	Collector supply voltage	Same system for collector currents;
V_C	Average dc voltage	also for base or emitter voltages
v_c	Ac component	and currents. Also applies to
v_C	Instantaneous value	drain, gate, and source of field-
V_c	RMS value of ac component	effect transistors
I_{CBO}	Collector cutoff current, emitter open	Reverse leakage current
BV_{CBO}	Breakdown voltage, collector to base, emitter open	Ambient temperature T_A is 25°C
h_{fe}	Small-signal forward-current transfer ratio in CE circuit	Same as ac beta for CE circuit
h_{ie}	Input resistance of CE circuit	Same as V_{in} for CE circuit

For the signal waveform of collector current in Fig. 2-7, give the letter symbol for the following values:

a. 4 mA as an instantaneous value
b. 1.27 mA as the rms value of the ac component
c. 3 mA at the Q point

2-7
GAIN CALCULATIONS

Examples can be taken from the load-line analysis in Fig. 2-7 for comparing the output to the input. We can calculate the gain in current, voltage, and power for the amplified signal.

Only ac values are considered, not the dc level, because the signal is represented just by the variations. The ac signal can be specified in rms, peak, or p-p values. However, the same measure must be used for both the output and input in order to compare the two.

It is generally best to use p-p values for gain calculations. The reason is that, if the signal is distorted, the opposite peaks will not be equal. Then there would be a question of which peak to use for the comparison.

P-P Amplitudes For a fluctuating dc waveform, subtract the minimum from the maximum value. The formula is

$$P\text{-}P \text{ amplitude} = \begin{cases} V_{max} - V_{min} \\ \text{or} \\ I_{max} - I_{min} \end{cases} \quad (2\text{-}4)$$

As an example, for the collector voltage in Fig. 2-7,

$$11.6 - 4.4 = 7.2 \text{ V p-p}$$

Also, the p-p collector current is

$$4.8 - 1.2 = 3.6 \text{ mA p-p}$$

Current Gain A_I The formula for A_I is

$$A_I = \frac{\text{output ac signal current}}{\text{input ac signal current}} \quad (2\text{-}5)$$

By using p-p values for the collector and base currents in Fig. 2-7, we have

$$A_I = \frac{3.6 \text{ mA}}{80 \ \mu A} = \frac{3600 \ \mu A}{80 \ \mu A}$$

$$A_I = 45$$

The same units, either milli- or microamperes, must be used for both currents. However, the gain factor has no units because the units in the ratio cancel.

Voltage Gain A_V The formula for A_V is

$$A_V = \frac{\text{output ac signal voltage}}{\text{input ac signal voltage}} \quad (2\text{-}6)$$

The p-p collector voltage in Fig. 2-7 is 7.2 V. For the base voltage, 100 mV p-p corresponds to the 80 μA of base current for the input signal. The values are derived from the input characteristics shown in Fig. 2-3. The gain calculations are

$$A_V = \frac{7.2 \text{ V}}{100 \text{ mV}} = \frac{7200 \text{ mV}}{100 \text{ mV}}$$

$$A_V = 72$$

An approximate formula that can be used for calculating voltage gain in the CE circuit, without a graphical analysis, is

$$A_V = \beta \times \frac{R_L}{r_{in}} \quad (2\text{-}7)$$

where R_L is the collector load, r_{in} is the dynamic ac resistance of the base-emitter junction, and β is the ac beta. Note that the voltage gain is beta multiplied by the ratio R_L/r_{in}. The higher R_L is, the greater the voltage gain is. For the values used here, R_L is 2000 Ω, r_{in} is 1250 Ω from Fig. 2-4, and the ac β is 50. Then

$$A_V = 50 \times \frac{2000}{1250} = 50 \times 1.6$$

$$A_V = 80$$

This value is a little higher than the voltage gain of 72 by graphical analysis, but the formula is a useful shortcut to the approximate amount of gain.

Power Gain A_P The A_P factor is the product of the current gain and voltage gain. The formula is

$$A_P = A_I \times A_V \qquad (2\text{-}8)$$

For the example here,

$$A_P = 45 \times 72 = 3240$$

Test Point Questions 2-7
(Answers on Page 42)

a. The signal output is 4 A p-p for a signal input of 200 mA p-p. What is the value of A_I?
b. The signal input is 600 μA p-p. The current gain A_I equals 100. What is the p-p output signal current?
c. The signal input is 30 mV p-p. The voltage gain A_V is 200. What is the output signal voltage?

2-8
POWER CALCULATIONS

By continuing with the same values for the load-line example in Fig. 2-7, we can determine the dc power dissipated at the collector and the amount of ac power output for the signal. The dc power dissipation should be less than the maximum rating for the transistor. The ac power output must be high enough to serve for the load in the next circuit. For an audio output stage driving a loudspeaker, for instance, the ac power determines how loud the sound will be.

Transistor Power Dissipation The transistor uses power from the dc supply to provide amplification of the ac signal. Just about all the power is dissipated at the collector junction. Therefore, the amount of dc power used is

$$P_{dc} = V_C \times I_C \qquad (2\text{-}9)$$

where V_C and I_C are average dc values for collector voltage and current at the Q point. For the values in Fig. 2-7, V_C is 8 V and I_C is 3mA. Then

$$P_{dc} = 8\text{ V} \times 3\text{ mA}$$

$$P_{dc} = 24 \text{ milliwatts (mW)}$$

The dc power dissipation has the relatively low value of 24 mW for this small-signal amplifier. However, consider a power output stage of an amplifier with $V_C = 6$ V and $I_C = 5$ A. Then

$$P_{dc} = 6\text{ V} \times 5\text{ A} = 30 \text{ watts (W)}$$

AC Power Output For a class A amplifier, the ac power can be calculated from the signal swings in collector current and voltage. The formula is

$$P_{ac} = \frac{(V_{max} - V_{min}) \times (I_{max} - I_{min})}{8} \qquad (2\text{-}10)$$

The divisor, 8, converts p-p to rms values for both V and I in the power calculations.

Division by 2 converts p-p to peak, and the divisor $1/\sqrt{2}$ converts peak to rms. Both V and I must be converted. The multiplying factor, then, is $1/2 \times 1/\sqrt{2} \times 1/2 \times 1/\sqrt{2}$, which is equal to 1/8.

For the ac signal swing in Fig. 2-7, the values are

$$P_{ac} = \frac{(11.6\text{ V} - 4.4\text{ V}) \times (4.8\text{ mA} - 1.2\text{ mA})}{8}$$

$$= \frac{7.2\text{ V} \times 3.6\text{ mA}}{8} = \frac{25.92\text{ mW}}{8}$$

$$P_{ac} = 3.24 \text{ mW}$$

The ac power output for the signal must be less than the dc power output because the amplifier cannot be 100 percent efficient. All the power for the ac signal output comes from the dc power supply.

Power Efficiency Power efficiency is the ratio of ac signal power output to dc power dissipation. The formula is

$$\text{Power efficiency, \%} = \frac{P_{ac}}{P_{dc}} \times 100 \quad \textbf{(2-11)}$$

For example, for the values of power already calculated for the small-signal class A amplifier, the ac power $P_{ac} = 3.24$ mW and $P_{dc} = 24$ mW,

$$\text{Power efficiency} = \frac{3.24 \text{ mW}}{24 \text{ mW}} \times 100$$

$$= 0.135 \times 100$$

$$\text{Power efficiency} = 13.5\%$$

Test Point Questions 2-8
(Answers on Page 42)

a. An amplifier has $V_C = 8$ V and $I_C = 2.4$ A. Calculate the dc power dissipation.
b. The same amplifier has p-p values of 12 V and 0.3 A for the signal swing in collector voltage and current. Calculate the ac power output.

2-9
TROUBLES IN THE
DC ELECTRODE VOLTAGES

When the ac signal is not amplified in the way it should be, the trouble is usually caused by dc voltages that are too high or too low at the transistor electrodes. A typical CE circuit is shown in Fig. 2-8 to illustrate the procedure of checking dc values for the amplifier.

To check voltages and currents power must be applied to the circuit. The dc values of a class A amplifier are approximately the same with or without ac signal. However, the dc values marked on a schematic diagram for commercial equipment are those that would be measured with no input signal. The ac signals are generally shown separately as oscilloscope waveforms. A VOM is preferable for making the measurements because it need not have one side at chassis ground.

Checking the Forward Bias Connect the VOM directly across the base and emitter terminals to measure V_{BE}. The value should be about 0.6 V for silicon or 0.2 V for germanium transistors. In Fig. 2-8, V_{BE} is $1.6 - 1.0 = 0.6$ V. Those values apply to most amplifiers. However, some pulse circuits may have a small reverse bias to cut I_C off until the input pulse drives the transistor into saturation.

When the reading for V_{BE} is zero, the junction is shorted. If V_{BE} is 0.8 V or higher, the junction is probably open.

Checking the Collector Current To check I_C, put the VOM between the collector and the supply voltage to read the $I_C R_L$ voltage drop. In Fig. 2-8, the VOM should read 4 V across the 2-kΩ R_L. Then I_C is equal to 4 V ÷ 2 kΩ = 2 mA.

If there is no voltage drop across R_L, the I_C must be zero. Either R_L is open or the transistor is cut off.

For the opposite trouble, excessive I_C makes V_{R_L} too high. Then V_C is very low or zero. Note that V_C is low when V_{R_L} is high.

Low I_C If I_C is zero or very low, the transistor may be open. However, check for correct forward bias and opens in the emitter circuit before making a replacement.

An important special case is R_L open. Then V_C is zero because there is no line to the supply voltage. I_C also is zero because there is no collector voltage.

High I_C To find out why I_C is too high, try shorting the emitter. That should cut I_C off because the forward bias is shorted. If I_C is still too high, the transistor probably has an internal short at the collector junction.

Checking the Emitter Circuit You can also check current by measuring the voltage drop across R_E in Fig. 2-8. Remember that V_E is $I_E R_E$ and also that I_E is $I_C + I_B$.

Make sure the resistance of R_E is normal. Emitter voltage V_E will read only a little higher than normal even if R_E is open. The reason is that the resistance of the voltmeter completes the circuit as a substitute for an open R_E when V_E is measured to ground.

Measuring I_C When the collector load is a coil or transformer primary, it may be better to measure I_C directly. You can open the collector circuit by cutting the foil of the printed circuit with a razor blade. Then put the leads of the VOM, set to read milliamperes or amperes, across the cut. The meter now reads current because it is in series with the collector circuit. After this test is finished bridge the cut with a spot of solder.

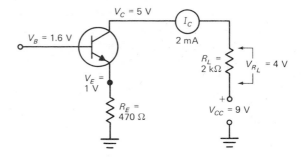

Fig. 2-8. Checking dc voltages and currents in a CE amplifier.

Test Point Questions 2-9
(Answers on Page 42)

Refer to Fig. 2-8.

a. With R_L open, what is the value of V_C?
b. With R_E open, what is the value of V_C?

SUMMARY

1. In order to amplify ac signal, the transistor amplifier must have the required dc operating voltages. They include reverse collector voltage and forward bias at the base.
2. For amplification of an ac signal, the variations in base current must vary the collector current. Then the varying voltage drops across R_L will result in variations of collector voltage.
3. For the CE circuit, the amplified signal voltage at the collector has inverted polarity.
4. Collector characteristic curves show V_{CE} plotted against I_C. A family of collector curves for the CE circuit is shown in Fig. 2-5 for different values of base current.
5. The beta (β) characteristic is the ratio of collector current to base current. Typical values of β are 10 to 300. β specifies the current gain for a CE amplifier.
6. The alpha (α) characteristic is the ratio of collector current to emitter current. A typical value is 0.98. It must be less than 1 because I_C must be less than I_E.
7. Graphical analysis with a load line drawn on the collector characteristic curves can determine all values of v_C and i_C for a specific R_L and V_{CC}. See Fig. 2-7.
8. In the letter symbols for transistors, capital letters such as V_C and I_C are used for average dc values. Small letters v and i are used for instantaneous changing values.

9. Current gain A_I is the ratio of output signal current to the input signal current. See Eq. 2-5.
10. Voltage gain A_V is the ratio of output signal voltage to the input signal voltage. See Eqs. 2-6 and 2-7.
11. Power gain A_P is the product $A_I \times A_V$. See Eq. 2-8.
12. The dc power dissipated at the collector is $V_C \times I_C$. See Eq. 2-9.
13. The ac power output is the product of the rms values for signal voltage and current. See Eq. 2-10.
14. When there are troubles in a transistor amplifier, first check the dc electrode voltages for the correct V_C and forward bias at the base.

SELF-EXAMINATION
(Answers at back of book)

1. In Fig. 2-1, what is V_{R_L}?
2. In Fig. 2-2, what is A_V?
3. In Fig. 2-3, what is the phase relation between the waveforms in Fig. 2-3b and f?
4. In Fig. 2-4, what is I_B for V_{BE} of 0.7 V?
5. In Fig. 2-5, what is I_C for V_C of 8 V and I_B of 160 μA?
6. In Fig. 2-7, what are V_C, I_C, and I_B at the Q point?
7. In Fig. 2-7, what is the p-p signal swing in v_C, i_C, and i_B?
8. Calculate the dc power dissipated at the collector for quiescent values of 2 A for I_C and 5 V for V_C.
9. Calculate ac power output for the p-p values of 4.3 A for i_C and 10 V for v_C.
10. In Fig. 2-8, what is V_C when I_C is zero?

ESSAY QUESTIONS

1. List the three dc electrode voltages for the junction transistor. Do the same for the FET.
2. What is meant by quiescent values for V and I?
3. Show an input characteristic curve like Fig. 2-4 but for a germanium transistor with typical bias of 0.2 V.
4. Show waveforms similar to those in Fig. 2-3b and f with the same input signal but with a voltage gain of 80.
5. What is meant by phase inversion of the signal in a CE amplifier?
6. Define beta (β). Why is it more than 1?
7. Define alpha (α). Why is it less than 1?
8. What is meant by r_{in} for a transistor in the CE circuit?
9. Compare the definitions for the following pairs of letter symbols: V_C and V_{CC}; V_D and V_{DD}; I_C and i_C; I_D and i_D; V_B and v_B; ac β and dc β; h_{fe} and ac β.
10. Give the formulas for voltage, current, and power gain.

11. Give the formula for dc power dissipation at the collector.
12. Give the formula for ac power output from an amplifier.
13. Give the formula for power efficiency in an amplifier.
14. Describe briefly two possible troubles in the dc values for a transistor amplifier.

PROBLEMS
(Answers to odd-numbered problems at back of book)

1. Calculate the dc power dissipation for $V_C = 5$ V and $I_C = 4$ A.
2. Repeat Problem 1 for $I_C = 4$ mA.
3. Calculate the ac power output with p-p signal values of 7 V for v_C and 5 A for i_C.
4. Repeat Problem 3 with 5 mA for i_C.
5. Calculate the efficiency with 20 W dc power dissipation and 6.5 W ac power output.
6. Calculate the efficiency with 250 mW dc power dissipation and 45 mW ac power output.
7. A change of 2 mA in i_B changes i_C by 120 mA. Calculate the ac beta.
8. A transistor has ac beta of 60. For a change of 2 mA in i_B, what is the change in i_C?
9. Refer to the collector characteristic curves in Fig. 2-9. Calculate the dc beta at a V_C of 3 V and an I_B of 1.5 mA.
10. V_E is at -8 V, the V_B is -7.4 V and the collector is at dc ground potential. Determine V_{BE} and V_{CE}.

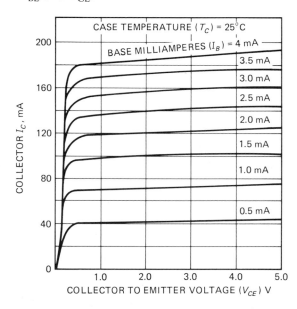

Fig. 2-9. Collector characteristic curves for a silicon medium-power NPN transistor. (RCA)

11. For the curves shown in Fig. 2-9 draw the load line for $V_{CC} = 4$ V and $R_L = 25$ Ω. Give all values as shown in Fig. 2-7, including dc power, ac power output and efficiency. Use I_B of 1.5 mA for V_{BE} of 0.6 V and Δi_B of $\pm$ 1 mA for Δv_B of $\pm$ 0.08 V.

SPECIAL QUESTIONS

1. Give two general requirements for any type of amplifying device.
2. Give three comparisons between the load-line analysis for a junction transistor in Fig. 2-7 and the same method for an FET.
3. By using the collector characteristics in Fig. 2-9, draw a forward-transfer curve for the transistor. Plot I_C against I_B, and use a constant V_C of 2 V.
4. Explain how it is possible to use a negative dc supply of -9 V for NPN transistors.

ANSWERS TO TEST POINT QUESTIONS

2-1	a. 3 mA	2-5	a. 100 μA	2-8	a. 19.2 W		
	b. 0.6 V		b. 4.8 mA		b. 3.6 W		
2-2	a. ± 50 mV		c. 4.4 V	2-9	a. 0 V		
	b. ± 40 μA		d. 4 mA and 6 V		b. 9 V		
	c. ± 1.8 mA	2-6	a. i_C				
	d. ∓ 3.6 V		b. I_c				
2-3	a. 20 μA		c. I_{CQ}				
	b. 7 mA	2-7	a. 20				
2-4	a. 50		b. 60 mA				
	b. 0.98		c. 6 V p-p				
	c. 200 mA						

Chapter 3
Combinations of Amplifiers

In most applications, one amplifier stage cannot supply enough signal output because the desired signal is usually very small at the input. For instance, the RF signal at the antenna of a radio receiver is generally in microvolts or millivolts. Also, audio signal from a microphone, phonograph, or tape recorder is in the millivolt range. The voltage or current needed to operate a loudspeaker is, however, much greater than the signal input to the amplifier. The louder the sound we want to hear, the greater the audio power output needed.

Between the low-level signal input and the high-level output, two, three, or more amplifier stages are required. A single stage that operates with a low-level signal does not have enough output power. A stage that can supply enough power needs a relatively large signal input. Different methods of combining amplifier stages are described in the following topics:

3-1
AMPLIFIERS IN CASCADE

When there are junction transistors in the CE circuit, the amplified output from the collector drives the base electrode of the next stage. In Fig. 3-1a, the idea is to build up the signal for enough base current i_B to drive the power output stage $Q3$. The base voltage for all the stages should vary by only 0.1 V, or less, for minimum distortion. However, the transistors are chosen for the required amount of collector current i_C. A transistor with more i_C in the output generally has more i_B for the input. A higher i_B means that more signal drive is needed to vary the base current.

Power Output Stage Transistor $Q3$ in Fig. 3-1a provides signal variations in i_C equal to ±2 A in this example. It needs a base signal i_B of ±100 mA. That value for the input is one-twentieth of the output, assuming a β value of 20 for the current-transfer ratio. The power output stage is the final amplifier, which drives the load instead of another amplifier.

The load generally requires relatively high signal current or signal voltage. As two examples, a loudspeaker may need 2 to 5 A of audio signal current through the voice coil and a picture tube may need 100 V of video signal for grid-cathode voltage.

Those values for the signal must be provided with minimum distortion.

Preamplifier For the opposite case, $Q1$ in Fig. 3-1a is the preamplifier for low-level input from the source of a signal. This stage is shown with a base-signal input of ±20 μA, which is small for i_B. The output signal is ±2 mA for i_C. As a result, $Q1$ has enough signal output at the collector to drive the base of $Q2$.

Distortion is generally not a problem in the preamplifier because of the small signal. However, the preamplifier must have very low hum and noise. A low-level signal can easily be masked by interference.

Driver In the middle between $Q3$ and $Q1$ is the driver stage $Q2$. This amplifier provides enough i_C to drive the base input circuit of the power output stage. The preamplifier can drive $Q2$, but $Q1$ does not have enough output power to drive $Q3$.

Cascade Multiplication of Current Gain A_I In Fig. 3-1a, the values of current gain are 100 for $Q1$, 50 for $Q2$, and 20 for $Q3$. The overall gain from the input at $Q1$ to the output from $Q3$ is the product of those values. Then

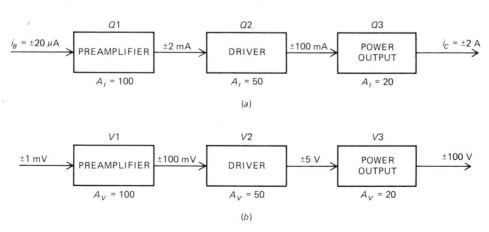

Fig. 3-1. Block diagram of three amplifier stages in cascade. (a) Three cascaded junction transistors. Current gain (A_I) is shown for each stage. (b) Three cascaded vacuum tube amplifiers. Voltage gain (A_V) is shown for each stage.

$$\text{Total } A_I = A_{I_1} \times A_{I_2} \times A_{I_3}$$

$$= 100 \times 50 \times 20$$

$$\text{Total } A_I = 100{,}000$$

The reason for the multiplication is that each stage amplifies the amplified signal from the preceding stage.

Cascade Multiplication of Voltage Gain

A_V The signal drive of vacuum-tube amplifiers is control-grid voltage. Cascaded stages are then used to build up the amount of voltage for the grids of successive stages (see Fig. 3-1b). The overall voltage gain is

$$\text{Total } A_V = A_{V_1} \times A_{V_2} \times A_{V_3}$$

$$= 100 \times 50 \times 20$$

$$A_V = 100{,}000$$

The idea of cascaded voltage gain with tubes applies to field-effect transistors also, but usually with lower voltages. With an FET, the input signal drive is variations in gate voltage for the input.

Circuit for Two CE Stages in Cascade
Figure 3-2 shows how the collector of $Q1$ drives the base of $Q2$ for two RC-coupled amplifiers. Each stage uses an NPN transistor in the common-emitter circuit. Emitter bias is provided by R_E for stabilization, with fixed forward bias at the base.

For cascaded gain, $Q2$ amplifies the amplified output of $Q1$. The original input signal is coupled by C_1 to the base of $Q1$. The amplified output of $Q1$ in the collector circuit is coupled by C_2 to the base of $Q2$. The signal—the output from the collector of $Q1$—is amplified again by $Q2$. The total cascaded gain for the two stages results in a strong signal for the output coupled by C_3 to the next circuit.

Cascaded Polarity Inversions of Signal Voltage
The signal waveforms shown in Fig. 3-2 indicate how each stage inverts the polarity, or shifts the phase by 180°. The polarity inversions are for signal voltage, not current.

Consider the phase inversion in $Q1$, which is NPN. The ac signal for v_{in} has positive and negative half-cycles. The positive polarity of v_{in} increases the base current. The result is more collector current. However, the collector voltage decreases because of a larger voltage drop across R_L. Therefore, increasing the positive voltage for v_{in} causes a decreasing positive voltage at the collector. The change is equivalent to a negative change in signal voltage for v_C. The result is that the positive half-cycle of v_{in} at the base corresponds to the negative half-cycle of v_C at the collector.

The phase inversion applies to $Q2$ also. There the negative polarity of v_1 into the base circuit

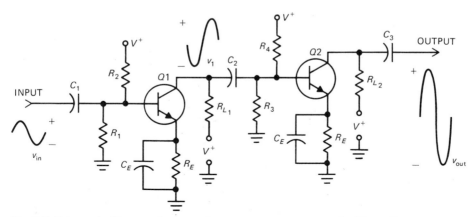

Fig. 3-2. Schematic diagram of two common-emitter stages in cascade. The collector output of $Q1$ drives the base input of $Q2$.

decreases the amount of i_B. Now the collector current decreases. As a result, the positive v_C for $Q2$ increases. The reason for this is that there is less voltage drop across R_L. Therefore, the negative half-cycle of the input of $Q2$ corresponds to the positive half-cycle of the amplified output signal v_{out}.

Note that v_{out} from the two cascaded stages has the same polarity as the original signal v_{in}. Two inversions of 180° total 360°, so the output signal is back in phase with the original input signal.

In general, when there is an even number of cascaded stages (2, 4, 6, etc.) the signal output is not inverted from the input. When the number of stages is odd (1, 3, 5, etc.), the output signal is inverted from the input. It should be noted that polarity inversions apply to PNP transistors also.

The phase of the ac input signal does not matter for most amplifiers, because the gain is the same. However, some types of clipping circuits require a specific polarity for the signal to be clipped properly.

In another application, the phase of an ac audio signal for a loudspeaker does not affect speaker operation. A special case, though, is the phasing of loudspeakers for stereo. Also, a picture tube for television requires a specific polarity of video signal to reproduce the image in the correct shades of white or black.

Test Point Questions 3-1
(Answers on Page 63)

a. Three cascaded stages have gains of 30, 20, and 10. What is the overall gain?
b. How many cascaded stages of CE amplifiers will result in polarity inversion of the input signal: two or three?

3-2
DARLINGTON PAIR

The Darlington pair circuit has two emitter followers in cascade. An emitter follower is the common-collector circuit. In order to review the CC circuit, just one stage is shown in Fig. 3-3. An

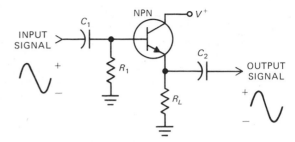

Fig. 3-3. The emitter-follower circuit.

input signal is applied to the base, and the output is taken from the emitter. The load impedance R_L is in the emitter circuit, instead of the collector circuit.

There is no phase inversion, because the positive and negative peaks of the signal occur at the same time as the positive and negative peaks of the input voltage. The voltage gain is slightly less than 1, but the current gain can be high because I_E is much more than I_B. An emitter follower provides high input impedance at the base and low output impedance from the emitter circuit.

The advantage of the Darlington pair shown in Fig. 3-4 is higher input impedance, lower output impedance, and more current gain as compared with a single emitter follower. The reason for this improvement is that the overall gain of the pair is the product of the gains of the individual cascaded stages.

In Fig. 3-4a an input signal is shown applied to the base of $Q1$. The output signal at the emitter is directly coupled to the base of $Q2$. This stage provides more current gain for the output signal from the emitter. The collector terminals for $Q1$ and $Q2$ are connected together to form the dc supply voltage terminal. Also, the base of $Q1$ must have forward bias voltage, but it is not shown here.

A typical Darlington unit is shown in Fig. 3-4b. Transistors $Q1$ and $Q2$ are actually one chip. Only three terminals are needed for the connections as shown in Fig. 3-4c. Terminal B is the base of $Q1$. Signal input is applied there by the external coupling capacitor C_1. Terminal E is the emitter of $Q2$. The load R_L is connected here for the output signal.

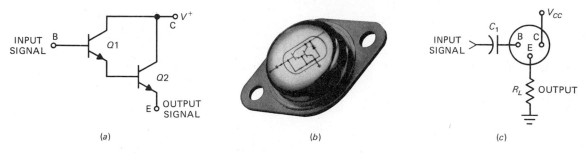

Fig. 3-4. Darlington pair consisting of two emitter followers in cascade. (a) Circuit with emitter of Q1 connected to base of Q2. (b) Device package with just three terminals. (c) Terminal connections.

The collector supply voltage is connected to terminal C. It should be noted that either of two connections can be used for V_{CE}. When R_L is returned to ground in the emitter circuit, as shown here, V^+ is used for the collector. Alternatively, the collector can be grounded but with R_L returned to the negative supply voltage for the emitter. Those polarities are for NPN transistors. The Darlington circuit can use two NPNs or PNPs or one of each.

Test Point Questions 3-2
(Answers on Page 63)

Answer True or False.

a. In the emitter follower, R_L is in the collector circuit.
b. In a Darlington pair, the emitter of Q1 is dc-coupled to the base of Q2.

3-3
PARALLEL AND SERIES STAGES

Just as resistors are connected in parallel or series to meet different current and voltage needs, so too can transistors be connected in parallel or series for a required application. In general, parallel connections divide the current but series connections divide the dc supply voltage.

Parallel Transistors In Fig. 3-5 all electrodes of Q1 are shown joined with the same terminals of Q2. Then the B, E, and C terminals are used

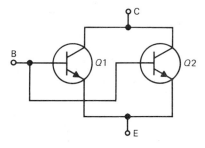

Fig. 3-5. Two transistors connected in parallel. All electrodes of Q1 are joined with like electrodes of Q2.

as one transistor, for signal and supply voltage, in any circuit configuration. However, the parallel combination has twice the current rating of one transistor, assuming the transistors are identical. The reason is that each transistor supplies one-half the output current. Actually, three or more transistors could be connected in parallel in the same way.

Parallel transistors would be used where a single transistor having the required power rating for one transistor might be too expensive or perhaps not be available. In an audio power output stage, for example, two 25-W transistors could cost less than one 50-W transistor.

Series Transistors In Fig. 3-6, Q1 and Q2 divide the dc supply of 48 V for their collector voltage. The two transistors are in series with the dc supply voltage because the emitter of Q1 is joined with the collector of Q2. Assuming they have the same current rating, Q1 and Q2 can be considered as equal resistances across the 48-V supply, or Q1

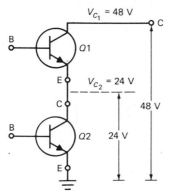

Fig. 3-6. Two transistors connected in series for their collector voltage. Transistors $Q1$ and $Q2$ divide 48 V from the dc supply.

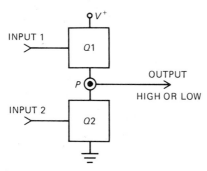

Fig. 3-7. Totem-pole connections for $Q1$ and $Q2$ in digital logic circuit.

serves as a voltage-dropping resistor for the collector of $Q2$.

Consider the collector voltage on $Q1$. The 48 V for V_{C_1} is to chassis ground. However, the emitter is at 24 V. Therefore, V_{CE} is $48 - 24 = 24$ V. As for $Q2$, the V_{C_2} of 24 V is also equal to V_{CE}. The reason is that the $Q2$ emitter is grounded.

As a result, both $Q1$ and $Q2$ have an effective reverse voltage of 24 V for the collector. This series arrangement is called *stacking*; $Q1$ and $Q2$ are stacked for the dc collector voltage.

It should be noted that separate connections are available to the base for $Q1$ and $Q2$. Each can have its own input signal with the required amount of base bias.

Totem Pole The name *totem pole* is used for a circuit with two stacked transistors in digital logic

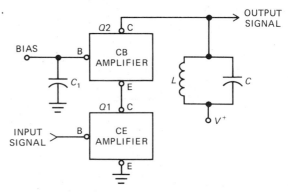

Fig. 3-8. Cascode circuit consisting of the common emitter amplifier directly coupled to the common-base amplifier.

applications. The block diagram in Fig. 3-7 shows $Q1$ and $Q2$ stacked for V^+.

A HIGH- or LOW-level output is available at point P, between the two transistors. Conduction in $Q2$ grounds P to pull the output voltage down to the LOW level.

Cascode Circuit The name *cascode* is similar to "cascade," but as Fig. 3-8 shows, the cascode circuit has a common-emitter stage directly coupled to a common-base stage. Transistors $Q1$ and $Q2$ are stacked for V^+.

An amplified signal output from the collector of $Q1$ drives the emitter of $Q2$. The base of $Q2$ is grounded for an ac signal with a bypass capacitor. This cascode circuit is used with broadband RF amplifiers for high gain with good stability.

Test Point Questions 3-3
(Answers on Page 63)

a. Is the current output increased by connecting transistors in series or in parallel?

b. In Fig. 3-6, what value of V_B is needed for 0.6 V forward bias on $Q1$?

3-4
PUSH-PULL AMPLIFIER

The push-pull amplifier circuit uses two transistors or tubes for opposite half-cycles of the ac signal.

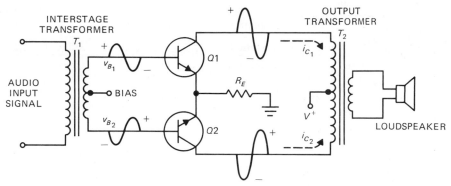

Fig. 3-9. Push-pull circuit for an audio output stage using NPN transistors.

In the example shown in Fig. 3-9, two NPN transistors are used in an audio power output stage driving a loudspeaker. Push-pull is also used for RF amplifiers.

The push-pull operation means that when the current in one stage is increasing, the current in the other stage is decreasing. However, the opposing changes are made to reinforce each other in the output load instead of canceling one another. The push-pull circuit is said to be *double-ended*. *Single-ended* just means that the amplifier is not push-pull. That is, there is just one output terminal for the signal with respect to the chassis ground.

A push-pull audio amplifier can operate class B for high efficiency, because each stage provides opposite half-cycles of the ac audio signal. However, class A operation also is used for minimum distortion. Class AB operation for good power and efficiency is very common. Push-pull operation is the best way to use two transistors for high power output with little distortion. Most audio power output circuits are push-pull.

Push-Pull Circuit In Fig. 3-9 the interstage transformer T_1 has a center-tapped secondary. It supplies equal and opposite ac signals to drive the base circuits for Q1 and Q2, which are CE amplifiers. Dc bias at the center tap provides forward voltage for both transistors.

The output transformer T_2 also has a center tap connected to V^+ for the collector voltage. Transistors Q1 and Q2 each uses one-half the primary

winding for collector current. The secondary winding feeds current to the loudspeaker without a center tap. In the primary, the three leads are usually color-coded red for V^+ and blue for each of the two collectors.

The Q1 and Q2 emitters are connected together, and there is one resistor R_E for emitter bias. A bypass capacitor may be used across R_E. However, the bypass is not needed in class A operation because the push-pull currents cancel variations in I_E.

Push-Pull Operation The equal and opposite variations of collector current i_C in class A operation are shown in Fig. 3-9. Note that i_{C_1} flows down through the top half of T_2 and returns to V^+ through the center tap. However, i_{C_2} flows up through the other half of T_2. (The i_C directions shown are for electron flow.) These two signal currents are 180° out of phase. When one is increasing, the other is decreasing because the opposite polarities of the input signals to the bases are 180° out of phase.

The push-pull effects in the secondary of the output transformer of the loudspeaker are additive. To see that effect in terms of transformer action, assume that i_{C_1} flowing down in the primary of T_2 has a counterclockwise magnetic field. Then i_{C_2} has a clockwise field because it flows up with the same direction of winding. The two fields induce voltages of opposite polarities in the secondary when the variations of i_{C_1} and i_{C_2} are the same. The fact is, though, that the push-pull variations

Table 3-1
Push-Pull Operation for Circuit in Fig. 3-9

Current	Direction	Magnetic Field	Variation	Induced Voltage
i_{C_1}	Down	Counterclockwise	Increase	+
			Decrease	−
i_{C_2}	Up	Clockwise	Decrease	+
			Increase	−

of signal current are opposite. When i_{C_1} is increasing because of more positive base voltage, i_{C_2} is decreasing because of less positive base voltage. On the next half-cycle, the variations are reversed but the changes in i_{C_1} and i_{C_2} are still opposite.

Remember that a counterclockwise magnetic field that is expanding with increased current has the same flux variations as a clockwise field collapsing with decreased current. Each will produce the same polarity of induced voltage. As a result, the push-pull operation allows the opposite collector currents to add in producing secondary voltage in the output transformer. Those points are summarized in Table 3-1.

Phase Splitter In the phase splitter circuit one stage is used to provide two equal and opposite output voltages to drive a push-pull amplifier. The purpose is to eliminate the push-pull input transformer.

Figure 3-10 shows a typical circuit. The phase splitter divides the output load resistance between the collector and emitter branches. The output signals from collector and emitter have opposite polarities. With equal load resistances, the two signals have the same amplitude but are 180° out of phase.

Push-Pull Advantages and Disadvantages
The push-pull output circuit cancels all components of i_C that have the same phase in the two transistors. This effect really applies to everything but the push-pull input signal. An example of cancellation is hum from the V^+ supply. The hum has the same phase in both transistors; therefore, it is canceled in the push-pull output.

Another advantage of the push-pull circuit is a reduction in the nonlinear amplitude distortion caused by the output signal not having the same relative amplitudes as the input signal. Such distortion is always a problem in amplifiers with large drives required for high power output. In push-pull operation, though, when one stage is near saturation in the output circuit, the other is near cutoff. The result is symmetrical peaks of output current.

Lack of symmetry in the amplitudes is called *second-harmonic distortion*, because it is equivalent to adding new frequencies at the second harmonic of the fundamental. The symmetry in the push-pull output can be considered to eliminate it. The advantage applies only to distortion generated in the amplifier itself. Any distortion that is part of the input is amplified as a push-pull signal.

A problem with class B push-pull operation is *crossover distortion* near cutoff; nonlinear amplification results when the stage crosses over to conduction. In fact, both stages may be off for a short time before one stage comes on. However, crossover distortion is reduced with class AB operation, and especially so with class A.

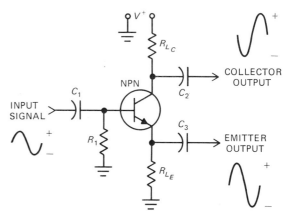

Fig. 3-10. Phase splitter circuit used to provide input to a push-pull circuit.

It is important that the two transistors in a push-pull stage have matched characteristics. Otherwise, the advantages of balance and symmetry in the output currents cannot be obtained.

Answer True or False.

a. Push-pull signals are equal and opposite in phase.
b. Class AB operation cannot be used for a push-pull audio power output stage.
c. A push-pull output transformer has a center-tapped primary.

3-5
STACKED SERIES OUTPUT

The circuit shown in Fig. 3-11 has the input transformer T_1 to supply push-pull drive, but the output transformer is eliminated. Instead, the coupling capacitor C_C is used for the audio output.

Transistors $Q1$ and $Q2$ are stacked in series for V^+. Both are NPN transistors. The emitter of $Q1$ at the top is directly connected to the collector of $Q2$. The junction at P is used for taking off the amplified audio signal.

In the input circuit, the push-pull signal from T_1 is connected to the base of $Q1$ and $Q2$. Instead of a tapped secondary, two separate windings are necessary. The split windings are across terminals 1 and 2 and 3 and 4. It is not possible to use a grounded center tap here because only the $Q2$ emitter is grounded and the emitter of $Q1$ is part of the output circuit. Resistors R_1 to R_4 provide the required forward bias voltage at the base of $Q1$ and $Q2$.

Although there is no output transformer, the amplified audio signal drives the loudspeaker by the charge and discharge currents of C_C. This capacitor is of the electrolytic type, and it has the high value of 470 μF for good low-frequency response. The positive side of C_C is at P.

Note that point P has the emitter current i_{E_1} of $Q1$ and the collector current i_{C_2} of $Q2$. With NPN transistors, an increase in i_{E_1} for $Q1$ makes P more positive. Consider the direction as electron flow returning to the emitter. With P made more positive, C_C can take on charge.

However, more i_{C_2} for $Q2$ has the opposite effect on C_C. The increase makes P more negative, or less positive, which allows C_C to discharge.

With a push-pull signal, $Q1$ and $Q2$ work in opposite directions. When i_{E_1} increases to make P more positive, i_{C_2} decreases to make P less negative (or also more positive). This is when the

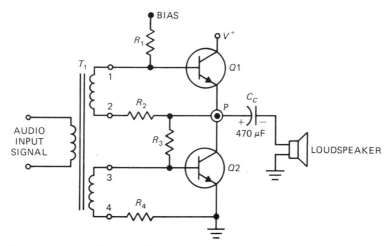

Fig. 3-11. A transformer with a split-secondary winding provides the input for a push-pull amplifier. The series-output connection does not need a push-pull output transformer.

loudspeaker has the charging current for C_C. On the next half-cycle, i_{E_1} decreases and i_{C_2} increases. Then both transistors make P less positive, which allows C_C to produce discharge current through the loudspeaker. As a result, the charge and discharge currents of C_C produce the audio signal output in the loudspeaker. This circuit with C_C for the loudspeaker is sometimes called an *OTL* circuit, meaning it is output transformerless.

Test Point Questions 3-5
(Answers on Page 63)

Refer to Fig. 3-11.

a. Does T_1 supply push-pull or single-ended input?

b. Is C_C used to provide an input or an output signal?

3-6
COMPLEMENTARY SYMMETRY

A PNP transistor is used with an NPN transistor to provide complementary current variations for the same input signal. As one current increases, the other decreases. The purpose is to eliminate the need for a push-pull drive. Furthermore, a stacked series-output circuit can be used, which eliminates the output transformer. As a result, the typical complementary-symmetry audio output circuit in Fig. 3-12 is completely transformerless.

Input Circuit In the circuit of Fig. 3-12 the audio signal is coupled to both $Q1$ and $Q2$ by C_1. The single-ended input delivers the same polarity to both transistors, but one is PNP and the other NPN. Assume the audio input signal increases positively or in a positive direction. The positive base voltage for $Q2$, which is NPN, increases i_{C_1}. However, more positive base voltage for the PNP transistor $Q1$ decreases i_{C_2} for this transistor.

On the next half-cycle, the negative signal makes the base less positive for both transistors. Then i_{C_1} decreases in the NPN transistor but i_{C_2} increases in the PNP transistor. The result is push-pull collector current for the output signal, even

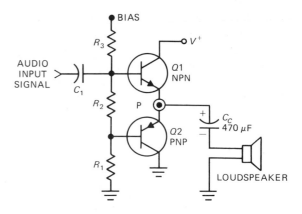

Fig. 3-12. Complementary symmetry with NPN and PNP transistors for push-pull operation without input or output transformers.

though the circuit has single-ended input signal at the base of $Q1$ and $Q2$. Resistors R_1 to R_3 are used for forward bias in the base circuits.

Output Circuit In Fig. 3-12 transistors $Q1$ and $Q2$ are stacked in series for V^+. The two emitters are connected together. Part of V^+ supplies positive V_{CE} for $Q1$. However, the other part of V^+ is the emitter voltage on $Q2$. The positive V_E corresponds to negative collector voltage for the PNP transistor.

The series-output circuit provides the audio signal to the loudspeaker by the charge and discharge currents of C_C. More I_E in the NPN transistor $Q1$ makes point P more positive and allows C_C to charge. More I_E in the PNP transistor $Q2$ makes point P less positive, which means C_C can discharge. However, the emitter currents are in push-pull: when one increases, the other decreases. As a result, the opposite half-cycles of the signal cause the charge and discharge of C_C to produce the load current in the loudspeaker.

Quasi-Complementary Symmetry The name *quasi-complementary symmetry* is applied to an audio circuit in which the driver stages are NPN and PNP but both power output stages are NPN. *Quasi* means similar to but not exactly the same as. The purpose is to avoid the higher cost of a high-power PNP transistor in the output by using the lower-cost NPN type instead. Also, it is easier to match the operating characteristics of two NPN power output transistors.

In quasi-complementary symmetry, the PNP and NPN drivers have the same single-ended input signal to the bases of both transistors. However, they supply the push-pull signal for the power output circuit.

Complementary Logic Circuits The complementary effect of P and N polarities is often applied to FETs. As shown in Fig. 3-13, the idea is to combine a P-channel FET with an N-channel FET. Transistors $Q1$ and $Q2$ are enhancement-type IGFETs or MOSFETs on an IC chip.

The two transistors are stacked in series for V^+, with the drain electrode of $Q1$ connected to the drain of $Q2$. Note that the source of $Q1$ is shown at the top. This FET has a P-channel, which needs negative drain voltage. However, V^+ at the source makes the drain negative with respect to the source. At the bottom, transistor $Q2$ with an N-channel has positive drain voltage.

The purpose of the circuit is to invert the pulses at the input. A HIGH or LOW level at the input is inverted to LOW or HIGH at the output. This use of opposite channel polarities is abbreviated to CMOS or COS/MOS, for complementary symmetry with metal-oxide semiconductors.

Test Point Questions 3-6
(Answers on Page 63)

Answer True or False for the circuit in Fig. 3-12.

a. Capacitor C_1 couples single-ended input signal for both $Q1$ and $Q2$.
b. The $Q1$ is an NPN transistor, but $Q2$ is PNP.
c. Transistor $Q2$ has a negative V_E from the V^+ supply.

3-7
DIFFERENTIAL AMPLIFIER

The circuit shown in Fig. 3-14 is the basic amplifier for linear IC chips. Also called a *differential pair*,

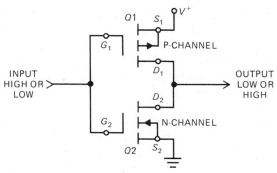

Fig. 3-13. MOSFET complementary symmetry with P and N channels, called *CMOS* or *COS/MOS*. This application is for a digital logic circuit.

the two transistors operate like a push-pull amplifier. The term *differential* refers to the difference of potential between two points in a balanced circuit. Balance in the transistors and resistances is a feature of integrated circuits.

In Fig. 3-14, $Q1$ and $Q2$ are NPN transistors in a differential pair; they are dc-coupled between the two emitters. Transistor $Q3$ is used as a constant-current source for emitter bias on both $Q1$ and $Q2$. They are balanced with equal values of R_L from one V^+ supply. The $-V_{EE}$ supply is used for forward bias at both emitters.

Input signals can be applied to the base of $Q1$ and the base of $Q2$. An amplified output signal

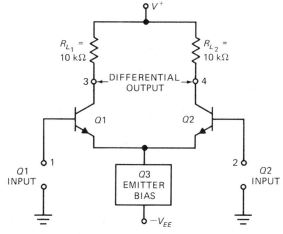

Fig. 3-14. Basic circuit of a differential amplifier. Transistors $Q1$ and $Q2$ are balanced amplifiers on an IC chip. Transistor $Q3$ is a constant-current source for emitter bias on $Q1$ and $Q2$.

can be taken either from one collector to ground or from one collector to the other collector.

When a signal is applied to only one base and the other base is grounded, the input is single-ended. The differential input is the push-pull signal to the base of both Q1 and Q2 between points 1 and 2 in the diagram.

When the amplified signal is taken from just one collector, the output is single-ended. The differential output is the signal from both collectors between points 3 and 4 in the diagram.

The combinations of input and output signals can be summarized as follows:

1. *Differential output.* The signal voltage between points 3 and 4, Fig. 3-14, is the difference between the two collector voltages. The difference is zero when the voltages are equal and of the same polarity. However, the two signal voltages add when they have opposite polarities.
2. *Differential input.* Two signal voltages of opposite polarity are connected to the base input terminals at points 1 and 2. The input is a push-pull signal.
3. *Single-ended input.* Either terminal 1 or terminal 2 can be used alone for the input signal, with respect to the chassis ground.
4. *Single-ended output.* The signal voltage is taken from the collector to chassis ground. The collector for either Q1 or Q2 can be used, or both collectors can be used. From both collectors to ground, the two signals provide a push-pull output.

Amplification of a Differential Signal   We can take an example of differential input and output voltages, assuming a gain of 100. Each stage operates as a common-emitter amplifier with input signal to the base. Let the input voltage to the Q1 base at point 1 vary by ±1 mV. The signal is amplified and inverted by Q1 to provide a signal of −100 mV in the collector circuit at point 3. Also, the opposite stage Q2 has an input signal of −1 mV for the base at point 2. The amplified output from Q2 at point 4 is then +100 mV. Both outputs are the same values but with opposite polarities.

The emitter voltage for both Q1 and Q2 is constant because the opposing signals cancel one another in the same return path for emitter current circuit. The only function of Q3 is to supply a constant bias current that is the same for Q1 and Q2. For balanced amplification, each stage must have the same bias. Actually, a resistance of about 2 kΩ can be used instead of Q3.

The differential output is the difference between the collector signal voltages at points 3 and 4. The difference in the output for the present example is 100 mV − (−100 mV), which is equal to 200 mV. For the input signal, the differential voltage between points 1 and 2 is 1 mV − (−1 mV), which is equal to 2 mV. Therefore, the voltage gain of the differential amplifier is

$$\text{Voltage Gain} = \frac{200 \text{ mV}}{2 \text{ mV}}$$
$$= 100$$

The gain achieved by differential operation is a result of two circuit conditions:

1. The output signal between the two collectors is equal to the differences in collector potential.
2. The collector voltages vary in opposite direction.

The differential gain is the same as for a single stage. The circuit can also amplify single-ended input signals of any phase.

Common-Mode Rejection   The advantage of the differential pair is that it does not amplify input signals applied to Q1 and Q2 that are in phase. This is called the *common mode.* When points 1 and 2 in the input are at the same potential, the difference voltage equals zero.

As an example, any variations in voltage from the power supply would be applied to both stages in the common mode; they include changes in dc supply voltage and hum or ac ripple. These common-mode variations are balanced out.

In short, the differential pair amplifies differential signals but rejects common-mode signals. How well the common mode is rejected depends on the balance of the two stages.

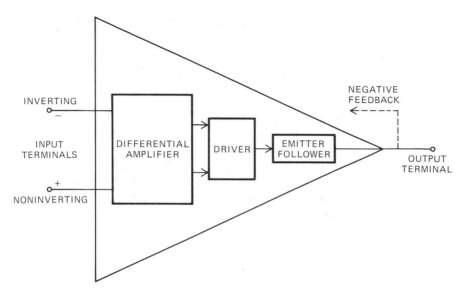

Fig. 3-15. Block diagram of circuits in an operational amplifier (op amp) IC chip.

Offset Voltage The offset voltage is the amount of differential voltage output without any input. Ideally, the offset should be zero. However, the stages are not perfectly balanced in gain or collector current. Also, temperature variations and stray signals can cause offset.

Test Point Questions 3-7
(Answers on Page 63)

Refer to Fig. 3-14.

a. Which terminals are used for differential input signals?
b. Which terminals are used for differential output signals?

3-8
THE OPERATIONAL AMPLIFIER (OP AMP)

The operational amplifier is the main application of the differential pair. It is usually packaged as an integrated circuit. A simplified block diagram of the stages in the IC chip for an op amp is shown in Fig. 3-15, and a flatpack unit with pin connections is shown in Fig. 3-16. Type letters LM in-

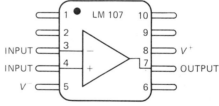

Fig. 3-16. Glass flat-pack IC unit LM 107 op amp showing pin connections. The size of the unit is 0.275 in square. (National Semiconductor Corporation).

dicate a linear, monolithic IC unit. Linear applications for amplifiers correspond to analog circuits, which include just about everything but digital pulse circuits. "Monolithic construction" means that the entire integrated circuit is formed on one silicon chip. The IC transistors can be the NPN and PNP junction types or FETs.

The name *operational amplifier* is derived from the fact that the amplifier was originally used to perform electronically the mathematical operations of addition, subtraction, integration, and differentiation. However, the op amp is so versatile that its use has been extended to other types of circuits. Some examples of its applications are audio amplifiers, RF amplifiers, waveshaping circuits, voltage regulators, sum or difference amplifiers, and special consumer ICs for radio and television receivers.

The op amp is actually a combination of amplifier stages. In Fig. 3-15, the signal is applied to the input terminals of the differential pair. The amplified output is coupled to the driver. This section has the required gain and provides a single-ended output to drive the emitter follower. It feeds the output terminal.

The minus (−) terminal of the input is called the *inverting terminal* because the signal applied there will be 180° out of phase with the amplified output. The plus (+) terminal is the *noninverting terminal*, because the amplified output is in phase with the input.

Note that the op amp has two input terminals for a signal but only one output terminal. The input impedance into the differential pair is high. However the input signal is usually single-ended, either to the (+) or (−) terminal, and the other end is returned to chassis ground. At the output terminal, its impedance is low because it comes from the emitter circuit of the emitter follower.

The op amp uses an external negative-feedback loop, from the output terminal to the negative input terminal. Negative feedback means that the signal being fed back is 180° out of phase with the input signal applied at the feedback point. The gain is reduced with negative feedback, because the feedback cancels part of the input signal. However, the distortion also is reduced. Most important, the negative feedback makes the amplifier stable and prevents it from breaking into oscillations. Also, the bandwidth is increased. Furthermore, the negative feedback can be controlled to set the gain and the frequency response of the op amp. The negative feedback is always applied to the minus (−) terminal of the input.

Without the external negative feedback loop, the *open-loop voltage gain* of the op amp unit itself is very high. A typical gain value is 100,000. However, the frequency response is very limited. With negative feedback, though, the *closed-loop gain* is reduced, ranging from 10 to 1000, approximately. The gain in an actual circuit depends on the component values in the feedback network.

Inverting Amplifier   In the inverting amplifier (Fig. 3-17), the input signal is applied to the minus (−) terminal and the plus (+) terminal is grounded. The result is an amplified output of opposite polarity, as shown by the waveshapes. Note that the feedback network also is connected to the negative input terminal at the junction of R_2 and R_1.

As an example of calculating the voltage gain, assume that R_2 is 500 kΩ and R_1 is 50 kΩ. Also, let the input signal voltage be 0.4 V, or 400 mV. For this circuit, the amount of amplified output voltage can be calculated as

$$V_{\text{out}} = \frac{R_2}{R_1} \times V_{\text{in}} \qquad (3\text{-}1)$$

$$= \frac{500 \text{ k}\Omega}{50 \text{ k}\Omega} \times 0.4 \text{ V}$$

$$= 10 \times 0.4 \text{ V}$$

$$V_{\text{out}} = 4 \text{ V}$$

The input signal voltage is multiplied by a factor equal to the ratio R_2/R_1. The ratio, 10, is the actual closed-loop gain for the op amp.

Noninverting Amplifier   In the noninverting amplifier (Fig. 3-18), the input signal is applied to the (+) terminal and the (−) terminal is returned to chassis ground through R_1. Now the amplified output is in phase with the input signal, as shown by the waveshapes.

Again assume R_2 is 500 kΩ, and R_1 is 50 kΩ, and the input signal is 0.4 V. For this circuit, the amount of amplified output voltage can be calculated as

$$V_{\text{out}} = \frac{R_1 + R_2}{R_1} \times V_{\text{in}} \qquad (3\text{-}2)$$

$$= \frac{550 \text{ k}\Omega}{50 \text{ k}\Omega} \times 0.4 \text{ V} = 11 \times 0.4 \text{ V}$$

$$V_{\text{out}} = 4.4 \text{ V}$$

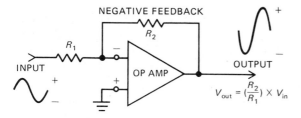

Fig. 3-17. The op amp as an inverting amplifier.

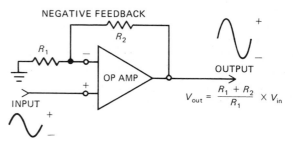

Fig. 3-18. The op amp as a noninverting amplifier.

The gain is slightly more than that of the inverting amplifier because the feedback is not connected to the same terminal as the input signal. Also, the input resistance is higher for this circuit.

Summing Amplifier Figure 3-19 is the circuit for the op amp summing amplifier, or adder. The inverting circuit is used here, but the summing can also be done with the noninverting circuit. In Fig. 3-19, three input signals are applied to the ($-$) terminal through resistors R_1, R_2, and R_3. The amount of negative feedback voltage is divided by each resistor in series with R_4 in the feedback path. As a result, the output voltage can be calculated as

$$V_{out} = -\left[\left(\frac{R_4}{R_1} \times V_{in}\right) + \left(\frac{R_4}{R_2} \times V_{in}\right)\right.$$
$$\left. + \left(\frac{R_4}{R_3} \times V_{in}\right)\right] \qquad (3\text{-}3)$$

With $R_1 = R_2 = R_3 = R_4$ the gain is 1 in each signal path. Then,

$$V_{out} = (1 \times 0.4\text{ V}) + (1 \times 0.4\text{ V})$$
$$+ (1 \times 0.4\text{ V})$$
$$V_{out} = 1.2\text{ V}$$

The output of 1.2 V is equal to the sum of the three input voltages of 0.4 V each. Note that the value of 3 kΩ for R_5 at the ($+$) terminal is made equal to the parallel combination of three 9-kΩ resistors at the ($-$) terminal. That value is used to balance the input to the two terminals of the differential amplifier in the op amp.

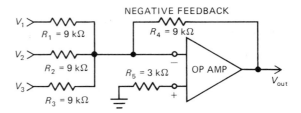

Fig. 3-19. The op amp as a summing amplifier, or adder, using the inverting circuit.

Gain-Bandwidth Product The typical frequency response of an op amp is from direct current, or 0 Hz, to more than 1 MHz. However, because of internal shunt capacitances, the gain drops off sharply as the frequency is increased. It is useful to specify the gain-bandwidth product (GBP), since it has a constant value for a particular op amp unit. The formula is

$$\text{GBP} = A_V \times f \qquad (3\text{-}4)$$

where A_V is the voltage gain at a particular frequency f, in hertz. As an example, with A_V of 1000 at 1000 Hz,

$$\text{GBP} = 1000 \times 1000\text{ Hz}$$
$$= 1{,}000{,}000\text{ Hz}$$
$$\text{GBP} = 1\text{ MHz}$$

The frequency response of an op amp with a constant GBP of 1 MHz is shown in Fig. 3-20. As additional examples, for the same unit, the gain is 100 for a frequency of 10 kHz; also, the gain is

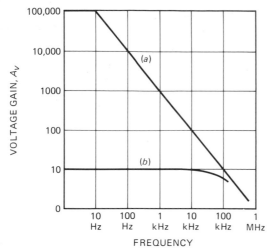

Fig. 3-20. Frequency response curve of a typical op amp. (*a*) Open-loop gain for gain-bandwidth product (GBP) of 1 MHz. (*b*) Closed-loop gain with a feedback of 1:10.

reduced to 10 for the higher frequency of 100 kHz. The values in the feedback network determine the gain of the op amp.

RC Compensation For increased stability, the op amp needs an *RC* network to compensate for internal phase shift. The compensation may be internal or external. A common value for the compensating capacitor is 30 pF, which reduces the GBP from 1 MHz to 0.1 MHz.

Input and Output Impedances A typical value for the input impedance of a differential pair is 1 MΩ at the (+) terminal. At the (−) terminal, the input impedance is lower because of the negative feedback loop. The high input impedance is desirable because it does not load down the circuit feeding into the op amp.

A typical values for the low output impedance from the emitter follower is 75 Ω or less. Low output impedance is desirable to provide power output to the load. Furthermore, a Darlington pair can be used at the input to increase the impedance and at the output to reduce the impedance.

CMMR This abbreviation stands for *common-mode rejection ratio.* Input signals that are in phase

with the differential pair are said to be in the common mode. The CMMR compares the gain of the differential input signal with the gain of the common-mode signal.

Slew Rate The term *slew* refers to the rate of change of the output voltage. As an example, the slew rate of 1 volt per microsecond (V/μs) means that the output voltage can change in amplitude by 1 V in 1 μs. A fast slew rate is desirable for wide-band amplifiers, especially those with nonsinusoidal signal waveshapes.

DC Supply Voltage Typical values of dc supply voltages are ± 15 to ± 22 V. Note that both positive and negative voltages are needed from a dual, symmetrical power supply (see Chap. 6, on Power Supplies). The V^+ is needed for the collector voltage, and the V^- is for the emitter supply voltage, assuming NPN transistors.

The dc load current drawn from the power supply for an op amp is generally 1 to 2 mA. A typical power rating is 500 mW.

Test Point Questions 3-8
(Answers on Page 63)

a. In an op amp, is a differential pair used for the input or for the output signal?
b. Is an emitter follower used for the input or the output signal?
c. Is the amplifier shown in Fig. 3-17 an inverting or a noninverting circuit?
d. Is negative feedback applied to the inverting or noninverting input terminal?

3-9
THE *RC* DECOUPLING FILTER
WITH CASCADED AMPLIFIERS

All the stages of an amplifier generally have a common dc supply line for V^+, bias, or both. Under this condition one or more stages may need an *RC* decoupling filter, such as R_1C_1 or R_2C_2 in Fig. 3-21. The filter is in the V^+ line.

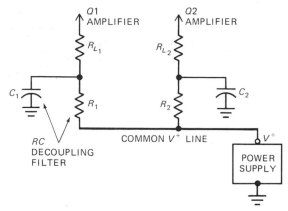

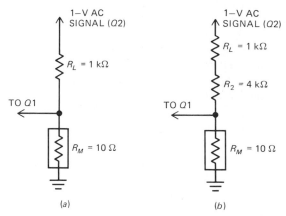

Fig. 3-21. Example of RC decoupling filters in a common V^+ line supplying cascaded amplifier stages.

Fig. 3-22. Function of R_2 as the series resistance in a decoupling filter to isolate R_L from the common power supply output resistance R_M. (a) Circuit without R_2. (b) R_2 added in series to reduce the signal voltage across R_M.

Decoupling means to isolate or separate two circuits to minimize the effects of the ac signal between them. The higher the amplifier gain, the more important is the decoupling.

When the stages are not decoupled, feedback can cause loss of gain, distortion, or oscillations. As an example of what happens, a small part of the ac signal from the last amplifier can be coupled back to the first amplifier through the common impedance of the power supply. The impedance provides mutual coupling between all the stages.

Both the R and C of the decoupling filter are needed, because each has its own function. Basically, the decoupling is a question of series and parallel connections. The R is in series to isolate each stage from the common impedance. The C is in parallel with R to bypass the ac signal so that the decoupling R is not part of the amplifier's ac load impedance.

Series R for Isolation The effect of R_1 or R_2 in the decoupling filter is illustrated in Fig. 3-22. In Fig. 3-22a the circuit has just the 1-kΩ load resistance, R_{L_2} for Q2, without any decoupling. In Fig. 3-22b, the series R_2 is added to isolate the ac signal of Q2 from the power supply. The R_M of 10 Ω represents the output resistance of the common power supply. The subscript M indicates mutual coupling.

For the values given, the voltage divider with R_L and R_M shown in Fig. 3-22a results in 0.01 V

of ac signal across R_M. That value is $1/100$ of the 1-V signal from Q2, because R_M is approximately $10/1000$ or $1/100$ of the total R in the divider. This ac signal of 0.01 V can be coupled to Q1 through the common line. The value of 0.01 V may seem small, but it is amplified in Q1 and Q2. Even smaller amounts of feedback can be a problem.

In Fig. 3-22b, however, the series R_2 of 4 kΩ is added. The total R in the divider becomes 5000 Ω, approximately. Now R_M is only $10/5000$ of the total R, and the ac signal across R_M also is only $10/5000$ of the 1-V signal. The value of $10/5000$ is $1/500$ or 0.002. The feedback signal of 0.002 V is much smaller than the 0.05-V ac signal without the decoupling resistor.

The higher the resistance of R_2, the better the isolation of the ac signal. However, R_2 cannot be too large, because it reduces the amount of V^+ for the Q2 amplifier. The same idea applies to the isolation resistance of R_1 for amplifier Q1 in Fig. 3-21.

Parallel C for Bypassing The need for C_2 to bypass R_2 in the decoupling filter is illustrated in Fig. 3-23. In Fig 3-23a, the series R_2 is used but is not bypassed. As a result, R_2 just becomes part of the load resistance. Instead of 1 kΩ, the effective load is 4 kΩ + 1 kΩ = 5 kΩ.

The voltage divider effect with R_M is the same as in Fig. 3-22b, but the higher load resistance allows the amplifier to provide more signal. Also, the signal may be distorted with too much R_L. Furthermore, with a tuned load impedance, R_2 should not be part of the resonant circuit. The decoupling for R_2 is not accomplished, therefore, without its bypass C_2. The capacitance of C_2 depends on whether the amplifier is for AF or RF signal. For bypassing, the reactance X_{C_2} should be one-tenth of R_2 or less at the lowest frequency.

In Fig. 3-23b, C_2 in parallel with R_2 allows the amplified output signal to pass through R_L but not R_2 which is bypassed. The ac signal path is through R_L and C_2 back to the common electrode of the amplifier. Now R_2 can isolate R_L from R_M because R_2 is not part of the load resistance for the ac signal. The same idea applies to the bypass C_1 across the isolating resistor R_1 for amplifier $Q1$ in Fig. 3-21.

In summary, then, in an RC decoupling filter or network, the series R provides isolation, but it must be bypassed by a parallel C for the ac signal.

It should be noted that troubles in an RC decoupling circuit can cause greater problems than just the lack of isolation. Consider these possibilities:

1. An open in R_1 or R_2 means no V^+ for $Q1$ or $Q2$. The amplifier will not operate at all.
2. A short in C_1 or C_2 also results in no V^+. The entire supply voltage will be across R_1 or R_2, and the resistor may burn out.
3. An open in C_1 or C_2 adds R_1 or R_2 to the load

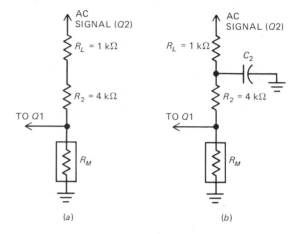

Fig. 3-23. Function of C_2 as the bypass capacitor in a decoupling filter to make R_L the only load for the ac signal. (a) Circuit without C_2. (b) C_2 added to bypass R_2 for the ac signal.

impedance of the amplifier. With a tuned load impedance, the gain is reduced drastically and there is little resonant response because the isolation resistance becomes part of the tuned circuit.

Test Point Questions 3-9
(Answers on Page 63)

Refer to Fig. 3-21.

a. Is the decoupling resistor for $Q2$, R_2, or R_{L_2}?
b. Is 10 μF or 47 pF a typical value for C_2 in an audio amplifier?

SUMMARY

1. When CE amplifiers are in cascade, the collector output signal of one stage drives the base input of the next stage. Cascading multiplies the gain.
2. The Darlington pair has two emitter followers in cascade with dc coupling.
3. The cascode circuit has a CE amplifier dc-coupled to a CB amplifier.
4. Stacked V^+ for amplifier stages means the stages are connected in series for the dc supply voltage.
5. A push-pull amplifier uses two stages to operate on opposite half-cycles of the

signal. See Fig. 3-9. The input signals are equal but 180° out of phase. The amplified output signals are also 180° out of phase.

6. In complementary symmetry, PNP and NPN transistors are used for push-pull amplification, but with a single-ended input and output.

7. Quasi-complementary symmetry means the driver stages are PNP and NPN but both output stages are NPN.

8. The differential amplifier uses two transistors for a push-pull input and a differential output between the two collectors. See Fig. 3-14. Either input terminal can also be single-ended, and either collector also can be used for a single-ended output. Integrated circuits are generally used for balance in the differential pair.

9. The operational amplifier, or op amp, includes a differential pair for the input signal, additional high-gain amplifiers, and an emitter follower for the output signal.

10. Negative feedback is applied from the output of the op amp to the input to increase bandwidth while reducing gain. The gain of the op amp circuit depends on the amount of negative feedback.

11. The input signal to the inverting terminal of the op amp produces an amplified output signal that is 180° out of phase. The input signal to the noninverting terminal is amplified without any phase reversal.

12. An *RC* decoupling filter is generally needed between stages of a cascaded amplifier to provide isolation from a common line. See Fig. 3-21. The series *R* isolates the ac signal from the common impedance. The parallel *C* bypasses the ac signal around *R*.

SELF-EXAMINATION
(Answers at back of book)

Choose (a), (b), (c), or (d).

1. Each of two cascaded stages has a voltage gain of 40. The overall gain is (a) 40, (b) 80, (c) 400, (d) 1600.

2. The Darlington pair consists of the following two stages: (a) CE and CC, (b) both CC, (c) both CE, (d) CE and CB.

3. Which of the following combinations has no phase inversion of the signal? (a) Two CE stages, (b) CE and CC stages, (c) three CE stages, (d) CE stage and emitter follower.

4. The circuit consisting of two transistors connected in series with the dc supply voltage is called (a) push-pull, (b) differential pair, (c) stacked V^+, (d) complementary symmetry.

5. Complementary symmetry uses two transistors that are (a) both NPN, (b) both PNP, (c) PNP and NPN, (d) both FET.

6. Which of the following circuits can operate class AB for audio power output? (a) Emitter follower, (b) cascade, (c) push-pull, (d) Darlington pair.
7. In Fig. 3-9 the audio signal is coupled to the loudspeaker by (a) T_1, (b) T_2, (c) R_E, (d) capacitive coupling.
8. Which of the following circuits uses complementary symmetry? (a) Fig. 3-2, (b) Fig. 3-9, (c) Fig. 3-11, (d) Fig. 3-12.
9. For the differential pair in Fig. 3-14, which two terminals are used for the differential output? (a) 1 and 2, (b) 3 and 4, (c) 1 and 3, (d) 2 and 4.
10. For the op amp in Fig. 3-15, negative feedback from the output is connected to the (a) plus $(+)$ input, (b) minus $(-)$ input, (c) emitter follower, (d) driver stage.
11. For the op amp in Fig. 3-17, with 50 kΩ for R_2 and 1 kΩ for R_1, the gain is (a) 50, (b) 1000, (c) 50,000, (d) infinite.
12. For the RC decoupling filters in Fig. 3-21 the isolation for Q2 is provided by (a) R_2, (b) R_{L_2}, (c) R_{L_1}, (d) C_1.

ESSAY QUESTIONS

1. How are two CE stages in cascade connected for an ac signal and for their dc supply voltage?
2. Show the circuit for two NPN transistors connected in series for their dc supply voltage.
3. Give two features of cascaded amplifiers.
4. Give three features of the push-pull amplifier.
5. Define the following circuits: **a.** Darlington pair, **b.** cascode amplifier, **c.** totem pole, **d.** stacked V^+.
6. Give two examples of cascaded amplifier stages that do not invert the ac input signal.
7. Show the waveforms for two ac signals that are inverted with respect to each other.
8. Draw the circuit for a push-pull audio amplifier with center-tapped input and output transformers.
9. Define complementary symmetry and quasi-complementary symmetry in audio amplifier circuits.
10. Draw the circuit for an audio output circuit using complementary symmetry.
11. Give two features of the differential amplifier.
12. Give the functions for R and C in a decoupling filter.
13. What are the three main sections of an op amp?
14. Name the three signal terminals in an op amp and the function of each.
15. Define negative feedback.
16. Define gain-bandwidth product.
17. Give five applications for op amps.
18. What is the difference between cascade and cascode amplifiers?
19. Define the following specifications for a differential amplifier or op amp: **a.** CMMR, **b.** slew rate, **c.** offset current or voltage.

PROBLEMS
(Answers to odd-numbered problems at back of book)

1. Three cascaded stages have A_V values of 40, 30, and 10. What is the overall voltage gain?
2. Three cascaded stages have A_I values of 40, 30, and 10. What is the overall current gain?
3. Calculate C_2 for bypassing R_2 of 4 kΩ at 50 Hz if $R_2 = 4$ kΩ. (See Fig. 3-21.)
4. Calculate C_2 in Problem 3 at 50 MHz.
5. Refer to Fig. 3-22b. If R_2 is increased to 9 kΩ, what will be the ac signal voltage across R_M?
6. For an op amp with GBP = 0.1 MHz, what is the gain for a 10-kHz bandwidth?
7. In Fig. 3-18, assume 50 kΩ for R_2 and 1 kΩ for R_1. **a.** Calculate the voltage gain. **b.** What is the bandwidth for an op amp with a GBP of 1 MHz?

SPECIAL QUESTIONS

1. Is it possible to have two op amps in one IC unit?
2. Give two examples of the need for *RC* decoupling.
3. Draw the circuit for an op amp adder with two input signals that uses the noninverting input.
4. What is meant by a dual symmetrical power supply?

ANSWERS TO TEST POINT QUESTIONS

3-1	**a.** 6000	**3-5**	**a.** Push-pull	**3-8**	**a.** Input		
	b. Three		**b.** Output		**b.** Output		
3-2	**a.** F	**3-6**	**a.** T		**c.** Inverting		
	b. T		**b.** T		**d.** Inverting		
3-3	**a.** Parallel		**c.** F	**3-9**	**a.** R_2		
	b. 24.6 V	**3-7**	**a.** 1 and 2		**b.** 10 μF		
3-4	**a.** T		**b.** 3 and 4				
	b. F						
	c. T						

Chapter 4
Audio Circuits

The word *audio* is Latin for "I hear." Audio signals correspond to sound waves in the range of frequencies that the ear can perceive as audible information. The range of audio frequencies is generally considered to be 16 to 16,000 Hz. The equipment and circuits used for audio signals are described here. Included are microphones, loudspeakers, phonograph records, magnetic tape recording and audio amplifiers.

Audio circuits are among the most common applications of electronics. They are used in radio transmitters and receivers. A transmitter needs a source of audio signal and audio amplifiers to supply the modulating information that is broadcast on the RF carrier wave. A receiver needs an audio amplifier for the detected signal and a loudspeaker to reproduce the sound. More details of the different types of audio equipment and circuits are given in the following topics:

4-1
SOUND WAVES AND AUDIO FREQUENCIES

Sound is a wave motion of varying pressure in a material substance such as a solid, liquid, or gas. The pressure variations result from the mechanical vibrations of a source producing the sound. Common examples are a vibrating metal reed and a plucked string. As shown in Fig. 4-1a, the vibrating reed produces varying values of pressure in the surrounding air. A *compression* is a point of maximum air pressure in the sound wave; minimum pressure is a *rarefaction*. Since air is compressible, the points of compression and rarefaction move outward from the vibrating reed and propagate sound waves in all directions. The sound waves can be detected by the human ear, in which they produce the sensation of hearing.

The velocity of sound waves is approximately 1130 feet per second (ft/s) [344.43 meters per second (m/s)] in dry air at a temperature of 20°C. In metals and liquids the velocity is much greater. There cannot be any sound in a vacuum.

Frequency of the Sound Waves In the example used here, the reed is vibrating at the rate of 1000 cycles per second (Hz). That rate is the frequency of the resulting sound waves, as shown in Fig. 4-1b. The waveform is a graph of pressure variations in the surrounding air, at any one point, changing with respect to time. One cycle of compression and rarefaction takes 1 millisecond (ms), which is the *period*. The frequency is $1/T$ or $1/1$ ms, which equals 1000 Hz.

Wavelength of the Sound Waves In Fig. 4-1c the variations in air pressure are shown at different distances starting at the source and radiating into the surrounding air. At any one time, different points have different pressures. One wavelength is the distance encompassing a complete cycle of pressure variations.

The wavelength of any wave motion depends on the frequency of the variations and the velocity of propagation. Specifically, the wavelength of sound waves with a velocity of 1130 ft/s is

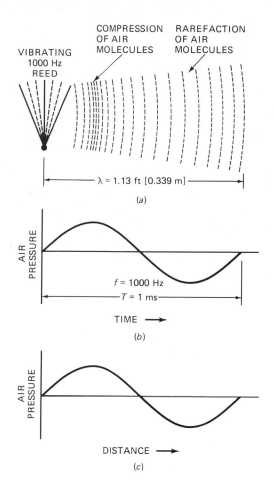

Fig. 4-1. Sound waves produced by a vibrating reed. (a) Compression and rarefaction of air molecules, propagating sound in all directions. (b) A cycle of variations in air pressure, at one point in space, but changing with respect to time. The period T is 1 ms, and f is 1000 Hz for this example. (c) A cycle of variations in air pressure, at a given time, with respect to distance from the source.

$$\lambda = \frac{1130 \text{ ft/s}}{f} \qquad (4\text{-}1)$$

where λ (the Greek letter lambda) is the wavelength in feet and f is the frequency of the wave in hertz.

For the example in Fig. 4-1,

$$\lambda = \frac{1130 \text{ ft/s}}{1000 \text{ Hz}} = 1.13 \text{ ft}$$

Loudness   The term *loudness* describes how the ear perceives the amplitude of sound waves. Greater amplitude means a louder sound. In the sound wave, more amplitude means greater changes in air pressure. In a corresponding electrical audio signal, more amplitude means greater changes in voltage or current. The loudest sound has about 100 times more amplitude than the threshold of audibility.

Tone and Pitch   Both *tone* and *pitch* are used to describe the effect of different frequencies, although pitch is applied mainly to sound waves. The range of audible frequencies is approximately 16 Hz to 16 kHz. Below 16 Hz, the sound is more a matter of feeling than hearing. Above 16 kHz, the ear cannot respond to the high-frequency variations.

High frequencies, as in the sound of a piccolo, are *treble tones.* At the opposite extreme, low frequencies such as the deep sound of an organ are *bass tones.* In audio equipment, the treble and bass controls adjust the frequency response.

The wide range of different sounds can be seen from the illustration in Fig. 4-2. For instance, the 88 keys of the piano cover the frequency range of 30 to 4100 Hz, approximately. The high-pitched piccolo produces treble sound at about 4600 Hz. The bass tuba produces 42 to 330 Hz.

A man's speaking voice produces sound with frequencies around 130 Hz; a woman's, an average of 250 Hz. A high-pitched soprano can produce 1170 Hz.

With those frequencies as typical, what produces the higher frequencies of up to 16,000 Hz? The answer is that any source produces sound waves that have overtones, or harmonics, which are multiples of the fundamental frequency.

Harmonic Frequencies   Consider a vibrating string as a source of sound. It might produce sound waves with a frequency of 500 Hz, with the entire string length vibrating at that frequency. That rate is the *fundamental frequency;* it is the lowest frequency the string can produce. Since the string is not perfectly rigid, however, shorter sections are also vibrating to produce additional frequencies. The frequency is higher for shorter lengths of the string; therefore, the vibrating sections produce frequencies higher than the fundamental. Those higher frequencies are called *harmonics.*

Furthermore, the harmonic frequencies are exact multiples of the fundamental. One-half of the string length vibrates at twice the fundamental frequency on the second harmonic; one-third of the string length produces the third harmonic. We could consider the vibrating string as producing an unlimited number of multiple harmonic frequencies. However, the amplitude of each harmonic decreases in inverse proportion to the harmonic number. Therefore, it is usually not necessary to consider more than 10 to 20 harmonics.

The higher harmonic frequencies must be included in the sound, which otherwise would not have its characteristic *timbre* or quality. It is the harmonics that make one source of sound different from another source although they are both producing the same fundamental frequency.

The Octave   The octave is a unit representing a span of frequencies having a ratio of 2 to 1. One octave above 500 Hz is 1000 Hz, and one octave above 1000 Hz is 2000 Hz. The reason for the name is that eight successive tones in the musical scale have a 2:1 frequency interval.

Test Point Questions 4-1
(Answers on Page 98)

When a tuning fork vibrates at 400 Hz:

a. What is the fundamental frequency of the sound?
b. What is the second harmonic frequency?
c. Calculate the wavelength of the fundamental frequency.

4-2
AUDIO SIGNAL

Referring to Fig. 4-3, the microphone at the left converts sound waves into a corresponding elec-

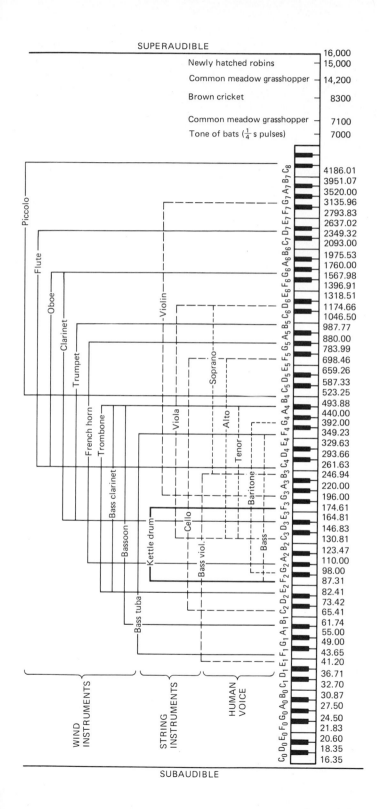

Fig. 4-2. Spectrum of frequencies for audible sound waves. The range is 16 to 16,000 Hz. (*Electronics*)

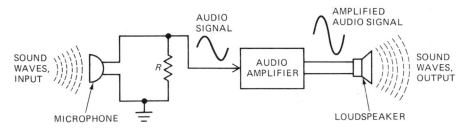

Fig. 4-3. Sound waves converted by a microphone to an electrical audio signal for amplification and then converted back to sound waves by a loudspeaker.

trical audio signal. The objective is to have variations in voltage and current that can be amplified. The frequency of the audio signal is the same as that of the sound waves, including harmonic components.

The amplitude of ac variations in the audio signal is increased in magnitude by the amplifier. Transistors in cascade can provide voltage and current gain of 100, 1000, or even 1,000,000.

After enough amplification, the audio signal is coupled to the loudspeaker. The speaker's vibrating cone converts the electrical variations back into sound waves. We have the original sound again, but with an increase in amplitude, or loudness. A faint whisper can be amplified enough to be reproduced as a loud sound, even in a large space such as a theatre.

Comparison of Audio Signals and Sound Waves

In Fig. 4-4a the variations are shown for a sound wave with the frequency of 1000 Hz. The signal is a variation in air pressure. The velocity of propagation is relatively low, at 1130 ft/s, because the disturbances must move through the physical medium of air.

In Fig. 4-4b, the corresponding audio signal at 1000 Hz includes variations in voltage V and current I. The electrical signal moves through a wire conductor with the speed of light, which is 3×10^{10} centimeters per second (cm/s). Now the electric and magnetic fields of the signal are propagated, not sound waves. We usually reserve the name *audio signal* for the variations of V and I in the audible range of frequencies, and the sound itself can be called a sonic signal.

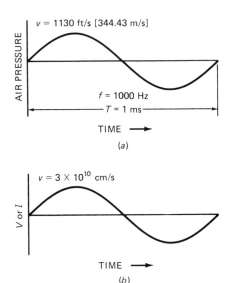

Fig. 4-4. Comparison of electrical audio signal with sound waves. (a) Sound variations of air pressure. The v is the velocity of sound waves in air. (b) Signal variations of voltage or current. The v is the velocity of current in a wire conductor; it is equal to the speed of the light.

Electromechanical Transducers

Microphones and loudspeakers are two common examples of electromechanical transducers. A transducer is able to convert one form of energy into another. The loudspeaker changes electric energy in the form of audio signal current into mechanical energy as sound waves of varying air pressure. The conversion is accomplished by means of a voice coil attached to a vibrating cone. A microphone converts sound waves into an electrical audio signal; its function is opposite that of the loudspeaker.

Frequency Range of AF Signals Most audio equipment does not operate over the full range of audible frequencies (16 to 16,000 Hz). The reason is that extremely low and high frequencies are more difficult to process in amplifiers and transducers. Furthermore, the full frequency range is often unnecessary. Telephones generally use the restricted range of 250 to 2750 Hz, and yet the speech is intelligible. For music and more natural speech, though, an AF range of about 250 to 8000 Hz provides much better quality. Phonograph records and tape recorders generally do not have frequencies higher than 15 kHz. In general, 50 to 15,000 Hz can be considered a full range of frequencies for audio equipment. This range is used with high fidelity broadcasting in the commercial FM radio band.

It should be noted that extending the frequency range is not always desirable. When very low frequencies are amplified, it is more difficult to reduce the effect of 60-Hz hum interference from the ac power line. Also, the dc power supply must have a very stable output voltage. At the opposite extreme, when very high frequencies are amplified, it is more difficult to reduce the effect of noise interference. Finally, transducers usually have a very low output at the high and low ends of their frequency response.

Stereophonic Sound When received by our ears sound has a directional effect to the left and right in space. The sound may be louder on one side than on the other side. The same effect in the reproduced audio signal can be provided by a system of *stereophonic sound,* or just *stereo.* Two separate left and right audio channels are required. Each channel provides the difference in sound that the left and right ears perceive when we listen to the original sound. The stereo effect gives a sense of direction to the sound.

Quadraphonic Sound The quadraphonic system takes the directional effect of stereo one step further. There are four audio signals. Two signals provide the usual stereo effect left and right from the front; the other two, left and right from the back. Four loudspeakers are necessary.

4-3
LOUDSPEAKERS

Typical loudspeaker construction is illustrated in Fig. 4-5. The speaker is a *dynamic speaker;* it is so called because it has a small, lightweight voice coil that can easily move in and out. Attached to the voice coil is a cone made of stiff paper or cloth. As the voice coil moves with audio signal current, the cone vibrates and produces sound.

The voice coil has about 20 turns of fine wire wrapped around a cardboard form that is typically 1 inch (in) [2.54 cm] in diameter. The coil is positioned in the air gap of the fixed magnetic field; it fits over the center of the field magnet. The magnet of a permanent-magnet (PM) speaker provides a steady field as shown in Fig. 4-6. In an electromagnetic (EM) speaker, a field coil with direct current supplies the steady magnetic flux.

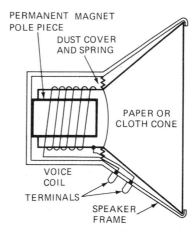

Fig. 4-5. Construction of a dynamic loudspeaker.

Although EM speakers are not used anymore they may still be found in older radio and audio equipment.

When audio signal current flows through the voice coil, its varying magnetic field reacts with the steady flux of the field magnet. The result is motor action, which moves the voice coil in accordance with the variations in audio signal. The motion of the horizontal coil shown in Fig. 4-5 is left and right. As the coil is attracted and repelled, it moves in and out. The attached cone vibrates like a piston; it compresses and rarefies the air to produce sound waves. The sound corresponds to the variations in the signal current in the voice coil. Typical cone diameters are 3, 5, 8, 10, 12, and 15 in.

The two leads for the voice coil are connected with flexible braided wire to stationary terminals on the speaker frame. A permanent-magnet (PM) speaker needs only those two connections for the audio signal, because the magnet supplies the field flux. Practically all loudspeakers now are the PM type; permanent magnets range from 2 ounces (oz) [56.7 grams (g)] for a small speaker to more than 5 pounds (lb) [2.27 kilograms (kg)] for the larger speakers. The better the speaker, the bigger the magnet, especially for good low-frequency response.

A speaker designed for low audio frequencies is called a *woofer*; a small speaker for high frequencies is called a *tweeter*. The *coaxial speaker* shown in Fig. 4-6 combines a woofer and tweeter on one frame. The frequency range is generally below 500 Hz for a woofer and above 8000 Hz for a tweeter. The midrange for loudspeakers can be considered to be 500 to 8000 Hz.

Loudspeaker Impedance Rating It is important that the loudspeaker impedance be matched to the impedance of the audio amplifier for power efficiency and low distortion. Transistor amplifiers are designed to operate into a speaker load in the range of 4 to 16 Ω.

The manufacturer specifies the impedance at 4, 8, or 16 Ω for most loudspeakers. That value is mainly the inductive reactance of the voice coil

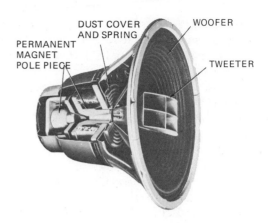

Fig. 4-6. Coaxial loudspeaker combining woofer and tweeter in one unit.   (*Jensen Mfg. Co.*)

at a specific test frequency of 400 Hz. For an 8-Ω reactance at 400 Hz, the inductance of the voice coil is 3.2 millihenrys (mH).

The dc resistance of the voice coil, as measured with an ohmmeter, is usually just one or two ohms. In many cases, the ac impedance can be approximated at about four times the dc resistance. If the ohmmeter reads infinity, the voice coil must be open.

Power Rating The manufacturer specifies the power rating of a loudspeaker to guard against feeding too much current into the voice coil. A possible result is that the coil would burn open or the paper cone would be torn off its support. A loudspeaker takes a relatively large amount of audio signal current. As an example, 20 W of power in 4 Ω corresponds to a current of 2.24 amperes (A) for the loudspeaker.

Typical power ratings are less than 1 W for small speakers to more than 50 W. It is important to realize that those are maximum speaker ratings. The amount of sound power the speaker actually produces depends on the amount of signal power supplied by the audio amplifier. A 50-W speaker can be operating with less than 1 W at low volume settings.

The speaker and amplifier power ratings should be about the same. A little higher rating for the speaker is acceptable but the speaker must not have a lower rating. A low speaker rating may result in damage to the speaker at high volume. On the other hand, it would not be practical to drive a 50-W speaker with a 1-W amplifier.

The efficiency of loudspeakers is generally 5 to 10 percent in converting audio signal in the voice coil to sound energy from the cone. For 10 W of audio power and 5 percent efficiency, as an example, the sound output is $10 \times 0.05 = 0.5$ W. That amount of sound power is enough, however, to fill a large living room with very loud sound.

Horn Loudspeakers The horn loudspeaker uses a small cone on a vibrating diaphragm as the driver and a flared horn for better acoustical loading. Acoustical loading involves the impedance of the surrounding air into which the speaker is working. The advantage of proper loading is higher efficiency, which can be 30 to 50 percent. Piezoelectric horn tweeters can be used for high-frequency signals up to 30 kHz.

Headphones The purpose of headphones is to concentrate sound at the ears while blocking extraneous background sounds. A typical pair of headphones, or a headset, is shown in Fig. 4-7. It contains two small dynamic speakers with separate leads for stereo. Maximum power rating is generally less than 500 mW, and the impedance is 8 Ω. Actually, the stereo separation with headphones is so great that headphones can sound better than loudspeakers.

Loudspeaker Baffle The baffle is an enclosure for the loudspeaker. The purpose is to prevent changes in air pressure at the front of the cone from being canceled by opposite changes at the back. That is why a loudspeaker sounds much better in an enclosure. Besides its use as a mounting for the speaker, any enclosure serves as a baffle. An enclosure should have unequal dimensions in a ratio such as 2:3:4 for depth, width, and height to minimize internal resonances.

An infinite baffle can be achieved by mounting the loudspeaker in a wall so that it plays into one room while its back is in a completely separate

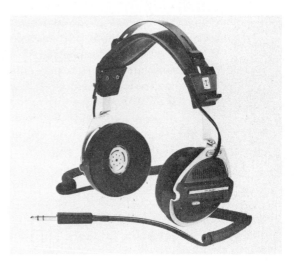

Fig. 4-7. Typical pair of headphones or headset. (*Koss*)

room. A more practical method is to have the loudspeaker completely sealed in its enclosure. Such an arrangement, called an *air-suspension loudspeaker,* is generally used for high-fidelity equipment. The enclosed air acts as a spring load on the speaker itself, which has a soft suspension. Inside the cabinet, damping material such as fiberglass insulation absorbs sound energy from the back of the speaker. Therefore, the speaker radiates only from the front. The enclosure must be very rigid, because even the slightest displacement of the sides cancels part of the baffle effect. Just to indicate the problem, it would be best to mount the speaker in a concrete wall to prevent vibrations of the baffle.

A *bass reflex enclosure* also has a sealed back, but an opening or *port* is cut out of the front. This enclosure is also called a *vented baffle.*

Multiple Loudspeakers The general principle is to connect multiple loudspeakers either in series or parallel, as shown in Fig. 4-8, or in series-parallel combinations. In Fig. 4-8a the two 8-Ω speakers in series have a total impedance of 16 Ω across the amplifier output. In Fig. 4-8b the two 8-Ω speakers in parallel have an equivalent impedance of 4 Ω. The choice of a circuit depends on the impedance that best matches the amplifier output.

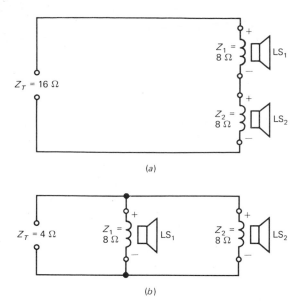

Fig. 4-8. Connecting multiple loudspeakers, (*a*) in series and (*b*) in parallel.

The series circuit uses less wire; but if one speaker opens, the others cannot operate.

Keep in mind that the series and parallel combinations only divide the power output. The speakers cannot have more audio power than the power supplied by the amplifier.

It is preferable to use multiple speakers with the same impedance. Otherwise, the distribution of power will be unequal, which may be a problem. In a series circuit, the higher impedance uses more power and the other uses less. The distribution is reversed when loudspeakers are in parallel, because then the lower impedance uses more power.

Speaker Wire The conductor used to connect the output of the amplifier to the loudspeakers is called *speaker wire*. Assuming the length of speaker wire is less than about 50 ft [15.24 m], almost any two-conductor cord or cable can be used. The current is usually in the range of amperes, so No. 18 gage or larger wire is necessary for minimum *IR* drop in the line. Ordinary lamp cord, called *zip cord*, is suitable. A type of zip cord with clear plastic insulation is often sold as speaker wire. Even

twin-lead television wire is convenient to use, because it is flat and easy to hide.

Shielding against interference is generally no problem because the loudspeaker circuit has low impedance and there is no amplification. Because of the low impedance, the capacitance of the line has a negligible effect on high-frequency response.

Speaker wire is usually color-coded so that both ends of the same wire can be easily identified. Color coding is often done by tinning one wire and leaving the other wire copper color.

Loudspeaker Phasing A loudspeaker is an ac device that operates properly whether or not the polarity of the amplifier output matches the polarity of the speaker terminals. When there is only one speaker, phase relationships have no meaning. When there are two or more speakers, however, it may be necessary to phase the speakers so that all the cones move in the same direction when driven by signals that are in phase. In stereophonic sound, the two speakers must be phased to get the stereo effect. The easiest way to phase two speakers is to follow the polarity markings on both the amplifier output terminals (usually marked *right speaker* and *left speaker*) and the speaker terminals. For each channel of a stereo system, the positive (+) terminal of the amplifier must be connected to the (+) terminal of the speaker. If the terminals are not marked, incorrect phasing can usually be heard by listening to a monophonic broadcast or signal. Out-of-phase speakers will produce greatly reduced volume since the compression and rarefaction portions of the sound waves will be bucking one another rather than aiding. By reversing the two terminals of one of the speakers, the two sound waves will be in phase and the volume will increase noticeably. When multiple speakers are not close together, the phasing usually does not matter so much because the room acoustics have a greater affect on the sound.

Frequency-Divider Crossover Network When woofer and tweeter loudspeakers are used, the frequencies for each are supplied by an *LC* network, as shown in Fig. 4-9. The filter provides high audio

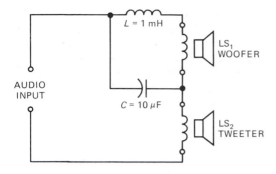

Fig. 4-9. Frequency divider or crossover network for woofer and tweeter loudspeakers with 8-Ω impedance. Crossover frequency is 1600 Hz.

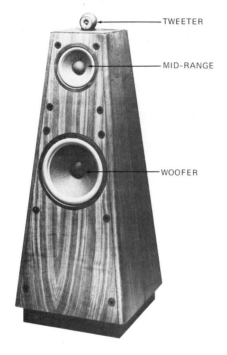

Fig. 4-10. Three-way speaker system in sealed enclosure. (*Epicure Products, Inc.*)

frequencies for the tweeter through the 10-μF series capacitor. Low frequencies have more capacitive reactance. The 1-mH series choke L couples low audio frequencies to the woofer. High frequencies have more inductive reactance. It should be noted that C must be a nonpolarized capacitor, because there is no dc voltage in the frequency-divider network.

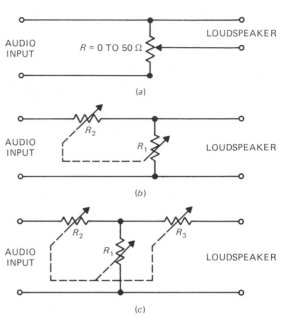

Fig. 4-11. Controlling the volume of a loudspeaker: (*a*) potentiometer, (*b*) L pad for constant impedance on one side of the network. (*c*) T pad for constant impedance on both sides of the network.

At the crossover frequency, where $X_L = X_C$, both speakers have the same amount of audio signal. That value for each speaker is less than the audio input, but together the speakers produce normal output. The crossover frequency is generally about 1600 Hz for a two-way system. The L and C values in Fig. 4-9 are based on this crossover frequency with 8-Ω speakers.

A three-way system uses a woofer, midrange speaker, and tweeter, as shown in Fig. 4-10. The divider network then requires three filters. Typical crossover frequencies are 800 and 5000 Hz.

Loudspeaker Level Controls Loudspeaker level controls are variable resistors that can be used to adjust the volume of a remote loudspeaker or balance the levels from the divider network with a two- or three-way system. The level adjustment of the tweeter is often called a *brilliance control*. The adjustment of the midrange speaker is called a *presence control*.

One way to adjust level is to use just the 50-Ω potentiometer shown in Fig. 4-11*a*. However, the circuit does not provide a constant impedance. The L pad in Fig. 4-11b has two variable resistors on a common shaft: R_2 in series and R_1 in parallel to keep a constant impedance on one side of the network. In Fig. 4-11c the T pad has two series resistances to maintain constant impedance on both sides of the network.

Test Point Questions 4-3
(Answers on Page 98)

a. Are the two terminals on a loudspeaker for the voice coil or field magnet?
b. Is a typical loudspeaker impedance 8 or 400 Ω?
c. The loudspeaker enclosure serves as a baffle for low frequencies. True or false?
d. What is the total impedance of two 4-Ω speakers in series?

4-4
MICROPHONES

The microphone converts sound energy into an electrical audio signal. A dynamic microphone uses the same principle as a dynamic loudspeaker, but in reverse. In fact, if you talk into a small speaker, an audio signal output can be taken from the voice coil terminals. That is the method used for "intercom" systems, in which you talk and listen with the same unit. For better frequency response and directional characteristics, though, there are many specialized types of microphone. They include the dynamic, crystal or ceramic, condenser or capacitor, and the carbon types. They can be in miniature form as a lapel microphone, tie-pin, or throat microphone.

Magnetic Microphones   The magnetic, or dynamic, microphone is probably the most common type (Fig. 4-12). It has a small voice coil with many turns of wire as fine as No. 48 gage. A permanent magnet supplies the steady field flux.

A diaphragm corresponds to the paper cone in a dynamic loudspeaker. When sound waves strike it, the diaphragm moves the coil in and out. The motion is only a few thousands of an inch, but a

Fig. 4-12. Typical dynamic microphone.   (*Shure Brothers, Inc.*)

small signal current is induced in the voice coil and becomes the audio output.

The frequency response of dynamic microphones is generally about 50 to 15,000 Hz. This range is considered good for high quality reproduction. Impedance of the coil is 4 to 150 Ω. The output signal level is about 1 mV.

Crystal and Ceramic Microphones   The crystal and ceramic microphones depend on the generation of voltage by a mechanical stress on the transducer element. This is called the *piezoelectric effect*. Rochelle salt crystals are a natural material that exhibits this property, but synthetic ceramic materials provide similar results and have better immunity to heat and humidity. The crystal or ceramic microphone usually costs less than the magnetic microphone and has higher impedance (50 to 100 kΩ) and greater output voltage. The quality can be good, but professional microphones are generally the magnetic type.

Condenser Microphones   In a condenser microphone a movable plate serves as one side of a capacitor (formerly called a condenser). The space between the plates varies with changes in the sound input. This in turn varies the charge and discharge current which is the audio output signal.

Dc voltage is needed to charge the capacitor. However, an *electret microphone* has a permanently charged plate. The frequency response of condenser microphones is good, but the output signal level is very low.

Carbon Microphone The carbon microphone contains a small enclosure packed with carbon granules. The sound input compresses the granules and changes the resistance across the enclosure. A dc voltage is used in the circuit to produce a current. The variations in current due to the variations in resistance, become the audio output signal. The output signal is high due to the dc voltage. However the frequency response is limited. The resistance of the carbon is about 100 Ω. Carbon microphones are generally used only where speech is being transmitted, such as in telephones.

Microphone Sensitivity The best microphones usually produce the smallest amount of signal. Typical output is about 1 mV.

Microphone sensitivity is defined as audio output for a standard sound pressure at a frequency of 1000 Hz. The output level is specified in *decibels* below the reference level of 1 V. Decibels (dB) are units for comparing two voltage or power levels as the logarithm of their ratio; they are explained in Chap. 5.

The ratio of decibel units to the 1-V reference is indicated as dBV. Since all microphones have outputs much less than 1 V, their sensitivity rating is in negative numbers of dBV. The negative sign means that the output is less than the 1-V reference value. For instance, an output of 1 mV, or 0.001 V, is equal to −60 dBV. It is important to realize that the larger the −dBV rating the smaller the amount of output voltage.

Each increase of −6 dBV corresponds to one-half the voltage. As an example, −66 dBV is equal to 0.5 mV, or one-half of 1 mV.

Directional Response of Microphones Typical directional patterns of microphone response are shown in Fig. 4-13. An omnidirectional microphone picks up sound from all directions, as shown

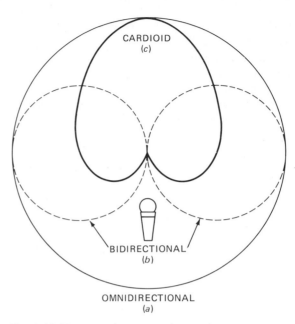

Fig. 4-13. Directional patterns of microphone response. Cardioid pattern *c* is unidirectional.

by the circular pattern *a*. The bidirectional response *b* picks up sound from two opposite directions. The unidirectional response *c* is called a *cardioid* pattern because it has the shape of a heart.

A cardioid microphone is often the preferred type. It picks up sound mainly from the front, some from the sides, but very little from the back. The directivity is helpful in reducing unwanted feedback from the loudspeaker in a public address system. The audio feedback from loudspeaker to microphone makes the amplifier oscillate, which produces a loud howl.

Microphone Plugs and Cables Microphone cable must be shielded to prevent pickup of interference. The microphone signal level is very low; the impedance is high; and the signal is amplified. Therefore, any unwanted signal introduced through the microphone cable can find its way to the amplifier output.

Three types of connector commonly used with microphone cables are shown in Fig. 4-14. The plugs shown in Fig. 4-14*a* and 4-14*b* are for shielded line with one inner conductor. This

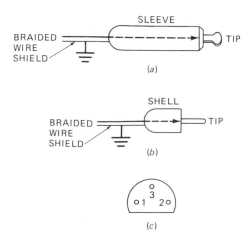

Fig. 4-14. Plugs for microphone cable: (a) telephone type, (b) RCA phono type, (c) Cannon 3-pin type.

"high" side of the line connects to the tip on the plug. The braided wire shield connects to the sleeve or barrel. That part of the plug is connected to chassis ground on the amplifier jack. In the plug, the tip is insulated from the sleeve or barrel.

The three-pin connector shown in Fig. 4-14c is for shielded two-wire cable. Two conductors inside the cable are enclosed in the shield. This cable provides better shielding.

An open in the cable or plug is a common problem with microphone connections. An open in the "high" side means no audio signal. When there is an open in the ground side, the line is sure to pick up interference. The reason is that the ungrounded, or "high," side then has very high impedance because it is "floating" without a ground return.

If there is any trouble with hum or interference from radio signals, a check for an open ground should be made. Even a small amount of 60-Hz hum picked up from the ac power line can be amplified enough to produce a loud hum or buzz. A radio signal can be picked up by an ungrounded line acting as an antenna. The unwanted signal can then be amplified and produce an audio output.

Wireless Microphone The wireless microphone is used sometimes to avoid the nuisance of

cables, especially when the person using the microphone must move around. The method is to include a small radio transmitter in the microphone assembly. One type of unit transmits an FM signal for distances up to 150 ft. The signal is picked up by an FM receiver and amplified.

Test Point Questions 4-4
(Answers on Page 98)

a. Is the typical output of a microphone 1 mV or 1 V?
b. Which is similar in operation to a speaker, the magnetic or the carbon microphone?
c. Has a cardioid microphone unidirectional or omnidirectional response?
d. Which must be shielded, microphone or speaker cable?

4-5
PHONOGRAPH RECORDS AND PICKUPS

In function, a phono pickup or cartridge is similar to a microphone; each is a source of audio signal. Both are electromechanical transducers; they depend on similar principles for converting mechanical energy into electric energy. As the stylus or needle in the pickup tracks the grooves of a phonograph record, the physical vibrations are converted to audio output.

The main types of phono cartridges use the magnetic or piezoelectric principles, as microphones do. However, the output of a phono pickup is much greater than that of a microphone because of the greater mechanical motion. Typical audio output levels are 5 to 10 mV for magnetic phono pickups and 200 to 1000 mV for the crystal or ceramic type. The frequency response is generally about 30 to 16,000 Hz for both types. A high-quality magnetic pickup can have even better response (Fig. 4-15).

Magnetic Phono Pickup for Stereo The construction and connections of the magnetic phono pickup for stereo are shown in Fig. 4-16. Practically all records and pickups are made with separate left and right audio channels for stereophonic reproduction. The record is cut with two audio signals in the grooves; the two are on opposite sides at a 45° angle.

Construction of a magnetic cartridge is shown in Fig. 4-16a. As the stylus moves the yoke, signal current is induced in the coils. Here the yoke is a movable iron core. In some types, the magnetic flux is fixed and the coils move. Either way, the motion produces electromagnetic induction for audio signal output.

In Fig. 4-16b, the cartridge is shown mounted in a holder or shell. Screws or spring clips can be used for the mounting. The four output pins, two

Fig. 4-15. Typical high-quality magnetic phono cartridge with attached record-cleaning brush. (*Pickering and Company*)

for right signal and two for left signal, connect to the four leads for audio output. The stylus or needle usually slips into the yoke with a friction fit.

In Fig. 4-16c, the left and right phono output plugs connect to the audio amplifier. Shielded wire is used; there is a common ground on the amplifier chassis. The individual ground lead with a spade lug is chassis-ground on the record player; it includes the shield in the pickup.

Note the following color code for stereo phono connections:

Red   Right channel, high side
Green   Right channel, low side
White   Left channel, high side
Blue   Left channel, low side

A typical magnetic phono cartridge is shown in Fig. 4-15. The purpose of the small brush is to clear dirt and dust from the record groove before the stylus passes over the groove.

Ceramic Phono Pickup for Stereo   As shown in Fig. 4-17, in a ceramic stereo cartridge two ceramic transducers are attached to a Y-shaped yoke. One ceramic element is for the right audio channel, and the other is for the left. When the stylus moves in the groove of the record, its arm twists the yoke and the ceramic elements. Because

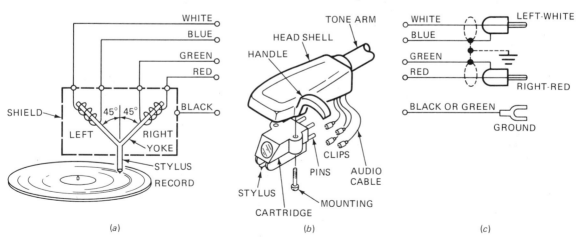

Fig. 4-16. Magnetic stereo cartridge for phonograph records. (*a*) Construction of cartridge with output leads. (*b*) Cartridge mounted in shell on tone arm. (*c*) Left and right phono plugs connect to phono jacks on the amplifier. (*From A. J. Wells, Audio Servicing, McGraw Hill, New York, 1980*)

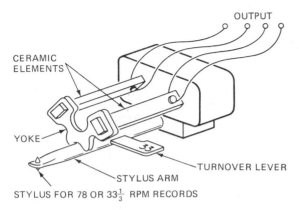

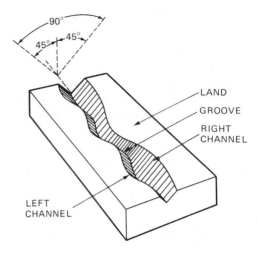

Fig. 4-17. Construction of ceramic stereo cartridge. Note turnover lever with two needles for 78 or 33-1/3 rpm records.

Fig. 4-18. Stereo record groove in 45°-45° system. Right and left audio signals are on opposite walls.

of the piezoelectric effect, that motion is converted to audio signal output.

Note the turnover lever for a choice of stylus. A 3-mil stylus is used for the older 78-rpm records. However, the fine-groove 33⅓-rpm records now in general use require a stylus diameter of 0.7 mil or less. One mil is 0.001 in.

Phonograph Records The record, or disk, has narrow grooves with variations that correspond to the audio signal. The grooves spiral into the center of the record. Actually, the record is a method of storing the information cut into the grooves. A master metal disk is made first, and then copies are pressed out by machine.

The old shellac records were made for a speed of 78 revolutions per minute (rpm) on the turntable. Now fine-groove vinyl records are used; they operate at either 33⅓ or 45 rpm. They have about 250 grooves per inch. The advantages are less surface noise, or scratch, and longer playing time. The 33⅓ rpm records have diameters of 10 or 12 in. They are also called long-play (LP) records. A 12-in record at 33⅓ rpm has a playing time of about 18 to 22 minutes (min).

The standard hole in the center of the record is ¼ in to fit over the spindle on the record player. However, 45-rpm records have a 1½ center hole, which requires either a separate spindle or an adapter.

The audio signal can be cut into the record grooves by using lateral (horizontal) variations or vertical (hill-and-dale) ones. For cutting stereo records, both approaches are used. As shown in Fig. 4-18, opposite walls of the groove are at 45° to a vertical plane, making them 90° or perpendicular to each other. This is called the 45°-45° system.

The right channel is recorded on one side of the groove and the left channel on the other. As the needle rides in the groove, the lateral and vertical forces produce diagonal motions for the phono pickup. The horizontal and vertical components are separated by the pickup for left and right audio signals.

The Stylus Figure 4-19 shows how the stylus, or needle, rides in the grooves of the record. The stylus must be very small to follow the variations. The diameter of the stylus for stereo microgroove records is typically 0.5 mil, or 0.0005 in [0.0013 millimeter (mm)]. Different stylus shapes are shown in Fig. 4-20. The conical stylus is most common, but the elliptical and shabata shapes provide better high-frequency response. The latter measure 0.3 mil in the short diameter.

The material used for the stylus is osmium metal, natural or synthetic sapphire, or diamond. A dia-

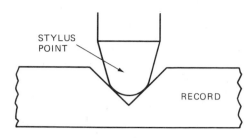

Fig. 4-19. Stylus resting in the groove of a record.

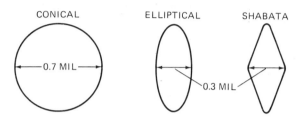

Fig. 4-20. Shapes of stylus tips.

Fig. 4-21. Typical record changer. (*Zenith Radio Corp.*)

mond stylus is best because it lasts the longest. Record wear dulls the stylus. The stylus can be cleaned with a soft brush, but the brushing should be from back to front so the force on the yoke is minimal.

Record Changers A typical record changer is shown in Fig. 4-21. Six to eight records can be stacked on the center spindle. At the end of each play, the tone arm lifts and moves clear of the record area, the next record drops, and the tone arm descends to the record to play again. The better units are also called an *automatic turntable,* because the audio quality is considered almost as good as the best manual turntable.

Speed Setting The standard speeds of phonograph turntables are 78, 45, 33⅓, and 16 rpm. The 16-rpm speed is used for extra long play on speech records. The 78-rpm and 16-rpm speeds are usually omitted from phonographs designed to play music recordings.

Tracking Force The weight of the tone arm and cartridge assembly must provide enough pressure to keep the stylus riding smoothly in the groove.

This pressure is called *tracking force.* The tracking force adjustment sets the weight of the tone arm assembly (including the cartridge) on the record for best tracking at the lightest weight for minimal record wear. Either a spring or a counterweight is used to oppose the weight of the tone arm and cartridge. Actually, the adjustment can balance the tone arm so that it stays horizontal without touching the record. The desired adjustment is a little more than that balanced position by the amount of weight recommended for the tracking force of the cartridge.

Magnetic cartridges usually require only ¾ to 3 grams (g) for accurate tracking. Ceramic cartridges need up to 8 g. The best tracking force to use is about 10 percent more than the minimum recommended by the manufacturer. The adjustment applies to manual and automatic turntables, but low-cost equipment may not have a tracking force adjustment control.

Antiskating Adjustment "Skating" refers to the tendency of the tone arm to move inward and apply more force on the inside edge of the record grooves. If the stylus should become clear of the record groove, the tone arm will skim across the top of the record toward the spindle. An antiskating control applies a slight counteracting outward pressure. Usually, the adjustment is set to the same number on the knob as the tracking force of the cartridge. Not all record players have an antiskating adjustment.

Test Point Questions 4-5
(Answers on Page 98)

a. Is the typical output of a magnetic phono cartridge 5 mV or 5 V?
b. Do microgroove records play at $33\frac{1}{3}$ or 78 rpm?
c. Are the right and left audio signals in a stereo record at 45° or 180°?
d. Is a typical tracking force for magnetic phono pickups 2 or 20 g?

4-6
MAGNETIC TAPE RECORDING

The magnetic tape recording offers the advantages of reduced surface noise and better stereo separation. There is practically no wear on the tape. Tape recordings are easily made with relatively inexpensive equipment. The same equipment can be used to play the recordings, erase them if desired, and reuse the same tape over and over again.

As shown in Fig. 4-22, the basis of tape recording involves magnetizing very small particles of a magnetic coating on the surface of a tape as the tape passes through a thin air gap of an electromagnet (the recording head). While it is in the gap, the tape is a continuation of the magnetic path.

During recording, a signal is applied to the coil in the head. The tape is then magnetized with the same variations as the signal. During playback, the magnetic variations on the tape induce an output signal in the coil.

One head can serve for both recording and playback, or separate heads can be used. As typical values, the head needs about 30 mV of audio signal for recording and induces about 1 mV on playback.

The main factors in tape recording are the magnetic properties of the tape, the tape speed in passing the head gap, and the gap width. Higher tape speed allows more high-frequency response but re-

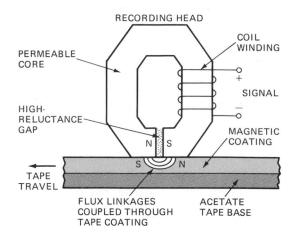

Fig. 4-22. Magnetic tape recording system. (*From G. P. McGinty, Video-Cassette Recorders, McGraw-Hill, New York, 1979*)

Fig. 4-23. A typical record-playback head for magnetic tape. Height is 1/2 in [1.27 cm]. (*Shure Brothers Inc.*)

duces the playing time. A smaller head gap also extends the response for high frequencies. A typical gap is only 6 to 9 mils [0.006 to 0.009 in] for audio tape recorders, and it is even less for video tape recorders. Actually, the gap is just a thin insulator of non-magnetic material in the ferromagnetic core of the head. A record-playback head for magnetic tape is shown in Fig. 4-23.

AC Bias for Magnetic Tape The ac bias technique is used in audio tape recorders to minimize distortion. The reason can be seen from the magnetization curve in Fig. 4-24. The curve is a plot

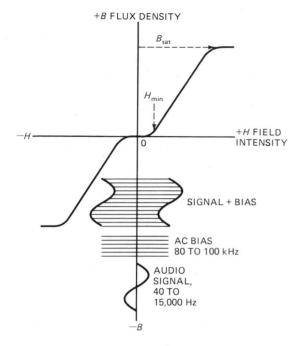

+B FLUX DENSITY

B_{sat}

H_{min}

−H +H FIELD INTENSITY

0

SIGNAL + BIAS

AC BIAS
80 TO 100 kHz

AUDIO
SIGNAL,
40 TO
15,000 Hz

−B

Fig. 4-24. The use of ac bias in audio tape recording to reduce distortion. The ac bias is also used to erase any signals (recording) from the tape.

of the flux density B in the magnetic coating on the tape against magnetic field intensity H in the gap, which is produced by current in the head. There is an offset from zero because the magnetic dipoles in the tape have inertia. No magnetic induction is produced until the field intensity reaches the minimum value necessary to organize the random dipoles in one direction. The amount of ac bias is just enough to exceed the minimum H. Typical values are 3 to 6 mA peak. The frequency of the ac bias current is 80 to 100 kHz.

The audio signal and ac bias are combined. As a result, the signal rides on the peaks of the ac bias. Then the audio recording variations use only the linear part of the B-H magnetization curve.

Tape Erase The ac bias current is also used to erase audio signal variations from the tape. If you use the record mode but do not apply an audio signal, any previous signal on the tape will be erased. The reason is that the ac bias itself has an

average value of zero, which causes the magnetic dipoles in the tape to assume a random orientation.

When you record audio signal, the added bias current automatically erases any previous signal. As a result, the new signal is recorded on a clean tape. That is why magnetic tape can be reused almost indefinitely. When there is a separate erase head, more bias current can be used for better erasing.

Types of Tape A coating of very small ferromagnetic particles with a binder material is put on a plastic base, which may be acetate or polyester (Mylar). The Mylar base is stronger and does not stretch. Thickness of the tape is approximately 0.001 in. The width is generally ¼ in [6.35 mm] for reel-to-reel tape and about ⅛-in [3.12-mm] tape in cassettes.

The magnetic coating of typical tape is ferric oxide, which chemically is the same as iron rust. An improved coating uses chromium dioxide (CrO_2) for better high-frequency response and higher signal-to-noise ratio. The chromium coating costs more, but its advantages are important at the low tape speed for cassette recorders. Another improved coating uses cobalt.

Chromium and cobalt tapes need more ac bias current because of their higher coercivity compared with iron. More field intensity is required to reduce the magnetic flux to zero.

Open-Reel Recorders Open-reel recorders feed tape from one reel to another. The recording tape is wound on a single reel. To play or record the end of the tape is threaded onto a second reel, called the *take-up reel*. Fast forward and reverse action is provided to find any position on the tape and to rewind in minimum time. Open-reel recorders generally use ¼-in tape operating at 7½ inches per second (in/s) [19.05 cm/s]. The reel is usually 7 in [17.78 cm] in diameter.

Cassette Recorders As shown in Fig. 4-25, a cassette is a miniature reel-to-reel system, except that the tape is already threaded in both reels. Cassette recorders provide fast forward or reverse

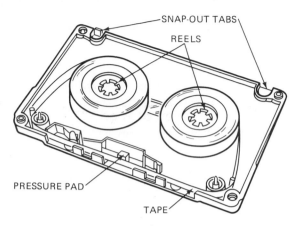

Fig. 4-25. Standard tape cassette. The dimensions of the cassette are 4 in by 2-1/2 in.

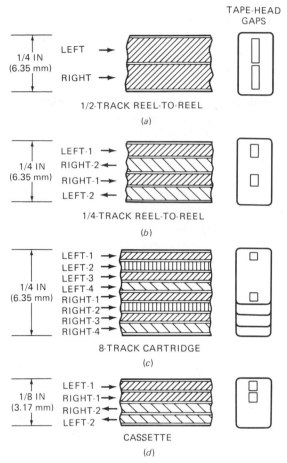

Fig. 4-26. Placement of head gaps for different stereo tape formats. (*From A. J. Wells, Audio Servicing, McGraw-Hill, New York, 1980*)

operation. Speed is 1⅞ in/s [47.63 mm/s] with ⅛-in [3.2-mm] tape. Some dual speed cassette recorders also operate at 3¾ in/s. The lower speed allows more playing time but makes extension of the high-frequency response more difficult. Still, improved tapes and a smaller head gap enable the cassette recorders to provide results almost as good as those of open-reel recorders. Cassette tape lengths provide up to two hours of playing time on two channels (one hour per channel). Playing times are specified as C-30, C-60, C-90, etc., indicating 15, 30, and 45 min playing time per channel respectively.

Note the snap-out tabs at both ends of the back of the cassette. When removed, the holes they leave prevent accidental erasing of prerecorded cassettes. The tape machine has a rod that extends into the hole to prevent operation in the recording mode. There is a hole at each end so that either one can be used when the cassette is turned over. If you want to record, though, the hole must be covered.

Eight-Track Recorders The eight-track recorder uses a cartridge that has only one reel with an endless loop of ¼-in [6.35-mm] tape. The speed is 3¾-in/s [9.53 cm/s]. Playing time is generally 40 to 90 min for all the tracks in succession, without any turnover. The eight-track cartridge is popular in compact stereo systems because any one of

the four pairs of tracks can be selected for program selection.

Stereo Tape Formats The way the head gaps are placed to provide separate audio channels on the tape is shown in Fig. 4-26 for different stereo tape formats. Note that turnover of the tape is necessary with ¼-track reels and cassettes but not for ½-track reels and the eight-track cartridge. The top two tracks of a cassette pass the head gaps in one direction. When the cartridge is turned over, the bottom two tracks become the top two tracks to pass the head gaps.

Dolby Noise-Reduction System Tape noise is a background hiss that may be heard when low-level audio signals are recorded. It is caused by random variations in the magnetization of the tape. The Dolby system is a means of reducing it.

Higher audio frequencies are usually relatively weak because they are mostly harmonics of the lower fundamental frequencies. In the Dolby system, weak audio signals above 500 Hz are expanded in amplitude for recording. The increase allows the signal to swamp out the noise. In the playback, these frequencies are compressed by the same amount to provide a resultant output with uniform frequency response. The decrease in signal amplitude also decreases tape noise. In the complete process, tape noise can be reduced by more than 50 percent.

Tape Troubles A dirty tape head is a common cause of weak signal and excessive tape noise. The head accumulates dust and loose oxide particles by the tape rubbing against the gap. As a result, a space is created between the gap and the tape, which decreases the magnetic flux. To remedy this condition the head can be cleaned with a soft cotton swab moistened with alcohol or a commercial fluid.

In small cassette recorders, the drive system generally uses a thin rubber belt from the motor to the reel. The drive is only on one reel, which pulls the other reel by the tape. The belt may become stretched or may break. When the tape speed is too low, check for a loose drive belt. When the reels do not turn at all, the belt may be broken.

Test Point Questions 4-6
(Answers on Page 98)

Answer True or False.

a. A typical gap for an audio head is 0.007 in.
b. The frequency for ac bias in audio recording is usually 60 Hz.
c. Cassette recorders generally use ¼-in half-track tape.
d. Chromium dioxide tape needs more ac bias than ferric oxide tape.
e. A dirty head can cause excessive tape noise.

4-7
AUDIO PREAMPLIFIER CIRCUIT

The preamplifier is used for very low levels of audio signal, such as the output of a magnetic microphone, phono pickup, or tape head. The typical audio input signal is 1 to 5 mV. A preamp circuit consists of RC-coupled amplifiers, as shown in Fig. 4-27. Because of the small signal, however, special attention is required to reduce hum and noise. A shielded lead is used for the input to avoid hum

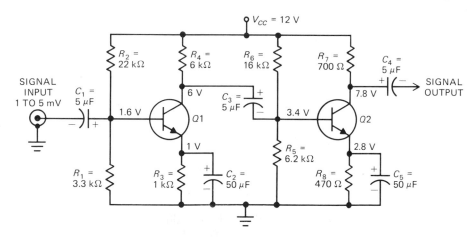

Fig. 4-27. Two cascaded CE stages for an audio preamplifier.

pickup. Also, the collector voltage and emitter current are low to reduce random noise generated in the transistor. Class A operation is necessary for minimum amplitude distortion.

In Fig. 4-27, $Q1$ and $Q2$ are common-emitter amplifiers in cascade. The input signal from the shielded line is coupled by C_1 to the base of $Q1$. The amplified output of $Q1$ is coupled by C_3 to the base of $Q2$. Finally, C_4 couples the collector output of $Q2$ to the next circuit, which would usually include driver and power output stages for the loudspeaker. The overall voltage gain of the preamplifier is 1000 to 5000.

All the audio coupling capacitors are $5\text{-}\mu\text{F}$ electrolytics. R_4 is the collector load for $Q1$, with R_7 for $Q2$. Both stages use emitter self-bias for stabilization and a voltage divider for base bias. Each stage has a net V_{BE} of 0.6 V for class A bias on silicon transistors.

Note the low values of collector voltage. The V_{CE} for $Q1$ is $6 - 1 = 5$ V. The stage operates with I_C of 1 mA. Its I_B for bias current in 20 μA.

For $Q2$, the V_{CE} is $7.8 - 2.8 = 5$ V. The stage operates with I_C of 6 mA. Its I_B for bias current is 100 μA.

Test Point Questions 4-7
(Answers on Page 98)

Refer to Fig. 4-27.

a. Is the R_1C_1 coupling circuit for $Q1$ or $Q2$?
b. What value of C is used for the audio coupling capacitors?
c. What is the V_{BE} bias for $Q2$?

4-8
AUDIO POWER OUTPUT STAGE

The audio output circuit is a power amplifier to supply enough audio signal current for the loudspeaker. An example of a single-ended amplifier with a transformer-coupled output is shown in Fig. 4-28. $Q1$ is the driver that swings the base current of the output stage $Q2$. Single-ended audio amplifiers must operate class A to prevent excessive distortion.

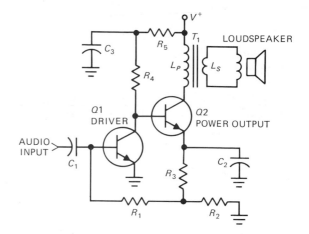

Fig. 4-28. Circuit for driver and audio output stages.

For more audio power, though, a combination of two transistors is used for the output stage. These circuits are push-pull amplifiers operating class B and AB for higher efficiency than class A.

Push-Pull Audio Output Two transistors amplify opposite half-cycles of the audio signal. Push-pull operation is the best way to use the two for maximum audio power with the least distortion. For an example of a push-pull audio output stage, refer to Fig. 3-9 on page 49. A center-tapped transformer is used to shift the output signals 180° out of phase with the input signals.

Complementary-Symmetry Output The complementary-symmetry circuit uses a PNP and a NPN transistor to provide push-pull output signals without the need for a center-tapped transformer in the input. An example is shown in Fig. 3-12 on page 52.

Quasi-Complementary Symmetry In the quasi-complementary circuit the driver stage, instead of the power output stage, uses complementary symmetry. The advantage is lower cost, since complementary transistors with a lower power rating can be used.

Stacked V^+ with Series Output An example of stacked V^+ with series output is shown in Fig.

3-11 on page 51. The circuit uses a push-pull input signal. However, a push-pull output is produced without the need for a center-tapped output transformer.

Circuit for Class A Driver and Output Stages In the circuit of Fig. 4-28, $Q1$ amplifies an audio signal from a preamplifier to drive the power output stage $Q2$. Both are CE amplifiers. Direct coupling is used from the collector of $Q1$ to the base of $Q2$. Note that part of the emitter bias on $Q2$ is used, through R_1, for the base bias on $Q1$.

The collector load resistance for $Q1$ is R_4. For the power output stage, though, the collector load of $Q2$ is the primary impedance of the audio output transformer T_1.

In the V^+ line for collector voltage, $Q2$ has practically the full dc supply voltage through the primary winding of T_1. Collector voltage for $Q1$ is lower because of the drop across R_5. The R_5C_3 network is a decoupling filter to isolate $Q1$ from the dc supply. The power stage can have the total unfiltered V^+ from the power supply, though, since there is no further amplification of any hum voltage. More V^+ allows more power output.

Audio Output Transformer In Fig. 4-28 T_1 is a step-down transformer. Its function is to match the low impedance of the loudspeaker's voice coil to the higher impedance needed for the collector of the output transistor. This impedance match results in more audio output and less distortion.

In addition, a separate secondary winding isolates the loudspeaker from the dc collector voltage. The ac audio signal is induced in the secondary winding without the need for a coupling capacitor. If the secondary is not grounded, the loudspeaker also will be isolated from any ground connections on the amplifier chassis. However, one side of L_S is often grounded in order to save extra wiring.

Since the audio transformer has an iron core with practically unity coupling, the power delivered to the secondary L_S equals the power supplied by the primary L_P. Therefore, L_P can provide the required Z_L with relatively high voltage and current compared with the low-impedance voice coil. For instance, assume the loudspeaker impedance is 4 Ω. If the speaker were connected directly in the collector circuit, there would be little audio output because 4 Ω is too low for the collector load impedance.

The turns ratio of the transformer needed for the impedance match is

$$\frac{N_S}{N_P} = \sqrt{\frac{Z_S}{Z_P}} \qquad (4\text{-}2)$$

where N is the number of turns. The turns ratio is equal to the square root of the impedance ratio. As an example, let the required Z_P for the collector be 200 Ω with Z_S of 4 Ω for the loudspeaker. Then

$$\frac{N_S}{N_P} = \sqrt{\frac{Z_S}{Z_P}}$$
$$= \sqrt{\frac{4}{200}} = \sqrt{\frac{1}{50}}$$
$$\frac{N_S}{N_P} = \frac{1}{7.07}$$

This answer means that the step-down transformer should have approximately seven turns of L_P for each turn of L_S.

Ratings for Maximum Audio Power Output Different methods are commonly used to specify audio power output. This can cause confusion because the various methods do not produce the same numerical value for the same amplifier. An example is the difference between continuous rms power and the rating for music power. The music power is specified for the purpose of indicating how the amplifier can supply peaks of power similar to those of a live performance. In terms of ratings, the music power is a higher number, generally by 20 to 50 percent, more than the rms power rating.

Furthermore, there are two methods of specifying music power. By the Electronic Industries Association (EIA) rating method, the power is measured after the sudden application of a transient signal with total harmonic distortion at 5 percent or less. By the Institute of High Fidelity (IHF) method, the transient power is measured with distortion of 1 percent or less. The IHF method gives rating numbers about 25 percent lower than the EIA method gives.

The EIA music-power rating is generally used for consoles in a complete package, the load requirements of which are known by the manufacturer. Separate high-fidelity audio amplifier units are rated by both the IHF method and continuous power output.

In addition, the output power can be specified for either a 4- or 8-Ω loudspeaker load impedance. The output power at 4 Ω is usually about 20 percent greater than at 8 Ω.

Test Point Questions 4-8
(Answers on Page 98)

a. Is a single-ended audio output stage usually operated at class A, B, or C?
b. In Fig. 4-28, which component is the emitter bypass for Q2?
c. What turns ratio is required for matching 8 Ω to 200 Ω?
d. Which is a higher number, the EIA music-power rating or continuous rms power output for the same audio amplifier?

4-9
TYPES OF DISTORTION

When the waveform of the amplified output signal is not exactly the same as that of the input signal, distortion has been introduced in the amplifier. Amplitude distortion means that the relative amplitudes in the signal have been changed. In frequency distortion, the relative gain is not the same for different frequencies. For high-fidelity audio, it is the absence of distortion that makes possible a reproduction close to the original, live sound.

Amplitude Distortion Amplitude distortion is produced by operating the amplifier over the non-

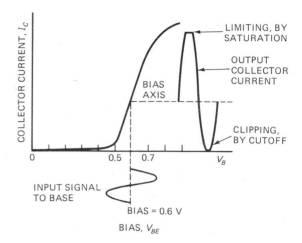

Fig. 4-29. Nonlinear amplitude distortion in a collector output signal produced by an excessive input signal causing overload.

linear part of the transfer characteristic of the amplifier. In extreme cases, the sound can be rough, hoarse, or garbled. Possible causes are low supply voltage, a weak amplifier, incorrect bias, or too much signal for the bias. Amplitude distortion is a problem mainly in power amplifiers because they handle larger signals.

Overload Distortion *Overload* distortion occurs when the input signal, or drive, is excessive. The result is very bad amplitude distortion. An example is shown in Fig. 4-29 for the transfer characteristic curve of a silicon NPN transistor. The bias V_{BE} for class A operation is 0.6 V. Linear operation on the curve includes only ± 0.1 V, approximately, around the bias. Actually, the transfer curve is perfectly linear only for a swing of about ± 0.05 V.

Below 0.5 V, the I_C approaches zero for cutoff. Above 0.7 V, the I_C becomes saturated. The excessive input signal drives V_{BE} too far toward saturation and cutoff. As a result, the sine-wave input signal I_C becomes flattened at the top and bottom.

The three main factors affecting overload distortion are:

1. Signal changes near saturation are reduced or eliminated in the output.

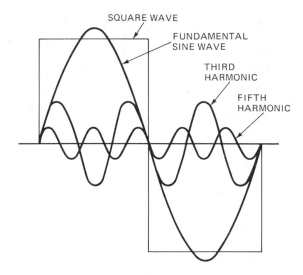

SQUARE WAVE

FUNDAMENTAL
SINE WAVE

THIRD
HARMONIC

FIFTH
HARMONIC

Fig. 4-30. How a sine wave flattened into a square wave corresponds to harmonic distortion.

2. Signal changes near cutoff are reduced or eliminated in the output.
3. The dc level of the output signal shifts.

In a class A amplifier, the dc level should stay the same with or without signal. In the distorted output signal of I_C in Fig. 4-29, though, note that the bias axis is not at the center of the waveform. The difference between the bias axis and the center axis of the waveform is the shift in dc level. Any shift of dc level in a class A amplifier is a measure of the amount of amplitude distortion.

Harmonic Distortion Actually, amplitude distortion is equivalent to harmonic distortion. Changing the relative amplitudes is the same as introducing harmonic components not present in the input signal. This idea of harmonic frequencies in a nonsinusoidal waveform is illustrated in Fig. 4-30. Consider the fundamental shown to be the original sine-wave input signal. Also, assume the distorted output is flattened to become a square wave. A square wave is composed of a fundamental sine wave at the same frequency plus odd-harmonic frequency components. Those harmonics were not present in the original sine-wave input signal. Therefore, the nonlinear amplification has introduced harmonic distortion in the form of new harmonic frequencies.

Nonlinear amplification or its resultant harmonic distortion is produced mainly in the audio output stage. Typical values are 1 to 5 percent harmonic distortion at full power output. Using less drive reduces the distortion to a great extent, but results in less output.

Intermodulation Distortion Another result of amplitude distortion is that the harmonics introduced in the amplifier can combine with each other or with the original frequencies to produce new frequencies that are not harmonics of the fundamental. This effect is called *intermodulation distortion,* because the frequencies combine with each other. Intermodulation is the reason for the rough, unpleasant sound of amplitude distortion, because that distortion is not harmonically related to the signal.

Frequency Distortion Frequency distortion results when the gain of the amplifier varies with frequency. Speech and music have many different frequency components. They correspond to the fundamental and harmonic frequencies produced by the sound source. Probably all the different frequencies have different amplitudes. However, the amplifier should provide the same gain for all the frequencies so that the frequencies will have the same relative amplitudes in the output. Then the original character of the sound will be maintained.

Generally, the problem is to amplify the very low and high audio frequencies. As an example, the frequency response curve for an *RC*-coupled audio amplifier is shown in Fig. 4-31. At the low-frequency end, the series coupling capacitors cause insufficient response, or a *droop* in the output. At the high-frequency end, the response curve has a *rolloff* because of the bypassing effect of shunt capacitance.

Insufficient output at low audio frequencies means weak bass in the sound. At the opposite extreme, insufficient treble is caused by a weak output at the high audio frequencies. The uniform gain at the center of the curve is called *flat response.*

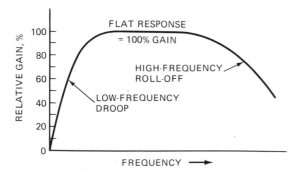

Fig. 4-31. Frequency distortion means the gain is not the same for all audio frequencies. The response curve shown is for an *RC*-coupled amplifier.

The absence of frequency distortion is important in the reproduction of music, which has a wide frequency range. Even in speech reproduction, the articulation or clarity of a voice improves with better frequency response. Conversational speech includes frequencies in the range of about 60 to 8000 Hz approximately.

Test Point Questions 4-9
(Answers on Page 98)

Answer True or False.

a. Amplitude distortion causes harmonic distortion.
b. Figure 4-29 shows an example of overload distortion.
c. In a class A amplifier, a shift in dc level with a signal input indicates there is no amplitude distortion.
d. High-frequency rolloff in an audio amplifier causes weak bass response.

4-10
NEGATIVE FEEDBACK

The negative feedback technique is probably the most important method of reducing distortion in amplifiers. "Feedback" means coupling part of the amplified output signal back to the input.

When the feedback is in phase with the input signal, the result is positive *feedback*, or *regenera-*

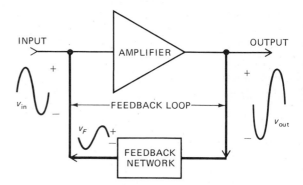

Fig. 4-32. Using negative feedback in an amplifier.

tion. Positive feedback is used in oscillator circuits. It increases gain, but the amplifier becomes unstable and has a tendency to oscillate. When the feedback is out of phase with the input signal it is *negative feedback, inverse feedback,* or *degeneration.*

Negative feedback reduces the amplifier gain because part of the input signal is canceled. However, all forms of distortion are reduced in about the same proportion as the loss of gain. It is usually not difficult to obtain as much gain as necessary, but distortion is always a problem. Negative feedback is commonly used to improve the quality of audio amplifiers, therefore, and especially to reduce the distortion of the power output stage. In addition, the amplifier is more stable with negative feedback.

The general idea of using negative feedback on an amplifier is illustrated in Fig. 4-32. The amplified output signal of an amplifier is out of phase with the input. Then part of the output can be coupled to the input for negative feedback. This circuit is the *feedback loop.* Note that the feedback signal V_F is negative when the input signal V_{in} is positive.

The amount of feedback is determined by the feedback network. That circuit can also vary the feedback for different frequencies. The amount of amplifier gain with feedback is called the *closed-loop gain.* Without feedback the gain is called *open-loop gain.*

The reason why negative feedback reduces distortion is because there is partial cancellation of the out-of-phase signals. As an example, suppose that a stage amplifies some amplitudes more than others because of nonlinear operation. The parts of the signal that are amplified too much provide more negative feedback voltage. The result is that there is more cancellation of these signals.

Suppose also that some frequencies have more relative gain than they should have. Such signal frequencies produce more negative feedback. The result is that there is more cancellation for those frequencies than for others with less gain.

The overall effect of the cancellation due to negative feedback, then, is to provide more uniform response in the amplifier. Both amplitude and frequency distortion are reduced.

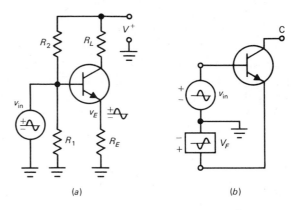

(a)   (b)

Fig. 4-33. An unbypassed emitter resistor produces negative feedback. (a) Circuit. (b) The feedback signal in series-opposition with the input signal.

Unbypassed Emitter Resistor for Negative Feedback

In the common-emitter amplifier circuit with an emitter resistor R_E for self-bias, the resistor is usually bypassed by a large capacitance. The bypass capacitor allows the emitter voltage to be a steady dc voltage for bias without any ac signal variations. When R_E is not bypassed, however, the ac component of the emitter voltage provides a signal for negative feedback. This degeneration in the emitter circuit reduces amplifier gain but also cancels distortion.

In Fig. 4-33, R_E is not bypassed in order to provide negative feedback. For the circuit in Fig. 4-33a, note that the feedback voltage V_E is shown in phase with the input signal V_{in}. However, the two are really out of phase. They cancel for the total base-emitter voltage, as in Fig. 4-33b. When V_E goes positive, it reduces the positive base voltage with respect to emitter. The potential difference between the two signals is the net V_{BE}. The higher the resistance of R_E the greater the amount of negative feedback.

The circuit in Fig. 4-33 is considered to be *current feedback* because it depends on I_E, which is mainly the collector current. It is also called *series feedback* because V_F is in series with V_{in}.

Negative Feedback from the Collector Voltage

In Fig. 4-34a the collector signal voltage is shown to be fed back through R_F to the base input

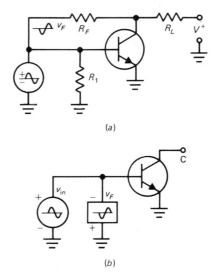

(a)

(b)

Fig. 4-34. Negative feedback from the collector voltage. (a) Circuit using R_F in the feedback loop. (b) Feedback signal in parallel with the input signal.

circuit. Since the amplifier inverts the signal V_F is out of phase with V_{in}.

The value of R_F, as a voltage divider with R_1, determines the amount of feedback. A higher resistance for R_F means less negative feedback. The equivalent circuit in Fig. 4-34b shows V_F in parallel with V_{in} to cancel part of the input signal.

The circuit shown in Fig. 4-34 is considered to be *voltage feedback* because it depends on the collector signal voltage. It is also called *parallel feedback* because V_F is in parallel with V_{in}.

Multistage Feedback The feedback loop can extend around more than one stage. Then the feedback is even more effective because of the cascaded gain. It is still necessary, however, for the negative feedback to oppose the input signal at the point where the feedback is injected. A practical system is to include the loudspeaker voice coil in the feedback loop, because the audio output circuit generally has the most distortion.

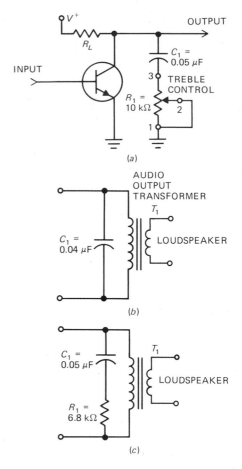

(a)

(b)

(c)

Fig. 4-35. Treble-cut circuits for audio tone control. (a) Variable *RC* filter. (b) Bypass capacitor across the primary of an output transformer. (c) *RC* filter across the transformer primary.

Test Point Questions 4-10
(Answers on Page 98)

a. Is distortion in an amplifier reduced by positive or by negative feedback?
b. In Fig. 4-33a, does increasing R_E provide less or more negative feedback?
c. In Fig. 4-34a, does the feedback loop include R_F or R_L?

4-11
TONE CONTROLS

The relative output at different audio frequencies determines the tone of the reproduced sound. Therefore, the tone can be controlled by varying the frequency response. Treble tone means more response at high audio frequencies, usually above 3000 Hz. Bass sound has more low audio frequencies, from about 300 Hz down. In many cases, bass tone is provided simply by reducing the gain for high frequencies. This method is called *treble cut.* The relative response emphasizes the bass, but at the expense of high-frequency response.

Treble-Cut Circuits The general treble-cut method is to use a shunt bypass capacitor to reduce the output at high audio frequencies. Typical values of C are 0.01 to 0.15 μF. Probably the most popular circuit is the *RC* filter shown in Fig. 4-35a. The tone control R_1 is in series with the bypass C_1. As the variable arm of R_1 is moved closer to terminal 3, the series resistance is re-

duced. Then C_1 can bypass more of the high audio frequencies to provide more treble cut, or more relative bass response. At the opposite end of R_1, with the variable arm at terminal 1, the series resistance reduces the bypass effect of C_1. That position is for treble response. In some circuits, an additional fixed R may be used in series to limit the reduction in volume.

In the circuit of Fig. 4-35b, the 0.04-μF bypass capacitor is simply connected across the primary of the output transformer. That method is often used to improve the thin sound of a small loudspeaker.

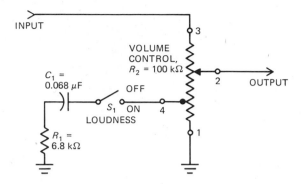

Fig. 4-36. Tone-compensated volume control.

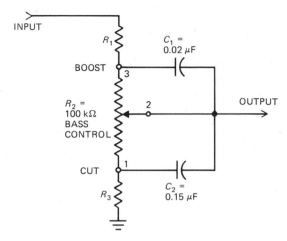

Fig. 4-37. Bass-boost circuit for audio tone control.

In Fig. 4-35c the R_1C_1 filter is shown connected across the primary. Not only is C_1 a tone control bypass but R_1 helps to maintain a uniform load impedance across the output circuit for different frequencies.

Tone-Compensated Volume Control   The ear is less sensitive to bass frequencies than treble frequencies at low volume. That may be why many people like to play music loud: it seems to sound better with more bass response. To increase bass tones of low volume, the volume control may have an extra tap to introduce a treble-cut filter into the circuit.

An example is shown in Fig. 4-36; a loudness switch connects the R_1C_1 filter to terminal 4. The filter is effective only at low-volume settings when the variable arm is near terminal 4. At high-volume settings, the R between terminals 4 and 3 minimizes the effect of the filter. Such a volume control, with provision for reducing the treble response, is sometimes called a *loudness control.* The loudness switch S_1 is provided to disconnect C_1 and R_1 if desired, because many people do not like the sound with a treble cut.

Bass-Boost Circuit   In the circuit shown in Fig. 4-37, R_2 is used to vary the low-frequency response without cutting the treble. Two coupling capacitors, C_1 and C_2, are used. The larger the coupling capacitor the better the low-frequency response.

When R_2 is varied, either C_1 or C_2 becomes more effective in coupling the audio signal to the next stage. When R_2 extends to terminal 3 at the top, C_1 is effectively shorted across the terminal 2 and the output. Then the larger value of C_2 is used for coupling. This condition results in maximum bass response. When the control is in position 1 the smaller C_1 is the coupling capacitor since C_2 is shorted across terminal 1 and the output.

It should be noted that the bass-boost circuit shown in Fig. 4-37 can be combined with the treble-cut circuit shown in Fig. 4-35a. Then the equipment features separate bass and treble controls, which is really the best way to adjust the tone. The sound is usually more natural without too much boost or cut at either end of the frequency response curve. The bass and treble response should be boosted or cut together.

Selective Negative Feedback for Tone Control   The selective negative feedback method is very effective in controlling the audio amplifier response at different frequencies. When more negative feedback is used for high audio frequencies, the result is less gain than at the low frequencies. Varying the amount of the feedback controls the treble response. For bass control, the amount of negative feedback is varied for the low audio frequencies.

RC filters can be used in series or shunt combinations to select the frequencies for more negative feedback. An example is shown in the dual stereo amplifier of Fig. 4-38 with bass-boost controls.

Test Point Questions 4-11
(Answers on Page 98)

a. In Fig. 4-35*a*, does terminal 1 or 3 on R_1 provide more cut in the treble response?

b. In Fig. 4-37, does C_1 or C_2 provide more bass response?

c. Is a tone-compensated loudness control used for high or low volume?

4-12
STEREOPHONIC AUDIO EQUIPMENT

Two separate amplifiers with loudspeakers are necessary to reproduce the audio signals individually for the left and right channels. A distance of 6 to 8 ft [1.83 to 2.44 m] between speakers is recommended to achieve a stereo effect. Stereo reproduction gives a sense of direction to the sound that eventually reaches the ears.

An example of duplicate audio amplifiers for stereo is shown in Fig. 4-38. The complete circuits for both amplifiers are in one integrated circuit (IC) unit, except for the components shown outside the dashed area. Maximum audio output is 7 W per channel with $V^+ = 35$ V as dc supply voltage for this IC package.

Note the balance control at the input. It is used to proportion the two audio-signal inputs. The control is usually at the center for balance. However, more signal can be supplied to either channel if necessary to compensate for the room acoustics.

The two bass-boost controls are on one shaft so that the tone can be adjusted for both channels together. Similarly, two volume controls for the audio input signals would be ganged, but they are not shown here.

The two inputs for left and right channels can be supplied by the following sources for audio signals:

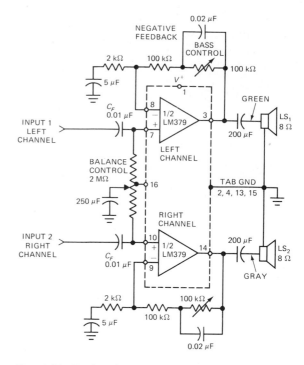

Fig. 4-38. Dual audio amplifiers for stereo using integrated circuit package LM 379. Audio output is 7 W per channel. Components outside the dashed area are not in the IC package. Volume controls are not shown. (*National Semiconductor Corporation*).

Stereo Records All phonograph records now are made with left and right audio signals for stereo cut into the groove on opposite walls at a 45° angle. The stereo pickup provides left and right audio signals for the phono output. The system is illustrated in Fig. 4-16.

Stereo Tape Two record-playback tracks are used for separate left and right audio signals. The tape formats for stereo are shown in Fig. 4-26.

Microphone Input Two microphones are used for stereo. In addition, microphone preamplifiers may be needed for the low-level output of magnetic microphones.

FM Radio Practically all stations in the commercial FM radio broadcast band of 88 to 108 MHz broadcast in stereo. Two audio signals are transmitted on one RF carrier for stereo receivers. The details of the stereo broadcasting system are explained in Chap. 17.

AM Radio Several systems of AM stereo broadcasting have been developed for the 540 to 1600 kHz band. Commercial AM stereo is likely to find its widest use in automobile radios.

Test Point Questions 4-12
(Answers on Page 98)

Answer True or False.

a. Phonograph records, magnetic tape, and FM radio broadcasts are all sources of stereo audio signals.
b. In Fig. 4-38, the balance control adjusts the tone for both channels.

4-13
AUDIO SECTION OF A RADIO RECEIVER

Any radio receiver has an audio section. A stereo receiver includes separate left and right audio amplifiers. In many cases, especially with high-fidelity receivers, the audio section contains provisions for external connections, as shown in Fig. 4-39. As a result, the audio part of the receiver can be used by itself as a separate stereo audio amplifier. Audio power output can range from 15 to 100 W per channel. A *tuner* has the RF circuits of a receiver but not the audio power output needed for a loudspeaker.

In Fig. 4-39a, the four pairs of jacks can be used for four stereo signals. The selector switch on the front panel of the receiver chooses one, as shown in Fig. 4-39b. Each pair of jacks consists of a left and right channel. Most jacks located at the rear of the receiver take the RCA type of shielded phono plug.

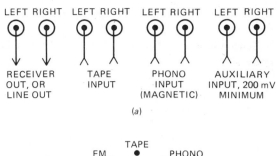

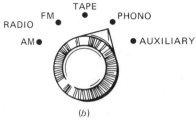

Fig. 4-39. (*a*) An example of jacks at the rear of a stereo receiver for audio input and output. (*b*) Selector switch on the front panel of the receiver. The switch is shown in phono position.

Audio Output from Receiver At the radio position of the selector switch, the detected audio signal from the RF section of the receiver is connected to the audio section. That position allows the normal operation of a radio receiver for either FM or AM. Furthermore, the audio signal can be taken at the terminals labeled RECEIVER OUT OR LINE OUT. The purpose may be to feed a tape recorder. Here the level is about 500 mV of audio signal output.

Audio Inputs The other three sets of terminals shown in Fig. 4-39 are for audio signal input. When they are used, the radio section of the receiver is disabled by the selector switch.

The tape input terminals connect to the output of a magnetic tape deck. The required audio input signal is about 200 mV. A *tape deck* includes the mechanisms for magnetic tape and some electronic circuits but it does not have the audio power output needed for a loudspeaker.

The phono input terminals are connected to the output of a phono cartridge. In this case the terminals are for a magnetic phono cartridge with a typical signal level of 5 to 10 mV.

The auxiliary input is for other input signals with relatively high levels of 200 to 500 mV. The input terminals can be used for a ceramic phono cartridge and possibly a ceramic microphone. Also, when the receiver does not have tape-input connections, the auxiliary position can be used for playback of a tape deck.

No inputs are shown in Fig. 4-39 for microphones. Receivers generally do not have microphone input connections because additional preamplifiers might be necessary. A separate audio amplifier unit, however, does have provisions for microphones. It should be noted that the combination of a microphone, audio amplifier, and loudspeaker forms a *public-address,* or *PA system.* The sound volume of a PA system depends on the power ratings of the amplifier and speakers.

For monophonic sources of audio signal, either the left or right input jack is used. The receiver can be switched to mono operation with both audio channels operating in parallel for the one audio signal.

Equalization   The term *equalization* refers to compensation for the nonlinear frequency response of phonograph records, magnetic tape, and microphones. The purpose is to provide flat frequency response in the overall process of recording and playback. *RC* networks are generally used for the equalizing circuits. It should be noted that the audio input jacks include the equalizing networks needed for each source of audio signal.

RIAA Characteristic   When a phonograph record is cut, low frequencies are attenuated and high frequencies are boosted to improve the signal-to-noise ratio. In playback, the opposite compensation is needed. The Record Industry Association of America (RIAA) standard equalization curve specifies the required frequency response.

Test Point Questions 4-13
(Answers on Page 98)

In Fig. 4-39, which jacks are used for the following:

a. Recording a radio program on tape
b. Playing back the output of a cassette tape deck
c. Amplifying the output of a magnetic phono cartridge

SUMMARY

1. Sound is a wave motion of varying air pressures. The velocity of sound in air is approximately 1130 ft/s [344.4 m/s].
2. An audio signal is an electrical variation of voltage or current. Its velocity is the same as that of electromagnetic waves: 3×10^{10} cm/s.
3. The amplitude of sound waves or audio signal determines volume.
4. The frequency of sound waves determines pitch, or tone. The AF range is 16 to 16,000 Hz. Low frequencies provide bass tone; high frequencies provide treble tone.
5. Harmonic frequencies are exact multiples of the fundamental frequency.
6. An octave is a range of 2:1 in frequencies.
7. A loudspeaker is an electromechanical transducer that produces sound output from an audio signal. The voice coil moves in a steady magnetic field provided by a permanent field magnet. The cone vibrates to produce sound waves.
8. A microphone is the opposite of a loudspeaker. Sound input to the microphone generates an audio signal. Common microphones are the dynamic or magnetic type and the crystal or ceramic type.

9. A phonograph pickup produces an audio signal output from the variations in the grooves of a record. Common types are the magnetic and ceramic phono pickups.
10. Magnetic tape is coated with tiny particles of iron oxide that can be magnetized. In the recording mode, the tape is magnetized as it passes through the air gap of an electromagnet (the recording head). The coil of the head is energized by the audio signal current. In the playback mode, the magnetized head induces a current in the coil as the tape travels pass the gap.
11. An audio preamplifier is a low-level amplifier.
12. An audio output stage is a power amplifier used to supply enough signal current for the loudspeaker. The power output stage can be single-ended class A or push-pull operating class B or AB.
13. The main types of distortion are amplitude distortion, which results from nonlinear amplification, and frequency distortion, which is caused by non-uniform gain for different frequencies.
14. Negative feedback is the coupling of part of the amplified output signal back to the input. In negative feedback, the output signal is out of phase with the input signal. The gain is lowered, but negative feedback is generally used to reduce distortion.
15. *RC* networks can be used to adjust the frequency response for tone control. A treble control varies the output at high audio frequencies. A bass control varies the output at low audio frequencies.
16. In stereophonic sound, individual left and right channels are used for a directional effect.

SELF-EXAMINATION
(Answers at back of book)

Choose (a), (b), (c), or (d).

1. The third-harmonic of 400 Hz is (a) 300 Hz, (b) 400 Hz, (c) 800 Hz, (d) 1200 Hz.
2. Treble tone corresponds to (a) high frequencies, (b) low frequencies, (c) high amplitudes, (d) low amplitudes.
3. Typical loudspeaker impedance rating is (a) 1 Ω, (b) 8 Ω, (c) 100 Ω, (d) 5000 Ω.
4. Typical output from a magnetic microphone is (a) 1 mV, (b) 500 mV, (c) 1 V, (d) 2 V.
5. The 45°-45° system refers to stereo for (a) magnetic tape recorders, (b) phonograph records, (c) crystal microphones, (d) loudspeakers.
6. The tape speed for cassette tape recorders, in inches per second, is (a) 1⅞, (b) 3¾, (c) 7, (d) 15.
7. Which of the following would apply to the audio power output stage? (a) Low-level signal; (b) complementary symmetry; (c) class C operation; (d) very little distortion.

8. To match 8 Ω to 200 Ω an audio output transformer must have a turns ratio of (a) 1:1, (b) 1:5, (c) 1:8, (d) 1:400.
9. Negative feedback is used in audio amplifiers to (a) increase gain, (b) increase distortion, (c) reduce distortion, (d) operate class C.
10. Less capacitive reactance in shunt with an audio amplifier has the effect of (a) bass cut, (b) treble boost, (c) treble cut, (d) more negative feedback.
11. The auxiliary input jacks of a radio receiver require audio output of (a) 0.1 to 0.5 mV, (b) 1 to 5 mV, (c) 200 to 500 mV, (d) 5 to 10 V.
12. Two 4-Ω speakers in series have a total impedance of (a) 2 Ω, (b) 4 Ω, (c) 8 Ω, (d) 16 Ω.

ESSAY QUESTIONS

1. **a.** What is the difference between sound waves and audio signals? **b.** Give the range of audio frequencies.
2. Describe briefly three types of electromechanical transducers used in audio systems.
3. List the fundamental frequencies up to the fifth harmonic for a signal at **a.** 400 Hz and **b.** 1000 Hz.
4. Define the following: PM loudspeaker, woofer, tweeter, crossover network, audio preamplifier, equalization, dynamic microphone, and public-address system.
5. Give two common values for loudspeaker impedance and power input.
6. Name two types of microphones and list two special features of each. What is the output voltage of each?
7. Name two types of phonograph pickup and list two special features of each. What is the output voltage of each?
8. Compare a cardioid microphone with a nondirectional microphone.
9. Give two requirements of a stereophonic audio system.
10. How is the 45°-45° system used in phonograph records?
11. Give two speeds for phonograph records, in revolutions per minute.
12. Give two speeds for magnetic tape, in inches per second.
13. State two functions for the ac bias used in magnetic tape recording.
14. Describe briefly two formats for stereo tape recording.
15. What is the advantage of the Dolby tape recording system?
16. Give two differences between tape cassettes and eight-track cartridges.
17. Show an example of a crossover network for loudspeakers.
18. Show an example of a level control for a loudspeaker.
19. Refer to the audio preamplifier circuit in Fig. 4-27. **a.** What is the signal input? **b.** What is the bias V_{BE} for both stages? **c.** Give the functions of C_1, C_3, and C_4. **d.** Give the functions of R_4 and R_7.
20. Name three types of circuits used for the audio output stage.
21. Give two functions for the audio output transformer.
22. Refer to the audio amplifier circuit in Fig. 4-28. **a.** What is the class of

operation of Q1 and Q2? **b.** How does Q1 obtain base bias? **c.** Is negative feedback used in this circuit? Explain your answer.

23. Give one advantage and one disadvantage of negative feedback in audio amplifiers.
24. Describe briefly three types of distortion in audio amplifiers.
25. Give two possible causes of amplitude distortion.
26. Show an example of negative feedback in an audio amplifier circuit.
27. Show an example of a treble-cut circuit for a tone control.
28. Give three examples of stereo signals for audio input.
29. Give the uses for the four pairs of audio jacks at the back of the receiver shown in Fig. 4-39.
30. **a.** How does an FM receiver differ from an FM tuner? **b.** What is meant by a stereo tape deck?

PROBLEMS
(Answers to odd-numbered problems at back of book)

1. Calculate a half-wavelength for sound at 40 and 16,000 Hz.
2. What frequency is two octaves above 300 Hz?
3. The frequency of 12,800 Hz is how many octaves above 50 Hz?
4. Calculate the turns ratio for an output transformer matching 8 Ω in the secondary to 40 Ω in the primary.
5. Calculate the turns ratio for a driver transformer matching the 140-Ω secondary to 6000 Ω in the primary.
6. Show how to connect four 8-Ω loudspeakers for a combined impedance equal to 8 Ω.
7. Referring to the preamplifier circuit in Fig. 4-27, calculate the collector current for Q1 and Q2.
8. Calculate the reactance of a 0.15-μF coupling capacitor at 50 Hz.
9. Referring to the treble-cut tone control in Fig. 4-35a, calculate the reactance of the 0.05-μF C_1 at 2, 4, and 8 kHz.
10. Referring to the bass-boost tone control in Fig. 4-37, calculate the reactance for C_1 and C_2 at 80 Hz.

SPECIAL QUESTIONS

1. Describe briefly two examples of audio equipment that you have used.
2. Which do you think is better, phonograph records or tape recordings? Explain why.
3. As tape equipment, would you prefer open-reel, cassette, or cartridge? Explain why.

4. Show in block diagram form how you would hook up: **a.** A stereo record player with audio amplifiers and loudspeaker. **b.** A stereo cassette deck with audio amplifiers and loudspeakers. **c.** A monophonic public-address system with microphone, audio amplifier, and four speakers in separate locations.

ANSWERS TO TEST POINT QUESTIONS

4-1	**a.** 400 Hz		**4-5**	**a.** 5 mV		**4-9**	**a.** T
	b. 800 Hz			**b.** 33⅓ rpm			**b.** T
	c. 2.8 ft			**c.** 45°			**c.** F
4-2	**a.** T			**d.** 2 g			**d.** F
	b. F		**4-6**	**a.** T		**4-10**	**a.** Negative
	c. T			**b.** F			**b.** More
4-3	**a.** Voice coil			**c.** F			**c.** R_F
	b. 8 Ω			**d.** T		**4-11**	**a.** 3
	c. T			**e.** T			**b.** C_2
	d. 8 Ω		**4-7**	**a.** Q1			**c.** Low
4-4	**a.** 1 mV			**b.** 5 μF		**4-12**	**a.** T
	b. Magnetic			**c.** 0.6 V			**b.** F
	c. Unidirectional		**4-8**	**a.** A		**4-13**	**a.** Line out
	d. Microphone			**b.** C_2			**b.** Tape input
				c. 1:5			**c.** Phono input
				d. EIA			

Decibel (dB) Units

Have you ever noticed that the ear is more sensitive to a change in sound intensity at low- than at high-volume levels? For instance, a 1-W increase of power output from 2 W to 3 W sounds much louder, but the same change from 10 W to 11 W is not so noticeable. The listener's impression of an increase or decrease in loudness depends on the ratio of the two powers. A ratio of 3 W:2 W is 1.5, or a 50 percent increase. That is greater than the ratio of 11 W:10 W, or 1.1, which is only 10 percent more. It is the ratio that determines the change of loudness.

That is why decibel values are based on the ratio of two power levels. Furthermore, the logarithm of the power ratio is used. The reason is to have smaller numbers that compress the extremes of small and large values. (See Appendix D for an explanation of logarithms.) When a power ratio is 1000, its logarithm is 3 in common logarithms to base 10; the common logarithms to base 10 are used for dB calculations.

Originally, the logarithmic unit was defined for audio measurements simply as the *bel,* which is equal to log (P_2/P_1). The unit generally used, though, is the decibel, which is equal to one-tenth of a bel. The decibel unit is abbreviated dB. A change in audio level of 1 dB is just perceptible to the ear. Although it was derived for audio, the dB unit is also commonly used for RF signals. The details of dB calculations are explained in the following topics:

5-1   Power Ratios and dB Units
5-2   Voltage Ratios and dB Units
5-3   Decibel Reference Levels
5-4   Common Decibel Values
5-5   Converting Decibels to Power or Voltage
5-6   Decibel Tables
5-7   Adding and Subtracting Decibel Units
5-8   Decibels and Loudness Levels

5-1
POWER RATIOS AND dB UNITS

The formula for comparing two values of power in decibel units is

$$dB = 10 \times \log \frac{P_2}{P_1} \qquad (5\text{-}1)$$

To calculate the number of dB, the following method can be used:

1. Reduce the ratio P_2/P_1 to one number. Be sure to use the same units for P_2 and P_1 in the ratio. As an example, let P_2 be 1 W and P_1 1 mW. Change 1 W for P_2 into 1000 mW. Then the ratio P_2/P_1 is 1000 mW/1 mW = 1000. That number is needed for the ratio in order to find $\log (P_2/P_1)$.
2. Always let P_2 in the numerator be the higher power; then the ratio must be more than 1. This procedure eliminates the problem of working with the negative logarithms of fractions.
3. Find the logarithm of the ratio P_2/P_1. Here 1000, or 10^3, has the log of 3.
4. Multiply the log by the factor 10 to calculate the number of decibels. Here we have 10 × 3 = 30 dB.

Stop with this answer. The result is $+30$ dB for a gain from 1 mW to 1000 mW, or 1 W. No antilogarithms are necessary because the decibel is meant to be a logarithmic unit.

For a decrease from 1000 mW to 1 mW, the loss would be -30 dB. The dB value is the same for an equal gain or loss. That is why the numerator P_2 in Eq. 5-1 can always be made the larger number in the ratio P_2/P_1.

The given problem will determine whether it is a dB gain or loss. More output than the input results in a gain. A loss means the output is less than the input. Amplifiers generally have a dB gain. Attenuation by resistance, as in a transmission line, results in a dB loss.

Example 5-1 What is the dB gain for an increase of power level from 13 to 26 W?

Answer

$$dB = 10 \times \log \frac{P_2}{P_1} = 10 \times \log \frac{26\ W}{13\ W}$$

$$dB = 10 \times \log 2$$

From a log table or calculator, log 2 = 0.3. Then

$$dB = 10 \times 0.3 = 3$$

The answer is 3 dB for a power increase ratio of 2, or double the power.

Example 5-2 What is the dB loss for a decrease of power level from 8 W to 1 W? Use the larger value of 8 W for P_2 in the numerator.

Answer

$$dB = 10 \times \log \frac{P_2}{P_1} = 10 \times \log \frac{8\ W}{1\ W}$$

$$dB = 10 \times \log 8$$

From a log table or calculator, 0.9 is the log of 8. Then

$$dB = 10 \times 0.9 = 9\ dB$$

However, the answer is -9 dB; the minus sign indicates the loss in power level.

Example 5-3 What is the dB gain of an amplifier that has 200-mW input and 2-W output?

Answer First change 2 W to 2000 mW. Then

$$dB = 10 \times \log \frac{P_2}{P_1} = 10 \times \log \frac{2000\ mW}{200\ mW}$$

$$dB = 10 \times \log 10$$

Since 1 is the log of 10,

$$dB = 10 \times 1 = 10$$

Note that 10 times the power also happens to be a gain of 10 dB. This example can also be solved if the units had been converted to watts rather than milliwatts.

No impedance values are needed to calculate a dB gain or loss in power level. The reason is that the effect of a low or high impedance is included in the value of power.

Test Point Questions 5-1
(Answers on Page 113)

a. An amplifier has a 7-mW input and a 14-mW output. How much is the dB gain?
b. A resistive attenuator reduces a 10-mW input to 5 mW. What is the dB loss?

5-2
VOLTAGE RATIOS AND dB UNITS

Remember that power is I^2R or V^2/R. Since dB units are used with ac signal levels, not dc values, Z is used instead of R. Then $P = I^2Z$ or V^2/Z. The dB formula for voltage ratios can be derived from the power ratio, as follows:

$$dB = 10 \log \frac{P_2}{P_1}$$

Substitute V^2/Z for P; then

$$dB = 10 \log \frac{V_2^2/Z_2}{V_1^2/Z_1}$$

To simplify this equation, double the coefficient 10 and take the square root of the log value. The dB value remains the same because these two operations cancel each other. Then

$$dB = 20 \log \frac{V_2/\sqrt{Z_2}}{V_1/\sqrt{Z_1}}$$

which upon further simplification becomes

$$dB = 20 \log \frac{V_2 \sqrt{Z_1}}{V_1 \sqrt{Z_2}}$$

Note that Z_1 and Z_2 become inverted as they are changed from denominators to numerators. For the case of $Z_1 = Z_2$

$$dB = 20 \times \log \frac{V_2}{V_1} \qquad (5\text{-}2)$$

This is the form in which voltage ratios are generally converted to dB units. However, the formula applies only for V_2 and V_1 across the same value of Z.

Just as in the power formula, always use V_2 for the larger voltage in the ratio to avoid ratios less than 1 and their negative logarithms. Again, the problem will determine whether it is a gain for $+dB$ or loss for $-dB$.

The coefficient is 20 instead of 10 because the square of V corresponds to P. When a number is squared, its logarithm is doubled. As a result, dB values for voltage ratios will always be twice the dB values for the same power ratio. For instance, double the power is a gain of 3 dB, but double the voltage is a 6-dB gain.

Formula 5-2 can be used for calculating dB values for V ratios as long as both voltages are across the same Z. That can be assumed for the problems here, unless stated otherwise.

Example 5-4 What is the dB gain for an increase of voltage from 13 to 26 mV?

Answer

$$dB = 20 \times \log \frac{V_2}{V_1}$$
$$= 20 \times \log \frac{26 \text{ mV}}{13 \text{ mV}}$$
$$= 20 \times \log 2$$
$$dB = 20 \times 0.3 = 6$$

The answer is 6 dB for double the voltage here, compared with 3-dB gain in Example 5-1 for double the power.

Example 5-5 What is the dB loss for an attenuation of 12 to 3 mV? Use the larger value of 12 mV in the numerator for V_2.

Answer

$$dB = 20 \times \log \frac{V_2}{V_1}$$

$$= 20 \times \log \frac{12 \text{ mV}}{3 \text{ mV}}$$

$$dB = 20 \times \log 4$$

From a log table or calculator, 0.6 is the log of 4. Then

$$dB = 20 \times 0.6 = 12$$

The answer is -12 dB, with the minus sign used to indicate the loss in voltage level.

Remember that Formula 5-2 is for two voltages across the same Z. When Z_2 and Z_1 are different, the voltage ratio must be multiplied by the correction factor $\sqrt{Z_1} / \sqrt{Z_2}$. That can also be stated as $\sqrt{Z_1 / Z_2}$. The next example illustrates use of the correction factor.

Example 5-6 What is the dB loss when a V_2 of 15 mV across 75 Ω drops to 3 mV for V_1 across 300 Ω?

Answer Since the input and output impedances are not equal, the correction factor $\sqrt{Z_1/Z_2}$ must be used.

$$dB = 20 \times \log \left(\frac{V_2}{V_1} \times \sqrt{\frac{Z_1}{Z_2}} \right)$$

$$= 20 \times \log \left(\frac{15 \text{ mV}}{3 \text{ mV}} \times \sqrt{\frac{300 \ \Omega}{75 \ \Omega}} \right)$$

$$= 20 \times \log (5 \times \sqrt{4})$$

$$= 20 \times \log (5 \times 2) = 20 \times \log 10$$

$$dB = 20 \times 1 = 20$$

This answer is -20 dB for the loss in voltage level. Note that the log is found only after the voltage ratio has been multiplied by the correction factor.

Current Ratios and dB Units Although it is seldom used, the dB formula for currents can be derived as

$$dB = 20 \times \log \left(\frac{I_2}{I_1} \times \sqrt{\frac{Z_2}{Z_1}} \right) \qquad \text{(5-3)}$$

This is similar to the voltage formula. However, note that Z_2 and Z_1 are not inverted here. The reason is that I^2R or I^2Z for power does not have an inverse relation as in V^2/Z.

Test Point Questions 5-2
(Answers on Page 113)

a. An amplifier has a 7-mV input and a 14-mV output. What is the dB gain in voltage?
b. An attenuator has an 8-mV input and a 2-mV output. What is the dB loss in voltage?

5-3
DECIBEL REFERENCE LEVELS

When only one value of P or V is converted to dB units, a reference level for the other value is assumed. There must be two values for a dB comparison. Several different references are in common use, but the value being used can be determined by the abbreviation, as follows:

dB   = 6 mW (0.006 W) reference in 500 Ω

dBm   = 1 mW (0.001 W) reference in 600 Ω

dBmV = 1 mV (0.001 V) reference across 75 Ω

The m indicates the 1-mW reference; mV indicates the 1-mV reference. However, the 6-mW reference is indicated just as dB.

The 6-mW Reference Any power level can be compared with the 6-mW reference by the formula:

$$dB = 10 \times \log \frac{P}{0.006 \text{ W}} \qquad \text{(5-4)}$$

As an example, the power level of 12 mW, or 0.012 W, is 3 dB more than the reference, since 12 mW is double 6 mW. Remember to keep the value of P and the reference level in the same units of either milliwatts or watts.

Example 5-7 An audio amplifier has an audio output of 24 W. What is the dB output?

Answer

$$dB = 10 \times \log \frac{24 \text{ W}}{0.006 \text{ W}} = 10 \times \log 4000$$

From a log table or electronic calculator, 3.6 is the log of 4000. Then

$$dB = 10 \times 3.6 = 36$$

This output is 36 dB, above the standard reference of 6 mW.

When the P to be converted is more than 6 mW, let P be the numerator. However, when P is less than 6 mW, make it the denominator in order to have a ratio greater than 1. The answer will be negative dB, however for a power less than the reference.

Example 5-8 Convert 3 mW to dB units.

Answer

$$dB = 10 \times \log \frac{6 \text{ mW}}{3 \text{ mW}}$$

$$= 10 \times \log 2$$

$$dB = 10 \times 0.3 = 3$$

This answer is -3 dB, however, because P is less than the 6-mW reference. It does not matter whether P is input or output.

When we compare input and output levels, a negative dB value means a loss with less output than the input. However, there is no question of a gain or loss in comparing one level to the ref-

erence. In that case, the negative dB value means only that the level is less than the reference.

The reference of 0.006 W in 500 Ω is so common that it is often indicated on ac voltmeters with 0 dB at 1.73 V. That voltage can be derived from the formula $P = V^2/Z$ as follows

$$V = \sqrt{P \times Z}$$

$$= \sqrt{0.006 \times 500}$$

$$= \sqrt{3} = 1.73 \text{ V}$$

The value of 1.73 V is another way to state the reference level of 6 mW in 500 Ω. The dB scale for the ac voltmeter shown in Fig. 5-1 is calibrated that way.

The 1-mW Reference The 1-mW reference is generally used for telephone service and studio equipment for radio broadcast stations. The formula is

$$dBm = 10 \times \log \frac{P}{1 \text{ mW}} \qquad \textbf{(5-5)}$$

Example 5-9 Calculate the dBm level for a 20 mW audio signal.

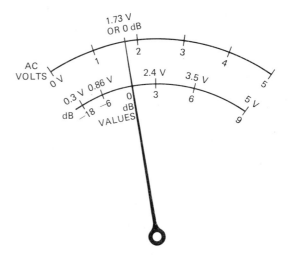

Fig. 5-1. Typical dB scale on an ac voltmeter. Zero dB is at 1.73 V.

Answer

$$dBm = 10 \times \log \frac{20 \text{ mW}}{1 \text{ mW}} = 10 \times \log 20$$

From a log table or calculator, 1.3 is the log of 20. Then

$$dBm = 10 \times 1.3 = 13$$

This answer is $+13$ dBm because P is more than the reference.

The 1-mV Reference The 1-mV standard is used for RF signal voltage on coaxial transmission lines. Such lines usually have an impedance of 75 Ω, which is Z for the reference. The formula is

$$dBmV = 20 \times \log \frac{V}{1 \text{ mV}} \qquad (5\text{-}6)$$

Here the multiplying factor is 20 instead of 10, because the dbmV values are for a voltage ratio. It should be noted that 50-Ω coaxial cable also is very common, but the dBmV reference is for 75 Ω line.

> **Example 5-10** Calculate the dBmV level for an antenna signal of 20 mV on the 75-Ω coaxial cable in a cable television system.
>
> **Answer**
>
> $$dBmV = 20 \times \log \frac{20 \text{ mV}}{1 \text{ mV}}$$
>
> $$= 20 \times \log 20 = 20 \times 1.3$$
>
> $$dBmV = 26$$

The answer is $+26$ dBmV because V is more than the reference. Note that 1.3 is multiplied by 20 for a voltage ratio, instead of 10 for a power ratio. No correction factor is necessary, because both voltages are across 75 Ω.

The VU Unit For audio measurements in radio broadcasting, the reference of 1 mW in 600 Ω is used for defining the volume unit (VU). Its main application is for an ac voltmeter calibrated in VU to monitor the audio modulation. The VU meter has standardized characteristics at 1000 Hz to indicate relative volume levels for the complex waveforms in voice and music signals.

Comparison of Positive, Negative, and Zero Decibel Values For any standard reference that is used,

+ decibels = level more than the reference

− decibels = level less than the reference

0 decibels = level equal to the reference

For negative decibel values, a higher number means less signal because it is further below the reference level.

It is important to remember that 0 decibels is not a zero level; it is just equal to the reference. Specifically, for the three reference levels:

$$0 \text{ dB} \quad = 6 \text{ mW}$$

$$0 \text{ dBm} \quad = 1 \text{ mW}$$

$$0 \text{ dBmV} = 1 \text{ mV}$$

Test Point Questions 5-3
(Answers on Page 113)

a. Convert 12 mW to dB
b. Convert 2 mW to dBm
c. Convert 2 mV to dBmV

5-4
COMMON DECIBEL VALUES

In Table 5-1 are listed a few power ratios and their dB values that are worth memorizing because they can be used as shortcuts in calculations. Note that the same ratios inverted apply for $+$dB and $-$dB. For instance, double the power is 3 dB, but one-half power is -3 dB. Those values result from the fact that log 2 is about 0.3 and $10 \times 0.3 = 3$.

Table 5-1	Common Decibel Values for Power Ratios
Power Ratio	**Decibels**
100	20
10	10
2	3
1.26	1
1	0
½	−3
¹⁄₁₀	−10
¹⁄₁₀₀	−20

Table 5-2	Common Decibel Values for Voltage Ratios
Voltage Ratio	**Decibels**
100	40
10	20
2	6
1.4	3
1.12	1
1	0
½	−6
¹⁄₁₀	−20
¹⁄₁₀₀	−40

Another useful ratio is 10 times or one-tenth the power. This power ratio is equal to 10 in dB units because log 10 is 1 and $10 \times 1 = 10$. Therefore, 10 times the power is 10 dB, but one-tenth the power is -10 dB.

The power ratio of 1.26 also is listed because it corresponds to 1 dB. That much change in audio power is just perceptible to the ear.

Shortcuts in dB for Power   With just the values given, many calculations can be made quickly. The technique involves dividing any ratio into factors of 2×10 and adding the corresponding values of 3 and 10 dB. Remember that adding the logarithms of numbers corresponds to multiplying the numbers.

Example 5-11   Calculate the dB value for a power ratio of 400.

Answer   The number, in factors of 2×10, is:

$$2 \times 2 \times 10 \times 10 = 400$$

The corresponding dB values are:

$$3 + 3 + 10 + 10 = 26 \text{ dB}$$

Therefore, the power ratio of 400 is equal to 26 dB. This example can be made shorter by using the factors 4×100 for 6 and 20 dB in power.

In general, each increase of twice the power ratio adds 3 dB. Doubling the power ratio of 400

to 800 increases the dB value to $26 + 3 = 29$ dB instead of 26 dB.

For each increase of 10 times in the power ratio, add 10 dB. An increase in the power ratio from 400 to 4000 increases the dB to $26 + 10 = 36$ dB.

Those changes apply to a decrease also. One-half the power ratio, from 400 to 200, decreases the dB value to $26 - 3 = 23$ dB. Also, one-tenth the power ratio, from 400 to 40, decreases the dB value to $26 - 10 = 16$ dB.

Shortcuts in dB for Voltage   In Table 5-2 are listed the same ratios as in Table 5-1, but for voltages instead of powers. The corresponding dB values are twice those for the same power ratio. For instance, double the voltage is 6 instead of 3 dB. Also 10 times the voltage is 20 dB instead of 10. The reason is that the dB voltage formula has the multiplying factor of 20 instead of 10.

The voltage ratio of 1.4 is listed in Table 5-2 in order to give the value for 3 dB. Note that a 3-dB gain in voltage is 1.4, or $\sqrt{2}$, times more, instead of 2 times more, than a power ratio. Also, 1 dB up is a voltage gain of 1.12.

The same shortcut methods for factors of 2×10 in a voltage ratio can be used. However, the corresponding dB values are 6 and 20 dB for voltage ratios.

Example 5-12   Calculate the dB value for a voltage ratio of 400.

Answer The number, in factors of 2×10, is

$$2 \times 2 \times 10 \times 10 = 400$$

The corresponding dB values are

$$6 + 6 + 20 + 20 = 52 \text{ dB}$$

Therefore, the voltage ratio of 400 is equal to 52 dB. Note that the 52 dB in voltage is twice the 26 dB in power found in Example 5-11 for the same ratio of 400.

In general, each increase of twice the voltage ratio adds 6 dB. Doubling the voltage ratio to 800 from 400 increases the value to $52 + 6 = 58$ dB.

For each increase of 10 times the voltage ratio, we can add 20 dB. A voltage ratio increase from 400 to 4000 increases the dB value to $52 + 20 = 72$ dB.

These shortcut methods apply to dB, dBm, and dBmV for gains or losses. The reason is that the technique just converts basic ratios to their corresponding decibel values.

Test Point Questions 5-4
(Answers on Page 113)

a. What is the dB value for a power ratio of 1000?
b. What is the voltage level for 6 dBmV?
c. What is the dB value for a power ratio of 4?

5-5
CONVERTING DECIBELS TO POWER OR VOLTAGE

The shortcuts can be used both to convert from voltage or power to dB and from dB to voltage or power. To convert a power or voltage ratio, the factors 2×10 are used. However, to convert a dB value, we use the corresponding terms 3 and 10 dB for power or 6 and 20 dB for voltage. Those conversions can be summarized as follows:

$$20 \text{ as a voltage or power ratio} = 2 \times 10$$

13 dB in power = 3 dB + 10 dB
26 dB in voltage = 6 dB + 20 dB

The methods apply to dB, dBm, or dBmV. To convert to an absolute value in watts or volts, remember that the references are 6 mW for dB, 1 mW for dBm, and 1 mV for dBmV.

Example 5-13 Convert a signal of 26 dBmV into the corresponding voltage level.

Answer The reference for dBmV is 1 mV. To simplify the decibel level,

$$6 \text{ dBmV} + 20 \text{ dBmV} = 26 \text{ dBmV}$$

The corresponding voltage ratios are

$$2 \times 10 = 20$$

This means the 26-dBmV level is 20 times the reference of 1 mV. Then

$$V = 20 \times 1 \text{ mV} = 20 \text{ mV}$$

Example 5-14 Convert an audio signal level of 26 dB to power, in watts.

Answer The reference power for dB units is 6 mW. To simplify 26 dB in power,

$$3 \text{ dB} + 3 \text{ dB} + 10 \text{ dB} + 10 \text{ dB} = 26 \text{ dB}$$

The corresponding power ratios are

$$2 \times 2 \times 10 \times 10 = 400$$

This means the power level is 400 times the reference of 6 mW. Then

$$P = 400 \times 6 \text{ mW} = 2400 \text{ mW or } 2.4 \text{ W}$$

The solution can be made shorter by using the factors 4×100 for 6 dB + 20 dB.

Example 5-15 Convert an audio signal level of -26 dB to power, in watts.

Answer Since 26 dB is 400 times more than 6 mW, from the preceding example, -26 dB is less than 6 mV by the factor 1/400. Then

$$P = 6 \text{ mW} \times \frac{1}{400} = \frac{6}{400} \text{ mW}$$

$$P = 0.015 \text{ mW}$$

Conversion Formula for Converting dB Units into Power It is not always convenient to use the shortcut methods. Then it becomes necessary to use antilogarithms. (How to find the value of an antilogarithm is explained in Appendix D.) The required formula in terms of the 0.006-W reference can be derived as follows:

$$\text{dB} = 10 \times \log \frac{P}{0.006 \text{ W}}$$

Dividing both sides by 10

$$\frac{\text{dB}}{10} = \log \frac{P}{0.006 \text{ W}}$$

Taking the antilog of both sides in order to eliminate the log factor

$$\text{Antilog} \frac{\text{dB}}{10} = \frac{P}{0.006 \text{ W}}$$

solving for P

$$P = 0.006 \text{ W} \times \text{antilog} \left(\frac{\text{dB}}{10}\right) \quad \textbf{(5-7)}$$

In using the formula, P is the desired power to be calculated, in watts, from a dB level. The procedure can be as follows:

1. First divide the stated number of dB by 10.
2. Find the antilog of the quotient. (Use a table of logs or a calculator.)
3. Finally, multiply the antilog by 0.006 W.

Example 5-16 Convert 26 dB of audio output into the corresponding power level.

Answer Here the reference is 0.006 W. Substituting in Formula 5-7 gives

$$P = 0.006 \text{ W} \times \text{antilog} \frac{26}{10}$$

$$= 0.006 \text{ W} \times \text{antilog } 2.6$$

The antilog of 2.6 is 400. Then

$$P = 0.006 \text{ W} \times 400$$

$$P = 2.4 \text{ W}$$

Example 5-16 is the same as Example 5-14 done by shortcut methods, and the answer is the same, 2.4 W. However, the advantage of Formula 5-7 is that it can be used for any values, including numbers that are not easy to simplify.

To convert dBm with the 1-mW reference, the formula is

$$P = 1 \text{ mW} \times \text{antilog} \left(\frac{\text{dBm}}{10}\right) \quad \textbf{(5-8)}$$

This answer for P is in mW units.

If we want to convert dBmV with the 1-mV reference, the formula is

$$V = 1 \text{ mV} \times \text{antilog} \left(\frac{\text{dBmV}}{20}\right) \quad \textbf{(5-9)}$$

Note that, in the voltage conversion, the number of decibels is divided by 20 instead of 10. This answer for V is in mV units.

Test Point Questions 5-5
(Answers on Page 113)

By using the 6-mW reference, convert the following into mW:

a. 9 dB
b. 0 dB
c. -9 dB
d. 13 dB

5-6
DECIBEL TABLES

In order to eliminate any calculations at all, a table of decibel values such as Table 5-3 can be used. Exact values are listed there, as calculated from logarithms. The first vertical column has the dB values. In the next column the corresponding power ratios are listed. Next is the column for voltages and current ratios. All the ratio values are divided into separate columns for gains (+dB) and losses (−dB).

To change dB into a ratio, find the dB value in the first column. Then look across that row for the corresponding ratio. As an example, 7 dB near the bottom of the first column corresponds to a power gain of 5.01 or to a voltage gain of 2.24. Note that the square of 2.24 is equal to 5. Those values show that the ratios for power are the squares of the voltage ratios for the same number of decibels.

For a loss, the −7 dB corresponds to a power ratio of 0.199, or approximately 0.2, which is equal to 1/5. The −7-dB loss in voltage is a ratio of 0.447, which is equal to 1/2.24. Those values show that the ratios for dB losses are the reciprocals of ratios for gains for the same number of decibels.

To change a power or voltage ratio into a dB unit, look for the ratio in the correct column and then read across to the left for the dB value. As an example, the power ratio of 5.62 corresponds to 7.5 dB at the bottom of the first column at the left.

Values more than 7.5 dB are continued in the table with another set of columns to the right. These values go up to 170 dB.

Test Point Questions 5-6
(Answers on Page 113)

From the decibel table, Table 5-3.

a. What power gain does 7 dB represent?
b. What power gain does 35 dB represent?
c. A power gain of 39.8 is how many decibels?
d. A voltage gain of 562 is how many decibels?

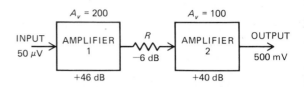

Fig. 5-2. Adding and subtracting dB levels.

5-7
ADDING AND SUBTRACTING DECIBEL UNITS

The fact that the decibel is a logarithmic unit means that cascaded values of gain are added, instead of multiplied, and decibel losses are subtracted. An example is illustrated in Fig. 5-2. The first amplifier has a voltage gain of 200, which is 46 dB. Then the signal is attenuated by a resistance that reduces the level by one-half. The loss is −6 dB. Finally, the last amplifier has a voltage gain of 100, or 40 dB.

Now we can calculate the overall gain from input to output. That value in voltage gain A_V is

$$200 \times 1/2 \times 100 = 10{,}000$$

In dB units, the overall gain is

$$46 \text{ dB} - 6 \text{ dB} + 40 \text{ dB} = 80 \text{ dB}$$

The output of 500 mV checks with the amount of gain in either voltage or dB units. In voltage, the input of 50 μV is multiplied by 10,000. Thus the output is 10,000 × 50 μV = 500,000 μV, which equals 500 mV. In dB units, the 80 dB is a voltage ratio of 10,000, which makes the output 10,000 × 50 μV = 500,000 μV or 500 mV.

Test Point Questions 5-7
(Answers on Page 113)

a. What is the overall gain for the following cascaded gains: 13 dB, −6 dB, 10 dB, and 20 dB? Give your answer in dB.
b. What is the output voltage for a gain of 7 dBmV in cascade with a loss of −7 dBmV?

Table 5-3
Decibel Table

dB	Power Ratio Gain	Power Ratio Loss	Voltage and Current Ratio* Gain	Voltage and Current Ratio* Loss	dB	Power Ratio Gain	Power Ratio Loss	Voltage and Current Ratio* Gain	Voltage and Current Ratio* Loss
0.1	1.02	0.977	1.01	0.989	8.0	6.31	0.158	2.51	0.398
0.2	1.05	0.955	1.02	0.977	8.5	7.08	0.141	2.66	0.376
0.3	1.07	0.933	1.03	0.966	9.0	7.94	0.126	2.82	0.355
0.4	1.10	0.912	1.05	0.955	9.5	8.91	0.112	2.98	0.335
0.5	1.12	0.891	1.06	0.944	10.0	10.00	0.100	3.16	0.316
0.6	1.15	0.871	1.07	0.933	11.0	12.6	0.079	3.55	0.282
0.7	1.17	0.851	1.08	0.923	12.0	15.8	0.063	3.98	0.251
0.8	1.20	0.832	1.10	0.912	13.0	19.9	0.050	4.47	0.224
0.9	1.23	0.813	1.11	0.902	14.0	25.1	0.040	5.01	0.199
1.0	1.26	0.794	1.12	0.891	15.0	31.6	0.032	5.62	0.178
1.1	1.29	0.776	1.13	0.881	16.0	39.8	0.025	6.31	0.158
1.2	1.32	0.759	1.15	0.871	17.0	50.1	0.020	7.08	0.141
1.3	1.35	0.741	1.16	0.861	18.0	63.1	0.016	7.94	0.126
1.4	1.38	0.724	1.17	0.851	19.0	79.4	0.013	8.91	0.112
1.5	1.41	0.708	1.19	0.841	20.0	100.0	0.010	10.00	0.100
1.6	1.44	0.692	1.20	0.832	25.0	3.16×10^2	3.16×10^{-3}	17.8	0.056
1.7	1.48	0.676	1.22	0.822	30.0	10^3	10^{-3}	31.6	0.032
1.8	1.51	0.661	1.23	0.813	35.0	3.16×10^3	3.16×10^{-4}	56.2	0.018
1.9	1.55	0.646	1.24	0.803	40.0	10^4	10^{-4}	100.0	0.010
2.0	1.58	0.631	1.26	0.794	45.0	3.16×10^4	3.16×10^{-5}	177.8	0.006
2.2	1.66	0.603	1.29	0.776	50.0	10^5	10^{-5}	316	0.003
2.4	1.74	0.575	1.32	0.759	55.0	3.16×10^5	3.16×10^{-6}	562	0.002
2.6	1.82	0.550	1.35	0.741	60.0	10^6	10^{-6}	1,000	0.001
2.8	1.90	0.525	1.38	0.724	65.0	3.16×10^6	3.16×10^{-7}	1,770	0.0006
3.0	1.99	0.501	1.41	0.708	70.0	10^7	10^{-7}	3,160	0.0003
3.2	2.09	0.479	1.44	0.692	75.0	3.16×10^7	3.16×10^{-8}	5,620	0.0002
3.4	2.19	0.457	1.48	0.676	80.0	10^8	10^{-8}	10,000	0.0001
3.6	2.29	0.436	1.51	0.661	85.0	3.16×10^8	3.16×10^{-9}	17,800	0.00006
3.8	2.40	0.417	1.55	0.646	90.0	10^9	10^{-9}	31,600	0.00003
4.0	2.51	0.398	1.58	0.631	95.0	3.16×10^9	3.16×10^{-10}	56,200	0.00002
4.2	2.63	0.380	1.62	0.617	100.0	10^{10}	10^{-10}	100,000	0.00001
4.4	2.75	0.363	1.66	0.603	105.0	3.16×10^{10}	3.16×10^{-11}	178,000	0.000006
4.6	2.88	0.347	1.70	0.589	110.0	10^{11}	10^{-11}	316,000	0.000003
4.8	3.02	0.331	1.74	0.575	115.0	3.16×10^{11}	3.16×10^{-12}	562,000	0.000002
5.0	3.16	0.316	1.78	0.562	120.0	10^{12}	10^{-12}	1,000,000	0.000001
5.5	3.55	0.282	1.88	0.531	130.0	10^{13}	10^{-13}	3.16×10^6	3.16×10^{-7}
6.0	3.98	0.251	1.99	0.501	140.0	10^{14}	10^{-14}	10^7	10^{-7}
6.5	4.47	0.224	2.11	0.473	150.0	10^{15}	10^{-15}	3.16×10^7	3.16×10^{-8}
7.0	5.01	0.199	2.24	0.447	160.0	10^{16}	10^{-16}	10^8	10^{-8}
7.5	5.62	0.178	2.37	0.422	170.0	10^{17}	10^{-17}	3.16×10^8	3.16×10^{-9}

*For equal Z.

5-8
DECIBELS AND LOUDNESS LEVELS

Sound waves consist of variations in air pressure. The loudness of the sound depends on the amount of air pressure and the corresponding power intensity at the ear, measured in units of watts per square centimeter (W/cm^2). The minimum intensity normally heard is 10^{-16} W/cm^2. That value is defined as zero decibels, a reference level for the threshold of audibility. In sound intensity, 1 dB above the threshold is called a *phon* unit.

Sound has loudness levels ranging from 0 dB up to approximately 130 dB, which is the threshold of pain. A decibel meter for measuring levels of sound intensity is shown in Fig. 5-3.

In Table 5-4 are listed some common sounds and their loudness levels. Note that whispering or very soft music is a sound level of 10 to 30 dB. Normal conversation or background music is about 60 dB. Very loud music is 100 dB.

It is interesting to note that about 25 W of electric power into a loudspeaker produces an intensity level of 100 dB, which is very loud sound. This conversion assumes a loudspeaker efficiency of about 1 percent.

An important feature of sound is that the ear is most sensitive to audio frequencies in the middle range of 500 to 5000 Hz. The audible response is shown by the graph in Fig. 5-4. The frequencies of 500 to 5000 Hz in the graph have the lowest values of sound intensity. Therefore, they are au-

dible with the least sound power, or they have the lowest threshold of audibility.

The curve of Fig. 5-4 is drawn on semilog paper. Linear spacing is used for the vertical axis, but the horizontal axis has logarithmic spacing. The frequencies increase in multiples of 10 in four groups or cycles of values. From left to right, the groups of values are in tens, hundreds, thousands, and ten thousands. The logarithmic spacing is generally used for frequency response curves. The purpose is to compress a wide range of values while still showing details of the middle frequencies.

The ear is most insensitive to audio frequencies

Fig. 5-3. A dB meter for measuring levels of sound intensity. (*VIZ Manufacturing Company*)

Table 5-4
Relative Loudness Levels

Type of Sound	Power at Ear, W/cm^2	Decibel Level
Threshold of pain	10^{-3}	130
Pneumatic chipper	10^{-4}	120
Thunder	10^{-5}	110
Very loud music	10^{-5}	110
Subway train	10^{-6}	100
Loud music	10^{-8}	80
Conversation	10^{-10}	60
Background music	10^{-10}	60
Soft music	10^{-13}	30
Whispering	10^{-14}	20
Threshold of hearing	10^{-16}	0

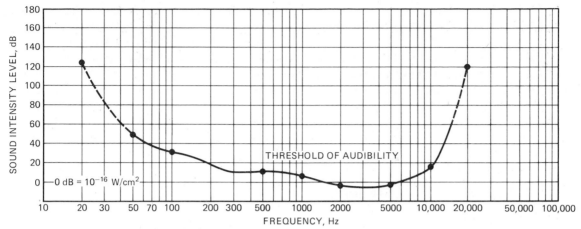

Fig. 5-4. Relative audibility for audio frequencies of 20 to 20,000 Hz. The ear is most sensitive to 500 to 5,000 Hz. The dashed curve below 50 Hz and above 15 kHz is for frequencies inaudible to many people.

below 50 Hz and above 15 kHz. Those values are shown by the dashed part of the curve. The poor audibility for low audio frequencies is the reason why the bass tone often sounds too weak at low volume, when played through an amplifier. To hear more of the low frequencies, we can turn up the volume or the bass response can be increased.

Test Point Questions 5-8
(Answers on Page 113)

a. In Fig. 5-4, is logarithmic spacing used for the values of dB or frequencies?
b. Which of the following dB values is an average sound level? 0, 60, 120.

SUMMARY

1. For power ratios, dB = 10 log (P_2/P_1), where P_2 is the higher power. A gain is +dB; a loss is −dB.

2. Common dB values to memorize are 3 dB for double power and 10 dB for 10 times the power. For losses, a power ratio of ½ is −3 dB and 1/10 is −10 dB.

3. For voltage ratios, dB = 20 log (V_2/V_1), where V_2 is the higher voltage. If V_2 and V_1 are not across the same Z, the voltage ratio must be multiplied by the correction factor $\sqrt{Z_1/Z_2}$.

4. Common dB values to memorize are 6 dB for double voltage and 20 dB for 10 times the voltage. The corresponding losses are −6 dB for the voltage ratio of ½ and −20 dB for 1/10 voltage ratio.

5. A common reference power for dB is 6 mW, or 0.006 W, in 500 Ω. Then 6 mW = 0 dB. Also, 3 dB is 12 mW and −3 dB is 3 mW.

6. Another common reference power is 1 mW, or 0.001 W, in 600 Ω. This decibel reference is indicated as dBm instead of dB. Then 1 mW = 0 dBm. Also, 3 dBm is 2 mW and −3 dBm is 0.5 mW.

7. A common reference voltage for decibels is 1 mV across 75 Ω, indicated as dBmV. Then 1 mV = 0 dBmV. Also, 6 dBmV is 2 mV and −6 dBmV is 0.5 mV.

8. Because the decibel is a logarithmic unit, decibel gains or losses in cascade are added or subtracted instead of being multiplied or divided.
9. In terms of levels of sound intensity, zero decibels is defined as the air pressure of 10^{-16} W/cm². That value is the threshold of audibility. Loudness levels range from 0 to 130 dB and average values are 40 to 70 dB.
10. The ear is most sensitive to audio frequencies of 500 to 5000 Hz, approximately.

SELF-EXAMINATION
(Answers at back of book)

Fill in the blanks with the proper values.

1. Double the power is a gain of _____ dB.
2. Ten times the power is a gain of _____ dB.
3. Twenty times the power is a gain of _____ dB.
4. Double the voltage is a gain of _____ dB.
5. Ten times the voltage is a gain of _____ dB.
6. One-half the power is a loss of _____ dB.
7. One-tenth the voltage is a loss of _____ dB.
8. A power level of 6 dB is _____ mW.
9. A power level of 6 dBm is _____ mW.
10. A voltage level of 6 dBmV is _____ mV.
11. A gain of 7 dB in cascade with 11 dB is a total of _____ dB.
12. A voltage level of 200 mV is _____ dBmV.

ESSAY QUESTIONS

1. Give the dB formulas for **a.** power ratios and **b.** voltage ratios across the same value of impedance.
2. Give the formulas for calculating decibels from the following reference levels. **a.** 6 mW, **b.** 1 mW, **c.** 1 mV.
3. What are the reference levels for dB, dBm, and dBmV?
4. Give the power level, in milliwatts, for **a.** 0 dB and **b.** 0 dBm.
5. Give two applications in which a power ratio results in $-dB$ instead of $+dB$ because of a loss instead of gain.
6. Why are dB gain values in cascade added instead of being multiplied?
7. Give the level of power for zero decibels of sound intensity at the threshold of audibility.
8. Which of the following audio frequencies, in hertz, is easiest to hear: 50, 1000, 10,000, or 15,000?
9. Derive the dB formula 5-2 for a voltage ratio from the dB power formula by using the substitution $P = V^2/Z$.
10. Derive the dB formula 5-3 for a current ratio from the dB power formula by using the substitution $P = I^2Z$.

PROBLEMS

(Answers to odd-numbered problems at back of book)

1. Calculate the dB gain or loss for the following: **a.** 2 mW input and 200 mW output, **b.** 5 mV input and 500 mV output, **c.** 18 mV input and 9 mV output, **d.** 10 mV input and 7.07 mV output.
2. Calculate the dB gain for 2 mW input and 280 mW output.
3. Calculate the dB gain for 2 mV input and 280 mV output.
4. Calculate the dB loss for 80 mV input and 10 mV output.
5. Calculate the dB gain for 30 mW input and 240 mW output.
6. Calculate the dB gain for 2 W input and 90 W output.
7. Give the power level for the following: **a.** 13 dBm, **b.** -6 dBm, **c.** 40 dBmV, **d.** -13 dB.
8. What is the overall gain for the following cascaded gains: $+17$, $+12$, -9, -9, and -2 dB?
9. Calculate the dB value for a power ratio of $(10^{-8} \text{ W/cm}^2) \div (10^{-15} \text{ W/cm}^2)$.

SPECIAL QUESTIONS

1. What is an advantage of using dB units?
2. Give three examples of dB units used in applications other than audio work.
3. Explain what is meant by four-cycle semilog graph paper, as in Fig. 5-4.

ANSWERS TO TEST POINT QUESTIONS

5-1	**a.** 3 dB	5-4	**a.** 30 dB	5-6	**a.** 5.01		
	b. -3 dB		**b.** 2 mV		**b.** 3160		
5-2	**a.** 6 dB		**c.** 6 dB		**c.** 16		
	b. -12 dB	5-5	**a.** 48 mW		**d.** 55		
5-3	**a.** 3 dB		**b.** 6 mW	5-7	**a.** 37 dB		
	b. 3 dBm		**c.** 0.75 mW		**b.** 1 mV		
	c. 6 dBmV		**d.** 120 mW	5-8	**a.** Frequencies		
					b. 60 dB		

Chapter 6
Power Supplies

A power supply converts the ac input of the 60-Hz power line to dc output voltage. This V^+ supply is needed for the amplifiers in electronic equipment. Transistors require dc collector voltage and dc bias for the base. Vacuum-tube amplifiers need the dc supply for plate and screen voltages.

The main component in the power supply is the rectifier, which generally is a silicon diode. The diode conducts only when forward polarity of voltage is applied. An ac input that has positive and negative half-cycles is converted to a dc output with a constant polarity. Additional requirements of the power supply and methods of using diode rectifiers in different circuits are explained in the following topics:

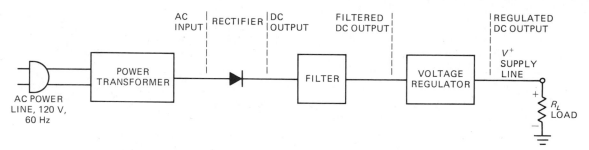

Fig. 6-1. Basic functions of a power supply from ac input to dc output.

6-1
BASIC FUNCTIONS IN A POWER SUPPLY

Basically, only a rectifier is needed to change the ac input to a dc output. Filter capacitors are also used, however, to remove the pulsating variations from the dc output. A dc voltage has a polarity, but it can still have changes in value.

In addition, a power transformer is often used to step up or step down the ac input voltage to the rectifier. The 120 V of the ac power line can be increased or decreased according to the turns ratio of the power transformer.

Finally, a voltage regulator may be used for the dc output. A regulator keeps the dc output voltage constant when the dc load current changes. Otherwise, the dc voltage would tend to decrease as the load current increased.

Those basic functions are illustrated by the block diagram in Fig. 6-1. The supply is shown with positive dc output for V^+. For negative dc output, the diode rectifier can be reversed, but V^+ is more common. Positive polarity is needed for collector voltage on NPN transistors and for plate and screen voltages on tubes.

Test Point Questions 6-1
(Answers on Page 135)

Refer to Fig. 6-1.

a. Is the dc output produced by the transformer or by the diode?
b. Is the ac input to the diode supplied by the transformer or by the filter?

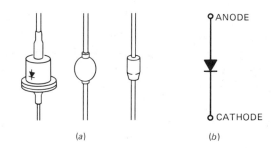

Fig. 6-2. (a) Silicon diode rectifiers, actual size. Packages shown from left to right are high-hat, pellet, and stick types. (b) Schematic symbol.

6-2
RECTIFIER DIODES

Silicon diodes are generally used for power supply rectification. Their maximum current rating ranges from 500 mA for small units, to more than 10 A. The advantage they offer is a very low internal voltage drop of approximately 1 V. In contrast, a typical vacuum-tube value is 18 V.

Note the schematic symbol in Fig. 6-2 for a semiconductor diode. The arrowhead marks the anode and the bar the cathode, which correspond to plate and cathode in a vacuum-tube diode. The symbol may be printed on the unit to indicate anode and cathode. Otherwise, a bar, stripe, or dot indicates the cathode. The cathode is used for positive dc output voltage, and the anode for the ac input.

Diode current flows only when the ac input provides forward voltage. The arrow shows the direction of hole current which consists of positive charges. Electron flow is in the opposite direction.

Forward voltage for the diode can be either positive at the anode or negative at the cathode. The ac input can supply one or the other on alternate half-cycles, but not both at the same time. A single diode, therefore, is a half-wave rectifier. It can conduct for only a half-cycle of the ac input.

Additional types of rectifier diodes include:

Selenium disk rectifier. This type also is a semiconductor, but it is much larger than a silicon diode.

Copper-oxide rectifier. This type is also called a *metallic rectifier.*

Vacuum-tube diode. This type has relatively low current ratings.

Gas-tube diode. This type has high current ratings.

Despite the different types of rectifiers, though, the silicon diode is used in most applications for converting an ac input to a dc output.

Peak Reverse Voltage The peak reverse voltage (PRV) rating is the maximum voltage that can be applied across the diode in the nonconducting or reverse direction. Too much reverse voltage can produce breakdown at the junction. The result will be a shorted diode.

The amount of reverse voltage across a diode in a typical rectifier circuit is about twice the dc output voltage. The reason why is illustrated in Fig. 6-3. When you consider the potential difference across the diode while the diode is not conducting, there are two voltages series-aiding. One is the positive dc voltage output at the cathode; the other

is the negative ac input voltage at the anode. For the example here, the dc output across the filter capacitor C_1 is 168 V. The negative peak of the input voltage is 168 V for an rms value of 120 V. Then $168 + 168 = 336$ V as the actual peak reverse voltage across the diode. The PRV rating should be greater. For silicon diodes, typical PRV ratings are 400 to 1000 V.

Testing the Diode Rectifier A semiconductor diode can easily be checked with an ohmmeter to see if it is open or shorted. The diode has only two leads. Out of the circuit, the diode resistance is measured in the forward direction and then the reverse direction based on the ohmmeter's polarity. R should be very low in the forward direction and practically infinite in the reverse direction of a silicon diode.

If R is low in both directions, the diode is shorted. If R is very high in both directions, the diode is open.

When checking the diode in the circuit with an ohmmeter, the power must be off and the filter capacitors discharged. One side of the diode should be disconnected to remove parallel paths.

Do *not* use low-power ohms for this test. The ohmmeter battery must supply enough voltage to turn on the junction for low R in the forward direction.

When there is an open rectifier in a power supply, voltage measurements will show normal ac voltage at the input terminal but no dc output at the other terminal. The ac input must be measured with an ac voltmeter, but the dc output is measured with a dc voltmeter.

Test Point Questions 6-2
(Answers on Page 135)

a. In the diode symbol, is the arrowhead cathode or anode?

b. Is forward voltage at the anode positive or negative?

c. Does a shorted diode read zero or infinite R in both directions with an ohmmeter?

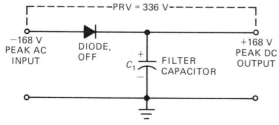

Fig. 6-3. Peak reverse voltage (PRV) across a diode rectifier on the nonconducting half-cycle.

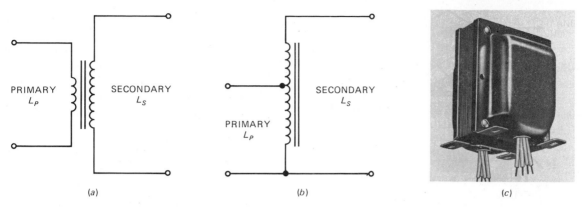

Fig. 6-4. Power transformers. (*a*) Two-winding transformer with isolated secondary winding L_S. (*b*) Autotransformer with L_S not isolated. (*c*) Typical multiwinding transformer.

6-3
THE POWER TRANSFORMER

The amount of ac voltage needed for the rectifier may be different from the 120 V of the power line in order to have the desired amount of dc output voltage. Then a power transformer such as in Fig. 6-4 is used. By mutual induction between L_P and L_S, the ac secondary voltage is increased or decreased in direct proportion to the turns ratio. As an example, when L_S has one-half the turns of L_P, the ac secondary voltage is one-half the ac primary voltage.

In Fig. 6-4*a*, L_P and L_S are shown as separate windings. Since L_P has no direct connection to L_S, it is an *isolated secondary.*

For the autotransformer in Fig. 6-4*b*, one winding is used for both L_P and L_S. The entire winding constitutes L_S. The tapped connection makes L_P just a part of the entire coil. Therefore, L_S is not isolated from L_P.

The point at which the coil is tapped determines the turns ratio of the autotransformer. This autotransformer is shown with more turns for L_S than L_P. Thus it is a step-up transformer. However, the connections can be reversed for a step down.

Line-Isolated Supply   A *line-isolated supply* is a supply with a power transformer that has an iso-

lated secondary. The primary is connected to the ac power line, but the rectifier circuit with V^+ output is in the secondary. The receiver chassis, which is usually the negative return for the dc output, then is a "cold" chassis because it is isolated from the ac power line.

Line-Connected Supply   With an autotransformer, or no power transformer at all, the receiver chassis can be "hot," meaning it is connected directly to the ac power line. If the chassis is connected to the high, or hot, side of the power line there is danger of an electric shock when the chassis or an exposed metal part is touched.

Remember that the ac power line has one side connected to earth ground and that the other side is the high connection. Suppose that the ac plug is oriented in the direction that connects the receiver chassis to the high side of the ac power. Touching the chassis is then the same as touching the ungrounded side of the 120-V power line. Furthermore, if two chassis with opposite power connections should touch each other, the result would be a short circuit across the ac power line.

To avoid that danger in a power supply that is not isolated, most equipment now uses a polarized ac plug with one blade wider than the other. Since the receptacle has one slot wider than the other the plug can be inserted in only one way. The wider blade is the ground side.

AC-DC Power Supply

The term *ac-dc power supply* is sometimes used for a supply that does not have a power transformer. With a transformer, a power supply can operate only from an ac source. Its function, however, is to supply dc output.

Without a transformer, the ac-dc supply can provide V^+ output either with an ac input or a dc input of the correct polarity for the rectifier diode. Actually, though, a dc input would hardly ever be used. The term "ac-dc power supply," then, simply means that no power transformer is used. Ac input is provided by direct connections to the power line. This type is definitely a line-connected power supply that can have a "hot" chassis.

Isolation Transformer

The isolation transformer has a 1:1 turns ratio, and it is used only to provide isolation from the ac power line. It should be used during the testing of equipment with a line-connected power supply. The equipment is connected to the isolation transformer, which is plugged into the ac power line. For test purposes, the isolation transformer can also have a variable turns ratio to provide ac voltages higher or lower than the line voltage.

Checking Transformer Windings with an Ohmmeter

The leads for the primary windings of power transformers are usually color-coded black. A step-up secondary winding is color-coded red. More details of the color coding are given in Appendix E. By using an ohmmeter, each winding can be tested for its normal resistance. The primary resistance is generally 5 to 50 Ω. The secondary winding resistance may be 1 to 400 Ω, depending on whether it is for voltage step-down or step-up. An open winding will cause the ohmmeter to read infinite resistance.

Test Point Questions 6-3
(Answers on Page 135)

Answer True or False.

a. The power transformer can be used instead of a diode rectifier.
b. An autotransformer provides isolation from the ac power line.

6-4
TYPES OF RECTIFIER CIRCUITS

Three popular types of rectifier circuits are shown in Fig. 6-5. They are:

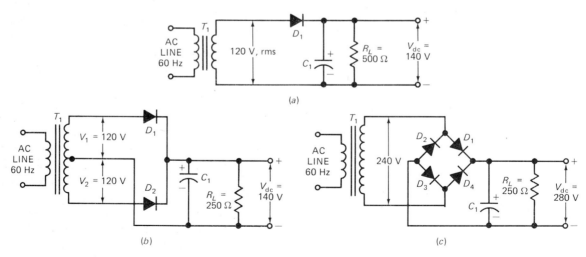

Fig. 6-5. Three basic types of rectifier circuits. (*a*) Half-wave with one diode. (*b*) Full-wave with center-tapped transformer and with two diodes. (*c*) Full-wave bridge with four diodes.

1. Half-wave rectifier. Only one diode is needed to conduct on one alternation of every cycle of the ac input.
2. Full-wave rectifier. The transformer has a center tap for the ac input. Two diodes are used to conduct on opposite half-cycles. Each diode supplies one-half the dc load current.
3. Full-wave bridge. This circuit uses four diodes in two pairs. The bridge circuit makes it possible to eliminate the center tap.

All the circuits are shown with an isolating power transformer T_1. However, the half-wave rectifier and full-wave bridge circuits can be connected directly to the ac power line. The full-wave circuit needs a transformer for the center tap.

In each case the top dc output terminal is positive. However, the polarity can be inverted to negative polarity by reversing the diodes.

DC Load Current   The rectifier is the source of dc voltage for load current for all the amplifiers connected in parallel to the power supply. Each stage is a parallel branch for direct current, mainly I_C. The total dc load current on the power supply is then essentially the sum of all the collector currents for the transistor amplifiers.

As an example, assume eight amplifiers with I_C values that add to 300 mA of total collector current for all the stages. The total could include four small-signal amplifiers with I_C of 5 mA in each stage, two driver stages with 40 mA, and two power output stages with 100 mA. The total is

$$4 \times 5 \text{ mA} = 20 \text{ mA}$$

$$2 \times 40 \text{ mA} = 80 \text{ mA}$$

$$2 \times 100 \text{ mA} = 200 \text{ mA}$$

$$\text{Total } I_C = 300 \text{ mA}$$

This total I_C is the load current I_L of the power supply.

The rectifier must conduct the I_L of 300 mA for all the amplifier stages in this example while supplying the required dc output voltage.

Furthermore, if we assume a dc output voltage of 140 V, the dc load current of 300 mA is equiv-

alent to a load resistance of 0.467 kΩ, or 467 Ω. The calculations are

$$R_L = \frac{V}{I_L} = \frac{140 \text{ V}}{300 \text{ mA}} = 0.467 \text{ k}\Omega$$

DC Voltage Output   In Fig. 6-5 some values are given to indicate the actual dc voltage output of a typical power supply with a filter capacitor C_1. In general, a half-wave rectifier with an ac input of 120 V has a dc output of 140 V, approximately. Those values are based on the following factors:

1. Dc load current of about 300 mA.
2. Silicon diode rectifier, which has an internal voltage drop less than 1 V.
3. Filter capacitor at the diode output terminal.

Remember that the 120 V for the ac input is an rms value; the peak value is 168 V. Furthermore, the dc ouput is more than the rms ac input because the filter capacitor can charge to the peak value. For any ac voltage input, the dc output is approximately 1.17 times the rms voltage of a silicon diode with a load current of about 300 mA.

As a result, the dc output voltage of the half-wave rectifier shown in Fig. 6-5a is 1.17 × 120 = 140 V, approximately. The dc output of the full-wave rectifier in Fig. 6-5b also is 140 V. The total secondary voltage is 240 V, but the center tap provides 120 V for the ac inputs V_1 and V_2 to each diode. The dc output voltage is the same as for one diode, but the two diodes can supply twice the load current.

The full-wave bridge of Fig. 6-5c uses the entire secondary voltage of 240 V as its ac input. Then the dc output voltage is 1.17 × 240 = 280 V, approximately. The circuit can supply double the dc output voltage of the circuit in Fig. 6-5b and double the dc load current of one diode. The extra capabilities come from using more diodes.

Half-Wave and Full-Wave Rectifiers   A half-wave rectifier uses only one half-cycle of the ac input for the dc output. Either the positive or the negative half-cycle can be used, depending on the diode connections. Positive voltage at the anode makes the diode conduct, and negative voltage at the cathode also is forward voltage.

A full-wave rectifier uses both half-cycles of the ac input for the dc output. At least two diodes are necessary.

Ripple Frequency The variations in rectified output are considered as an ac ripple voltage superimposed on the dc level. Assuming 60 Hz for the ac input, the dc output has either of the following:

1. 60-Hz ripple with a half-wave rectifier
2. 120-Hz ripple with a full-wave rectifier

The ripple frequency of the half-wave rectifier shown in Fig. 6-5a is 60 Hz. Only one-half of each cycle of the ac input is used to produce dc output. The ripple frequency of the full-wave rectifiers shown in Fig. 6-5b and 6-5c is 120 Hz because both half-cycles of the ac input are used.

A higher ripple frequency is easier to filter because smaller capacitors can be used to provide low reactance. The full-wave rectifier offers the advantage of better filtering, therefore, as compared with a half-wave rectifier.

Test Point Questions 6-4
(Answers on Page 135)

a. What is the filtered dc output voltage of the half-wave rectifier shown in Fig. 6-5a?
b. What is the ripple frequency of the full-wave rectifier shown in Fig. 6-5b?
c. What is the filtered dc output voltage of the bridge rectifier shown in Fig. 6-5c.

6-5
ANALYSIS OF HALF-WAVE RECTIFIER

Details of the half-wave rectifier circuit are illustrated in Figs. 6-6 to 6-8. The half-wave rectifier with a single diode is the basic circuit of all types of power supplies. Rectifier circuits with more diodes are just combinations of half-wave rectifier circuits.

Positive DC Output In Fig. 6-6a, rectifier $D1$ conducts when the ac input voltage makes the diode anode positive. Current can then flow in the dc output circuit. When the ac input is negative at the anode, the diode cannot conduct. Then there is no dc output across R_L. The dashed half-cycle in Fig. 6-6b, indicates the part of the ac input that is missing in the dc output. This waveform just shows the rectification without the effect of any filter capacitor.

The operation of this rectifier circuit can be summarized in the following steps:

1. The positive ac input provides forward voltage at the anode of $D1$.
2. Current can flow in only one direction. Electron flow is from cathode to anode, through L_S, and back to cathode through the load resistance R_L. The path is indicated by the arrows for I.
3. The resultant IR drop across R_L provides the dc output voltage for the load.
4. The polarity of dc output voltage across R_L is positive at the cathode side compared with the grounded side. Note that the electron flow through R_L is from minus to plus at the cathode terminal.
5. Assuming a 60-Hz ac input, the fluctuations in the dc output, that is, the ripple frequency, is 60 Hz.

The basic reason why the dc output voltage has positive polarity to ground is that the diode conducts only when the ac input is positive to ground. Then R_L is connected by the conducting diode to the voltage source only when the voltage is positive.

Remember that chassis ground is only a common connection. It is not necessarily the start or finish for current. Electrons that leave the diode cathode to supply the current must return to the cathode for a complete path.

Negative DC Output In Fig. 6-7a, rectifier $D1$ conducts when the ac input makes the diode cathode negative. Note that the diode is reversed, compared with Fig. 6-6. The ac input also is shown reversed to indicate that $D1$ conducts when the

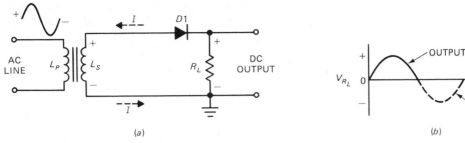

(a)   (b)

Fig. 6-6. Half-wave rectifier for positive dc output. (a) The circuit with I in the direction of electron flow. (b) Waveform of the rectified output without a filter capacitor.

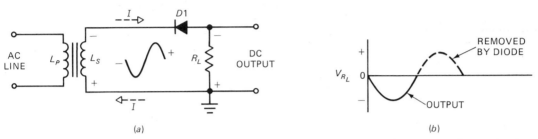

(a)   (b)

Fig. 6-7. Half-wave rectifier with diode inverted for negative dc output. (a) The circuit with I in the direction of electron flow. (b) Waveform of the rectified output without a filter capacitor.

high side of L_S is negative. Making the cathode negative is equivalent to making the anode positive. Both methods provide forward voltage for the diode.

The diode cannot conduct on the opposite half-cycle, when the ac input makes the $D1$ cathode positive. Then there is no dc output across R_L. The dashed half-cycle, Fig. 6-6b, indicates the part of the ac input that is missing in the dc output.

With $D1$ conducting in Fig. 6-7a, the electron flow can be considered from the top of L_S, from cathode to anode in the diode, through R_L, and returning to the grounded side of L_S as the source. As a result, the top of R_L is the negative side of V_{R_L}, and electron flow is into that side. The circuit for a negative dc output is sometimes called an *inverted power supply*.

Polarities for DC Output Voltage The fundamental way to consider polarities is that the dc output has the same polarity as the ac input voltage that makes the diode conduct. Chassis ground is only a common return which does not have any specific polarity.

The two polarity possibilities for a diode can be summarized briefly as follows:

1. Connecting the ac input to the anode produces a positive dc voltage output at the load in the cathode circuit.
2. Alternatively, connect the ac input to the cathode for negative dc voltage output at the load in the anode circuit.

The dc output polarity is determined by the ac input because that is the source of power. The rectifier is only a device that can make the connection between source and load each time the source repeats the same polarity. The rectifier action of passing through only one polarity is also called *commutation*.

Positive and Negative DC Output Voltages The circuit of Fig. 6-8 is called a *dual, symmetrical supply* because it provides equal and opposite dc voltage outputs. The type is needed in applications that require both V^+ and V^- for supply voltage, as is the case with operational amplifiers (op amps).

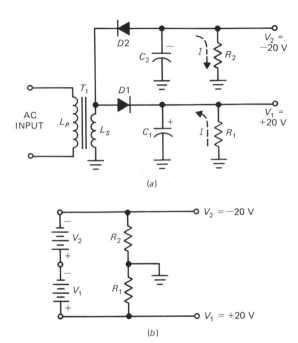

(a)

$V_2 = -20$ V

$V_1 = +20$ V

(b)

Fig. 6-8. Dual symmetrical power supply for equal and opposite dc output voltages. (a) Circuit with $D1$ and $D2$ in opposite polarities. (b) Equivalent circuit for dc output $V1$ and $V2$.

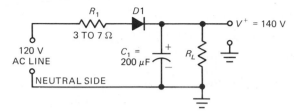

Fig. 6-9. Typical line-connected half-wave power supply with filter capacitor C_1.

Fig. 6-10. Electrolytic filter capacitor. The can is common negative. Height is 2 in.

Note, in Fig. 6-8a, that both $D1$ and $D2$ are connected to the same input voltage across L_S in the power transformer T_1 but that one diode is inverted with respect to the other. When an ac input voltage drives the anode of $D1$ positive, that diode conducts. The I is shown for electron flow. As a result, the dc output voltage V_1 at the cathode is positive across R_1. On the next half-cycle, negative ac voltage drives the cathode of $D2$ negative. That polarity at the cathode corresponds to positive polarity at the anode, which makes $D2$ conduct. Its dc output voltage V_2 at the anode is negative across R_2.

The equivalent circuit for the dual output voltages are shown in Fig. 6-8b. V_1 and V_2 are equal at 20 V, but V_2 is negative and V_1 is positive. The polarities are shown with respect to chassis ground.

Line-Connected Half-Wave Power Supply No power transformer is used for the case of a half-wave rectifier circuit that is not isolated from the ac power line as in Fig. 6-9. $D1$ is a silicon diode. C_1 is a filter capacitor that removes the 60-Hz ac ripple from the dc output. The value of 200 μF requires that C_1 be an electrolytic capacitor. In general, electrolytics must be used for filter capacitors in power supplies because of the large C. Remember that electrolytic capacitors such as the one in Fig. 6-10 must be connected in the correct polarity.

In the filtering action, the capacitor provides a dc output during the time the diode is not conducting. C_1 is an input filter capacitor because it is connected directly to the diode cathode, at the input to the dc side of the circuit.

R_1 in series with $D1$ is used as a *surge-protection resistor*. It limits the peak current through the diode when C_1 is charging. The resistance is generally 3 to 7 Ω with a power rating of 5 W. When a transformer is used, though, R_1 is not needed because L_S then provides series resistance.

This circuit has a positive dc output at the cathode for V^+, with an ac input applied to the anode.

Note that the V^- side is connected to the grounded or neutral side of the ac power line to reduce shock hazard.

Effects of the Filter Capacitor The filter capacitor C_1, Fig. 6-9, charges fast through the very low resistance of the diode when it conducts, but it discharges slowly through the load resistance when the diode is off. As a result, C_1 builds up charge to maintain the dc output voltage. In the process, the filter capacitor has the following effects:

1. The 60-Hz ac ripple is practically eliminated from the dc output.
2. The dc output voltage is maintained during the entire cycle of ac input, even when the diode is not conducting.
3. The dc voltage across C_1 puts reverse bias on the diode so that the diode can conduct only at the peak of the ac input. As a result, the diode is a peak rectifier.
4. C_1 can charge to the peak value of the ac input, up to 168 V. As a result, the dc output voltage can be more than the rms value of 120 V.

In summary, C_1 charges through the diode and discharges through the load. The filter capacitor is effectively the source of dc voltage.

Test Point Questions 6-5
(Answers on Page 135)

a. Does a diode require positive ac input voltage at the anode or cathode for conduction?
b. Does the diode of question **a** have positive dc output voltage at the anode or cathode?
c. Is the ripple frequency for a half-wave rectifier 60 or 120 Hz?

6-6
FULL-WAVE RECTIFIER

In the circuit shown in Fig. 6-11a two diodes are used to rectify both half-cycles of the ac input. Each diode is a half-wave rectifier, but the two have the common load R_L. Only one-half the total

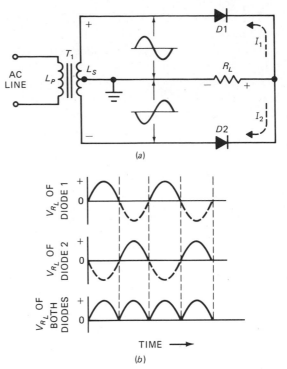

(a)

(b)

Fig. 6-11. Full-wave rectifier. (a) The circuit with I in the direction of electron flow. (b) Waveform of the rectified output without a filter capacitor. V_{RL} is at cathode.

secondary voltage, from each end of L_S to the grounded center tap, is used for each diode. Because of this requirement, a transformer with a center tap must be used. The rectifier waveforms are shown in Fig. 6-11b, without any filtering.

The opposite ends of a coil always have opposite polarities with respect to the coil center. When the top of L_S is positive, the bottom is negative, as shown in Fig. 6-11a. For that polarity of ac input, the anode of D1 is positive to make the diode conduct. On the next half-cycle, though, the ac input has opposite polarity. Then the bottom of L_S is positive, the anode of D2 is positive, and this diode conducts.

For conduction in either diode, the rectified current flows in the same direction through R_L. That direction is from left to right in the figure, as shown by the dashed arrows for I_1 and I_2 for electron flow. The dc output voltage is positive at the common cathodes of the diodes.

The waveforms in Fig. 6-11b show that the full-wave rectifier uses the full cycle of the ac input to produce a dc output. Only one diode conducts at a time, but the outputs are combined in R_L. When one diode is off, the other diode is conducting. As a result, the ripple frequency is 120 Hz. That is double the 60-Hz ac input because both alternations of each cycle are able to produce a dc output from the rectifier.

Full-Wave Operation The circuit can be analyzed in the following steps:

1. Both diodes have the same R_L for the common cathodes, returning to the center tap on L_S. Therefore, current in either diode must flow through R_L.
2. When the $D1$ anode is driven positive by the ac input, current flows through $D1$ and the top half of L_S to chassis ground and returns through R_L to cathode. That direction for I_1 is electron flow.
3. On the next half-cycle of the ac input, the $D2$ anode is positive. I_2 flows through $D2$ and the bottom half of L_S and returns through R_L to the $D2$ cathode.
4. Note that I_1 and I_2 flow through R_L in the same direction.
5. As a result, both diodes produce a dc output voltage in the same polarity, with the cathode side positive.

For a 60-Hz input, the ripple frequency is 120 Hz. Better filtering is possible because of the higher ripple frequency. Also, full-wave rectifiers can provide double the dc load current that one half-wave diode can provide.

DC Output Voltage without Filtering No filter capacitor is shown for the circuit in Fig. 6-11a to illustrate the basic rectifier waveforms. Consider the bottom waveform in Fig. 6-11b for the full-wave output. Its average dc value is 90 percent of the rms ac input. That percent is based on the ratio of average and rms values for a sine wave. Remember that rms is 0.707 of the peak value,

whereas average is 0.637 of the peak. The ratio 0.637/0.707 = 0.9, or 90 percent. For 120 V of rms ac input, then, the average dc output is 120 × 0.9 = 108 V.

For half-wave rectification, the unfiltered dc output is 0.9/2, or 0.45, of the rms ac input because alternate half-cycles are not rectified.

These values are for a rectified output without filtering. However, the filter capacitor increases the dc ouput voltage close to the peak value of the ac input.

Test Point Questions 6-6
(Answers on Page 135)

Refer to Fig. 6-11a.

a. For the polarities shown, is $D1$ or $D2$ conducting?

b. Is the ripple frequency of the dc output 60 Hz or 120 Hz?

6-7
FULL-WAVE BRIDGE RECTIFIER

The circuit of Fig. 6-12 is a full-wave rectifier that does not need a center-tapped transformer. As shown, all the secondary voltage is used for the dc output voltage.

T_1 is an isolation transformer here, but the bridge can be connected directly to the ac line. Also, the power transformer can be used to step the ac voltage up or down for the input to the bridge circuit.

Small bridge rectifiers with four diodes in each unit are shown in Fig. 6-13. The four leads correspond to terminals A, B, C, and G in Fig. 6-12. Two leads are for the ac input and two are for the dc output.

The details of conduction by the four diodes are shown in Fig. 6-14. In brief, the bridge allows two diodes to conduct in series with R_L between them on each half-cycle. The equivalent circuit in Fig. 6-14a shows R_L as the center arm of the bridge.

The right end of R_L at point C in Fig. 6-14 is

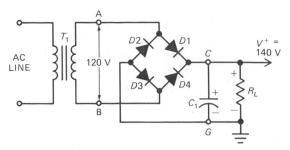

Fig. 6-12. Full-wave bridge rectifier with four diodes. This circuit does not need a center-tapped transformer.

Fig. 6-13. Bridge rectifiers with four diodes in each unit. The unit in the foreground has four pins for use with a socket. The other two have solder lugs as terminals. Size is ½ in square. (*General Instrument Corporation*)

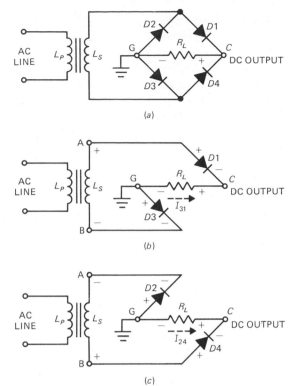

(a)

(b)

(c)

Fig. 6-14. How two of the four diodes in a bridge rectifier conduct on opposite half-cycles of the ac input. Current shown is electron flow. (a) The dc load R_L is in the center arm of the bridge. (b) When point A is positive, D1 and D3 conduct through R_L. D2 and D4 do not conduct. (c) When point B is positive, D2 and D4 conduct through R_L. D1 and D3 do not conduct.

connected to the cathodes of D1 and D4. At the other end of R_L, point G is chassis ground. As a result, point C at the cathode end of R_L has a positive output with respect to ground.

In Fig. 6-14b, conduction is shown for the ac input that makes point A positive with respect to B. Then the D1 anode is positive while the D3 cathode is negative. Both D1 and D3 conduct. The path for electron flow I_{31} is from the bottom of L_S at point B, from cathode to anode in D3, through R_L from cathode to anode in D1, and returning to point A at the top of L_S. The diodes D2 and D4 are not shown in this diagram because they are effectively out of the circuit with reverse voltage.

On the next half-cycle the ac input makes point A negative and point B positive, as shown in Fig.

6-14c. Then D2 and D4 conduct. The D4 anode is made positive while the D2 cathode is negative. The electron flow I_{24} is from the top of L_S, from cathode to anode in D2, through R_L, from cathode to anode in D4, and returning to point B at the bottom of L_S. Now the diodes D1 and D3 are effectively out of the circuit with reverse voltage.

Note that I is in the same direction through R_L for both half-cycles. The result is V^+ voltage at the end of R_L connected to the cathodes of D1 and D4. If all four diodes were inverted, the dc output at this point would be negative. The ripple frequency is 120 Hz with full-wave rectification.

a. How many diodes are needed in a bridge rectifier?

b. In Fig. 6-12, which diode conducts with $D1$?

6-8
VOLTAGE DOUBLERS

The amount of dc output voltage can be made twice the ac input by using series combinations of diodes and their filter capacitors. Typical voltage-doubler circuits are shown in Figs. 6-15 and 6-16. Voltage-tripler and -quadrupler circuits are also possible.

Full-Wave Doubler The full-wave doubler circuit, Fig. 6-15, can be analyzed as follows:

1. On the half-cycle when point A of the ac input makes the $D1$ anode positive, the diode conducts to charge C_1. The dc voltage across C_1 is positive at the cathode of $D1$.

2. On the next half-cycle, point A of the ac input makes the $D2$ cathode negative. Then that diode conducts to charge C_2. The dc voltage across C_2 is negative at the anode of $D2$.

3. The total dc output voltage is taken across C_1 and C_2 in series with respect to chassis ground. As a result, V_{dc} is twice the ac input voltage.

The ripple frequency is 120 Hz for the full-wave circuit, because both half-cycles of the ac input produce a dc output. Full-wave doublers are seldom used, however, because the circuit does not have a common ground connection for both the ac input and the dc output. The common return is important for reducing shock hazard and also stray pickup of hum from the power line.

Half-Wave Doubler The half-waver doubler circuit, Fig. 6-16, can be used as a line-connected power supply because one side of the ac input is common to the return for the dc output voltage. *Cascade doubler* is the usual name for this circuit, because the rectified output of $D1$ is used to boost

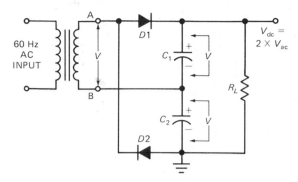

Fig. 6-15. Full-wave voltage doubler.

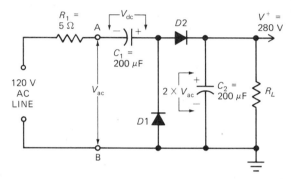

Fig. 6-16. Half-wave, or cascade, voltage doubler used as line-connected power supply.

the ac input to $D2$. The operation is as follows:

1. On the half-cycle when point B at the bottom of the ac input makes the $D1$ anode positive, the diode conducts to charge C_1. That dc voltage is positive at the cathode of $D1$.

2. On the next half-cycle of ac input, point A at the top becomes positive. It is important to note that the polarity of the dc voltage across C_1 now is series-aiding with the ac voltage input.

3. As a result, when point A is positive for the anode of $D2$, the ac voltage is boosted by the amount of V_{dc}. The ac voltage for $D2$ varies around the level of V_{dc} instead of the zero axis. Then the peak ac input to $D2$ is approximately twice the ac input to $D1$.

4. When $D2$ conducts to charge C_2, therefore, the dc output voltage is doubled. The V^+ of

280 V in Fig. 6-16 is twice the typical value of 140 V for a half-wave rectifier.

The surge-limiting resistor R_1 serves for both diodes. However, the ripple frequency is 60 Hz for half-wave rectification because only $D2$ produces the dc output voltage.

For higher voltage multiplication, two doublers can be connected in cascade for a quadrupler. In a voltage tripler, one doubler is combined with another half-wave diode. Voltage multipliers, however, have the disadvantage of poor voltage regulation. This effect means that the dc output voltage drops with more load current. Large filter capacitors are needed to help maintain the output voltage.

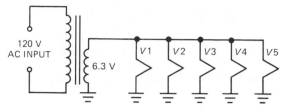

Fig. 6-17. Heaters connected in parallel across the 6.3-V transformer winding.

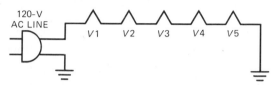

Fig. 6-18. Heaters connected in series across the 120-V ac power line.

Test Point Questions 6-8
(Answers on Page 135)

a. In Fig. 6-15, does diode $D1$ charge C_1 or C_2?
b. In Fig. 6-16, is the V^+ output produced by $D1$ or $D2$?

6-9
HEATER CIRCUITS FOR VACUUM TUBES

The heaters for equipment that uses tubes can be connected either in parallel or in series. Heaters for parallel circuits are usually rated at 6.3 V. The tube type number starts with the digit 6, as in 6GH8. A transformer is needed to step the 120 V of the ac line down to 6.3 V. All the heaters are in parallel across the heater winding, as shown in Fig. 6-17. Parallel heaters have the same voltage rating, but they can have different current ratings.

Series heaters are used with a line-connected supply, which does not have a power transformer. Then all the heaters are connected in series across the 120 V ac power line, as shown in Fig. 6-18. Typical heater ratings for a series connection are 5, 10, 12, 17, 25, and 35 V. Series heaters can use different voltages, but they must have the same current rating, usually 450 or 600 mA.

It is important to note that when one heater in a series string opens, no current can flow in the entire string. Then all of the heaters are disconnected and the equipment does not operate at all.

Test Point Questions 6-9
(Answers on Page 135)

a. What is the typical value of voltage for heaters connected in parallel?
b. What is the total voltage across a series heater string if the ac power line supplies 120 V?

6-10
FILTERS FOR POWER SUPPLIES

The dc output from a rectifier is undirectional. Without any filtering, however, the rectified output will still vary in amplitude. This fluctuating dc waveform corresponds to an ac ripple superimposed on the average dc level.

The ripple in the V^+ supply causes hum at 60 or 120 Hz. In radio receivers, the excessive ripple in the audio signal produces a constant low-pitched sound called *hum*. In television receivers, excessive hum ripple at 60 or 120 Hz in the video signal produces one or two pairs of dark and light horizontal bars. The bars usually drift slowly up or down the screen.

It is the function of the power-supply filter to remove the ac ripple component from the dc voltage output. Then the V^+ supply has a steady value. The filtering is accomplished by connecting shunt capacitors and series chokes in the output circuit of the rectifier. They function as smoothing components for the V^+ supply. An example of a filter capacitor can be seen by referring back to Fig. 6-10. A filter choke is shown in Fig. 6-19. Typical C values are 50 to 1000 μF, which requires electrolytic filter capacitors. L is 2 to 10 H for an iron-core filter choke.

The electrolytic filter capacitor must be connected in the correct polarity. Its voltage rating should be a little more than the dc output of the power supply. With too small a voltage rating for C, its dielectric film can be punctured, which short-circuits the capacitor. Too high a voltage rating also is not desirable for C because enough voltage must be applied to form the dielectric film.

How the Filter Functions A filter capacitor smooths the output voltage because C opposes any change in voltage. The filter capacitors must be connected in parallel with V^+. Otherwise, a series capacitor would block the dc output.

A filter choke smooths the output current because L opposes any variations in current. The coil is connected in series so that the dc output current must flow through it.

The filter capacitor is the main factor in reducing ripple. C charges through the low internal resistance of the conducting rectifier. The capacitor charges fast. However, the discharge path through the higher resistance of the load has a much longer RC time constant. Therefore, the capacitor discharges slowly. The slow discharge prevents the capacitor voltage from decreasing much before the capacitor is charged again by the rectifier. The net result is that the dc output voltage across the capacitor has a relatively constant level.

Percentage of Ripple The ripple factor is a measure of how effective the filter is in reducing the ac ripple in the dc output voltage. The formula is

$$\text{Ripple factor, \%} = \frac{V_{\text{ripple}}}{V_{\text{dc}}} \times 100 \quad \textbf{(6-1)}$$

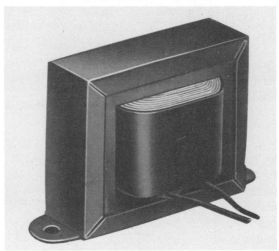

Fig. 6-19. Power-supply filter choke. Inductance L is 8 H, and R is 250 Ω. The choke shown is 3 in wide.

The ac ripple voltage is an rms value for either 60 or 120 Hz.

Example 6-1 A power supply with a dc output of 140 V has a 60-Hz ripple of 1.4 V. What is the percent ripple?

Answer

$$\text{Ripple factor, \%} = \frac{1.4\text{ V}}{140\text{ V}} \times 100$$

$$\text{Ripple factor, \%} = \frac{140}{140} = 1 \text{ percent}$$

The 1 percent ripple is a typical value just with a filter capacitor alone at the rectifier output. More filter components can reduce the ripple factor to 0.2 percent or less.

Capacitor-Input and Choke-Input Filters Filters of the capacitor-input and choke-input type are used in power supplies. The comparison is illustrated in Fig. 6-20. In Fig. 6-20a, note that the input filter capacitor C_1 is connected directly across the rectifier output. All the rectifier current can charge C_1 without flowing through L_1 or R_L. Then C_1 can charge to the peak value of the ac

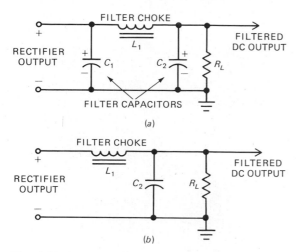

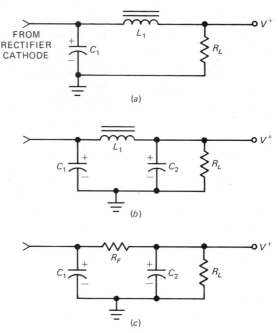

Fig. 6-20. Power-supply filters. (a) Capacitor-input circuit with C_1 at the rectifier output. (b) Choke-input circuit with L_1 at the rectifier output.

Fig. 6-21. Typical power-supply filters with capacitor input. (a) L type. (b) π type with series choke. (c) π type with series R.

input voltage. As a result, more dc output voltage can be produced as compared with the choke input filter shown in Fig. 6-20b. C_2 is the output filter capacitor used to provide additional filtering for the voltage across R_L.

In the choke-input filter arrangement, Fig. 6-20b, all the rectifier current must flow through L_1 connected directly to the rectifier. This filter has better voltage regulation with large values of load current. However, the capacitor-input filter is generally used in receivers because of its higher voltage output.

Typical filter circuits with capacitor input are shown in Fig. 6-21. The L type in Fig. 6-21a has just one capacitor C_1 and choke L_1. Most popular is the π type, Fig. 6-21b, with two capacitors to improve the filtering. C_1 is the input filter capacitor at the rectifier cathode, and C_2 is the output filter capacitor across the load. Usually, both C_1 and C_2 are in one unit with a common negative terminal. In Fig. 6-21c the filter resistor R_F is used instead of a choke to save space and money. R_F serves as a filter by providing a longer time constant for C_1 discharging through the load. However, the resistor has a higher IR voltage drop, which reduces the V^+ output compared with a filter choke.

6-11
VOLTAGE REGULATORS

The dc output voltage of a power supply tends to decrease when the load current increases. Also, the rms ac input level may vary up or down. Regulation in a power supply is used to keep the dc voltage output constant in spite of variations in either the dc load current or the ac input voltage.

Voltage regulation also improves the filtering. Three common types of voltage regulator circuits are used:

1. *Zener diode.* With the reverse breakdown voltage across a zener diode, the output voltage is constant for a wide range of current values. Typical zener diodes are shown in Fig. 6-22. Such zener regulators or voltage-reference diodes are often used in voltage ratings of 3 to 18 V. Series diodes are used for higher ratings.
2. *Voltage-regulating power transformer.* This is a special transformer designed to provide a constant ac input to the rectifier. Regulation is accomplished by saturation of the iron core.
3. *Feedback regulators.* In this type of circuit, a sample of the dc output is fed back to a stage that can control the amount of output voltage. When the sample indicates too little voltage, the output is increased. The output is lowered when the sample voltage is too high. An adjustment to maintain the dc output voltage at a specific level is usually provided.

One or more of these methods can be used in a *regulated power supply.* A typical commercial unit is shown in Fig. 6-23.

Regulation Factor The ability of the power supply to maintain a constant dc output voltage with variations in the load is specified as

Load regulation factor, %

$$= \frac{V_N - V_L}{V_N} \times 100 \qquad (6\text{-}2)$$

where V_N is the open circuit output with no load, and V_L is the full load output. It should be noted that the lower the percentage the better the voltage regulation. A lower value means less difference between the load and no-load voltages.

Example 6-2 The dc output voltage drops from 48 V with no load to 46 V at full load. What is the percent of load regulation?

Fig. 6-22. (a) Zener diodes. Color band or rounded end is cathode side. Length is ¾ in. (b) Symbol.

Fig. 6-23. Regulated power supply. Width is 14 in.   (*Kepco*)

Answer

Load regulation factor, %

$$= \frac{48 \text{ V} - 46 \text{ V}}{48 \text{ V}} \times 100$$

$$= \frac{200}{48} = 4.2 \text{ percent}$$

Actually, a regulated power supply can have a load regulation factor of less than 0.1 percent.

Zener Diode Regulator Circuit Note the schematic symbol for a zener diode in Fig. 6-24. Also, reverse bias is used with positive voltage at the cathode. This zener diode is rated for 12 V breakdown, 150 mA maximum current, and 10 W power dissipation. Small units are rated at 1 W or less.

Regulation is accomplished by the zener diode with the series-regulating resistor. R_S has the function of providing a voltage drop that varies with the amount of load current. However, the diode must have 12 V, or more, to operate in its breakdown mode to provide enough reverse current.

When the unregulated voltage tends to increase, because of less load current, the diode passes more

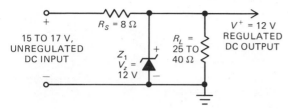

Fig. 6-24. Zener diode voltage regulator circuit. Note that 12 V across Z_1 is reverse voltage. R_S is series-regulating resistor.

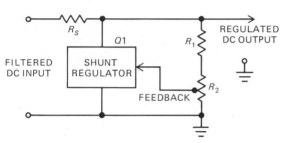

Fig. 6-25. Basic shunt regulator circuit with feedback.

current and a higher voltage drop appears across R_S. When there is a decrease in the unregulated voltage, R_S has a lower voltage drop. The net result is a constant dc output voltage of 12 V across the zener diode and the load R_L.

Feedback Regulator Circuits There are two types of feedback regulator circuits: shunt and series. In the circuit of Fig. 6-25, Q1 is a shunt regulator in parallel with the output. The amount of current shunted by Q1 affects the voltage across R_S. This *IR* drop determines the amount of dc output voltage.

The amount of current in Q1 is determined by the feedback; this sample of the dc output can increase or decrease the current in Q1. The result is a constant output voltage because of the feedback control of current in the shunt regulator.

In the circuit of Fig. 6-26, Q1 is in series with the dc output. The Q1 current and *IR* voltage drop determine the output voltage. A larger voltage drop across Q1 means less output. Less current in Q1 allows more output voltage. Actually, Q1 serves as a variable resistance in series with the load. In this function, it is called a *pass transistor*.

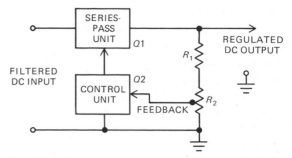

Fig. 6-26. Basic series-pass regulator circuit with feedback.

The amount of current in Q1 is determined by Q2, which controls the bias on Q1. Furthermore, Q2 has feedback as a sample of the output voltage. Then Q2 can control the effective resistance of Q1 to provide a constant dc output voltage. The series regulator is more efficient than the shunt regulator with its shunt current, which is waste current.

There are two types of series regulators, linear and switching. When the series-pass transistor Q1 operates continuously, the circuit is a linear voltage regulator.

When Q1 is switched between cutoff and saturation, the circuit is a *switching regulator*. In this circuit, Q1 can be a silicon controlled rectifier (SCR). The switching voltage, which is at a frequency much higher than 60 Hz, is provided by a separate switching circuit. This regulator circuit is much more efficient than linear regulation because Q1 is not on all the time.

An IC regulator unit is shown in Fig. 6-27. The small package contains the regulator circuits but not the power supply.

Test Point Questions 6-11
(Answers on Page 135)

a. Does the zener diode need forward or reverse voltage?

b. Is the zener diode used as a shunt or series regulator?

c. Should the load regulation factor be a high or low percentage?

6-12
POWER-SUPPLY TROUBLES

The power supply is a common source of trouble because the components operate with high current, high voltage, or both. High current produces heat, which can cause opens or shorts in the components. Typical troubles are as follows:

1. *Defective rectifier.* A single defective diode can result in no V^+ output, although the ac input is normal. Remember that V^+ is measured with a dc voltmeter, but an ac voltmeter is needed to measure the input. It should also be noted that, in a circuit with two or more rectifiers, one defective diode may just cause a reduced dc output.

2. *Open input filter capacitor.* This can cause the V^+ output to be low; also, there may be excessive hum. The filters are electrolytic capacitors. With years of use, they dry out and lose their capacity to hold a charge. Then the capacitor is effectively open.

3. *Shorted input filter capacitor.* This trouble results in no V^+ output with very high current in the rectifier and power transformer, which can burn out. If there is a fuse or circuit breaker, it will open.

4. *Open output filter capacitor.* This defect will have little effect on the V^+ output, but the hum will be more than normal.

5. *Shorted output filter capacitor.* This will cause excessive current in the rectifier circuit, including the filter choke or resistor. The choke or resistance may burn open.

6. *Open filter choke.* With this problem there is dc voltage at the rectifier side of the choke but no V^+ output for the load. The same result

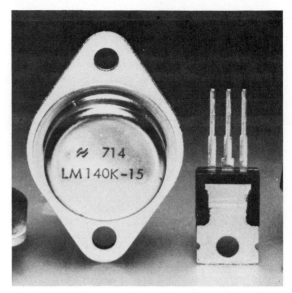

Fig. 6-27. Voltage regulator IC packages. The TO-3 metal can is at left (LM means linear monolithic construction). The TO-202 plastic package is at right. (*National Semiconductor Corporation*)

occurs when a filter resistor is used.

7. *Open heater in series string.* The equipment does not operate at all because there is no current in the series heater circuit.

8. *Excessive hum.* Assuming the V^+ output is close to normal, hum is only a filtering problem. In most cases, replacing the filter capacitors corrects the trouble.

Test Point Questions 6-12
(Answers on Page 135)

a. Is excessive hum caused by an open filter choke or capacitor?

b. Is the lack of a V^+ output caused by an open filter choke or capacitor?

SUMMARY

1. The main components in a power supply are the rectifier, which converts an ac input to a dc output, and the filter, which removes ac ripple from the rectified output. The ac input can be stepped up or down by a transformer. A regulator may also be used for a constant dc output voltage.

2. Silicon diodes are generally used for the rectifier function because of the low internal voltage drop of approximately 1 V.

3. The peak reverse voltage (PRV) across the diode rectifier on the nonconducting half-cycle is about twice the peak ac input voltage.
4. In a line-connected power supply, the ground side of the dc output circuit is not isolated from the ac power line.
5. A half-wave rectifier uses only one half-cycle of the ac input for the dc output. The ripple frequency is 60 Hz.
6. A full-wave rectifier uses both half-cycles of the ac input for the dc output. The ripple frequency is 120 Hz.
7. The ripple factor is the percent of ac ripple in the dc output. See Formula 6-1. The ripple should be very small to eliminate hum.
8. Electrolytic filter capacitors are used to remove the ripple. A capacitor input filter has C directly at the rectifier output terminal.
9. The load regulation factor is the percent of difference between the no-load and full-load output voltages. See Formula 6-2. This percent should be small.
10. The main types of rectifier circuits are summarized in Table 6-1.

Table 6-1
Rectifier Circuits

Type	Diagram	Diodes	Ripple, Hz	Notes
Half-wave	Fig. 6-6	1	60	Dc output voltage about 1.2 × rms ac input
Full-wave	Fig. 6-11	2	120	Needs power transformer with a center tap
Full-wave bridge	Fig. 6-12	4	120	Does not need a center tap
Cascade doubler	Fig. 6-16	2	60	Half-wave output

SELF-EXAMINATION
(Answers at back of book)

Match the numbered statements at the left with the lettered statements at the right.

1. Conducts in only one direction
2. Removes ac ripple
3. Uses four diodes
4. Autotransformer
5. Supplies an ac input voltage
6. Constant dc output voltage
7. Full-wave rectifier
8. Series heaters
9. Open filter choke
10. 60-Hz ripple

(a) Zener diode
(b) Power transformer
(c) Diode rectifier
(d) 120-Hz ripple
(e) Half-wave rectification
(f) Filter capacitors
(g) No V^+ output
(h) One circuit
(i) One winding
(j) Bridge rectifier

ESSAY QUESTIONS

1. What is the function of the power supply in a radio receiver?
2. Give the functions of the power transformer, diode rectifier, and filter in a power supply.
3. **a.** What is the advantage of a line-isolated power supply? **b.** Give two examples of a line-connected power supply. **c.** Why does an autotransformer *not* provide isolation from the power line?
4. Define the peak reverse voltage (PRV) rating for a diode rectifier.
5. Compare capacitor-input and choke-input filters.
6. Give typical values for the electrolytic filter capacitors and filter choke in a power supply.
7. Define ripple factor. Should it be a high or low value?
8. Define load regulation factor. Should the factor be a high or low value?
9. What is the ripple frequency for **a.** a half-wave rectifier and **b.** a full-wave rectifier?
10. Draw the circuit for a half-wave rectifier with a π-type capacitor input filter.
11. Draw the circuit for a full-wave center-tapped rectifier with a center-tapped transformer and an L-type choke input filter.
12. Draw the circuit for a full-wave bridge rectifier with a π-type filter using a resistor instead of the choke.
13. Describe briefly the operation of the two diodes in a cascade voltage doubler on opposite half cycles of the ac input.
14. Show how to connect a zener diode with reverse voltage across the diode.
15. What is meant by **a.** shunt regulator, **b.** series regulator, **c.** switching regulator?
16. Give the symptoms in hum and V^+ for the following troubles: **a.** open input filter capacitor, **b.** open filter choke, **c.** open heater in series string, **d.** open diode in half-wave rectifier circuit.

PROBLEMS

(Answers to odd-numbered problems at back of book)

1. The rms ripple voltage is 20 mV for a 15-V dc output. Calculate the ripple factor, in percent.
2. A filter choke has a 200-mA dc load current with a dc resistance of 80 Ω. Calculate its dc voltage drop.
3. The dc output voltage is 40 V at full load and 41 V without any load current. Calculate the load regulation factor, in percent.
4. A diode has a 120-V rms ac input and a 140-V dc output. What is the peak reverse voltage across the diode?
5. Calculate the capacitive reactance of a 100-μF capacitor at 60 and 120 Hz.

SPECIAL QUESTIONS

1. Name five types of electronic equipment that use a power supply other than batteries.

2. Give two different values of V^+ used in electronic equipment.
3. What is the battery voltage in a typical electronic calculator or portable radio?

ANSWERS TO TEST POINT QUESTIONS

6-1	a.	Diode	6-5	a.	Anode	6-9	a.	6.3 V
	b.	Transformer		b.	Cathode		b.	120 V
6-2	a.	Anode		c.	60 Hz	6-10	a.	T
	b.	Positive	6-6	a.	D1		b.	T
	c.	Zero		b.	120 Hz		c.	F
6-3	a.	F	6-7	a.	Four	6-11	a.	Reverse
	b.	F		b.	D3		b.	Shunt
6-4	a.	140 V	6-8	a.	C_1		c.	Low
	b.	120 Hz		b.	D2	6-12	a.	Capacitor
	c.	280 V					b.	Choke

Chapter 7
Thyristor Power Circuits

Thyristor is the name of a class of silicon semiconductors that can be switched on and off electronically to control relatively large amounts of current for motors and other electrical equipment. The silicon controlled rectifier (SCR) and the triac are examples of thyristors. The SCR is like the diode rectifier, but an additional gate electrode is used to trigger the start of conduction. The triac also has a gate electrode, but it controls both alternations of an ac wave. The name *thyristor* comes from the thyratron gas tube, which has the same function as a triggered electronic switch.

Additional types of thyristors are the unijunction transistor (UJT) and diac. They are used as trigger devices for the gate of an SCR or triac.

Thyristors are very useful in replacing mechanically controlled switches and relays. The construction and applications of thyristors are described in the following topics:

7-1   Four-Layer Construction of Thyristors
7-2   The Silicon Controlled Rectifier (SCR)
7-3   The Triac
7-4   Static Switching of Thyristors
7-5   SCR Characteristic Curves
7-6   Triac Characteristic Curves
7-7   The Diac
7-8   The Unijunction Transistor (UJT)
7-9   Unijunction Oscillator
7-10   Phase Control of Thyristors
7-11   Thyristor Motor Control
7-12   Radio-Frequency Interference (RFI)
7-13   Zero-Point Switching
7-14   How to Test Thyristors

7-1
FOUR-LAYER CONSTRUCTION OF THYRISTORS

Typical thyristors are shown in Fig. 7-1. Their construction is shown schematically in Fig. 7-2. Alternate doping produces alternate P- and N- layers of doped silicon. Three junctions, J_1, J_2, and J_3, are formed between the four layers. For an SCR, the P-layer at one end is the anode and the N-layer at the opposite end is the cathode.

Consider positive voltage applied to the anode and negative voltage to the cathode, as shown in Fig. 7-2b. Note that J_1 and J_3 now are forward-biased but J_2 in the middle has reverse bias. No current can flow between anode and cathode because of reverse bias at J_2. There is a small leakage current, but it is not forward current.

When the applied voltage has opposite polarity, as shown in Fig. 7-2c, J_2 now has forward bias but J_1 and J_3 are reverse-biased. Again no current can flow except for the leakage current.

The result is a high resistance through the four-layer diode in either direction. One of the junctions has reverse bias for either polarity of the applied voltage. The high resistance state is the OFF condition.

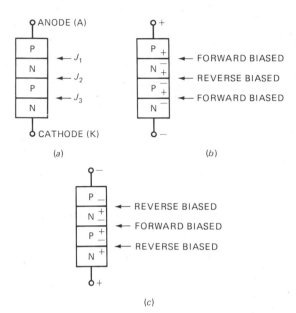

Fig. 7-2. Thyristor construction. (a) Four layers with three PN junctions. (b) Positive voltage applied to anode. (c) Negative voltage at anode.

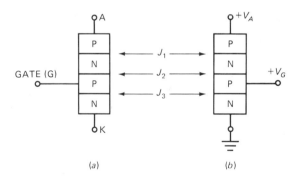

Fig. 7-3. Gate electrode for thyristors. (a) Connected to internal P-layer. (b) Positive voltage at anode and gate to turn thyristor on.

Gate Electrode The third terminal, the gate electrode, is connected to the inside P-layer (Fig. 7-3a). Now the gate can force the thyristor into conduction. There are two conditions, as shown in Fig. 7-3b.

1. Anode voltage positive
2. Gate voltage positive

Fig. 7-1. Typical thyristors. (a) Silicon controlled rectifier (SCR) rated for 25 A. Width is 1 in. (b) Triac rated for 6 A. Width is ¾ in.

The positive polarity applies to an SCR. After the SCR is turned on, it will remain on and permit current to flow from anode to cathode even after the gate voltage is removed. This ability to hold the current on is called *latching*.

Turn-on Positive voltage applied to the gate-cathode PN junction forward-biases J_1, and gate current readily flows. Charge carriers are injected into the region, as in any PN junction. Some of the charges drift to the reverse-biased J_2. There they allow the leakage current through that junction to increase. The increased leakage current adds to the gate current. The effect produces more carriers at the reverse-biased junction, and so on. This regenerative action very quickly turns the thyristor on. Once the thyristor is turned on, the gate current is not required to produce the charge carriers, because they are then produced by the main current stream.

Turn-off The gate current may be removed, but the thyristor remains on. In fact the thyristor *cannot* be turned off by removing the gate current. In order to turn the thyristor off, the main current stream must be reduced to near zero.

Two-Transistor Equivalent In Fig. 7-4a the dashed line divides the thyristor into two parts. Q1 at the left is PNP, and Q2 is NPN. The imaginary equivalent transistor circuit is shown in Fig. 7-4b. Note that V_G at the gate is the input

to the base of Q2. Also the collector output of Q_2 is the input to the base of Q1. Its collector output feeds the base of Q2. Since the output of each transistor is connected to the input of the other, the result is a regenerative feedback circuit. A positive pulse applied to the gate electrode produces the turn-on condition with forward current in the thyristor. The turn-on action is called *triggering* or *firing*.

7-2
THE SILICON CONTROLLED RECTIFIER (SCR)

In Fig. 7-5a the four-layer construction of the SCR is illustrated; its schematic symbol is shown in Fig. 7-5b. The outline drawing in Fig. 7-5c is for a high-power SCR rated at 25 A. Note that the gate electrode is the shorter terminal. In the symbol, the diode part shows that current can flow in only one direction through the SCR. A more specific name for the SCR is "reverse-blocking triode thyristor," meaning it does not conduct with negative voltage at the anode.

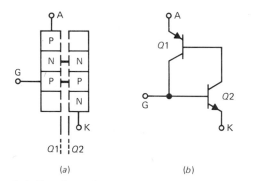

(a)   (b)

Fig. 7-4. Four-layer thyristor equivalent to two transistors. (a) Thyristor considered in two sections. (b) Equivalent circuit for Q1 and Q2.

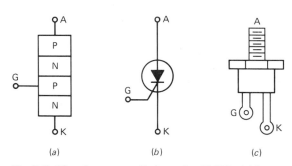

(a)   (b)   (c)

Fig. 7-5. The silcon controlled rectifier (SCR). (a) Layer arrangement. (b) Schematic symbol. (c) Outline drawing. The gate is the shorter terminal.

DC Control Circuit with an SCR The circuit in Fig. 7-6a allows the SCR to turn on a large value of dc load current in the anode-cathode path by means of a small voltage pulse at the gate. The waveforms are shown in Fig. 7-6b.

In Fig. 7-6a the SCR is shown connected in series with the 40-Ω load R_L and the 400-V dc supply through the switch S_1. Let S_1 be closed. The SCR does not conduct, however, because there is no gate pulse.

At the appropriate time, a 2-V gate pulse is applied to trigger the SCR into conduction. It conducts because it has a positive voltage at the anode. Then the SCR has very low resistance, with only about 1 V between anode and cathode. The current of approximately 10 A now flows in

the load circuit. That value is calculated as follows: 400 V/40 Ω = 10 A, neglecting the 1-V drop across the SCR.

Because the SCR resistance becomes very low, the amount of I_L is determined by R_L. Therefore, the load resistance must be high enough to limit I to a value less than the SCR rating.

Gate Voltage and Current Typical values of gate voltage and current are approximately 10 mA for the gate current and 1 to 2 V between gate and cathode. The minimum pulse duration to turn the SCR on is about 5 μs.

Turning the SCR Off When a thyristor is triggered into conduction, it stays on even after the gate pulse is removed, as shown for the SCR in Fig. 7-6b. To turn off the thyristor, the regenerative action of charge carriers between the layers must be stopped.

The turn-off requires that the anode-cathode current be reduced below a level called the *holding current*. That value is typically 3 to 10 mA for medium power ratings. Furthermore, the current must be below that level for at least 50 μs, although faster turn-offs are available. The turn-off process is also called *commutation*.

In a dc control circuit, the turn-off is usually accomplished by opening a switch in the load circuit.

AC Control Circuit with SCR An ac control circuit is shown in Fig. 7-7. Every time the ac voltage goes to zero in each ac cycle, the gate signal regains

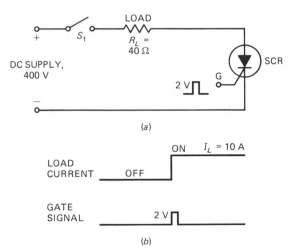

(a)

(b)

Fig. 7-6. SCR in a dc control circuit. (a) Circuit diagram. (b) Waveforms.

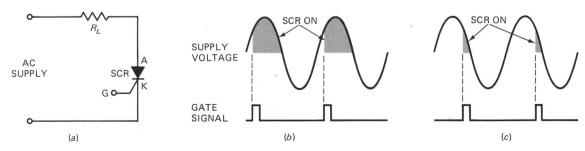

(a)   (b)   (c)

Fig. 7-7. SCR in an ac control circuit. (a) Circuit diagram. (b) Gate pulse at the start of the positive half-cycle of the ac supply. (c) Gate pulse later in the ac cycle.

control of the SCR. By controlling when the gate pulse arrives, the total amount of time during which the SCR allows load current to flow can be varied. In that way the thyristor circuit can control the average amount of power delivered to the load.

In the circuit of Fig. 7-7b, the signal occurs at the start of each positive half-cycle of the ac supply voltage. Then the SCR conducts for practically the entire half-cycle. In Fig. 7-7c, though, the gate pulse is timed to arrive later in the ac cycle. Then the SCR conduction is held to a minimum.

Test Point Questions 7-2
(Answers on Page 156)

a. How many electrodes has an SCR?
b. The gate is at -5 V and the anode voltage 500 V. Is the SCR on or off?
c. The gate is at 2 V and the anode is at 400 V. Is the SCR on or off?

7-3
THE TRIAC

Since the SCR conducts in only one direction, only one-half of the ac power input can be delivered to the load even when the gate signal is as early as possible. This limitation is overcome by the triac, which is another popular type of thyristor (Fig. 7-8). In external appearance the triac is similar to an SCR, but the triac is bidirectional. The internal layers and doping are so arranged that current can flow in either direction during both half-cycles of the ac input. Also, the gate signal can control each half-cycle.

From the symbol in Fig. 7-8a the triac can be viewed as two SCR's connected head to tail, but with one gate. This connection is also called *inverse-parallel.* The triac is equivalent to a bidirectional SCR.

The designations anode and cathode cannot be applied to a triac because the triac can conduct with either end positive. Instead, the end connections are main terminal 1 (MT1) and main terminal 2 (MT2). The difference is that the gate electrode internally is closer to MT1. Gate voltage

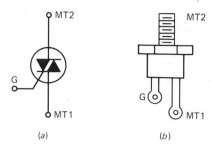

Fig. 7-8. The triac bidirectional thyristor. MT1 and MT2 are main terminals. (*a*) Schematic symbol. (*b*) Outline drawing.

and polarity are specified with respect to that terminal.

The bidirectional characteristics of a triac can be illustrated with opposite polarities of an applied voltage. When MT2 is positive and the gate trigger also is positive, the triac can conduct like the SCR. However, when MT1 and the gate are positive with respect to MT2, the triac can conduct with current in the opposite direction. In that way the triac can supply power to the load for both half-cycles of the ac input. Keep in mind that for a load such as a light bulb or a heater the direction of current does not matter.

Test Point Questions 7-3
(Answers on Page 156)

a. Which is bidirectional, the SCR or the triac?
b. Is the gate in a triac closer to MT1 or MT2?
c. Is gate voltage specified with respect to MT1 or MT2?

7-4
STATIC SWITCHING OF THYRISTORS

In static switching, the thyristor is either completely off or completely on without any phase control. In function it is similar to an electromechanical relay, but it is all-electronic and without switch contact problems. Also, electronic switching can be done at much higher frequencies than with relays.

Figure 7-9 illustrates the general idea of a static

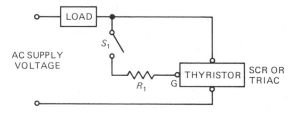

Fig. 7-9. Static switching of a supply voltage across a load through a thyristor.

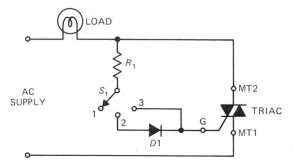

Fig. 7-10. Triac circuit for three-step power control with switch S_1.

switching circuit with either an SCR or a triac. The ac power input is switched by the thyristor in the return line of the load back to the ac supply voltage.

The switch S_1 applies gate voltage through R_1 from the load circuit. S_1 in the gate circuit switches only a few milliamperes of gate current even when the thyristor is controlling tens or hundreds of amperes.

When the gate circuit is off, with S_1 open, the thyristor is off. Then the load current is practically zero. Closing S_1 supplies gate current and turns the thyristor of Fig. 7-9 on. When an SCR is used, current flows through the load every other half-cycle. No current can flow when the anode is negative. However, that is when the gate regains control to fire the SCR on the next positive half-cycle.

When a triac is used, current flows continuously every half-cycle. When MT2 and the gate are positive, the polarity is similar to that of firing an SCR. When the polarity is opposite, current flows through the triac in the opposite direction.

Figure 7-10 shows a triac control circuit in which the amount of power delivered to a lamp can be varied in three steps. In position 1 of the switch, the gate circuit is open and the lamp is off. Then the gate does not have the voltage needed for firing the triac. In position 2, the diode $D1$ is used to provide gate voltage only in the positive polarity. The triac fires only on the positive half-cycles. As a result, the lamp is at approximately half its full brilliance. The operation every other half-cycle is like that of an SCR. However, in position 3, without $D1$, gate current flows every half-cycle. Then the triac conducts current continuously, so the lamp lights at full brilliance.

Test Point Questions 7-4
(Answers on Page 156)

a. With an SCR in Fig. 7-9, does S_1 control the gate or anode circuit?
b. In Fig. 7-10, is the triac operation like SCR operation in position 2 or 3 of S_1?
c. Through which terminals of the triac does the lamp current of Fig. 7-10 flow?

7-5
SCR CHARACTERISTIC CURVES

A typical volt-ampere characteristic of the anode-cathode curve is shown in Fig. 7-11. This curve is for zero gate current. The important effect of the gate trigger current is illustrated in Fig. 7-12.

Forward Anode Voltage and Current On the curve of forward anode voltage vs. current, $+V$ is positive voltage at the anode with respect to the cathode. The $+I$ is forward current. Its direction is hole current from anode to cathode. Electron flow is in the opposite direction.

Reverse Anode Voltage In the reverse direction, the SCR curve for $-V$ and $-I$ is the same as the curve of any reverse-biased diode. The $-I$ is only a few microamperes. However, when the reverse breakdown voltage is reached, the $-I$ increases sharply. If the current is not limited, it can destroy the junction.

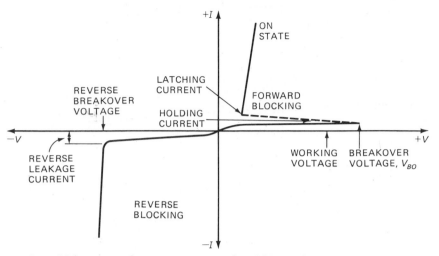

Fig. 7-11. Volt-ampere characteristic curve of an SCR anode-cathode circuit.

Breakover Voltage V$_{BO}$ The breakover voltage value is the $+V$ at which the forward current suddenly jumps to a high value of $+I$. The SCR is then ON with very low resistance. How much forward current flows is limited by the external load resistance in the anode circuit.

Holding and Latching Currents   The *holding current* is the $+I$ value necessary in the anode circuit to keep the SCR in conduction while it is ON.

The *latching current* is the $+I$ value needed to switch the SCR anode circuit ON from the OFF condition. This I is typically about three times more than the holding current. When the SCR is switched into conduction, the gate voltage must be on long enough for the anode current to reach the value for latching.

Working Voltage   In Fig. 7-11, note that the working voltage value is lower than the breakover voltage. The SCR is always operated below its breakover voltage for zero gate current. Then the SCR never turns on unless gate voltage is applied for triggering. In that way the timing of the gate pulses controls the turn-on of the SCR.

The condition with positive anode voltage but no breakover is called *forward blocking*. Although

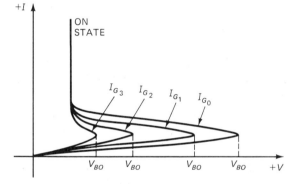

Fig. 7-12. Effect of gate current in lowering the breakover voltage of an SCR.

the SCR is OFF, it is ready to be triggered ON by the gate.

Gate-Current Control   Injecting gate current into the SCR lowers the breakover voltage, as shown in Fig. 7-12. Here I_{G_0} is for zero gate current. This situation is the same as that shown in Fig. 7-11, but the other examples in Fig. 7-12 are for increasing gate current. Note that, as gate current is increased, the breakover voltage is reduced.

When there is enough gate current, the breakover voltage becomes lower than the operating voltage or the forward blocking voltage of the SCR. That is how the SCR is used. The injection of gate current lowers the breakover voltage to a value below that of the applied voltage, thereby turning the SCR ON.

Note that the ON state is the same for all the different values of gate current in Fig. 7-12. The gate current triggers the SCR ON; but when the SCR conducts, the amount of forward current is determined by the anode circuit.

SCR Ratings   In the letter symbols given below, the subscript D is used for the OFF state with forward blocking, subscript R for reverse blocking, and subscript T for forward current. Examples are listed here with values for a medium-power SCR:

V_{DOM}	Maximum forward blocking anode voltage, with gate open (600 V)
V_{ROM}	Maximum reverse anode voltage (600 V)
I_T	Rated forward anode current (16 A or less)
V_T	Forward voltage drop across anode-cathode in conduction (1 V)
V_{GT}	Gate trigger voltage (3 V)
I_{GT}	Gate trigger current (20 mA)

The ratings for thyristors are generally given for sine-wave ac voltage at a frequency range of 50 to 400 Hz.

Test Point Questions 7-5
(Answers on Page 156)

a. Does more gate current lower or increase the breakover voltage?
b. Is V_{DOM} the forward breakover or the blocking voltage?
c. Is a typical value for forward anode current 5 mA or 10 A?
d. Is a typical gate trigger voltage 2 or 600 V?

7-6
TRIAC CHARACTERISTIC CURVES

Because a triac conducts on both half-cycles of an ac voltage, the operation is as illustrated with four quadrants in Fig. 7-13. Quadrant I at the upper

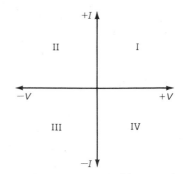

Fig. 7-13. The four quadrants used for triac characteristic curves.

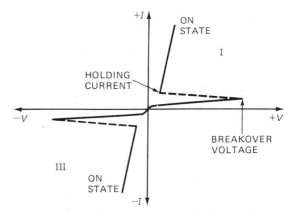

Fig. 7-14. Triac characteristic curve showing operation in quadrants I and III.

right is for positive voltage and current. In the opposite corner, quadrant III is for negative voltage and current. The triac operates in quadrants I and III, as shown by its characteristic curve in Fig. 7-14.

By definition, $+V$ means that MT2 is positive with respect to MT1. Also, $+I$ is the forward current between the main terminals with that polarity of applied voltage. These voltage and current values are in quadrant I.

The $-V$ means that MT1 is positive instead of MT2. Also, $-I$ is the forward current for that polarity of applied voltage.

Triac Operation   Figure 7-14 shows that the triac is switched ON just like an SCR, but the triac can break into conduction with positive or negative applied voltage. If the curve in quadrant III were folded over so that it was in quadrant IV, with − V coinciding with + V, it would be a mirror image of the curve in quadrant I. For either polarity, the gate can trigger the triac to conduct for the load connected in the main circuit of MT2 and MT1. While it is on, the triac has a very low resistance and a voltage drop of approximately 1 V.

Triac Triggering An SCR can be turned on with gate voltage only of positive polarity, but the triac can be triggered with positive or negative gate voltage. This dual polarity applies to triggering the breakover points in both quadrants I and III. Therefore, it is necessary to specify the trigger polarity as + or − in each quadrant.

The designation I(+) or III(−) means that the gate current is in the same direction as the main forward current for either quadrant. Opposite trigger polarities are possible, but more gate current is necessary. Therefore, the trigger polarities of I(+) and III(−) are generally used. Triggering in those modes is automatic when the gate voltage is taken from the ac line applied to the main terminals of the triac. Then the gate voltage follows the polarity of the MT voltage.

Test Point Questions 7-6
(Answers on Page 156)

Answer True or False.

a. A triac can conduct with +V at MT2 and positive gate voltage.
b. A triac can conduct with +V at MT1 and negative gate voltage.
c. The voltage drop across a triac or SCR in conduction is about 1 V.

7-7
THE DIAC

The thyristor known as the *diac* does not have a gate electrode. Instead, it is used as a trigger diode for the gate of either a triac or an SCR. The diac is a two-terminal, three-layer device; it is some-

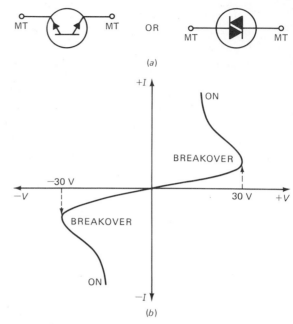

Fig. 7-15. The diac or bidirectional trigger diode. (*a*) Schematic symbol. (*b*) Characteristic curve.

times called a *bidirectional trigger diode*. Figure 7-15a shows two schematic symbols for the diac. The characteristic curve of the diac is shown in Fig. 7-15b.

Note that the diac operates in quadrants I and III as a triac does. However, conduction depends only on the voltage across its two terminals. A typical breakover voltage is 30 V in either direction. The diac is in the OFF state until its voltage exceeds the breakover value. Then the diac is in the ON state.

Diacs are used in power-control circuits for more effective control of the turn-on point for either the triac or SCR. Sometimes a neon bulb is used for the same function. It has a breakdown rating of 60 to 90 V, which is its ionizing potential.

Test Point Questions 7-7
(Answers on Page 156)

a. What is the breakover voltage for a diac, 30 or 300 V?
b. Which has a gate electrode, the diac or triac?

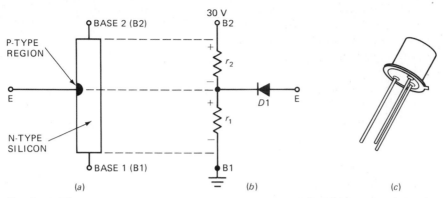

Fig. 7-16. The unijunction transistor (UJT). (*a*) Construction with two base electrodes and one emitter. There is no collector. (*b*) Base equivalent to two resistances r_1 and r_2. (*c*) Typical UJT. Diameter is 0.3 in.

7-8
THE UNIJUNCTION TRANSISTOR (UJT)

In spite of "transistor" in its name, the UJT is used as a thyristor. It has a unique construction that gives it a negative resistance characteristic for specific operating voltages. The sharp drop in R corresponds to a breakover point in I when the UJT is switching to the ON state. This operating characteristic makes the UJT very useful as a sawtooth oscillator and pulse generator for timing circuits used in triggering the SCR or triac.

Construction   The UJT consists of N-type silicon, as shown in Fig. 7-16*a*. Connections to the ends are called *base 2* and *base 1*. Then a PN junction is made by adding a P-type region near the middle of the silicon. That electrode is the emitter. The PN junction behaves as a diode. In fact, the name *double-base diode* is sometimes used for the UJT. The UJT is packaged as a small transistor with power ratings less than 1 W, as shown in Fig. 7-16*c*. The maximum emitter current is 8 to 20 mA. This corresponds to the gate current needed to turn the SCR or triac on.

Equivalent Circuit   The silicon bar serves as a resistance divided into two parts r_1 and r_2 by the PN junction (Fig. 7-16*b*). The total of the internal r_1 and r_2 is the interbase resistance, designated r_{BB}.

Its value is typically 5 to 10 kΩ. Also, r_1 is usually a little greater than r_2 because the emitter is a little closer to B2. The diode $D1$ is the equivalent of the PN junction, where the emitter meets the silicon channel.

Standoff Ratio   The ratio of r_1 to the total resistance of the channel is called the *standoff ratio.* The formula is

$$\eta = \frac{r_1}{r_1 + r_2} = \frac{r_1}{r_{BB}} \qquad \textbf{(7-1)}$$

where the Greek letter η (eta) is the standoff ratio. Typical values are 0.5 to 0.8. As an example, with r_1 of 5 kΩ and r_2 of 3 kΩ,

$$\eta = \frac{5}{5 + 3} = \frac{5}{8} = 0.625$$

The standoff ratio η has no units; it is a ratio of resistances.

Operating Characteristic   The UJT schematic symbol and volt-ampere characteristic curve are shown in Fig. 7-17. V_{BB} is the dc supply voltage applied across r_{BB} between base 1 and base 2. The subscript V is for valley-point current and voltage, and subscript P is for peak-point values. Between them, the curve shows negative resistance for the UJT.

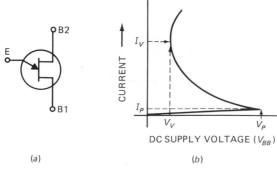

(a)

(b)

Fig. 7-17. (a) Schematic symbol of unijunction transistor. (b) Volt-ampere operating characteristic curve.

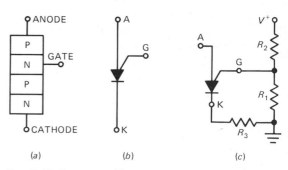

(a)   (b)   (c)

Fig. 7-18. Programmable unijunction transistor (PUT). (a) Internal P-N construction. (b) Schematic symbol. (c) R_1R_2 voltage divider sets the standoff voltage.

Referring back to Fig. 7-16b, consider $+30$ V applied to base 2, as shown. Part of the voltage is across r_1 as a voltage divider with r_2. Note that the polarity of V_1 makes the equivalent diode $D1$ cathode positive. Then the diode cannot conduct because of the standoff voltage. Remember that positive voltage at a diode cathode is reverse voltage.

A positive voltage applied to the emitter cannot produce current unless it is great enough to overcome the standoff voltage at the base plus the voltage drop across the $D1$ junction. The voltage needed is the peak-point voltage V_P. Its value is

$$V_P = V_{D1} + (\eta \times V_{BB})$$

where V_{D1} is the internal junction voltage and the quantity in parentheses is the standoff voltage across r_1. The value of V_{D1} can be taken as 0.7 V for a silicon junction. As an example, with a standoff ratio of 0.6 and V_{BB} of 30 V,

$$V_P = 0.7 + (0.6 \times 30)$$
$$= 0.7 + 18$$
$$V_P = 18.7 \text{ V}$$

The positive voltage needed at the emitter to trigger the UJT ON must be more than 18.7 V in this example.

When V_P is exceeded, the UJT fires for the ON state. Here it operates in its negative resistance region on the characteristic curve. The UJT suddenly conducts maximum emitter current. This effect is a typical breakover for a thyristor.

The conduction effectively short-circuits r_1 and causes the emitter voltage to drop near V_V, which is the valley voltage. Now the emitter voltage is not high enough to overcome the standoff voltage at the base and the UJT changes to the OFF state. A typical value is 2 V for the valley point.

Programmable Unijunction Transistor (PUT)   In a conventional UJT, the standoff ratio is determined by the internal construction. However, more flexible operation can be obtained by using the device shown in Fig. 7-18a. It has four layers and a separate gate electrode. The user can program the standoff characteristics by choosing appropriate values in the external circuit. Therefore, this thyristor is called a *programmable unijunction transistor* (PUT).

In the schematic symbol in Fig. 7-18b, the anode corresponds to the emitter of the UJT. The anode has an input voltage to turn the PUT on. The gate and cathode are comparable with the base electrodes. Although the symbol is similar to that of an SCR, the PUT has the gate closer to the anode. Also, the PUT is a relatively low-power device with anode current less than 150 mA.

The circuit in Fig. 7-18c shows how R_1R_2 serve as a voltage divider to set the standoff voltage. Their ratio determines V_{R_2} to bias the gate-anode junction OFF. Voltage applied to the anode must exceed that standoff voltage plus the internal junction voltage before current flows. Then the anode circuit breaks into the ON state. The purpose for R_3, a current-limiting resistor, is to improve the temperature stability of the circuit.

Answer True or False.

a. In the UJT, voltage is applied to the emitter to turn it ON.
b. A typical standoff ratio for the UJT is 0.6.
c. The anode of a PUT corresponds to the emitter of the UJT.

7-9
UNIJUNCTION OSCILLATOR

The basic circuit for both the UJT and the PUT oscillator is the sawtooth generator shown in Fig. 7-19. When power is applied, the circuit oscillates to produce a sawtooth voltage across C_S at the emitter terminal. Also, sharp voltage pulses are produced across R_1 at base 1. Those pulses are very useful for triggering the gate electrode of a triac or SCR.

Sawtooth Voltage A sawtooth voltage is also called a *ramp voltage* because it has a slow rise and fast drop. The action repeats at a regular rate to produce the sawtooth waveform.

The rise in sawtooth voltage across C_S is produced as C_S charges through R_S toward the V^+ of the dc supply. The UJT is OFF at the time. The drop in sawtooth voltage is produced as C_S discharges through the emitter circuit when the UJT conducts.

The charging is relatively slow because R_S in the charging circuit has higher resistance than the conducting UJT in the discharge circuit. In general, a sawtooth voltage is produced in two steps:

1. C charges slowly through a high R in a series path.
2. C discharges fast through a low R in a parallel path.

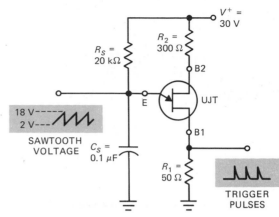

Fig. 7-19. UJT oscillator circuit. A sawtooth voltage is available at the emitter; sharp voltage pulses are available at B1.

Trigger Pulses The sharp voltage pulses at $B1$ are produced by current through R_1 when the UJT conducts. Actually, this I is the surge of C_S discharge current. The pulse voltage is equal to $I \times R_1$.

Oscillator Frequency One cycle of sawtooth voltage includes the charge and discharge of C_S. Since there is a pulse for every discharge, the pulse frequency is equal to the sawtooth frequency.

The period T of one cycle is determined by the time constant of C_S with R_S on charge, approximately. Discharge time is so short that it can be ignored. Remember that, for one RC time constant, C charges to approximately 63 percent of the applied voltage. Furthermore, when the standoff ratio of the UJT also is 0.63, the peak-point voltage will be reached in about one time constant. Then the frequency of the UJT sawtooth oscillator can be calculated approximately as $1/T$ or $1/RC$. For the 20 kΩ of R_S and 0.1 μF of C_S in Fig. 7-19, as examples:

$$f = \frac{1}{RC} = \frac{1}{20 \times 10^3 \times 0.1 \times 10^{-6}}$$

$$= \frac{10^6}{20 \times 0.1 \times 10^3}$$

$$f = \frac{1000}{2} = 500 \text{ Hz}$$

The 500 Hz is the frequency of the sawtooth voltage and the trigger pulses.

Test Point Questions 7-9
(Answers on Page 156)

Refer to Fig. 7-19.

a. Does the UJT fire at 18 or 2 V?
b. Does C_S charge through R_S *or* R_1?
c. Is the pulse output produced across R_1 or R_2?

7-10
PHASE CONTROL OF THYRISTORS

In their applications, thyristors are used to vary the average power delivered to the load. As examples, a thyristor can control the amount of light produced by a lamp, the speed of a motor, or the heat generated by a heater. Power control is accomplished by changing the instant of time in each ac cycle at which the thyristor is turned on.

Specifically, the phase of the ac voltage applied to the thyristor gate terminal is shifted to make the ac voltage lag the voltage applied to the load circuit. In that way, a fraction of each cycle is needed for the gate voltage to build up to the point at which it is able to turn the thyristor on. When the phase-angle delay is small, the gate triggers early in each cycle. The power delivered to the load then is near maximum. When the delay is longer, the gate triggers later in each cycle and less power is delivered to the load.

The required phase angle is obtained by taking the gate voltage across the capacitor in an *RC* phase-shifting network. In a series *RC* circuit, the voltage across *C* lags the applied voltage. The phase angle can be 0 to 90°. When *R* and X_C are equal, the angle is 45°.

Furthermore, additional delay results when a trigger such as a diac is used to supply the gate voltage for an SCR or triac. Time is needed for the diac to reach its breakdown voltage, which is typically 30 V. As a result of the *RC* phase shift and the trigger delay, the total angle of lag can be 30 to 150°. The delay can be varied by adjusting the series *R* in the *RC* phase-shift network.

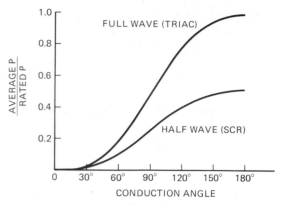

Fig. 7-20. Power delivered by a thyristor varies with the conduction angle.

Conduction Angle In each half-cycle, the thyristor conducts for 180° minus the angle of delay. The conduction angle then is between 150 and 30°. The greater the lag the smaller the conduction angle.

The curves in Fig. 7-20 show how the power delivered by the thyristor varies with the conduction angle. The power is in terms of average power compared with rated power. For the full half-cycle of 180°, the value 1.0 means full power. The curves show that varying the conduction angle between 30 and 150° allows almost the maximum possible variation from full to zero power.

Power-Control Circuit Figure 7-21 shows a circuit for an *RC* network to control the diac voltage that triggers the gate of a thyristor. Either a triac or SCR can be used. The *RC* values are typical for the 120-V 60-Hz ac power line. The current rating of the thyristor is selected to match the required load current.

When the thyristor is OFF, no current flows through the load. Then practically the entire line voltage is across the *RC* circuit. The capacitor charges toward the peak of the ac line voltage. The time it takes to reach the breakover voltage of the diac depends on the setting of *R*. When the voltage across the diac equals the breakover voltage, the diac conducts. Then the stored charge of *C* is injected into the gate circuit of the thyristor.

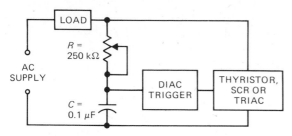

Fig. 7-21. Thyristor power-control circuit with an RC phase-shifting network.

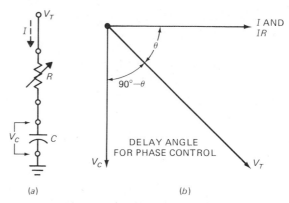

(a)   (b)

Fig. 7-22. Phase angles for RC network in Fig. 7-21. (a) RC circuit. (b) Phasor diagram showing V_C lagging the applied voltage V_T.

The injected gate current turns the thyristor on, which allows current to flow through the load.

Once turned on, the SCR or triac remains ON even though gate current ceases. However, when the ac load current drops below the holding current, the thyristor turns itself OFF and the cycle repeats. In an ac circuit, therefore, the thyristor is turned off at each zero current crossing, and the thyristor regains control. Load current will not start again until a new gate pulse is received.

If the thyristor in the circuit of Fig. 7-21 is an SCR, current flows for some part of *every other* half-cycle. If the device is a triac, then current flows for some part of all the half-cycles.

Phase Delay for V_C   The phasors with their angles are shown in Fig. 7-22 for the R and C of Fig. 7-21. The RC circuit has current I that leads the applied voltage V_T. This phase θ is the angle with a tangent equal to X_C/R. More X_C increases the angle θ. Higher R decreases θ.

However, the capacitor voltage V_C has an angle 90° lagging I for any capacitance. As a result, V_C lags behind the source voltage V_T. The angle of V_C with respect to V_T is 90° minus θ. For instance, when θ is 30°, the delay angle is 60°. More R reduces θ but the angle for V_C is increased.

dv/dt Rating   The dv/dt ratio indicates the "rate of rise in voltage" permissible for the thyristor. The four-layer construction has appreciable capacitance. As in any C, it charges in proportion to the rate of change of voltage. The faster the rate (dv/dt), the faster the charging time and the higher the value of charging current.

In a thyristor circuit, the voltage changes across the thyristor every time the thyristor is gated on or turned off. Whenever it turns off, its voltage jumps from about 1 V during conduction to the voltage of the ac supply at the instant of turn-off.

Depending on the inductance and capacitance in the circuit, the voltage can change at a rate of many volts per microsecond. At that rate of change of voltage, enough current may flow through the thyristor to turn the thyristor on again when the rest of the circuit is acting to turn it off. For that reason, thyristors are rated in terms of the maximum dv/dt they can sustain at turn-off time.

Snubber Network   The rate of rise of voltage across the thyristor can be limited simply by placing a series RC circuit across the thyristor, as shown in Fig. 7-21. The capacitor limits the dv/dt, and the resistor limits the current. This shunting RC circuit is referred to as a "snubber." In this circuit, the RC phase shift network for the diac also is a snubber for the SCR or triac.

Test Point Questions 7-10
(Answers on Page 156)

a. In Fig. 7-21, does the V_C trigger the diac or the SCR?

b. In Fig. 7-21, will less R increase or decrease the conduction angle of the thyristor?

7-11
THYRISTOR MOTOR CONTROL

Thyristor motor control includes control of speed and direction of rotation, clockwise or counterclockwise. Power tools, kitchen appliances, and industrial motors are among the many applications of thyristors in controlling motors.

Directional Rotation of AC Motors The rotational torque of an ac motor results from the stator's rotating magnetic field, which attracts the rotor field. A rotating field is produced by phase differences in the stator windings of the motor. Two stator windings are used in single phase motors to produce the phase difference.

The phase difference of the currents is obtained by a different resistance-reactance ratio in each winding. One way to get the difference is shown in Fig. 7-23a for the popular capacitor induction motor. The capacitor C for winding 2, marked L_2, produces the required phase shift.

Directional control is added by arranging the circuit as shown in Fig. 7-23b. With S_1 closed, winding L_1 is across the ac line but the phasing capacitor C is in series with winding L_2. That connection results in clockwise rotation. Opening S_1 and closing S_2 produces rotation in the counterclockwise direction.

If the motor is of any appreciable size, the switch contacts for S_1 and S_2 must be large enough to conduct the high current. It is here that a thyristor, like the SCR, is useful to conduct current for the field windings. Figure 7-23c shows how it does so. The anode circuit of SCR1 and SCR2 conducts the high current for the field windings. S_1 and S_2 can be small switches for only the relatively low gate current of about 50 mA. Note that R_1 and R_2 are current-limiting resistors for the gate circuits.

Control of Series Motors In the series motor the field winding and armature winding are connected in series. A commutator and brushes are used to feed current into the rotating armature. Because the armature current always has the same direction as the field current, the motor can operate on either ac or dc power. It is often called a *universal motor*. The characteristics are high torque at low speeds, reduced speed with more load current, and higher speed with reduced load. This type of motor is used in many appliances.

The series motor is very suitable for phase-shift speed control, as shown in Fig. 7-24. The circuit is similar to the *RC* phase-shift circuit, shown in Fig. 7-21, with a diac trigger and thyristor. In Fig. 7-24, the SCR feeds current to the series motor. Increasing R in the phase-shift network makes the SCR fire later in each half-cycle. The conduction angle is reduced for the SCR, then. It delivers less power to the motor, which reduces the speed. A circuit of this type can also be used to control the speed of split-phase ac induction motors.

Counter EMF Motor Control Thyristors can be used to provide constant motor speed under vary-

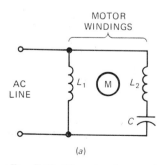

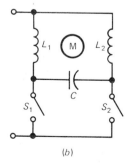

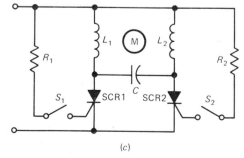

Fig. 7-23. Directional control of an ac motor. (*a*) Capacitor C connected for stator winding L_2. (*b*) C switched to either L_1 or L_2. (*c*) SCR1 and SCR2 used instead of S_1 and S_2.

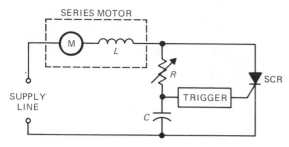

Fig. 7-24. RC phase shift circuit with an SCR for speed control of a series motor.

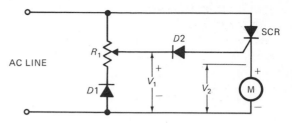

Fig. 7-25. Circuit for counter emf motor control.

ing load conditions. The method is to use as a feedback signal the counter emf generated in the rotating armature. The counter emf is proportional to the motor speed, and opposite in polarity to the applied voltage.

A typical control circuit is shown in Fig. 7-25. The SCR anode-cathode circuit supplies current for the motor. The voltage for the gate is the combination of V_1 from the potentiometer R_1 and V_2, which is the counter emf of the motor. Note that V_2 is plus at the SCR cathode. This polarity corresponds to negative at the gate. V_1 and V_2 are opposing because they tend to produce opposite directions of current in the gate circuit.

The circuit operates by comparing V_1 with V_2. When the motor speed is high, the counter emf V_2 is greater than V_1. The net voltage for the gate is then negative. The SCR cannot turn on, even at the peak of the forward half-cycle. When the motor speed is low, V_2 is less than V_1. Then the SCR is turned on for each half-cycle. The lower the counter emf, the earlier the SCR is turned on for each half-cycle of forward anode voltage. That action increases the motor speed. In a like manner the control circuit reduces speed when the motor is too fast.

Diode $D1$ is used to block gate current in the reverse direction. Also, $D2$ allows current in the R_1 branch only during the positive half-cycle of the applied ac voltage. In addition, a capacitor can be connected across R_1 for smoother control of the motor speed.

Answer True of False.

a. The motor in Fig. 7-23a rotates in a direction opposite to that of the motor in Figure 7-23b with S_2 closed.

b. In Fig. 7-24, the diac trigger supplies current to the motor field winding.

c. In Fig. 7-25, the SCR is OFF when V_2 is greater than V_1.

7-12
RADIO-FREQUENCY INTERFERENCE (RFI)

When a thyristor is gated ON, the current jumps quickly from zero to the load current determined by Ohm's law at the instant of turn-on. The change is called a *current step*. Steps of current produce electromagnetic radiation at all frequencies. Starting from the fundamental at 60 Hz, the harmonic frequencies extend well past the AM radio broadcast band of 535 to 1605 kHz. The high frequencies are generated because of the very high rate of change in the current.

The RF interference may be radiated directly or conducted into the ac power line. Either way, the broad range of frequencies can interfere with radio communications equipment. Direct radiation can be minimized by keeping the thyristor circuit compact and shielded. The shield enclosure should be grounded.

Filters help to minimize the line-conducted interference. Commercial filter packages are available, but an effective filter can be made with the

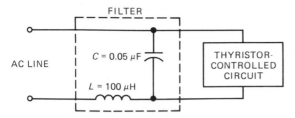

Fig. 7-26. Filter to reduce RF interference.

circuit shown in Fig. 7-26. L is an RF choke used to isolate the RF interference. It is connected in series with the ac line in either side or in both sides. The reactance is high at radio frequencies, but not at 60 Hz. C is an effective RF bypass capacitor across the ac line. Its reactance is low at radio frequencies, but not at 60 Hz. The entire filter itself may be shielded.

Test Point Questions 7-12
(Answers on Page 156)

Answer True or False.
Refer to Fig. 7-26.

a. L is an RF choke.
b. C is a bypass for 60 Hz.

7-13
ZERO-POINT SWITCHING

For minimum RFI, the ideal switch should turn on at the instant the voltage across it was zero and turn off when the current was zero. Since a thyristor normally turns off when its current goes through zero, the current steps are avoided for that condition. If the turn-on could be accomplished only when the ac supply voltage went through its zero axis, minimum RFI would be produced by the thyristor. The name for that method of power control is *zero-point or zero-voltage switching*. Actually, the turn-on is not exactly at 0 V, but near zero.

In zero-point switching, the thyristor is turned on for a whole number of half-cycles or complete cycles of the ac power supply. The half-cycles apply for an SCR and full cycles for a triac. Power control is accomplished by varying the number of cycles delivered to the load.

Figure 7-27 illustrates the zero-voltage method. The block diagram of the circuits in Fig. 7-27a shows that the thyristor is gated on from an AND gate. (Digital logic gates are explained in Chap. 8.) An AND gate will permit a signal to pass through it only if there is a signal present at all of its inputs. In this digital logic circuit, the two inputs to the AND gate are:

1. Zero-crossing pulses from a detector that samples the ac voltage. The input indicates when the ac input voltage is crossing the zero point.
2. Pulses from the power-control circuit. These pulses have a variable width, in time.

The pulse width from the control circuit determines the number of half-cycles or full cycles for turn-on of the thyristor. When both pulses appear

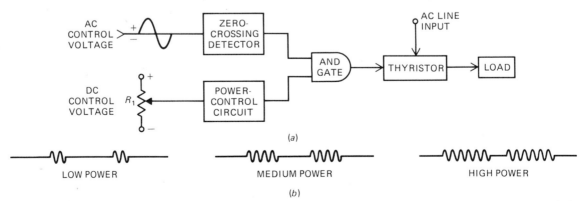

Fig. 7-27. Principle of zero-voltage switching for thyristor power control, using either a triac or SCR. (a) Block diagram of circuits. (b) Waveforms of thyristor output for the load. Full cycles shown for a triac.

at the inputs of the AND gate, a voltage appears at the output and becomes the input to the thyristor.

The waveforms for the control of average power to the load are shown in Fig. 7-27b with full cycles for a triac. Low average power means the triac is turned on for a short interval. For higher power, a longer turn-on period allows more cycles of ac input delivered to the load through the triac. In each case, though, turn-on takes place within a few volts of zero and turn-off occurs at zero current. As a result, the current steps are minimized to reduce the effect of generating RF interference.

Test Point Questions 7-13
(Answers on Page 156)

Answer True or False.

a. Zero-point switching is used with thyristors to minimize RF interference.
b. In Fig. 7-27b, the highest power corresponds to the maximum number of ac cycles.

7-14
HOW TO TEST THYRISTORS

The method of testing the SCR and triac is relatively simple. A circuit such as that of Fig. 7-28 is used to answer three questions:

1. Is the thyristor off until triggered by the gate?
2. Does the thyristor turn on when triggered?
3. Does the thyristor remain on after the trigger voltage is removed?

The lamp in Fig. 7-28 serves as a visual indicator of conduction, and it limits current through the

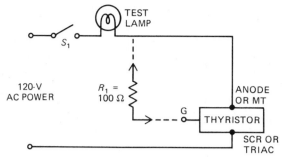

Fig. 7-28. Circuit for testing thyristors.

thyristor. As an example, a 100-W lamp is used as the load with a thyristor rated for 1 A.

When S_1 is closed, the lamp should not light. Voltage is applied to the anode or MT terminal, but there is no trigger voltage. If the lamp lights, the thyristor is shorted.

Next, touch R_1 momentarily between the gate and anode or MT terminal. The lamp should light and stay on. If it does not, the thyristor is open.

With an SCR, the lamp is on at half-brilliance because it is conducting only every other half-cycle. With a triac, the lamp is on at full brilliance. When S_1 is opened, the lamp should turn off.

Test Point Questions 7-14
(Answers on Page 156)

Refer to Fig. 7-28.

a. With S_1 closed but R_1 not connected, the bulb lights. Is the thyristor open, shorted, or good?
b. With S_1 closed and R_1 connected, the lamp does not light. Is the thyristor open, shorted, or good?

SUMMARY

1. Thyristors are solid-state switching devices with a four-layer PNPN structure. Breakover voltage makes the thyristor conduct for the ON state. No conduction is the OFF state.
2. The SCR has an anode, cathode, and a gate electrode to trigger conduction in the anode circuit. The SCR is triggered on by the gate current, but it can be turned off only by the anode circuit. The SCR conducts with positive voltage at the anode and gate.

3. The triac is a bidirectional SCR. The main terminal (MT) electrodes can conduct with either polarity. Also, the gate trigger voltage can be positive or negative.
4. Both the SCR and triac operate with an applied voltage below the breakover value. Then the gate trigger voltage lowers the breakover voltage to make the triac conduct. With ac voltage applied, the SCR conducts only on the positive half-cycles. The triac conducts on both half-cycles.
5. The breakover voltage of a thyristor is the forward V needed to start conduction. The holding current is the I needed to maintain the ON state. The latching current is the I needed to start full conduction from the OFF state.
6. Thyristors are used for power control by controlling the time in each ac cycle when the gate current produces triggering. Supplying the gate pulses early in the cycle provides maximum power to the load; later triggering results in less power.
7. Bidirectional triggers are two-terminal devices that break down to conduct at a specific V in either polarity. Diacs and neon lamps are examples. They are useful in control circuits for the gate of an SCR or triac.
8. The unijunction transistor (UJT) is a switching device that is commonly used to generate trigger pulses in timing circuits. Its electrodes are emitter, base 1, and base 2. The basic circuit is a sawtooth oscillator the frequency of which is set by an RC network for the emitter. See Fig. 7-19.
9. Phase control is used to delay the firing of a thyristor in order to vary the thyristor's conduction cycle. The phase shift is obtained with an RC network. Less delay for a greater conduction angle allows more power delivered to the load. This method is used to control motor speed.
10. Power control with thyristors requires a step jump in current once or twice in each cycle. The current steps produce radio-frequency interference (RFI). The RFI can be reduced with LC filters.
11. In zero-point or zero-voltage switching, the thyristor is gated on when the supply voltage goes through its zero axis, approximately. The purpose is to eliminate RF interference. Power control is obtained by varying the number of complete cycles, or half-cycles, of current for the load.
12. A triac or SCR can be tested with the lamp circuit in Fig. 7-28 to see if it is open or shorted.

SELF-EXAMINATION
(Answers at back of book)

Answer True or False.

1. Four-layer PNPN construction is used in both the SCR and the triac.
2. A thyristor is either on or off; it serves as an electronic switch.
3. The SCR has a gate electrode, but the triac has none.
4. The SCR anode conducts with either positive or negative voltage.

5. A gate voltage of 1 to 2 V can provide the gate current for triggering an SCR or triac.
6. Turning the gate voltage off automatically turns a thyristor off.
7. If an SCR does not turn on with anode voltage but no gate current, it must be defective.
8. A triac operates in quadrants I and III of its volt-ampere characteristic curve.
9. Injection of gate current reduces the breakover voltage of a thyristor.
10. Thyristors are always used with a supply voltage higher than the breakover voltage.
11. The voltage across a diac must never exceed the diac's breakover voltage.
12. The peak-point voltage of a UJT is less than the supply voltage.
13. In a UJT sawtooth generator, the frequency of oscillations depends on an RC time constant.
14. In phase control of thyristors, a large conduction angle allows more power to the load.
15. RFI can be eliminated from power-control circuits by filtering out the 60 Hz frequency.
16. Zero-point switching for thyristors is used to minimize RF interference.

ESSAY QUESTIONS

1. Give three examples of power-control circuits in which thyristors can be used.
2. Show the PNPN structure of the SCR. Draw the schematic symbol, and label the electrodes.
3. Draw the schematic symbol of the triac, and label the electrodes.
4. How does a triac differ from an SCR?
5. What is the function of the gate in a thyristor?
6. How does a diac differ from a triac?
7. Define breakover voltage, holding current, and latching current of an SCR.
8. Explain how an SCR is turned off.
9. Show the MT terminals of a triac connected to the ac supply line in series with a load R_L. What is connected to the gate?
10. Repeat question 9 for the anode and cathode of an SCR.
11. Explain how a diac is used with a triac.
12. Draw a circuit for testing the SCR or triac and explain how it is used.
13. Draw the volt-ampere characteristic curves for an SCR and a triac. Compare two characteristics of an SCR and a triac.
14. Draw the schematic symbols for an SCR, triac, diac, and UJT.
15. What is meant by the dv/dt rating of a thyristor? Why is it important?
16. How is an RC snubber circuit used?
17. For the UJT, define standoff ratio, standoff voltage, valley-point voltage, and current. Give a typical value for the valley voltage.
18. Draw the waveform for a sawtooth voltage and mark one cycle.
19. What is the difference between the UJT and PUT?
20. What is meant by zero-point switching? What is its advantage?

PROBLEMS

(Answers to odd-numbered problems at back of book)

1. Refer to Fig. 7-6a. What is the value of I_L through R_L when the SCR conducts?
2. What is the internal R of a thyristor conducting 10 A with 1 V across it?
3. A UJT has the base divided into r_1 of 6 kΩ and r_2 of 4 kΩ. Calculate the standoff ratio.
4. A sawtooth voltage has a period of 2 ms for one cycle. What is its frequency?
5. For a standoff ratio of 0.6 and a dc supply of 20 V, calculate **a.** standoff voltage and **b.** peak-point voltage.
6. Refer to Fig. 7-26. Calculate X_C for the 0.05-μF capacitor and X_L for the 100-μH RF choke at the power line frequency of 60 Hz.
7. In a UJT sawtooth generator, the peak-point voltage for the emitter is 12.7 V and the valley voltage is 2 V. What is the p-p voltage of the sawtooth waveform?
8. Calculate the reactance X_C of a 0.1-μF capacitor at 60 Hz.
9. Refer to Fig. 7-22. Calculate the phase-angle delay of V_C for the RC network at 60 Hz with R at **a.** 50 kΩ and **b.** 100 kΩ.

SPECIAL QUESTIONS

1. Give one application of thyristor power control.
2. Why is it that thyristors are not used for audio power amplifiers?
3. From a manufacturer's manual, give the values of three important ratings of either a triac or SCR.

ANSWERS TO TEST POINT QUESTIONS

7-1	**a.** Three	**7-5**	**a.** Lower	**7-9**	**a.** 18 V		
	b. Low		**b.** Blocking		**b.** R_S		
	c. P		**c.** 10 A		**c.** R_1		
7-2	**a.** Three		**d.** 2 V	**7-10**	**a.** Diac		
	b. Off	**7-6**	**a.** T		**b.** Decrease		
	c. On		**b.** T	**7-11**	**a.** T		
7-3	**a.** Triac		**c.** T		**b.** F		
	b. MT1	**7-7**	**a.** 30V		**c.** T		
	c. MT1		**b.** Triac	**7-12**	**a.** T		
7-4	**a.** Gate	**7-8**	**a.** T		**b.** F		
	b. 2		**b.** T	**7-13**	**a.** T		
	c. MT1 and MT2		**c.** T		**b.** T		
				7-14	**a.** Shorted		
					b. Open		

Chapter 8
Digital Electronics

In the decimal number system of counting by 10s the digits 0, 1, 2, 3, 4, 5, 6, 7, 8, and 9 are used. There are 10 digits including 0. In digital circuits, however, only the digits 0 and 1 are used. There are two digits, 0 and 1, in a system of binary numbers.

The reason for using binary numbers is that digital circuits operate with square-wave pulse waveforms, which have only two amplitudes. The pulses are either all on or all off. The ON, or HIGH, state can be represented by binary 1, and the OFF, or LOW, state as binary 0.

There are many applications of digital electronics in addition to the natural function of counting pulses. Even audio and video information can be converted into digital form by pulse sampling. The advantage of digital circuits is the excellent signal-to-noise ratio. More details of binary numbers and the types of circuits used in digital electronics are explained in the following topics:

8-1
DIGITAL AND ANALOG CIRCUITS

The world of electronics can be classified as either digital or analog circuits. Everyone has seen many digital devices with their numerical displays. A good example is the electronic calculator shown in Fig. 8-1a. The input from the keyboard uses the decimal digits 0 to 9. Inside the calculator, however, the information is processed in binary form; only the two digits 0 and 1 are used. The output at the display shows decimal numbers. An important part of most digital systems is coding and decoding of decimal and binary numbers.

An analog circuit is illustrated in Fig. 8-1b. The gradual turning of the potentiometer shaft is the analog input. The slow change in R causes a change of I in the circuit, as indicated by the analog meter at the right. The changes in current provide an analog signal. The word *analog* means there is some similarity or proportion in the signal variations as compared with the input.

Digital and analog signals are compared in Fig. 8-2. The sine wave in Fig. 8-2a is an analog signal; its values increase or decrease gradually, and the changes are smooth and continuous. Two other common examples of analog information are audio and video signals. In an audio signal the variations in V or I correspond to continuous changes in sound. In a video signal the amplitude variations correspond to changes in picture information.

The square wave in Fig. 8-2b is a digital signal. Its main characteristic is just two distinct amplitudes, such as 0 and $+5$ V in the illustration. A digital signal has no in-between values. The pulse is either all on or all off.

In many cases, the $+5$ V level is called the HIGH voltage. The HIGH amplitude is also referred to as *logic 1*, or just *1*. The reference of 0 V then becomes the LOW level, which is *logic 0*, or just *0*. This method of using the $+5$ V for logic 1 is a *positive logic system*.

A digital circuit is one that processes a digital signal. The signal itself is a train of pulses in which the HIGH and LOW levels represent the binary digits 1 and 0. An example is shown in Fig. 8-3. The

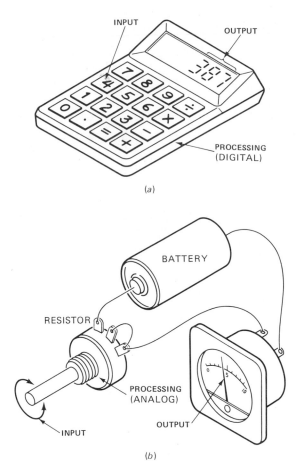

Fig. 8-1. Comparison of digital and analog systems. (a) Electronic digital calculator. (b) Example of an analog circuit.

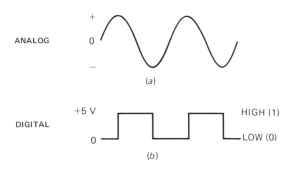

Fig. 8-2. (a) Sine wave as an example of an analog signal with continuous variations. (b) Square wave as an example of a digital signal that is at either HIGH or LOW amplitude.

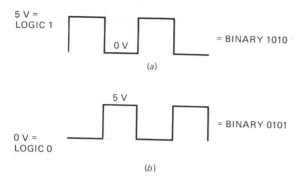

Fig. 8-3. (a) Digital signal with a train of pulses for binary information 1010. (b) Inverted signal corresponds to the complementary signal 0101.

5-V level represents logic 1, and 0 V is logic 0. The four levels in Fig. 8-3a correspond to binary 1010. In Fig. 8-3b, however, the pulse signal is inverted, or in opposite polarity. As a result, the signal has the opposite binary information, or 0101. The values of 1010 in Fig. 8-3a and 0101 in Fig. 8-3b are *complementary*, meaning that each has logic 1 where the other has logic 0. An interesting and useful principle of digital signals, then, is that phase inversion of the pulses provides complementary binary information.

Test Point Questions 8-1
(Answers on Page 187)

a. How many levels has a digital signal?
b. Is a pure sine wave a digital or an analog signal?
c. Is the HIGH voltage level of a digital signal 1 or 0 in positive logic?

8-2
ENCODING AND DECODING BINARY NUMBERS

Encoding means changing decimal numbers to a binary form. In *decoding* the digital information is converted back into decimal numbers. The coding process can also be used for letters and symbols. Methods of converting numbers are explained in the next section, but for now consider the system in the digital calculator illustrated in Fig. 8-4. The

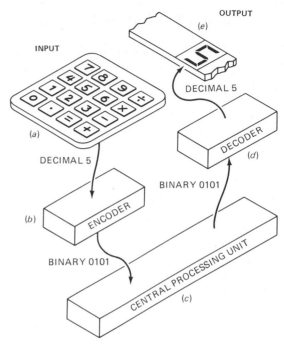

Fig. 8-4. Example of encoding and decoding in an electronic calculator. Steps (a) to (e) described in text.

letters **a** to **e** indicate the sequence of operations, as follows:

a. Here the input is the decimal number 5 punched in at the keyboard.
b. The encoder changes decimal 5 to digital form as the binary digits 0101.
c. The central processing unit contains digital logic circuits for the required calculations. Everything here is in binary form.
d. The decoder changes the binary signals back into decimal form.
e. The output display is in decimal form, showing the original number 5.

In an actual calculation, the central processing unit provides the answer in binary form to be converted into a decimal number at the output display.

Place Value in Binary Numbers Refer to the table of values in Fig. 8-5. Decimal numbers 0 to 15 are in the column at the left. Actually, 0 to 9 are the digits for decimal numbers, but 10 to 15

DECIMAL COUNT	BINARY COUNT			
	2^3 (EIGHTS)	2^2 (FOURS)	2^1 (TWOS)	2^0 (ONES)
0	0	0	0	0
1	0	0	0	1
2	0	0	1	0
3	0	0	1	1
4	0	1	0	0
5	0	1	0	1
6	0	1	1	0
7	0	1	1	1
8	1	0	0	0
9	1	0	0	1
10	1	0	1	0
11	1	0	1	1
12	1	1	0	0
13	1	1	0	1
14	1	1	1	0
15	1	1	1	1

Fig. 8-5. Binary equivalents for decimal numbers 0 to 15.

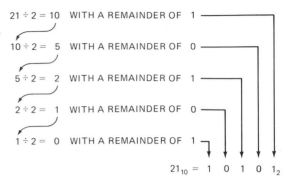

Fig. 8-6. Method of converting decimal numbers to binary numbers.

are also listed to include numbers with more than one decimal place. The decimal point, although not shown, is after the last digit at the right for whole numbers of 1 or more. The first place to the left of the decimal point gives the number of units indicated by the digit. The next place to the left is in 10s, then 100s, and so on, in multiples of 10 for higher decimal counts.

The binary numbers corresponding to decimal numbers 0 to 15 are also shown in Fig. 8-5. Only the two digits 0 and 1 are used. The binary place values are in powers of 2 starting with 1s, then 2s, 4s, 8s, and so on. A binary number also has an assumed binary point, after the last digit at the right, that corresponds to the decimal point.

Look at decimal 5 in the table with binary equivalent 0101_2. The subscript indicates the base 2 of binary numbers. This binary number is read as "zero, one, zero, one in base two," going from left to right. However, to make the binary count, go from right to left, starting from the binary point. The binary count for 0101 is $1 + 0 + 4 + 0 = 5$. As another example, the bottom number in the table is 15_{10} or 1111_2. The decimal equivalent of 1111_2 is calculated as $1 + 2 + 4 + 8 = 15$.

Decimal-to-Binary Conversion The divide-by-2 method for converting decimal numbers to their binary equivalent is shown in Fig. 8-6. In this example, 21_{10} is converted to its binary equiv-

alent. The first division is for the first place at the right, next to the binary point. This means the equivalent binary number will be developed from right to left.

Each division is for each place in the binary number. For each division, an exact answer without any remainder means the binary digit for that place is 0. When there is a remainder of 1, the binary digit is 1. The binary equivalent of 21_{10} is 10101_2 as calculated in Fig. 8-6. As a check

$$1 + 0 + 4 + 16 = 21.$$

Repeat the divisions until the quotient is zero.

Binary-to-Decimal Conversion The process of converting from binary to decimal just involves counting the place values for base 2. From right to left, they are 1, 2, 4, 8, 16, 32, 64, etc., in powers of 2. These values correspond to powers of 10 in decimal numbers. For instance, the 1000s column is 10^3 in base 10; it corresponds to the 8s column in base 2, or 2^3.

A 1 in a column means that the decimal equivalent of that column is to be added. A 0 means that the decimal value of that column is *not* added.

IC Encoders and Decoders Integrated circuits are generally used in encoders and decoders. A typical IC encoder is illustrated in Fig. 8-7. (In Fig. 8-7a the 74147 printed on the case is the device number similar to a vacuum tube number.)

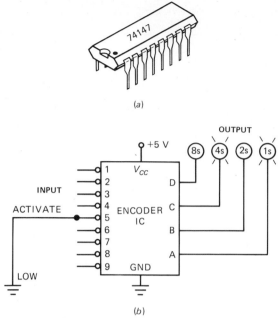

(a)

OUTPUT

Fig. 8-7. IC encoder used to produce a binary output. (a) Dual-in-line (DIP) package with 16 pins. (b) Block diagram with input and output connections. Only the bulbs for the 1s and 4s places are lit.

The whole IC unit is only about 1 in long. The 74147 translates a decimal input to its equivalent binary number at the output.

The block diagram in Fig. 8-7b shows input and output connections. The dc supply voltage to the transistor logic circuits inside is $+5$ V for V_{CC} with respect to chassis ground. There are nine input terminals. The small bubble or circle at each input indicates that a LOW voltage is needed to activate the input. The four outputs A, B, C, and D do not have the bubbles, which means they are active HIGH outputs. It takes a LOW to activate the input to produce a HIGH at the output. Then a HIGH output can light one of the indicator bulbs. It should be noted that a bulb when it lights, indicates only that binary place. The output is a binary signal. For the example here, input terminal 5 is activated with a LOW to make the encoder produce binary 0101 for the output.

The decoder illustrated in Fig. 8-8 converts the binary input to a special code that lights individual segments of the seven-segment display. The decimal display for each digit is shown in Fig. 8-8a.

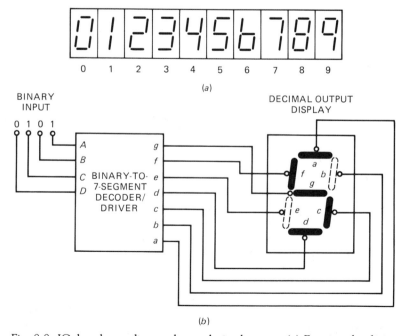

(a)

(b)

Fig. 8-8. IC decoder used to produce a decimal output. (a) Forming the digits 0 to 9 on a seven-segment display. (b) Block diagram showing how the binary input is converted to a decimal output that drives the display for each digit.

Individual segments are indicated as *a*, *b*, *c*, *d*, *e*, *f*, and *g*. One seven-segment unit can display any digit from 0 to 9. As an example, when all the segments light, the display is 8. For 5, segments *b* and *e* are off.

In the block diagram in Fig. 8-8*b* the binary input at terminals A, B, C, and D for 0101_2 is shown. The decoder changes that input signal to output signals that light segments *a*, *c*, *d*, *f*, and *g*, on the display. Then it shows decimal 5. The visual information of the output indicator is in decimal form converted from the binary information at the input of the decoder. Although the decoding here is for only one decimal place, the same procedure can be used for more places.

Test Point Questions 8-2
(Answers on Page 187)

a. What two digits are used in binary numbers?
b. Is a device that converts from decimal to binary numbers called an encoder or a decoder?
c. Convert decimal 19 to a binary number.
d. Convert binary 101010 to a decimal number.

8-3
BINARY LOGIC GATES

Logic gates are circuits the output voltage of which can be predicted from the conditions at the input. An example of their logical use, is illustrated in Fig. 8-9. The switches in this circuit are arranged as a logic AND function. The circuit in Fig. 8-9*a* has series switches for the input, which can be closed to light the bulb as the output. In itself, the series circuit is not actually a binary gate, but the logic is similar. For the series connections, it is logical to predict that the bulb cannot light unless both switch S_A and switch S_B are closed. If both are open, or just one is open, the bulb cannot light and there is no output. The actual gate circuits are made with transistors and diodes, usually in an IC package.

Truth Table It is useful to list the different logical possibilities, as in Fig. 8-9*b*. The list, called a *truth*

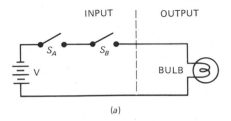

(a)

S_A	S_B	BULB
OPEN	OPEN	OFF
OPEN	CLOSED	OFF
CLOSED	OPEN	OFF
CLOSED	CLOSED	ON

(b)

Fig. 8-9. Series switches as an example of the logic AND function. (*a*) Both S_A and S_B must be closed to light the bulb. (*b*) Table listing the four possible combinations for the two switches and the resulting output for each case.

table, gives known facts for the input and admitted truths for the output. This truth table is for the series circuit in Fig. 8-9*a*. Whether the switches S_A and S_B are open or closed states the conditions for the input. Whether the bulb is on (lighted) or off (not lighted) describes the output.

There are four possible switch combinations: both off, S_A on and S_B off, S_B on and S_A off, and both on. Those conditions are listed in the four horizontal rows. As the truth table shows, only the input conditions in the bottom row, both switches on, allow the bulb to light.

The AND Gate The logic symbol and input-ouput combinations of the AND gate are shown in Fig. 8-10. The input terminals are for digital pulses. At any one time, a pulse is either at the HIGH or LOW level. The AND gate is defined as a circuit in which both inputs A and B must be HIGH in order to have HIGH output at Y. That function is similar to the series circuits with switches shown in Fig. 8-9.

When there are more than two inputs, all must be HIGH for a HIGH level of the output voltage. For that reason the AND gate is also called an *all gate*.

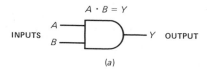

(a)

INPUTS		OUTPUT
A	B	Y
0	0	0
0	1	0
1	0	0
1	1	1

ALL ⟶

(b)

Fig. 8-10. The AND gate. (a) Logic symbol. (b) Truth table with binary 1 for HIGH level. All inputs must be 1 for output of 1.

A table showing all possible combinations of inputs and their respective outputs is called a *truth table*. The truth table (Fig. 8-10b) is shown for two inputs. Logic 1 indicates the HIGH level and 0 the LOW. Across the bottom row, the truth table of the AND function shows that all inputs must be at 1 to produce 1 at the output. Different circuits can be used to form an AND gate, but all have the same truth table.

Boolean Equations Boolean algebra provides a shorthand method of indicating logic functions. (The algebra is a special branch of mathematics formulated by George Boole for operations in logic.) The Boolean expression for the AND function is

$$A \cdot B = Y \qquad (8\text{-}1)$$

which is read as A and B equals output Y. The dot means AND in Boolean algebra. To save space, the dot may be omitted to make the Boolean expression $AB = Y$. The meaning remains the same: A and B equals output Y.

The OR Gate The logic symbol and truth table of the OR gate are shown in Fig. 8-11. The OR gate produces logic 1 in the output when any of the inputs is at 1. For that reason, it is also called the *any gate*.

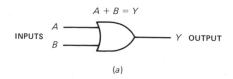

(a)

INPUTS		OUTPUT
A	B	Y
0	0	0
0	1	1
1	0	1
1	1	1

ANY

(b)

Fig. 8-11. The OR gate. (a) Logic symbol. (b) Truth table. Any input at 1 produces output of 1.

The OR gate also produces output at 1 when all of the inputs are at 1. However, the OR function is different from the AND function in that the output is at 1 when any single input is at 1. The OR function corresponds to the action of parallel switches for the inputs.

Note the difference between the logic symbol for the OR gate in Fig. 8-11a and that for the AND gate in Fig. 8-10a. The OR gate symbol has a curved line and the AND gate symbol a straight line at the input side.

The Boolean expression for the OR gate is

$$A + B = Y \qquad (8\text{-}2)$$

which is read as A or B equals output Y. The + sign means OR in Boolean algebra. When there are three inputs, the result is $A + B + C = Y$. No matter what the number of inputs, a 1 at any input causes the output to be at logic 1.

Inverter or NOT Function The logic symbol shown in Fig. 8-12a has a triangle for an amplifier with a bubble, or circle, at the output to indicate inversion. The inverter performs the NOT function, which changes the input to its opposite logic state. If the input is 0, the output will be 1; if the input is 1, the output will be 0. The inversion process is also called *negating* or *complementing*, because binary 1 and 0 are complements of each other.

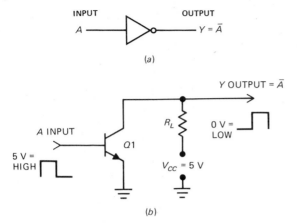

(a)

(b)

Fig. 8-12. (a) Logic symbol for an inverter. Note circle or bubble at inverted output. The bar over the A indicates inverted A, NOT A, or complement of A. (b) Common-emitter circuit as an inverter.

The Boolean expression for inversion, or the NOT function, is

$$A = \bar{A} \qquad (8\text{-}3)$$

which is read as input A equals output not A. The bar over the A means NOT, or the complement of that function.

The common-emitter amplifier circuit is an inverter. When pulses are the input signal, the amplified output has inverted polarity. (see Fig. 8-12b).

The NAND Gate The NAND gate combines the AND function with the NOT function. Actually, it is just an AND gate followed by an inverter. Its truth table and logic symbol are shown in Fig. 8-13. Note that the symbol has a bubble at the Y output to indicate inversion.

The truth table (Fig. 8-13b) compares the output for AND in the third column with NAND in the fourth column. The NAND outputs are just complements of the AND outputs. However, note the unique conditions for the bottom row, in which both inputs are at 1. Instead of output at 1, the NAND gate has output at 0. All inputs must be at 1 for 0 in the output.

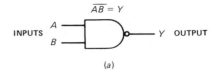

(a)

INPUTS		OUTPUT	
A	B	AND	NAND
0	0	0	1
0	1	0	1
1	0	0	1
1	1	1	0

(b)

Fig. 8-13. The NAND gate. (a) Logic symbol for NAND gate, with bubble at output. (b) Truth table for AND with NAND functions. Note that the NAND output is the complement of the AND output.

The Boolean expression for the NAND gate is

$$\overline{AB} = Y \qquad (8\text{-}4)$$

The NAND gate is like the AND gate, but the long overbar inverts the output of the NAND function. The process is called NANDing the inputs. Read the function as A NANDed with B equals output Y.

The NAND gate is very widely used. The circuit is considered the universal gate for logic design because NAND gates can be combined to produce any Boolean function.

The NOR Gate The NOR circuit is just an OR gate followed by an inverter. The truth table and logic symbol for the NOR gate are shown in Fig. 8-14. In the symbol, the bubble at the Y output indicates inversion.

In the truth table (Fig. 8-14b), any input of 1 makes the output 0 (LOW). That is just the opposite of the OR gate, in which any input of 1 makes the output 1.

The Boolean expression for the NOR gate is

$$\overline{A + B} = Y \qquad (8\text{-}5)$$

where the overbar indicates NORing the functions A and B. Read the expression as input A NORed with B equals output Y.

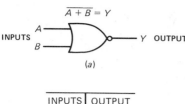

$$\overline{A + B} = Y$$

INPUTS — A, B — Y OUTPUT

(a)

INPUTS		OUTPUT
A	B	Y
0	0	1
0	1	0
1	0	0
1	1	0

(b)

Fig. 8-14. The NOR gate. (a) Logic symbol with bubble at output. (b) Truth table.

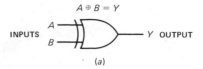

$$A \oplus B = Y$$

INPUTS — A, B — Y OUTPUT

(a)

INPUTS		OUTPUT
A	B	Y
0	0	0
0	1	1
1	0	1
1	1	0

(b)

Fig. 8-15. Exclusive-OR gate. (a) Logic symbol. (b) Truth table.

Exclusive-OR Gate   The name exclusive-OR gate is usually shortened to XOR gate. The truth table and logic symbol are shown in Fig. 8-15. Note in the symbol that an extra curved line is used at the input side.

The truth table of the XOR gate shows the output is HIGH (logic 1) when any, but not all, of the inputs is at 1. This exclusive feature eliminates a similarity to the OR gate. The XOR gate responds with a HIGH output only when an odd number of inputs is HIGH. When there is an even number of HIGH inputs, such as two or four, the output will always be LOW.

The Boolean expression for the XOR gate is

$$A \oplus B = Y \qquad (8\text{-}6)$$

Note the unique XOR symbol with a circle around the + symbol of OR. Read the expression as A exclusively ORed with B equals output Y.

Exclusive-NOR Gate   The exclusive-NOR circuit, abbreviated XNOR, is the last of the seven basic logic gates. The circuit is just an XOR gate followed by an inverter. The truth table and symbol for the XNOR gate are shown in Fig. 8-16. In the symbol, the bubble at the Y output indicates inversion of the XOR function.

The truth table of the XNOR gate is shown in Fig. 8-16b. The XNOR output is LOW (logic 0) when

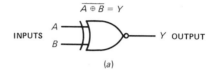

$$\overline{A \oplus B} = Y$$

INPUTS — A, B — Y OUTPUT

(a)

INPUTS		OUTPUT
A	B	Y
0	0	1
0	1	0
1	0	0
1	1	1

(b)

Fig. 8-16. Exclusive-NOR gate. (a) Logic symbol with bubble at output. (b) Truth table.

the inputs have an odd number of 1s. Note, in the output column, that Y is the complement of the Y output of the XOR gate.

The Boolean expression for the XNOR gate is

$$\overline{A \oplus B} = Y \qquad (8\text{-}7)$$

which is read as A exclusively NORed with B equals output Y.

Summary of Logic Gates   The seven basic gates are listed in Table 8-1, each with its symbol and Boolean expression.

Table 8-1
Types of Logic Gates

Name	Symbol	Boolean Equation
AND	A, B → Y	$A \cdot B = Y$
OR	A, B → Y	$A + B = Y$
INVERT	A → $\overline{A}$	$A = \overline{A}$
NAND	A, B → Y	$\overline{AB} = Y$
NOR	A, B → Y	$\overline{A + B} = Y$
XOR	A, B → Y	$A \oplus B = Y$
XNOR	A, B → Y	$\overline{A \oplus B} = Y$

Test Point Questions 8-3
(Answers on Page 187)

a. Give the usual name for an "all gate."
b. Which gate corresponds to the action of parallel switches for the input?
c. Which gate is formed by inverting the output of the AND gate?

8-4
COMBINATIONS OF LOGIC GATES

In the application of logic gates, a common procedure is to use the following steps:

1. Construct a truth table for the desired results.
2. Develop a Boolean expression from the truth table.
3. Construct a logic symbol diagram to match the Boolean expression.
4. Use the logic diagram for actual wiring of a digital circuit with standard IC units.

The procedure is illustrated in Fig. 8-17. In Fig. 8-17a is a truth table, and in Fig. 8-17b is the Boolean expression. The logic symbol circuit is given in Fig. 8-18. The logic gate connection shown in Fig. 8-18b is called the AND OR *pattern*. The AND gates are used to drive the OR gates at the output, in this example.

The truth table of Fig. 8-17a has three inputs, A, B, and C, with all of their possible combinations in eight horizontal lines. The desired output Y is to be HIGH at logic 1. Note that this output of logic 1 results only with the two input combinations for lines 5 and 8, shown by shaded areas. The desired Y output of 1 is just a starting point for this design.

Note that 1 and 0 are complementary, just like B and $\overline{B}$. For instance, when B is 1 then $\overline{B}$ is 0.

In line 5 of the truth table the combination of C AND not B AND not A activates the output Y

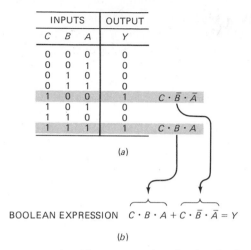

INPUTS			OUTPUT	
C	B	A	Y	
0	0	0	0	
0	0	1	0	
0	1	0	0	
0	1	1	0	
1	0	0	1	$C \cdot \bar{B} \cdot \bar{A}$
1	0	1	0	
1	1	0	0	
1	1	1	1	$C \cdot B \cdot A$

(a)

BOOLEAN EXPRESSION $\quad C \cdot B \cdot A + C \cdot \bar{B} \cdot \bar{A} = Y$

(b)

Fig. 8-17. Truth table in (a) used to develop Boolean function in (b).

with a 1. This Boolean expression is $C\overline{B}\overline{A} = Y$.

In line 8 of the truth table the input combination of C AND B AND A activates output Y with a 1. Here the Boolean expression is $CBA = Y$.

When the two Boolean expressions are combined, the result is

$$CBA + C\overline{B}\overline{A} = Y$$

This Boolean expression means that the two groups CBA and $C\overline{B}\overline{A}$ must be ORed together. The procedure is shown in Fig. 8-18b, where the two-input OR gate labeled 1 is used. Of the two inputs, one is for CBA and the other is for $C\overline{B}\overline{A}$.

The next question is how to provide the two inputs CBA and $C\overline{B}\overline{A}$. The answer is shown by the logic diagram in Fig. 8-18c for the CBA part alone and in Fig. 8-18d for $C\overline{B}\overline{A}$.

In Fig. 8-18c the three digital signals A, B, and C are ANDed; gate 2 is used for output of CBA. The output of the AND gate is connected to the bottom input terminal of the OR gate labeled 1.

In Fig. 8-18d another AND gate, labeled 3, is connected with the inverters labeled 4 and 5. The AND gate 3 supplies the other input of $C\overline{B}\overline{A}$ for the top input terminal of OR gate 1. Note that AND gate 3 has input signals of C, $\overline{B}$, and $\overline{C}$. The $\overline{B}$ input to gate 3 is supplied by inverter 4. The

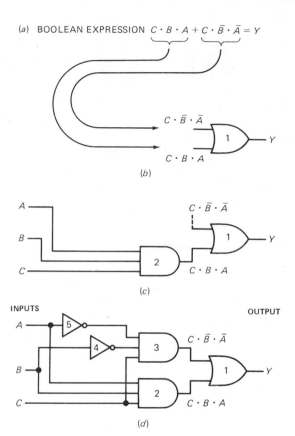

Fig. 8-18. Developing a logic symbol diagram. (a) Boolean expression. (b) OR gate number 1 for output. (c) AND gate number 2 for one input to OR gate. (d) Adding AND gate number 3 with inverters 4 and 5 for other input to OR gate.

$\overline{A}$ input to gate 3 is supplied by inverter 5.

The final result in Fig. 8-18d performs the Boolean function described by the truth table in Fig. 8-17. This logic circuit can actually be wired with standard IC units by using a two-input OR gate, a pair of three-input AND gates, and two inverters.

Test Point Questions 8-4
(Answers on Page 187)

a. In Fig. 8-18, is the OR gate or the AND gate nearest the output?
b. In Fig. 8-17, is the Boolean expression developed for HIGH or LOW output?

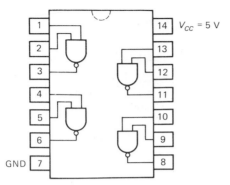

Fig. 8-19. IC package with four two-input NAND gates.

8-5
TYPES OF LOGIC CIRCUITS

Integrated circuit logic gates contain the properties of resistors, diodes, bipolar junction transistors, and the insulated-gate or metal-oxide semiconductor field-effect transistors (IGFET or MOSFET). The gates are usually packaged in groups of two or four in the same IC unit. Each gate can have from two to eight inputs. As an example, Fig. 8-19 shows the pin connections for a package with quad two-input NAND gates.

The integrated circuit logic gates are classified in families according to the principal functions used and how they are connected. Their common abbreviations are described next.

DTL The abbreviation DTL is for diode-transistor logic. The diodes are used for the AND or the OR gate, and junction transistors serve as inverters and buffer amplifiers. Actually, each diode can be made as the emitter-base junction of an integrated transistor. The basic DTL configuration is the NAND gate, the output of which is LOW when all inputs are HIGH.

TTL, or T²L The abbreviation TTL, or T²L, is for transistor-transistor logic. Instead of diodes for the gate, a special transistor with multiple emitters is used for the input signals. Space in the integrated circuit is thereby saved. The basic circuit configuration is the NAND gate.

ECL The abbreviation ECL is for emitter-coupled logic. Each input signal is applied to the base of separate transistors with a common emitter resistance R_E to form the gate. The switched output from the emitter is coupled to an emitter follower for output. The circuit is also called *current-mode logic* (CML), because the input transistors switch the current through R_E. The advantage of the method is high speed for the switching gates, because the transistors are not saturated. A switching time of less than 5 ns is typical.

Schottky TTL The Schottky TTL method is transistor-transistor logic, but a Schottky diode is used for clamping of the base-collector junction voltage to prevent saturation of the transistor. A Schottky diode is a special type that has no minority carriers and very low voltage drop in the forward direction. Those features mean fast switching between the ON and OFF states. Typical is a switching time of less than 5 ns, which is similar to that of ECL.

CMOS, or COS/MOS These abbreviations indicate complementary symmetry using P-channel and N-channel MOS transistors. The advantage is that no resistors are necessary, which allows very high density of gates on the IC chip with low power dissipation. The switching speed is about the same as that of saturated junction transistors in TTL.

The CMOS circuit for a two-input NAND gate is shown in Fig. 8-20. Four transistors are used in two pairs. $Q1$ and $Q2$ are P-channel, and $Q3$ and $Q4$ are N-channel.

Furthermore, $Q1$ and $Q2$ are connected in parallel. Either one conducting connects the Y output to the supply of 5 V. That action is *pulling up* the output for a high at Y. Then $Q3$ or $Q4$ is cut off, or both are cut off, to isolate Y from ground.

$Q3$ and $Q4$ are connected in series. Both must be conducting to connect the Y output to ground. That action is *pulling down* the output for a low at Y.

All four transistors are the enhancement type of IGFET. The drain current I_D is close to zero without any gate voltage V_G. Positive V_G increases I_D for conduction in an N-channel. The P-channel, however, is cut off. With the opposite polarity

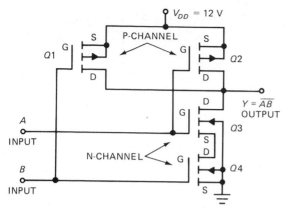

Fig. 8-20. Circuit for NAND gate using complementary P-channel and N-channel field-effect transistors.

of input, negative V_G cuts an N-channel off, but a P-channel conducts.

The A input is applied to the gate electrode of Q3 and Q2 with complementary channels. Also, the B input is applied to the complementary Q4 and Q1. When both inputs A and B are high, the N-channel pulldown transistors conduct to ground point Y. Also, the P-type pullup transistors are off. Therefore, Y is at ground.

When input A is high but B is low, Q4 cannot conduct. When input B is high but A is low, Q3 cannot conduct. When either Q3 or Q4 is in the OFF state, Y cannot be pulled down because Q3 and Q4 are in series.

When both A and B inputs are low, the P-channel pullup transistors conduct for a HIGH at Y. Then the N-channel pulldown transistors are OFF.

When any input or every input is at a LOW, the output is HIGH. Only when both inputs A and B are HIGH can the output Y be LOW. Therefore, this circuit implements the NAND function $\overline{AB} = Y$.

Test Point Questions 8-5
(Answers on Page 187)

a. What is the abbreviation for transistor-transistor logic circuits?

b. Is the logic circuit in Fig. 8-20 ECL or CMOS?

c. In Fig. 8-20, are the pulldown transistors Q1 and Q2 or Q3 and Q4?

8-6
ELECTRONIC ADDERS

When connected properly, logic gates can perform addition of binary numbers. As a review, a sample problem of binary addition, along with the equivalent decimal addition, is illustrated in Fig. 8-21. Remember that binary numbers are to base 2. The 2 is the base, or *radix*, of the number system. The base, or radix, of decimal numbers, is 10. The number of digits that can be used for counting is one less than the base. The decimal system uses digits 0 to 9, but binary numbers can use only the digits 0 and 1.

The rules of binary addition are

$$0 + 0 = 0$$
$$0 + 1 = 1$$
$$1 + 0 = 1$$

The problem comes in the carry, or overflow, for a place count. We cannot have $1 + 1 = 2$ in binary form because 2 is not a binary digit. Therefore, in binary arithmetic

$$1 + 1 = 0 \text{ with a carry of } 1$$

The *carry of 1* means add 1 in the next place to the left, which is the 2s column.

In binary arithmetic

$$1 + 1 + \text{carry-in of } 1 = 1 \text{ with carry-out of } 1$$

The carry-in of 1 at the left side of the equation comes from the preceding column, and the carry-

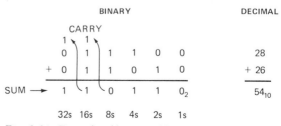

Fig. 8-21. Example of binary addition. Equivalent decimal addition is at the right.

out goes into the next place column. The place columns must be added from right to left from the assumed binary point.

This addition with a carry-in can occur for any column except the 1s place. For numbers greater than zero, that column can have a carry-out but not a carry-in.

We can apply those rules in Fig. 8-21. Starting with the 1s column at the right, $0 + 0 = 0$ for the sum shown. Then $0 + 1 = 1$. Next, $1 + 0 = 1$. So far in these three columns, there is no need for a carry because the sum does not exceed 1. In the fourth column, though, $1 + 1 = 0$ with a carry of 1. The carry of 1 keeps the sum correct by adding one more count in the next higher order of places. In the fifth column, the addition is really $1 + 1 + 1$ from the carry-in. The sum in that column, then, is 1 and the carry of 1. In the last column, the sum is $0 + 0 + 1$ for the carry, which equals 1.

The binary sum of all the place counts, from right to left, then, is

$$0 + 2 + 4 + 0 + 16 + 32 = 54$$

The sum of 54 is the same as $28 + 26 = 54$ in decimal numbers.

The Half-Adder Consider the 1s column in a binary addition problem. That column is the first place to the left of the binary point. An electronic adder circuit would need two inputs and two outputs. The two inputs are for the two digits to be added, either 0 or 1. One output terminal is for the sum of the two inputs. The other output is for the carry, if necessary. A value of 0 here means no carry, whereas 1 is the carry-out to the next place. This circuit is a half-adder; it has the symbol shown in Fig. 8-22a.

The half-adder would behave according to the truth table in Fig. 8-22b. Note the output columns for the sum and carry. Furthermore, the output columns can be produced by using two gates as follows:

1. The sum column is the output of the XOR gate. Remember that the exclusive OR gate has HIGH

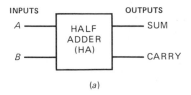

(a)

INPUTS		OUTPUTS	
B	A	SUM	CARRY
0	0	0	0
0	1	1	0
1	0	1	0
1	1	0	1
DIGITS TO BE ADDED		XOR	AND

(b)

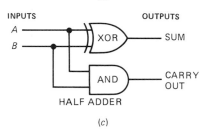

(c)

Fig. 8-22. The half-adder circuit. (a) Logic symbol. (b) Truth table. (c) Logic diagram using gates.

output when either input is HIGH but not when both inputs are the same.

2. The carry column is the output of the AND gate. Both inputs must be HIGH for there to be a HIGH in the output.

Figure 8-22c shows how the XOR and AND gates are connected. This circuit for the half-adder can be used for binary addition. However, it is only for the 1s place. The reason is that it has no input for a carry-in.

The Full-Adder As shown in Fig. 8-23, a full-adder is formed by using two half-adder circuits and an OR gate. A half-adder is indicated by the block labeled HA. Note the carry-in input, which requires the extra half-adder. The output of the OR gate forms the carry-out output. The full-adder is for binary addition in all places except the 1s place, which means the 2s, 4s, 8s, etc. Note the symbol Σ (sigma) for the sum.

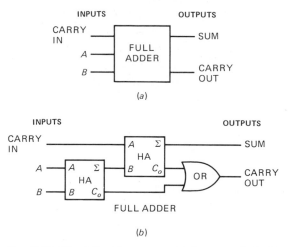

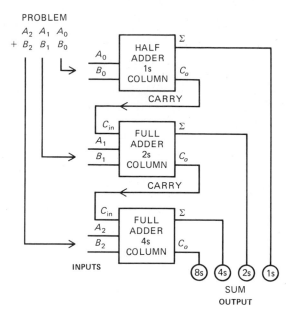

Fig. 8-23. The full-adder circuit. (a) Logic symbol. (b) Logic diagram using half-adders and an OR gate. Note the symbol Σ (sigma) for the sum.

Fig. 8-24. Logic diagram of a 3-bit parallel adder.

Parallel Adders   Half-adders and full-adders can be interconnected to form parallel adders for multiple-place columns. Figure 8-24 shows a half-adder with two full-adders connected to form a parallel adder. The circuit adds three binary-digit places. The term binary digit is usually referred to by its shortened name, *bit*. Therefore, the adder in Fig. 8-24 is called a 3-bit parallel adder. The binary numbers 0 and 1 are represented by A and B, with subscripts to indicate the place. The subscript digit is the exponent of 2 needed for the count in that place. Subscript 0 is for the units place, where A and B are 1 or 0; subscript 1 is for the 2s place, where A and B are 2 or 0; subscript 2 is for the 4s place, where A and B are 4 or 0.

The binary numbers from the 1s column of the addition problem are entered into the A and B inputs of the half-adder at the top of the diagram. Also, the bits A_1 and B_1 for the 2s column and A_2 and B_2 for the 4s column are entered into the two full-adders below. The sum of the two 3-bit numbers will appear almost instantly in the output at the lower right. Note the carry-out lines from each adder to the carry-in of the next adder. Those lines keep track of any carries in the addition. The 8s column is needed here for the carry-out of the 4s column.

Parallel adders are very common in digital circuits. More full-adders can be added to calculate the sum of longer binary numbers. Furthermore, subtraction can be performed with adder circuits by using complementary numbers. In the binary system, 0 and 1 are complements of each other. Also, multiplication and division can be performed by binary adders with special procedures called *algorithms*. An algorithm is a set of specific rules for solving a problem in a definite number of steps.

Test Point Questions 8-6
(Answers on Page 187)

a. Calculate the binary sum of 111010_2 and 11011_2.

b. What is the sum of 111010_2 and 11011_2 in decimal form?

c. Does the 8s column in binary addition need a half-adder or a full-adder?

d. How many full-adders would a 4-bit parallel adder contain?

8-7
FLIP-FLOPS

A logic gate can make a logical decision based on the immediate conditions at the input terminals. The gates do not normally have a memory characteristic, however, to retain the input data. Flip-flops, on the other hand, are circuits that do have the valuable feature of remembering. The reason is that a flip-flop circuit is bistable, which means that it can remain in either of two states with their output either HIGH or LOW. Then an input pulse can switch the output. After the flip-flop is switched, it is stable in that state until another pulse switches the circuit back to the original condition. The name *flip-flop* describes the ability of the circuit to change between two stable states.

A flip-flop circuit consists basically of two cross-coupled inverter stages. In Fig. 8-25, $Q1$ and $Q2$ represent two CE amplifiers. The collector output of $Q1$ drives the base input of $Q2$. Also, the output of $Q2$ is fed back to $Q1$ through the emitter-bias circuit with R_E and C_E for both stages.

The requirement is to have conduction in one stage cut off the other stage. Assume that first $Q1$ conducts. Then $Q1$ is on. Its collector voltage drops to drive the base of $Q2$ past cutoff. As a result, $Q2$ is cut off. For that stable state, $Q1$ is on and $Q2$ is off.

The outputs are the collector voltages. V_C for $Q1$ is LOW, close to zero, because of the $I_C R_L$ drop. However, V_C for $Q2$ is HIGH; it is equal to the 5 V of the dc supply because $Q2$ is not conducting.

Next, an input pulse can be applied to force the base of $Q2$ into conduction. Then $Q2$ has a drop in collector voltage that drives $Q1$ into cutoff. The feedback is through the voltage drop across R_E. As a result, the outputs are reversed. Now V_C for $Q1$ is HIGH and V_C for $Q2$ is LOW.

The two outputs are always opposite, or complementary in terms of logic 1 and 0. When one output is HIGH, the other must be LOW. The flip-flop is stable in either state. Only when an input pulse switches the action can the HIGH and LOW outputs be reversed.

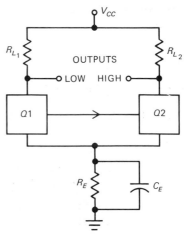

Fig. 8-25. Basic circuit for a flip-flop to provide two outputs at opposite HIGH and LOW levels. Either output terminal can be at HIGH or LOW.

R-S Flip-Flop The R-S type is the basic flip-flop logic circuit. The symbol is shown in Fig. 8-26a, with the abbreviation FF for any flip-flop. Its two inputs are called *set* (S) and *reset* (R). The two outputs are always complementary, as indicated by the standard symbols Q and $\overline{Q}$. The $\overline{Q}$ means *not* Q. When Q is at logic 1, $\overline{Q}$ must be 0. Also, if Q is at logic 0, $\overline{Q}$ must be 1.

Note the small bubbles for inversion at the S and R input terminals. The bubbles show that this FF has active LOW inputs. Logic 0 is required to activate the R or S input.

The exact behavior of a flip-flop is defined by its truth table, as in Fig. 8-26b. In the left column, the four modes of operation are:

1. *Prohibited mode.* This mode is not used because it drives both outputs HIGH. A flip-flop must operate with complementary outputs.
2. *Set mode.* Caused by activating the S input for normal Q output set to logic 1. In this FF the bubble at S means that a LOW here makes Q go HIGH.
3. *Reset mode.* Caused by activating the R input to reset or clear the Q output to logic 0. In this FF the bubble at R means a LOW here resets Q to make it LOW.
4. *Disabled mode.* This mode is the "do nothing" or "remembering" state, because there is no

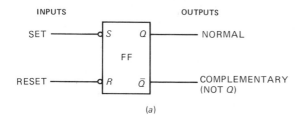

(a)

MODE OF OPERATION	INPUTS		OUTPUTS	
	S	R	Q	$\bar{Q}$
PROHIBITED	0	0	1	1
SET	0	1	1	0
RESET	1	0	0	1
DISABLED	1	1	NO CHANGE	

(b)

Fig. 8-26. The R-S flip-flop. (a) Logic symbol. (b) Truth table.

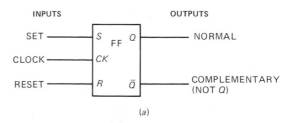

(a)

MODE OF OPERATION	INPUTS			OUTPUTS	
	CK*	S	R	Q	$\bar{Q}$
DISABLED	⊓	0	0	NO CHANGE	
RESET	⊓	0	1	0	1
SET	⊓	1	0	1	0
PROHIBITED	⊓	1	1	1	1

* ⊓ is positive clock pulse

(b)

Fig. 8-27. The clocked R-S flip-flop. (a) Logic symbol. (b) Truth table.

change. The outputs stay as they were, with Q either set or reset. In this FF the disabled mode results with both inputs at 1, because logic 0 is needed for switching. Because of the disabled mode, the flip-flop "remembers" the preceding state by remaining at that state until it is switched. That operation is possible because a flip-flop is a bistable circuit. It is stable in either the set or reset mode, and it can stay that way until it is switched to the opposite state.

The R-S flip-flop circuit is sometimes called an R-S latch. Latching means the circuit maintains one condition, in the disabled mode, until it is released by an S or R pulse. Latch circuits are used for temporary storage of digital information.

Clocked R-S Flip-Flop A clock is a circuit that generates periodic pulses at an exact frequency for the purpose of synchronizing digital circuits. The R-S flip-flop and logic gates described previously are asynchronous, because they do not have clocked input. The clocked R-S flip-flop shown in Fig. 8-27a is synchronous; it operates in step with a master clock for the system. This circuit has three inputs for set (S), reset (R), and the clock (CK) pulses.

Note that all three are active HIGH inputs; no bubbles are shown at the terminals. A HIGH is required to activate the input for switching the output.

The clock input times the set and reset operation. Because of the timing, the outputs wait to be set or reset until a clock pulse enters the CK input terminal.

The truth table is shown in Fig. 8-27b. The prohibited mode in the bottom row is not used. The set operation occurs with S at 1, because this FF has active HIGH inputs. Also, the reset operation occurs with R at 1. Whenever S or R is at 1, the other must be at 0. The disabled, or remembering, mode results when both inputs are at 0, because only one HIGH input is needed for switching the output.

The clock input has no effect when the FF is in the disabled state. However, the S and R pulses cannot activate the flip-flop unless a positive clock pulse is at the CK input at the same time that either S or R is HIGH.

Most flip-flops use some type of clocked operation. Correct timing is a problem in large systems because of unequal propagation delays. The clock

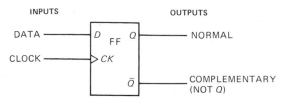

Fig. 8-28. Logic symbol for the D flip-flop.

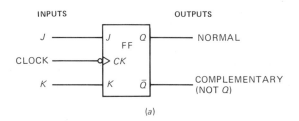

(a)

MODE OF OPERATION	INPUTS			OUTPUTS	
	$CK*$	J	K	Q	$\bar{Q}$
DISABLED	⊓	0	0	NO CHANGE	
RESET	⊓	0	1	0	1
SET	⊓	1	0	1	0
TOGGLE	⊓	1	1	TO OPPOSITE STATE	

* ⊓ is positive clock pulse

(b)

Fig. 8-29. The J-K flip-flop. (a) Logic symbol. (b) Truth table.

pulses provide synchronizing signal as a timing reference for all the pulses in the entire system.

D Flip-Flop The logic symbol for the D type of flip-flop is shown in Fig. 8-28. This FF has only one input terminal D (for data input) plus CK for clock pulses. The D type is also called a *delay flip-flop* because that is how it is used. Data at the D input is delayed from the Q output by one clock pulse.

The arrowhead > symbol next to the CK input terminal shows that the triggering of the flip-flop occurs on the positive edge of the clock. When that pulse goes from LOW to HIGH, data is transferred from the D input to the Q output. It should be noted, though, that different types of triggering can be used. Also, the trigger polarity can be indicated by showing the edge of the clock pulse sloping up or down.

J-K Flip-Flop The J-K circuit is considered the universal flip-flop because it can be used in many ways. The logic symbol is shown in Fig. 8-29a. There are two data input terminals, labeled J and K, and the clock input, CK. The bubble at the arrow shows that triggering of the FF occurs during the HIGH-to-LOW transition of the clock pulse.

The truth table in Fig. 8-29b lists four useful modes of operation for the J-K flip-flop:

1. Disabled mode. This is the memory state, as in an R-S flip-flop.
2. Reset mode. This is the same as in a clocked R-S flip-flop with active HIGH input.
3. Set mode. This is also the same as in a clocked R-S flip-flop with active HIGH input.
4. Toggle mode. The toggle action means that the normal Q output is switched between the

HIGH and LOW states at the repetition rate of the clock pulses.

The toggle mode is a new and useful function. The J-K flip-flop operates in that mode when both J and K inputs are HIGH at logic 1. Then each clock pulse toggles the outputs to switch to their opposite states. On repeated clock pulses, the Q output may go HIGH, LOW, HIGH, LOW, etc. Then Q has one complete cycle of pulse data for every two clock pulses. As a result, the toggle flip-flop is a binary divide-by-2 counter.

Most flip-flops are available in convenient IC packages for their many applications. Inside the IC chip, a flip-flop is usually formed by a combination of logic gates.

Test Point Questions 8-7
(Answers on Page 187)

a. What are the two outputs of a flip-flop called?
b. In the set mode, does Q go HIGH or LOW?
c. In the reset mode, does Q go HIGH or LOW?

8-8
DIGITAL COUNTERS

A *digital counter* consists of a group of flip-flops. Each flip-flop is usually the *J-K* type operating in the toggle mode. In this operation the *J* and *K* terminals are either open or tied to a HIGH voltage to remain at logic 1. The input pulses are applied to the *CK* terminal. As a result, the input toggles the output terminals *Q* and $\overline{Q}$ between logic levels 1 and 0.

The number of flip-flops equals the number of bits required in the final binary count. For instance, a 2-bit counter with two flip-flops counts up to binary 11, or decimal 3. A 4-bit counter with four flip-flops counts up to binary 1111, or decimal 15.

2-Bit Counter An example of a 2-bit counter with two *J-K* flip-flops is given in Fig. 8-30*a*. The bubble at the *CK* terminal shows triggering on the negative edge of the clock input pulses. Output from the first flip-flop, FFA, is available at terminal A as Q_A. The Q_A signal also supplies the input to drive the *CK* terminal of FFB. The Q_B output is available at terminal B. Only the *Q* outputs of the flip-flops are used here. The truth table is shown in Fig. 8-30*b*.

To analyze the counting circuit, we start with the clock pulses at 0 V and both flip-flops in the reset condition with Q_A and Q_B at logic 0. The waveforms are shown in Fig. 8-30*c*. On the clock pulse input, the first negative-going edge is labeled *one* to indicate the first trigger pulse for FFA. At this time Q_A goes from LOW to HIGH. The change does *not* trigger FFB because the *CK* input needs a negative-going change from HIGH to LOW.

The next trigger on the clock pulse toggles Q_A from HIGH to LOW. Now the Q_A output also triggers FFB, which toggles from LOW to HIGH. Output terminal Q_B is now HIGH.

The third clock trigger toggles Q_A back to HIGH. The change does not affect FFB. Finally, the fourth trigger pulse toggles Q_A to LOW again. At that time, Q_B also is triggered. In short, FFA toggles FFB only when Q_A goes from HIGH to LOW.

The four toggles on FFA and two toggles on FFB complete a cycle of operations for the 2-bit counter with its two flip-flops. At this time the counter has the flip-flops in the reset state with both Q_A and Q_B at logic 0, which is just like the start before any trigger pulse was applied. The same action is repeated for every four negative-going trigger pulses on the clock pulse input. The toggled values of Q_A and Q_B are tabulated in 8-30*b*, and the waveforms are shown in Fig. 8-30*c*.

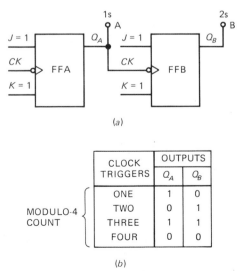

(a)

CLOCK TRIGGERS	OUTPUTS	
	Q_A	Q_B
ONE	1	0
TWO	0	1
THREE	1	1
FOUR	0	0

MODULO-4 COUNT

(b)

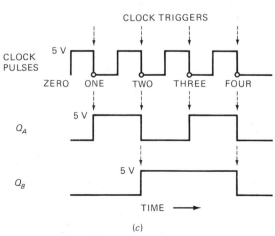

(c)

Fig. 8-30. A 2-bit binary counter. (*a*) Circuit with two *J-K* flip-flops. (*b*) Table showing FFA and FFB toggling between 0 and 1. (*c*) Clock input pulses and outputs at Q_A and Q_B terminals.

Modulo Number The circuit shown in Fig. 8-30 is a modulo-4 counter, because it has 4 counts to the reset state. As another example of modulo number, a modulo-16 counter has 16 counts. The modulus of a counter is the number of different states in the combined output during the counting sequence.

By adding more flip-flops, the modulus number can be increased in powers of 2: 2, 4, 8, 16, etc. A modulus of 16 needs 4 flip-flops for 16 counts of 0 to 15. Furthermore, the modulus can be changed from a binary number by using the J and K terminals. Common examples are modulo-5 and modulo-10. The modulo-10 is a decade counter for 0 to 9 in the decimal system.

The actual count is 1 less than the modulo number. For instance, a modulo-4 counts up to decimal 3, which equals binary 11.

Serial and Parallel Counters The counter shown in Fig. 8-30 is of the serial type, because the Q output of one flip-flop drives the clock input of the next. It is also called a *ripple counter*, because the count ripples through each flip-flop in succession. In a parallel counter the clock input pulses are applied to the CK input on all the flip-flops at the same time. The Q output of each FF drives a

J and K terminal of the next FF. The parallel type is also called a *synchronous counter*, whereas the series counter is asynchronous. The parallel counter is faster, because it can count at higher clock frequencies.

Up and Down Counters The circuit shown in Fig. 8-30 is an up counter, because each toggle increases the stored binary value by one bit. In the toggle table shown in Fig. 8-30c, note that the outputs in their binary places increase from 00 to 01 to 10 to 11.

In a down counter the count is decreased by one bit for each count. The down count is accomplished by using the $\overline{Q}$ (NOT Q) output of each FF to drive the following FF. The count itself, however, is still taken from the Q outputs.

Presettable Counter A presettable counter starts the count from a specific value instead of zero. As an example, a down counter can start from 4 and count down to 0 for reset.

Modulo-16 Ripple Counter See Fig. 8-31a for another example of a digital counter. Four flip-flops are used for four binary bits, and there are four binary places in the output. Note that FFA

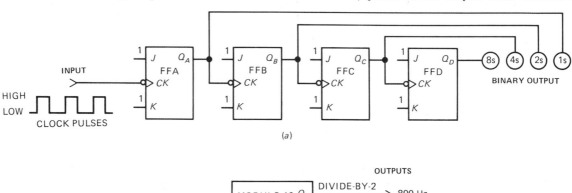

(a)

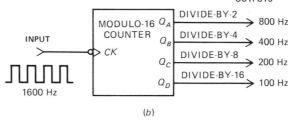

(b)

Fig. 8-31. (a) Logic diagram of modulo-16 counter with four flip-flops for 4 bits. (b) Same counter used as a frequency divider.

supplies information for the first place, next to the assumed binary point. The 4-bit binary counting sequence in the output is the same as shown before in Fig. 8-5. At the end of 16 clock input pulses (0 to 15), all the outputs Q_A, Q_B, Q_C, and Q_D are 1 for a binary output of 1111, which equals decimal 15.

Frequency Dividers The same modulo-16 counter as shown in Fig. 8-31a is shown in Fig. 8-31b for the application of a frequency divider. Specifically, the input pulses to the CK terminal form a square-wave signal with the frequency of 1600 Hz.

The first flip-flop FFA will divide the frequency by 2 and provide 800 Hz at the output terminal Q_A. That frequency is again divided by 2 to provide 400 Hz at Q_B. Also, Q_C has 200 Hz and Q_D has 100 Hz. The final output, at 100 Hz, results from the divide-by-16 action of the counter on the 1600-Hz input.

Such frequency division is a common use for counters. The electronic digital clock is based on this application.

Decimal Display for the Count A decimal display, a common application of the counter, is illustrated in Fig. 8-32. The first block at the left is a decade counter, which counts 0 to 9, but in binary form. The middle block is a decoder. It converts the count in BCD form to the code needed to drive each of the seven segments in the numerical display. The BCD is the abbreviation for binary-coded decimal; each decimal digit from 0 to 9 in each decimal place is represented as a binary word.

The combination of functional blocks as in Fig. 8-32 is common in digital electronics. The electronic counting can be done only by binary circuits. The binary data must be converted to decimal form to have a convenient way for reading out the information. Finally, the seven-segment display is a compact device for showing digits, but coded signals are needed to activate its segments.

Test Point Questions 8-8
(Answers on Page 187)

a. In Fig. 8-30, the circuit counts up to what number?
b. In Fig. 8-31a, the circuit counts up to what number?
c. In Fig. 8-30, does FFB toggle when FFA goes HIGH or LOW?
d. Do Figs. 8-30 and 8-31a show up or down counters?

8-9
SHIFT REGISTERS

Shifting action can be seen in a calculator display. To enter the number 25, the 2 key is pressed and released. The number 2 appears on the display. Next the 5 key is pressed and released. This causes the 2 to shift one place to the left and 25 appears on the display. The action shows that the shift register has memory in addition to the shifting action: the numbers were remembered even after

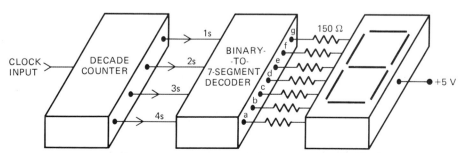

Fig. 8-32. Typical system for counter-decoder-decimal display. Letters a to g on the decoder correspond to the segments of the display.

the keys were released. Also, the 2 was shifted to the left to make place for the 5. An electronic calculator has many large registers in its complex digital circuits.

The circuit of a shift register consists basically of flip-flops, usually the D type with a *clear* or *reset* terminal, plus logic gates for special functions. One flip-flop is used for each bit in the binary data. The shifting means transferring the information in each FF from its Q output to the next FF.

The logic diagram for a 4-bit, shift-right register is shown in Fig. 8-33a. Four D-type flip-flops are used. All the CK inputs are wired together. Also, all the CLR terminals are connected to the clear or reset input. The normal Q outputs are connected to the indicators A, B, C, and D for each binary place. Also, the Q output of FFA drives the D input of FFB, which drives FFC to drive FFD. The serial load data is applied to the D-input terminal of FFA. *Serial loading* means loading a number one bit at a time.

The chart in Fig. 8-33b illustrates the operation. At the top the clear input is activated with a low. This action clears, or resets, the register to 0000.

The next sequence is serial loading of 0111 into the register one bit at a time. In the table note that the rows from top to bottom correspond to the load data 0111 from right to left. Four clock pulses will be needed to load this 0111 into the register.

Look at the top horizontal row for loading the first bit 1 in the data 0111. This input at the D input of FFA is transferred to the A output. Also, the logic 0 that was at A is transferred to B, shifting one place.

In the second horizontal row the next logic 1 in the load input makes the outputs shift again. The 1 that was at A shifts one place to B, and the input 1 is transferred to output A. This shift from A to B is necessary to make a place for the next bit in the load data of the input.

The shift-right occurs for every clock pulse. Finally, the register shows 0111 for the outputs, with

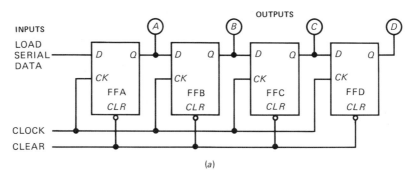

(a)

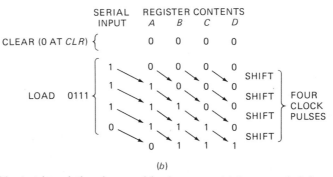

(b)

Fig. 8-33. A 4-bit, shift-right, serial-load register. (a) Logic-symbol diagram. (b) Clearing and loading sequence. The flip-flops are D type with a clear or reset terminal.

each in the correct place corresponding to A, B, C, and D.

In general, registers can be only 4 bits long or 16 or 32 bits long. They can shift left or right. The input can be serial-loaded or parallel-loaded with input data to all the flip-flops. Since shift registers are very common, they are available in convenient IC packages as a complete unit.

Test Point Questions 8-9
(Answers on Page 187)

Refer to the logic diagram in Fig. 8-33.

a. Is this a shift-right or a shift-left register?
b. Is the register loading serial or parallel?
c. Is the input loaded at terminal D of FFA or FFD?

8-10
SEMICONDUCTOR MEMORIES

The two main types of semiconductor memory are the *random-access memory* (RAM) and the *read-only memory* (ROM). The memory function is the storage of digital information as 1s and 0s. Essentially, in the *read mode* digital information is extracted from the memory and in the *write mode* information is put into the memory.

The RAM is also called a *read-write memory* because it can provide both functions, whereas the ROM provides only the read mode. The comparison is illustrated in Fig. 8-34.

A RAM is something like a shift register, but it will hold many more bits of data. The information can be put into the memory for storage. The input operation of the RAM in its write mode resembles that of a scratch pad. To extract the information from the RAM, the read mode is used. *Random access* means that any bit or word can be read at any time by means of the address information. The address identifies the bit or word.

The data extracted does not erase the stored information. However, the RAM is a temporary type of storage because the information is destroyed when the power is turned off. For that reason, it

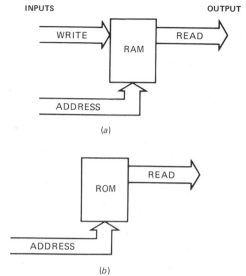

INPUTS   OUTPUT

(a)

(b)

Fig. 8-34. Input and output features of semiconductor memories. (a) Random-access-memory (RAM) for read and write functions. (b) Read-only-memory (ROM).

is also called a *volatile* memory. A memory of this type can use the flip-flop circuit for storage.

The ROM is a permanent type of memory; it contains memory cells that are permanently programmed by the manufacturer with a specific pattern of 1s and 0s. In some types, the ROM can be programmed by the user; it is then a programmable read-only memory (PROM).

Test Point Questions 8-10
(Answers on Page 187)

a. Has the RAM or the ROM both read and write mode capabilities?
b. Is extracting information from storage the read or the write mode?

8-11
MULTIPLEXERS AND DEMULTIPLEXERS

The devices known as multiplexers and demultiplexers are the electronic equivalents of a single-pole, many-position rotary switch. The basic idea of how an electronic multiplexer works is shown in Fig. 8-35a. Any one of many inputs can be selected by the control. The data at that input will be transferred to the output. Rapid rotation of the control will change the parallel-input word to serial data. Several bits taken together form a *word*. The word in Fig. 8-35a in parallel form is LHHLLL. The electronic multiplexer is also called a *data selector*.

In its operation, a demultiplexer is the opposite of a multiplexer. In Fig. 8-35b the demultiplexer is shown distributing a serial string of bits to the correct outputs. As a result, the serial data is changed to parallel data. The demultiplexer is also called a *distributor* or *decoder*.

An application of the two functions is in the transmission of data by telephone lines, coaxial cable, or radio. In the transmission, it is better to have the data in serial form, so only one line or channel is needed. At the receiving end, the data is reassembled as parallel data, back in its original form.

Test Point Questions 8-11
(Answers on Page 187)

a. Is input to a multiplexer serial or parallel data?
b. Is output from a demultiplexer serial or parallel data?

8-12
DIGITAL AND ANALOG CONVERTERS

Digital and analog converters change information from one form to the other. As an example, a digital voltmeter has an analog input that must be converted to digital form. This is a function of the analog-digital (A/D) converter. In the opposite case, a digital-analog (D/A) converter changes binary data back into analog form. We can say that the A/D converter encodes the analog values as binary numbers and the D/A converter decodes the data to provide the original analog information.

The function of an A/D converter is illustrated in Fig. 8-36. Note that each value of input voltage generates a specific binary output of 1s and 0s. However, the binary number itself does not equal

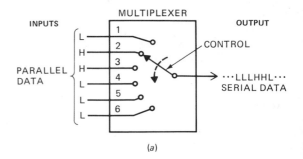

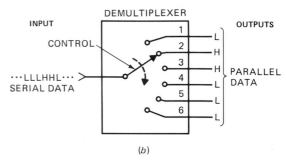

Fig. 8-35. (*a*) Multiplexer's similarity to a rotary switch. (*b*) demultiplexer's similarity to a rotary switch.

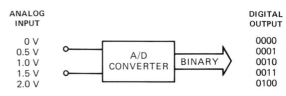

Fig. 8-36. Function of an analog-to-digital (A/D) converter.

the analog value. The scaling of the analog input is set by the design of the system. In the case of a digital voltmeter, the scaling factors determine the different voltage ranges.

The circuits for an A/D voltage converter include a sawtooth voltage generator to produce a ramp voltage as a reference and a comparator circuit. The comparator indicates which of its two input voltages is larger and, in turn, whether the output is binary 1 or 0. Then a gate and counter is used to provide the binary data. A comparator circuit is another application of the op amp.

The function of a digital-analog (D/A) converter is illustrated in Fig. 8-37. Each binary count at the input will generate a specific analog voltage

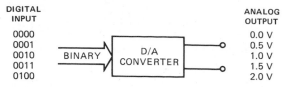

DIGITAL INPUT		ANALOG OUTPUT
0000		0.0 V
0001		0.5 V
0010	BINARY → D/A CONVERTER	1.0 V
0011		1.5 V
0100		2.0 V

Fig. 8-37. Function of a digital-to-analog (D/A) converter.

at the output. The scaling factor can be set for the system.

The circuits of a D/A converter consist mainly of two parts: a resistor network for the binary input and an op amp as a summing amplifier for the analog output. (A circuit for the op amp summing amplifier is shown in Fig. 3-19.) The resistors determine the amount of feedback for the op amp, which controls the amount of gain.

Test Point Questions 8-12
(Answers on Page 187)

a. Does a D/A or an A/D converter change analog voltage to binary data?

b. Which has binary input, the D/A or the A/D converter?

8-13
SEVEN-SEGMENT DISPLAYS

Seven-segment displays are generally used as the readout devices for digital calculators, clocks, cash registers, etc. As shown in Fig. 8-8, each digit is formed by energizing some or all of the seven segments. Most commonly used are the light-emitting diode (LED), liquid-crystal display (LCD), and fluorescent displays.

LED Display The display that emits a reddish glow is usually of the LED type. It is available in a complete package similar to that shown in Fig. 8-38a, which plugs into a standard DIP IC socket. The construction is illustrated in Fig. 8-38b. In this example a common-anode circuit is used; all the diodes are connected to a common pin to be connected to the +5-V supply. Note the letters a, b, c, d, e, f, and g clockwise from the top for each segment. A decimal point, marked D.P., is

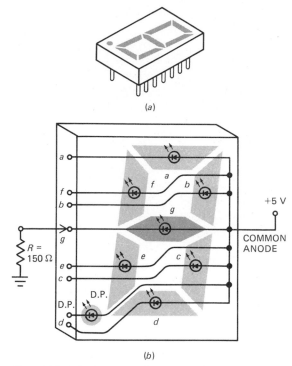

Fig. 8-38. (a) Commercial DIP unit for a LED seven-segment display. (b) Circuit to light segment g. This display has decimal point (D.P.).

also included here, but not all seven-segment displays contain a decimal point.

To light a segment, the cathode must be grounded through the 150-Ω R. The resistance limits current of the diode. Each segment operates as one diode, which requires forward voltage to emit light. Special compounds of gallium are used for the LED. Typical ratings are 1.2 V with a forward current of 20 mA. The radiation is in the wavelength for red, green, or yellow light.

In Fig. 8-38b, segment g is lit because its cathode is returned to ground through the 150-Ω R. Each segment and the decimal point could be tested in this way. To see if it lights, apply 5 V in series with 150 Ω.

The LED displays are also available in common-cathode form; then all the cathodes are connected to a common pin that returns to chassis ground. The 5 V is connected through 150 Ω to segments a to g to light each segment.

LCD Display   The liquid-crystal display operates much differently than the LED type. An LED generates light output, whereas an LCD controls light. The LCD needs light input to be seen, whereas the LED produces its own light. The main advantage of the liquid-crystal display is that it consumes an extremely small amount of power.

A diagram of a common LCD is shown in Fig. 8-39; the construction is that of a *field-effect LCD*. When energized, the LCD segment would appear black as compared with the rest of the shiny surface. Segment *e* is energized in the illustration. The rest of the segments would be nearly invisible.

The key to LCD operation is the *nematic fluid*, which is a liquid crystal. It is sandwiched between two glass plates. Voltage is applied across the nematic fluid, from the top metallized segment to the rear metallized backplane. Then the fluid transmits light differently and the energized segment of the display becomes visible.

Liquid-crystal displays are energized with ac voltage; in Fig. 8-39 the ac input is a 30-Hz square wave. Direct current will destroy this type of display.

LCDs that produce milky white characters also are available. This type is usually the *dynamic scattering LCD;* it uses a different nematic fluid. The field-effect LCD is probably more common. In some applications a lamp is provided for viewing the LCD in the dark.

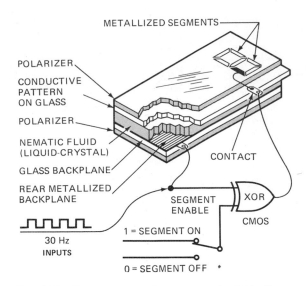

Fig. 8-39. Construction and operation of a field-effect-seven-segment LCD.

Test Point Questions 8-13
(Answers on Page 187)

a. Does the LCD or the LED emit a reddish glow?
b. Does the LCD or the LED consume the smaller amount of power?
c. Does the LCD or LED segment need 5 V with about 20 mA direct current?

8-14
THE CLOCK TIMER

Most digital systems need a clock to time the operation of the various system parts. An electronic clock is a timing device that generates a continuous series of pulses, as shown in Fig. 8-40*a*. The pulses provide a digital waveform in which HIGH and LOW values are repeated at a specific frequency.

The clock is essentially an oscillator circuit. The basic oscillator for pulse waveforms is the multivibrator (MV) circuit. It consists of two stages that alternate between conduction and cutoff. There are three types of MV circuits:

1. Monostable, or one-shot, multivibrator
2. Bistable multivibrator, or flip-flop
3. Astable, or free-running, multivibrator

Free-Running Multivibrator   An example of the astable, or free-running MV is shown in Fig. 8-40*b*; discrete transistors *Q*1 and *Q*2 are used. An astable MV has no stable state. The circuit oscillates continuously because conduction in one stage cuts off the other. A square-wave output can be taken from the collector of either *Q*1 or *Q*2, but with opposite polarities. V_C for the cutoff stage is at 5 V for the HIGH level; V_C for the stage that conducts is close to zero for the LOW level.

The pulse output may be symmetrical or not, depending on the *R* and *C* values in each stage. When one stage is cut off longer than the other, the output is asymmetrical.

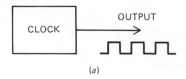

(a)

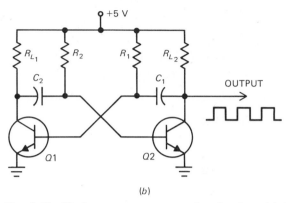

(b)

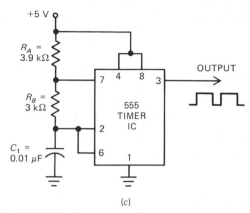

(c)

Fig. 8-40. Clock system to generate digital pulses. (a) Symbol. (b) Multivibrator using discrete components in a clock circuit. (c) Same clock circuit using a 555 IC timer.

IC Clock Timer   Figure 8-40c shows the popular 555 timer IC package connected with three external components, R_A, R_B, and C_1, to form the equivalent of an astable multivibrator for a clock circuit. The output is a square-wave pulse train. The frequency depends on the values for R_A, R_B, and C_1 according to the formula

$$f = \frac{1.44}{(R_A + 2R_B) \times C_1} \qquad (8\text{-}8)$$

Substituting the values in Fig. 8-40c gives us

$$f = \frac{1.44}{[3900 + (2 \times 3000)] \times 0.01 \times 10^{-6}}$$

The result is

$$f = 14.545 \times 10^3 \text{ Hz}$$
$$= 14.545 \text{ kHz}$$

Note that increasing R_A, R_B, or C_1 reduces the output frequency, because those values are in the denominator of the fractions for Formula 8-8.

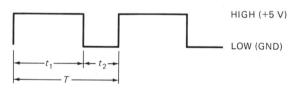

Fig. 8-41. Output waveform for the 555 IC timer.

More details of the square-wave output are shown in Fig. 8-41. The waveform is asymmetrical because t_1, when the pulse waveform is HIGH, is a longer time than t_2, when the pulse is LOW. The total time period T is equal to $t_1 + t_2$. The frequency of the clock determines T. Specifically, the frequency is equal to $1/T$.

Test Point Questions 8-14
(Answers on Page 187)

a. Can an astable or a bistable MV be used as a clock timer?

b. In Fig. 8-40c, does increasing R_B increase or decrease the clock frequency?

SUMMARY

1. In the decimal number system there are 10 digits, 0 to 9, and a base, or radix, of 10. In the binary number system there are just two digits, 0 and 1. and a base, or radix, of 2. Numbers in either base can be encoded for the other base. See Fig. 8-5.

2. Digital circuits operate with pulses that have either a HIGH voltage level for binary 1 or LOW level for binary 0 in a system of positive logic. Analog signals, such as sine waves, have continuous variations in amplitude.

3. Logic gates are the building blocks for all digital circuits. The basic gates are AND, OR, NOT, NAND, NOR, XOR, and XNOR. See Table 8-1 for the logic symbol and Boolean expression for each.

4. Electronic adders are constructed from logic gates to add binary numbers. A half-adder combines 0 and 1 with a carry of 1 to the next place if needed, but it has no terminal for carry-in. A full-adder has provision for carry-in and carry-out.

5. The flip-flop (FF) is a basic digital circuit with a memory characteristic that can be used to store information. The circuit is a bistable multivibrator that stays in one state until switched to the opposite state. The two outputs Q and not Q ($\overline{Q}$) are always opposite as HIGH and LOW logic levels.

6. The main types of flip-flop are the R-S, with set and reset input terminals, D, and J-K. Clock input pulses to time the switching make the flip-flop synchronous. The clocked J-K flip-flop is often used in the toggle mode, in which the Q and $\overline{Q}$ outputs are reversed by each clock input pulse.

7. A counter is a group of clocked flip-flops. In a serial counter, each FF output drives the next input. A parallel or synchronous counter has clock pulses into all the flip-flops.

8. Digital counter circuits are commonly used for frequency division of the clock-pulse input. A J-K flip-flop in the toggle mode automatically divides by 2 because the output pulses are at one-half the input frequency.

9. A shift register uses flip-flops and logic gates to transfer digits from one binary place to the next, either left or right.

10. The two basic types of semiconductor memory are RAM and ROM. The RAM, random-access memory, provides both the read and write functions. In the write mode, digital information is put into the memory; in the read mode, the information is taken out. The ROM, read-only memory, provides only the read function.

11. An electronic multiplexer or demultiplexer operates like a single-pole many-position rotary switch. The multiplexer selects data one at a time for serial output from the parallel lines at the input. The demultiplexer does the opposite; it changes serial data from the input to the parallel form at the output.

12. An analog-digital (A/D) converter changes analog signals to digital form. The circuit includes an op amp as a comparator to indicate when the analog signal is higher or lower than a reference voltage.

13. A digital-analog (D/A) converter changes digital signals to analog form. The circuit includes an op amp as a summing amplifier to form an analog voltage from the bits of digital information.

14. The seven-segment DIP IC is commonly used to display decimal digits. As shown in Figs. 8-8 and 8-38, each digit is formed by energizing some or all of the segments. The LED type needs forward voltage to emit light, usually red. The liquid-crystal display (LCD) controls the transmission of light through its fluid.

SELF-EXAMINATION
(Answers at back of book)

1. Is the LOW voltage on a digital-pulse signal called a logic 0 or a logic 1.
2. Does the binary number system use base 2 or base 10?
3. If any of the inputs are HIGH, will the output be HIGH for the AND or for the OR gate?
4. Must the AND gate have all or any of the inputs HIGH for a HIGH at the output?
5. Give the Boolean expression for the NOR gate.
6. Does inverting the output of the AND gate form a NAND or a NOR gate?
7. Does a half-adder include an AND gate with an OR or a XOR gate?
8. What timing pulses are needed for a synchronous device?
9. Is toggle operation used with a flip-flop or a gate circuit?
10. In the set mode, does the Q output of a flip-flop go HIGH or LOW?
11. Is the astable or bistable MV used as a clock generator?
12. How many flip-flops are needed for a 4-bit counter?
13. Is the memory device that has both read and write capabilities a RAM or a ROM?
14. Which is used as a data selector, the multiplexer or the demultiplexer?
15. Is the op amp used in A/D and D/A converters or in shift registers?
16. Which needs dc forward voltage to emit light, the LED or the LCD?
17. When all seven-segments of a display are energized, is the number shown 7 or 8?
18. Which uses field-effect transistors, the T^2L or the CMOS family of logic circuits?

ESSAY QUESTIONS

1. Explain the difference between analog and digital signals.
2. How would you say the number 1010_2?
3. In the number 1010_2 what is the count for the 8s place?
4. List the binary counting sequence that is equal to decimal counting from 0 to 20.
5. How can you tell from the schematic symbol in Fig. 8-7 that this encoder has active LOW inputs?
6. Give the logic symbol, Boolean expression, and truth table for **a.** an AND gate; **b.** an OR gate.

7. Give one difference between a gate and a flip-flop circuit.
8. What is meant by an inverter for digital circuits?
9. What does the overbar, as in $\overline{A}$, mean in a Boolean expression?
10. Give three ways to describe a logic problem on paper.
11. Give the number of half-adders and full-adders needed for an 8-bit parallel adder circuit.
12. Name three types of flip-flop circuits.
13. Why is a flip-flop also called a *latch circuit*?
14. Give three types of multivibrator circuits.
15. How are op amps used in D/A and A/D converters?
16. Give the functions of a multiplexer and a demultiplexer.
17. Compare the operations of LCD and LED seven-segment displays.
18. Explain what happens to the outputs of a flip-flop when in the toggle mode.
19. Name three types of diode and transistor circuits used for constructing IC logic packages.
20. What two important characteristics does a shift register exhibit?
21. What do the letters ROM mean?
22. Describe how each segment is made to light in an LED display.
23. Give the place values for the first five places in binary and digital numbers.
24. Define modulo number, ripple counter, decade counter, down counter, 4-bit word.
25. What is meant by the carry-in and carry-out for adders?

PROBLEMS
(Answers to odd-numbered problems at back of book)

1. Give the decimal equivalent of 111011_2.
2. Give the binary equivalent of decimal 80. Check your answer.
3. Give the binary equivalent of decimal 207.
4. Refer to Fig. 8-8. If the input is 0001_2, which segments on the display will have to be activated to show the decimal number?
5. Write the Boolean expression for the truth table in Fig. 8-42.

INPUTS			OUTPUT
C	B	A	Y
0	0	0	0
0	0	1	0
0	1	0	0
0	1	1	0
1	0	0	0
1	0	1	1
1	1	0	1
1	1	1	0

Fig. 8-42. Truth table for Problem 5.

6. Use the Boolean expression of Question 5 to draw a logic symbol diagram with AND, OR, and NOT gates.

7. A 3-bit ripple counter similar to the counter shown in Fig. 8-30 would have **a.** how many flip-flops, **b.** what modulus number, **c.** what binary count as maximum?
8. What is the binary sum of 101010_2 and 11011_2? Check your answer with decimal numbers.

SPECIAL QUESTIONS

1. Name five devices you have seen that use the seven-segment display for readout.
2. Name three types of digital and two types of analog test equipment.
3. What is meant by a truth table?
4. Compare the advantages and disadvantages of digital and analog signals.
5. Give the part number for three digital ICs.
6. If a number contains any of the digits 2, 3, 4, 5, 6, 7, 8, or 9, why is it not possible for it to be a binary number?
7. How could the fraction ½ be encoded in binary form?

ANSWERS TO TEST POINT QUESTIONS

8-1   **a**   Two
      b. Analog
      c. 1

8-2   **a.** 0 and 1
      b. Encoder
      c. 10011
      d. 42

8-3   **a.** AND gate
      b. OR
      c. NAND

8-4   **a.** OR
      b. HIGH

8-5   **a.** TTL or T^2L
      b. CMOS
      c. Q3 and Q4

8-6   **a.** 1010101_2
      b. 85
      c. Full-adder
      d. Three

8-7   **a.** Q and $\overline{Q}$
      b. HIGH
      c. LOW

8-8   **a.** binary 11, or decimal 3
      b. binary 1111, or decimal 15
      c. LOW
      d. Up

8-9   **a.** Shift-right
      b. Serial
      c. FFA

8-10   **a.** RAM
      b. Read

8-11   **a.** Parallel
      b. Parallel

8-12   **a** A/D
      b. D/A

8-13   **a.** LED
      b. LCD
      c. LED

8-14   **a.** Astable
      b. Decrease

Chapter 9
Digital and Linear IC Packages

An integrated circuit combines transistors, diodes, resistors, and small capacitors on a single IC chip, usually silicon. The advantage is a tremendous saving in space as compared with *discrete components,* which are separate units.

A digital IC package contains digital circuits such as the logic gates, flip-flops, and counters described in Chap. 8. Linear IC packages include everything else for applications in analog circuits. A few examples are audio amplifiers, IF amplifiers, AM and FM radios, and the chroma circuits of color TV receivers.

The IC units are fast replacing discrete components in all electronic equipment except for high-power applications. Digital IC packages are usually rated for 1 W or less. Linear IC units have power ratings of up to 10 W. The IC packages will probably continue to grow in use because of their small size and weight, low cost, low power consumption, high reliability, and improved performance. More information is given in the following sections.

9-1   IC Packages and Pin Connections
9-2   Inside an IC Package
9-3   Production of IC Chips
9-4   Types of Linear IC Circuits
9-5   Digital IC Families
9-6   Methods of Testing

9-1
IC PACKAGES AND PIN CONNECTIONS

Of the common types of IC packages and pin connections, the dual in-line package (DIP) shown in Fig. 9-1a and b is most popular. The 8-pin unit of Fig. 9-1a is sometimes called the *mini DIP*. Its length is ⅜ in. In Fig. 9-1b the DIP has 14 pins in a TO-116 package about ¾ in long. DIP packages are manufactured with either plastic or ceramic cases.

Note the position of pin 1 next to a notch or mark on the DIP. The pins are numbered counterclockwise from the notch when viewed from the top.

Some IC packages also come in the transistor-style metal can shown in Fig. 9-1c. The unit is an eight-lead TO-5 metal case. The flatpack IC package in Fig. 9-1d is an older style.

Test Point Questions 9-1
(Answers on Page 201)

a. How many pins are on the mini DIP?
b. Is the pin next to the notch on a DIP 1 or 5?

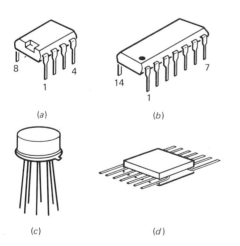

(a) (b)

(c) (d)

Fig. 9-1. IC packages and pin connections, top view. (*a*) 8-pin dual in-line package (DIP). Length is ⅜ in. (*b*) 14-pin DIP, ¾ in long. (*c*) Transistor style TO-5 metal can. Diameter is 0.35 in. (*d*) 14-pin flatpack about ¼ in square.

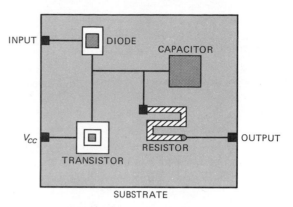

Fig. 9-2. Integration of transistors, diodes, *R* and *C* on silicon chip. The black lines and areas show conductor paths.

9-2
INSIDE AN IC PACKAGE

The circuits to be integrated are similar to those in which discrete components are used. However, there are some differences. The IC usually contains only transistors, diodes, and resistors. It is impractical to form inductors in an IC. Also, only very small capacitors, in the picofarad range, can be included. When inductors and larger values of C are needed, they are usually part of the circuitry external to the IC. In a schematic diagram of the circuit inside an IC, the transistors are not shown with circles because they are formed as part of the IC and not discrete components.

The method of integrating transistors, diodes, resistors, and capacitors on a silicon ship is illustrated in Fig. 9-2. In addition, a transistor can be used as a diode by connecting the collector to base. Typical C values are 3 to 30 pF. Typical R values are 100 Ω to 25 kΩ.

Basic IC construction of the 8-pin DIP is shown in Fig. 9-3 at the top. The actual integrated circuit is the small silicon chip sliced from the circular wafer shown in the lower part of the figure. The chip may be smaller than 0.1 in square, but it contains all the integrated circuits. Note the connecting wires from the chip to each of the eight pins on the IC package. The body of the DIP is made from either plastic or a ceramic material.

The tiny silicon IC chips are manufactured in a group as part of a silicon wafer, as shown in Fig. 9-3 at the bottom. The wafer has a diameter of 1.5 to 2 in, and it is 0.01 in thick. One wafer contains hundreds of individual chips. Impurities are introduced into the pure materials changing their properties. This process is called *doping*. After a complex process of producing the pattern of the integrated circuits with doped semiconductors, the wafer is tested and then diced into hundreds of chips. Each silicon chip is a complete integrated circuit. It is mounted in an IC package with leads from the pins connected to the IC chip. The actual chip is only a small part of the larger and heavier IC package.

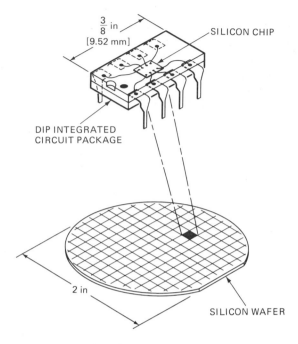

Fig. 9-3. Assembly of a silicon chip into an 8-pin DIP.

Test Point Questions 9-2
(Answers on Page 201)

a. Are transistors, resistors, or inductors most common inside an IC unit?
b. The circuit in Fig. 9-6b is on the single chip shown in Fig. 9-3. True or false?

9-3
PRODUCTION OF IC CHIPS

Bipolar integrated circuits are made with NPN or PNP transistors, usually the silicon NPN type. The field-effect transistor is formed in MOS integrated circuits. The complementary type with P and N channels is CMOS, or COS/MOS.

Bipolar IC Chips The method of fabrication of silicon NPN transistors is shown in Fig. 9-4. The symbol is shown in Fig. 9-4a, and Fig. 9-4b illustrates the doped layers and their junctions. This cross section of the chip shows that the silicon substrate is lightly doped with P-type impurity. The P-substrate is actually one chip of the silicon wafer shown in Fig. 9-3b.

With the doped substrate as a starting platform, the large N-type islands in Fig. 9-4b are formed by doping. The darker areas indicate more doping,

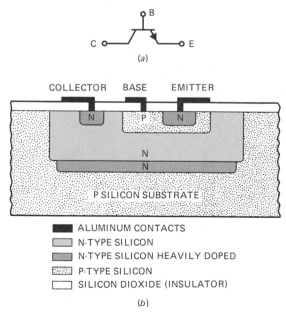

Fig. 9-4. Forming an NPN transistor for integrated circuits. (a) Schematic symbol. (b) Fabrication on silicon wafer with PN junctions by using bipolar technology.

often labeled N$^+$. These islands have the purpose of isolating the transistor from other transistors on the same substrate.

Next the smaller P-type island is formed for the base electrode of the transistor, and then N-type doping forms the emitter and collector. In the production process, the doping is limited to specific spots by the equivalent of windows for the vaporized impurity elements.

Then a layer of silicon dioxide is formed as an insulator over the entire unit. However, aluminum contacts go through for connections to the emitter, base, and collector. Those contacts would also connect other components on the chip much as the copper foil on a printed-circuit board does. Aluminum is used for the contacts because it makes an ohmic contact, without any junction voltage.

The cross section of a silicon chip shown in Fig. 9-4b is greatly magnified. Hundreds of bipolar transistors can be formed in this way on a tiny chip. The bipolar construction is used for both linear and digital IC chips.

MOS Integrated Circuits The MOS integrated circuit method is illustrated in Fig. 9-5. The schematic symbol in Fig. 9-5a is for the MOSFET or IGFET with an N-channel, as indicated by the arrow in from the source. Also, the broken channel line shows the enhancement type. With positive drain voltage for the N-channel, this FET will pass current from source to drain only when positive gate voltage is applied.

The cross section of the silicon chip is shown in Fig. 9-5b. First the substrate is lightly doped with P-type impurity. The substrate serves as a platform on which the electrodes are made. Next, two small N-type islands are formed with a narrow channel between them, also N-type. Then a silicon dioxide layer is formed for insulation. The aluminum connections extend through the oxide layer to contact the N-areas for source and drain. Note that the source electrode is connected internally to the substrate, as indicated by the black dot at the left.

However, the gate contact remains insulated from the channel. The result is the insulated-gate field-effect transistor, or IGFET. When positive

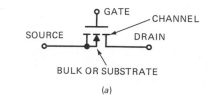

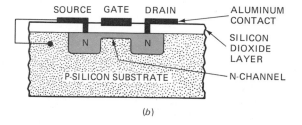

Fig. 9-5. Forming an insulated-gate FET for integrated circuits. (a) Schematic symbol for N-channel, enhancement type. (b) Fabrication on silicon wafer by using MOS technology.

voltage is applied to the gate, electrostatic induction produces more negative charges in the channel and allows electron flow from source to drain.

For the CMOS integrated circuits, the transistors are formed with complementary P and N channels. These transistors are connected together on the chip to meet opposite requirements for gate voltage V_G. An N-channel can be turned on by positive V_G, whereas a P-channel needs negative V_G to conduct. The CMOS type of integrated circuit has the advantage of extremely low power consumption.

IC Terminology Abbreviations are used to describe the number of individual circuits on one digital IC unit chip. They are:

SSI Small-scale integration; 10 gates or fewer
MSI Medium-scale integration; 11 to 99 gates
LSI Large-scale integration; 100 to 999 gates
VLSI Very large-scale integration; 1000 gates or more

To form chips with LSI and VLSI, MOS technology is used because such transistors take less space. Typically, 300 MOS transistors require the same area on a chip as six bipolar transistors. That area is equivalent to the space required for one discrete bipolar transistor.

The following terms specify the methods used to produce linear IC chips by bipolar technology:

Monolithic IC. All the components are formed on the silicon substrate. This is the construction method illustrated in Figs. 9-2 to 9-5.

Thin-film IC. The substrate is ceramic or glass. All the semiconductor components are deposited on this insulating platform.

Thick-film IC. Resistors and capacitors are formed on the substrate, but the transistors are added as discrete units.

Hybrid IC. Monolithic and thin-film construction are combined. Also, discrete transistors may be added when high power is needed.

Test Point Questions 9-3
(Answers on Page 201)

a. Which transistor takes less space on an IC chip, the NPN or the FET?
b. Which IC consumes less power, the CMOS or the bipolar type?
c. Is the IC shown in Fig. 9-4b an example of monolithic or hybrid construction?

9-4
TYPES OF LINEAR IC CIRCUITS

Linear IC units are used in analog circuits. Typical applications include the operational amplifier (op amp), voltage regulators, voltage comparators, A/D and D/A converters, and other circuits for audio, radio, and television equipment. The manufacturers provide IC handbooks, such as tube or transistor manuals that give diagrams, pin numbers, operating data, and typical uses. Special application notes also are usually available.

Voltage regulators maintain a constant output voltage with varying input voltage or changes in load current. The IC regulators are available for rated outputs of 5 V for digital circuits and 6, 8, 12, 15, 18, and 24 V. Output load current is generally rated at more than 1 A. The regulators are specified for either positive or negative dc voltage. Other circuits include dual audio amplifiers for stereo. Power ratings are 2 to 6 W with dc supply voltage of 20 to 40 V.

The IC timer is used in analog circuits for square-wave pulses and in digital circuits as a clock-pulse generator. A typical power rating is 600 mW, with 5 to 20 V for the dc supply.

Op Amp The op amp circuit is basic to most linear IC applications. (The op amp circuit with differential amplifiers is explained in Chap. 3.) Not only is the op amp a part of almost all linear IC functions, but it is available in its own IC package for many uses. Figure 9-6a shows the popular type number 741 op amp in an 8-pin DIP. The circuit is shown in Fig. 9-6b, and the maximum ratings are given in Fig. 9-6c.

The op amp circuit of Fig. 9-6b consists of differential amplifiers and driver stages for the emitter-follower output. Note the inverting and noninverting inputs at 2 and 3 for the amplified output at pin 6. A dual power supply is necessary with V^+ at pin 7 for the collector voltage and V^- at 4 for the emitter voltage. The two offset-null terminals can be used to balance the differential amplifier.

Maximum dc supply voltage is ± 18 V. Power dissipation is 500 mW, maximum. That power rating corresponds to approximately 14 mA for maximum I from the dc power supply with a potential difference of 36 V.

Transistor Array The transistor array contains several transistors in a single IC chip, as shown in Fig. 9-7. They take less space than discrete transistors. Also important is the fact that the transistors are closely matched in electrical and thermal characteristics. The matching is an advantage of diode arrays or resistor arrays, which are available in an IC package. The transistor array is used for interfacing between digital IC units and output lamps or other indicating devices.

The type number 3086 IC shown in Fig. 9-7 contains five silicon NPN transistors on a monolithic substrate. It is a 14-pin DIP. Transistors $Q5$, $Q4$, and $Q3$ are separate and have their own pin connections. $Q2$ and $Q1$ form a differential pair.

The transistors may be used separately like discrete units or as matched groups. Each transistor

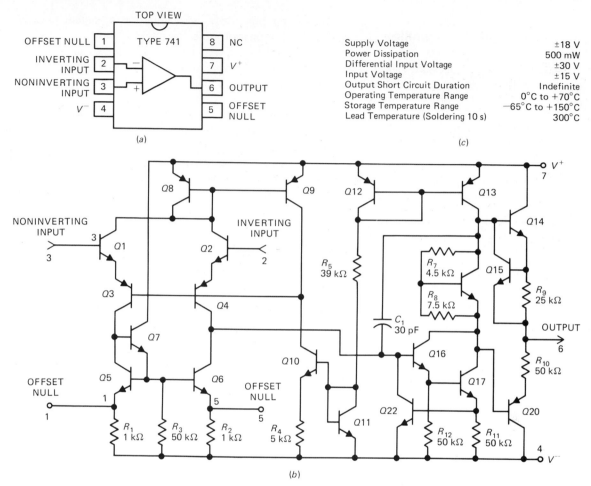

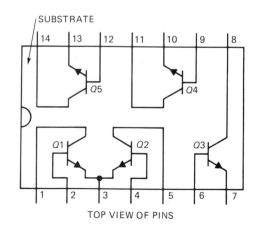

Fig. 9-6. The bipolar IC type 741 op amp. (a) 8-pin DIP. (b) Schematic diagram. (c) Maximum ratings.

has a beta (β) of 100, which is the current gain in a common-emitter circuit. Maximum dc supply voltage to the collector is 15 V, and 300 mW is the maximum power dissipation of each unit.

Test Point Questions 9-4
(Answers on Page 201)

Answer True or False.

a. The IC 741 op amp shown in Fig. 9-6 needs positive and negative dc supply voltage.

b. The IC 3086 transistor array shown in Fig. 9-7 has five differential pairs.

Fig. 9-7. Schematic diagram of IC type 3086 transistor array. (National Semiconductor Corporation)

Chapter 9/Digital and Linear IC Packages   **193**

9-5
Digital IC Families

Actually most IC packages are digital. There are so many of them that they are classified in families on the basis of the type of logic, such as TTL and ECL, or the bipolar and MOS methods of fabrication. Digital IC packages in the same family are compatible. That feature means they use the same V^+ and can be connected together easily. Each family may have its own data handbook or manual that gives diagrams, ratings, truth tables, and applications.

A popular group is the 7400 series of bipolar TTL packages. There are over 200 different types for practically every function in digital circuits. Information on a few types is shown in Figs. 9-8 and 9-9. Typical values for the IC units in this series are V_{CC} of 5 V, a logic HIGH of 3.5 V or more, and a logic LOW of 0.1 V or less.

Fan-out   The fan-out specifies the maximum number of additional gates that one gate can drive. Too many circuits cannot be added because of the increased load current. For the 7400 series of TTL circuits, the fan-out rating is 10.

Type 7400 Quad NAND Gate   See Fig. 9-8a. Inside the 14-pin DIP are four two-input NAND gates. The V_{CC} pin 14 is connected to +5 V, and the

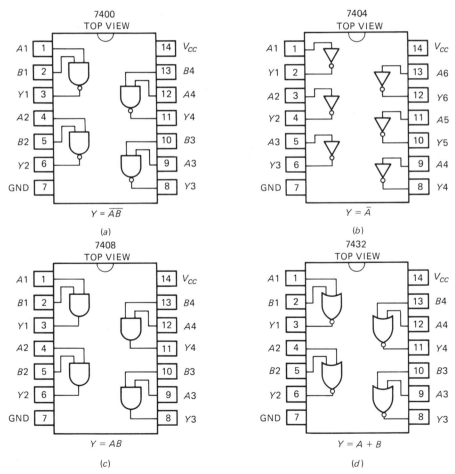

Fig. 9-8. Examples of 7400 series TTL digital IC packages. (a) Quad, two-input NAND gate. (b) Hex inverter containing six inverters. (c) Quad two-input AND gate. (d) Quad two-input NOR gate.

negative return connects to the ground pin 7. Those pins supply dc voltage for all the gates. However, there are separate pin connections for the input and output pulses. Each gate has the logic function $Y = \overline{A\,B}$.

Type 7404 Hex Inverter See Fig. 9-8b, in which the *hex* means "six." Each of the six inverters has the logic function $Y = \overline{A}$. However, all six operate from the same dc supply.

Type 7408 Quad AND Gate See Fig. 9-8c. Four separate two-input AND gates are powered from the same V_{CC} supply. Each AND gate has the logic function $Y = AB$.

Type 7432 Quad OR Gate See Fig. 9-8d. Four separate two-input OR gates are powered from the

same V_{CC} supply. Each OR gate has the logic function $Y = A + B$.

Type 7476 Dual J-K Flip-Flop See the wiring diagram in Fig. 9-9a, the truth table in Fig. 9-9b, and the maximum ratings in Fig. 9-9c. The unit has two separate J-K flip-flops FF1 and FF2 in a 16-pin DIP. The common V_{CC} connection is at pin 5, and the ground is at pin 13. Each FF has its own output and input terminals. Preset is the same as set, and clear is the same as reset. The bubbles for the clear and preset inputs indicate an active LOW.

The truth table (Fig. 9-9b) shows the input and output states. At the top, the three rows are for operation without clock pulses. Then J and K are not used. This operation is asynchronous. The bottom four rows are for synchronous operation with clock pulses. Data on inputs J and K will be

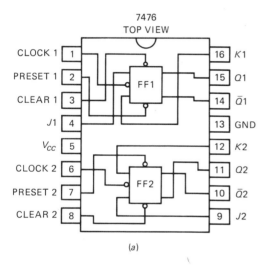

(a)

Supply Voltage V_{CC}	4.75 to 5.25 V
Logic Input 1 Voltage	2.0 V minimum
Logic Input 0 Voltage	0.8 V maximum
Logic Output 1 Voltage	2.4 V minimum
Logic Output 0 Voltage	0.4 V maximum
Maximum Clock Frequency	15 to 30 MHz

(c)

OPERATING MODE	INPUTS					OUTPUTS	
	PRESET	CLEAR	CLOCK	J	K	Q	$\bar{Q}$
Asynchronous Set	L	H	X	X	X	H	L
Asynchronous Reset (Clear)	H	L	X	X	X	L	H
Prohibited	L	L	X	X	X	H	H
Toggle	H	H	⊓	h	h	$\bar{q}$	q
Load 0 (Reset)	H	H	⊓	l	h	L	H
Load 1 (Set)	H	H	⊓	h	l	H	L
Hold (Disabled)	H	H	⊓	l	l	q	$\bar{q}$

H = HIGH voltage level steady state.
L = LOW voltage level steady state.
h = HIGH voltage level one setup time prior to the HIGH-to-LOW clock transition
l = LOW voltage level one setup time prior to the HIGH-to-LOW clock transition.
X = Don't care.
q = Lower case letters indicate the state of the referenced output prior to the HIGH-to-LOW clock transition.
⊓ Positive clock pulse.

(b)

Fig. 9-9. IC type 7476 J-K flip-flop. (a) Connections on 16-pin DIP. (b) Truth Table. (c) Absolute maximum ratings.

transferred to the output on the HIGH-to-LOW transition of the clock pulse. This type of trigger is indicated by the bubble on the FF symbol in Fig. 9-9a.

In the maximum ratings in Fig. 9-9c, note that as little as 2 V at an input is considered a logic 1, or HIGH. However, 3.5 V is typical. Also, a LOW can be as much as 0.8 V, but it is typically 0.1 V. These HIGH and LOW values are the same for all circuits in the TTL logic family.

MOS Families All the functions described for bipolar logic families can also be carried out with metal-oxide semiconductor (MOS) devices. They use field-effect transistors. The MOS families have many applications in which large-scale integration (LSI) is needed for over 100 gates in one package. The advantages are low power consumption and dense packaging.

The MOS families can be considered in the three groups.

PMOS	P-channel
NMOS	N-channel
CMOS, or	Complementary P and
COS/MOS	N channels

The CMOS, or COS/MOS, type is most common. Typical drain supply voltage V_{DD} is 12 V.

Since the IGFET is an electrostatic device, the MOS packages must be handled with care to prevent buildup of static charge. Typical input resistance of the insulated gate is 10^{12} Ω. Breakdown voltage of the silicon dioxide insulator is about 100 V. MOS devices usually have internal protective diodes to prevent arcing, but excessive gate voltage can short-circuit the gate electrode to the channel.

The Microprocessor The microprocessor application is a growing use of large-scale integration with MOS technology. A microprocessor (μP) is an IC that usually performs all the functions of the central processing unit (CPU) of a computer. Applications of microprocessors include personal computers for home use, traffic-control systems, numerically controlled machine tools, and automatic diagnostic testing of automobile engines.

Test Point Questions 9-5
(Answers on Page 201)

a. In the TTL family is a logic 1 usually 1 V or 3.5 V?
b. Can the V_{DD} for a CMOS digital package be 12 or 50 V?

9-6
METHODS OF TESTING

Testing procedures for circuits that contain IC units will be discussed for the timer-counter shown in Fig. 9-10a. The waveform of pulses is shown in Fig. 9-10b. This circuit has digital information in the 7476 dual J-K flip-flop. The 555 timer is a linear IC unit, but it produces clock pulses for the 7476 IC circuit. At the input side of the timer, R_A, R_B, and C_1 are discrete components. They can be tested with an ohmmeter. The pulse waveforms can be observed with an oscilloscope. (Oscilloscopes are described in Chap. 18.) Also, the HIGH and LOW levels in the flip-flops can be checked with the logic probe shown in Fig. 9-11.

Circuit Analysis The 555 timer operates as an astable multivibrator to produce a train of square-wave pulses at point A in Fig. 9-10a. This waveform is the clock input to FF1. The 7476 is wired as a 2-bit counter. Clock input to FF2 is supplied by FF1. The J and K terminals for both flip-flops are held HIGH at 5 V so that the clock input can toggle the Q outputs. The set and reset terminals are not used here. Also, the $\overline{Q}$ terminals are not used.

Point B after FF1 is the divide-by-2 output, or 1s place for the count. This point is Q1 of FF1. Point C at FF2 is the divide-by-4 output, or 2s place. Note that the Q1 output of FF1 is also the clock input for FF2.

Consider the waveforms shown in Fig. 9-10b. At the start, FF1 and FF2 are cleared, or reset, by a clock pulse from the timer. Then Q1 and Q2 for both flip-flops are LOW at logic 0, so that the binary count starts at 00. On the HIGH-to-LOW transision of clock pulse one, FF1 toggles from LOW to HIGH.

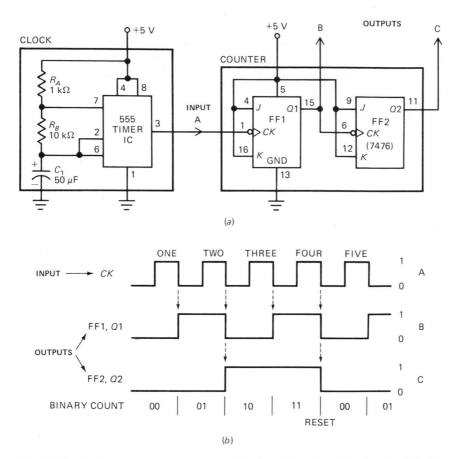

Fig. 9-10. Clock-counter system using a 555 timer IC and a 7476 flip-flop IC. (*a*) Wiring diagram. (*b*) Waveform diagram.

Then $Q1$ goes HIGH for logic 1. The count is now binary 01, as shown below the clock waveform in Fig. 9-10*b*. FF2 does not toggle because its clock input needs a change to LOW, not to HIGH.

Both flip-flops are stable with $Q1$ HIGH and $Q2$ LOW until clock pulse two has a falling edge to toggle $Q1$ to LOW. Now this output toggles FF2 to make $Q2$ HIGH. The resulting binary count is 10 for the $Q2$ and $Q1$ outputs. Remember that the 1s place is to the right, next to an assumed binary point.

The HIGH-to-LOW transition on clock pulse three makes $Q1$ toggle to HIGH. However, $Q2$ remains at its HIGH stage. Now the binary count is 11.

On the falling edge of clock pulse four, $Q1$ is toggled to LOW. That change also toggles $Q2$ to

LOW. Then the binary count changes to 00, as at the start. Clock pulse five starts the counting sequence again.

Oscilloscope Waveforms It should be common practice to check the operation of a digital circuit like that of Fig. 9-10 with an oscilloscope to display the pulses. Usually a dual-trace oscilloscope would be used to show two waveforms at the same time.

Points A and B can be compared. At B, the waveform for $Q1$ should show one square-wave cycle for every two cycles at A. Also, the edges of the pulses at B must coincide with the falling edges of the pulses at A. Another comparison is between the waveforms at B and C. The square wave at C has one-half the frequency at B.

Chapter 9/Digital and Linear IC Packages   **197**

Logic Probe As shown in Fig. 9-11, the logic probe can be used to check the operation of a digital circuit. The two alligator clips are connected to the power supply. Then the needlelike tip of the logic probe is touched to any part of the circuit under test. The LED indicators light to give the logic state. This probe can be used to check logic levels in DTL, TTL, or CMOS circuits. The switch selects the logic-family threshold levels. Frequency response is up to 10 MHz. The high input impedance is 300 kΩ.

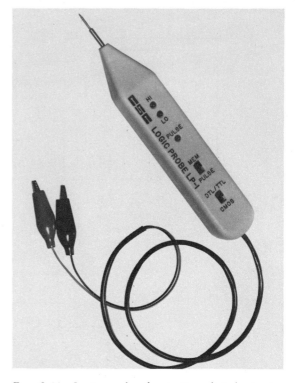

Fig. 9-11. Logic probe for testing digital circuits. (*Global Specialties Corporation*)

Test Point Questions 9-6
(Answers on Page 201)

Answer True or False.

a. A dual-trace oscilloscope cannot show two waveforms at the same time.

b. The logic probe shown in Fig. 9-11 must be connected to the power supply of the circuit under test.

c. In the system of Fig. 9-10, the clock input terminal of FF2 is toggled by the Q1 output of FF1.

SUMMARY

1. The two main classes of integrated circuits are linear ICs and digital ICs.

2. The most common IC outlines are the 8-, 14-, and 16-pin dual-in-line packages (DIP), square flatpack, and TO-5 transistor outline with a metal case. The pins of the DIP are numbered counterclockwise from the notch or indent when viewed from the top.

3. The operational amplifier (op amp) is the basic circuit for linear IC packages, including voltage regulators, A/D and D/A converters, voltage comparators, and audio, radio, and TV circuits. Transistor arrays and diode arrays also are available.

4. In bipolar IC construction, the silicon chip is used to form many NPN transistors. Resistors and small capacitors, but not inductors, can also be integrated.

5. In MOS IC construction, the silicon chip is used to form many field-effect transistors. CMOS, or COS/MOS, has complementary N and P channels.

6. Digital IC packages are classified in families according to the type of logic, such as TTL or ECL with bipolar transistors or MOS with field-effect transistors. The popular TTL family has 5 V for the dc supply, 3.5 V as a logic HIGH, and 0.1 V for a LOW.

7. Fan-out refers to the number of additional gates that a gate can drive.
8. Digital circuits can be tested with an oscilloscope to display the pulse waveforms or a logic probe to check HIGH and LOW levels.

SELF-EXAMINATION
(Answers at back of book)

1. Is the operational amplifier (op amp) used in linear or digital IC packages?
2. Does a transistor or an inductor take less space on a silicon IC chip?
3. Is the 14-pin DIP more or less common than the flatpack IC package?
4. In MOS fabrication, is the insulating layer of the gate electrode silicon dioxide or plastic?
5. When both P and N channels occur in MOS technology, is the chip PMOS, NMOS, or CMOS?
6. Is the most popular bipolar family of digital IC chips TTL, DTL, or CMOS?
7. Does the 555 timer produce a continuous string of pulses operating in an astable or bistable mode?
8. Is the 7400 series of digital IC packages part of the DTL, T^{2}L, or CMOS family?
9. Is the maximum fan-out of the 7400 series of digital IC gates 10 or 100?
10. In Fig. 9-7, does the IC array consist of NPN or PNP transistors?
11. Are the IC transistors formed with MOS technology similar to NPN and PNP transistors or the IGFET?
12. In the 7400 series of bipolar digital IC packages, is 3.5 V a HIGH or LOW?
13. Would a digital IC with two flip-flops be classified as SSI or LSI?
14. In the 7476 dual J-K flip-flop, is each triggered by the transition to HIGH or to LOW on the clock pulse?
15. Does ECL stand for emitter-coupled logic with bipolar transistors or MOS-FETS?
16. Can pulse waveforms be compared with the oscilloscope or the logic probe?
17. Are the connecting leads in an IC chip aluminum or silicon?
18. Which use the least power, TTL, ECL, or CMOS chips?
19. Which is an electrostatic device, the MOSFET or NPN and PNP bipolar transistors?
20. In the Fig. 9-10 system, is the clock-pulse input frequency divided by 4 at terminal A, B, or C?

ESSAY QUESTIONS

1. Give at least five types of circuits in linear IC packages.
2. Give at least eight types of digital circuits for IC packages.
3. Give at least three families of digital IC packages.
4. Name four types of components generally formed in IC chips.
5. What is meant by discrete components used with an IC chip?

6. Draw the top view of a 14-pin DIP. Show the notch and pins 1 to 14.
7. Give the meanings of the abbreviations SSI, MSI, LSI, and VLSI.
8. Give the meanings of the abbreviations DTL, TTL, T^2L, and ECL.
9. Give the meanings of CMOS, COS/MOS, PMOS, MOSFET, and IGFET.
10. What is the typical supply voltage with logic HIGH and LOW values for the TTL family of bipolar digital IC packages.
11. List at least five items of information that can be found on an IC data sheet.
12. Give descriptions for the following digital IC packages: 7400, 7404, 7408, 7432, 7474, 7476. Use a manufacturer's manual if necessary.
13. Explain how a logic probe would be connected to the circuit in Fig. 9-10 to check the output of the divide-by-4 counter.
14. Name all seven terminals on each J-K flip-flop in Fig. 9-9.
15. List the components of the IC transistor array in Fig. 9-7. Give two advantages of IC transistors over discrete transistors.
16. What are D/A and A/D converters?
17. **a.** Give at least three features of the operational amplifier (op amp) circuit.
    b. Give at least five applications of the op amp circuit.

PROBLEMS
(Answers to odd-numbered problems at back of book)

1. Figure 9-12 shows the bottom view of an 8-pin mini DIP. Give the numbers for the pins labeled *a*, *b*, *c*, and *d*.
2. Refer to the circuit in Fig. 9-13 with a 555 timer and two LED units. Will the diodes light when the timer output at pin 3 is at logic 1 or at logic 0?
3. Calculate the internal resistance of each LED in problem 2 with pin 3 of the timer at 3.3 V.
4. Refer to the dual *J-K* flip-flops shown in Fig. 9-9. When an input is at 2.6 V, is that a logic 1 or 0?

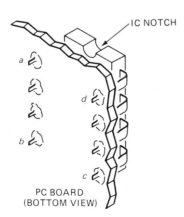

Fig. 9-12. Bottom view of DIP for Problem 1.

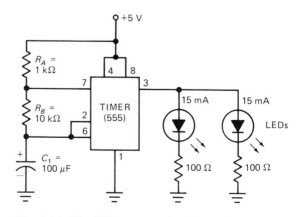

Fig. 9-13. The 555 timer wired in the astable mode for Problem 3.

5. Refer to the 555 timer and 2-bit counter in Fig. 9-11. After clock pulse 6 on the waveform diagram, give **a.** the binary count for B and C, **b.** the logic state of Q1 on FF1, **c.** the logic state of Q2 on FF2.

SPECIAL QUESTIONS

1. Name at least three companies that manufacture linear and digital IC packages.
2. List at least three handbooks or data manuals for IC packages.
3. Which do you think have more applications, digital or linear IC packages?

ANSWERS TO TEST POINT QUESTIONS

9-1	**a.**	8	**9-4**	**a.**	T
	b.	1		**b.**	F
9-2	**a.**	Transistors	**9-5**	**a.**	3.5 V
	b.	True		**b.**	5 V
9-3	**a.**	FET	**9-6**	**a.**	F
	b.	CMOS		**b.**	T
	c.	LSI		**c.**	T
	d.	Monolithic			

Chapter 10
Tuned RF Amplifiers

At radio frequencies above 100 kHz, approximately, an amplifier stage is generally tuned to a specific frequency by means of a resonant circuit. The tuned circuit can supply input signal to the amplifier, or it can serve as the output load impedance, or it can be used for both functions. The purpose is to provide maximum gain only for the frequencies at and near resonance.

The reason why a tuned RF amplifier is so useful is that its gain depends on the resonant response of a tuned circuit. Generally, inductance L and capacitance C are used for the resonance. Either L or C can be varied to tune the LC circuit for maximum amplifier gain at the desired frequency. The tuning action is the means by which a radio or television receiver selects the desired station from all the others broadcasting at different carrier frequencies. The types of tuned RF amplifiers and additional details of the resonant response are described in the following topics:

10-1   Functions of the RF Amplifier
10-2   Typical RF Amplifier Stage
10-3   Single-Tuned Circuit Response
10-4   Shunt Damping Resistor
10-5   Single-Tuned Transformer Coupling
10-6   Double-Tuned Transformers
10-7   Stagger Tuning
10-8   Wave Traps
10-9   Neutralization of RF Amplifiers
10-10   Mechanical and Crystal Filters
10-11   Random Noise

10-1
FUNCTIONS OF THE RF AMPLIFIER

The idea of an RF stage that is tuned to amplify only the desired signal is illustrated in Fig. 10-1. Coming into the amplifier are three bands of signals each 10 kHz wide. A band is used here, and not just a single frequency, because the signal is modulated. These examples are for an AM signal with a bandwidth of ±5 kHz produced by 5 kHz modulation on the RF carrier wave.

One band of frequencies is centered around 900 kHz; the next is 995 to 1005 kHz for a 1000-kHz signal; and the third band is centered around 1100 kHz. The amplitude indicated is 1 mV for all three RF signals into the amplifier.

The amplifier is tuned by the LC circuit resonant at 1000 kHz. Therefore, the stage amplifies the 1000-kHz signal the best. Assume a voltage gain of 100. The amplifier output then is 100×1 mV $= 100$ mV, but only for the 1000-kHz signal. There is little gain for the other signals because they are too far off the resonant frequency of the tuned circuit.

The gain of a tuned RF amplifier results from the high impedance of the parallel-resonant LC circuit. At the resonant frequency of 1000 kHz, the output load impedance for the amplifier is maximum. For the RF signal at 900 kHz, the impedance of the tuned circuit is very low because the frequencies are far below resonance. Also, the impedance is low for 1100 kHz, which is far above resonance.

Examples of RF coils and capacitors for the tuned LC circuit are shown in Fig. 10-2. The L of the coil (Fig. 10-2a) can be adjusted by turning the screw to move the ferrite core. The mica trimmer capacitors across each coil (Fig. 10-2b) can be varied to adjust C. In both cases, the LC circuit can be tuned to resonance at the desired frequency.

RF Response Curve   In Fig. 10-3 amplifier gain, on the vertical axis, is plotted against frequencies, on the horizontal axis. The resultant curve around f_r in the middle is the response curve of the tuned RF amplifier. Essentially, it is the same as the res-

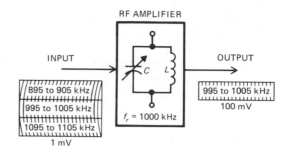

Fig. 10-1. Function of a tuned RF amplifier stage. In this example, the amplifier is tuned to 1000 kHz for a frequency range of 995 to 1005 kHz.

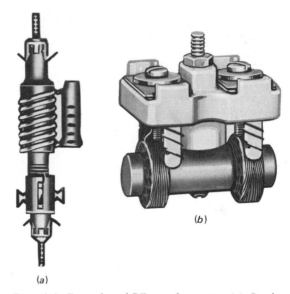

Fig. 10-2. Examples of RF tuned circuits. (a) Single-tuned circuit with L and C for 17 MHz. The L of the coil is adjusted by the screw. (b) Double-tuned transformer with L and C for 455 kHz. Adjustable trimmer capacitors are at the top.

onance curve of the LC circuit. Only frequencies at f_r and close to resonance are in the *passband* of the amplifier, meaning that they are amplified with close to 100 percent response.

The passband is considered to include frequencies with a response of 70.7 percent or more. Signal frequencies too far below the passband, or too far above it, are not amplified to any extent. Those signals can be present in the input but not in the amplified output because the frequencies do not produce a resonant response in the LC circuit. The passband of a tuned RF amplifier is like a window. The desired signal at the resonant frequency can pass through, but other signal frequencies are blocked from the amplified output.

For specific examples of relative response and actual gain values, assume that 100 percent response is a gain of 100 when f_r is at 1000 kHz. Those values correspond to the RF amplifier example of Fig. 10-1. Then, with 1-mV input signal, the amplified output is 100 mV for 1000 kHz at f_r. Within the passband, 70.7 percent response provides output of 70.7 mV for frequencies close to resonance. The 70.7 values are for the two *edge frequencies*, one below and the other above f_r. Between f_r and the edge frequencies, the gain values are between 70.7 and 100.

The width of the passband depends on the Q of the tuned circuit. A higher Q means a narrow bandwidth, or a sharp response. A wider passband results from a lower Q.

Selectivity From the response curve in Fig. 10-3, it looks as though an undesired signal could be close to the passband but not in it. Would such a signal come through? It depends on the selectivity, which is the ability to reject such frequencies close to the passband while allowing the desired signal to be amplified.

An example of "perfect" selectivity is shown in Fig. 10-4. This response curve has vertical sides, or *skirts*. Any frequencies not in the passband have no gain at all. Of course, it is not possible to have so perfect a response, but modern filter circuits can approach the ideal. Resonant circuits are actually tuned filters.

The selectivity of a response curve is indicated by the slope of the skirts. The slope is measured by comparing the bandwidth at the edge frequencies with the bandwidth lower down on the curve. See the example in Fig. 10-5. Each bandwidth is determined by subtracting the lower from the higher frequency at two points with the same gain.

The top of the curve is marked as 0 dB gain to represent maximum gain for 100 percent response. Less response is shown in units of −dB to indicate less gain than the maximum.

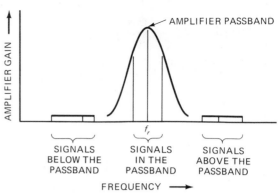

Fig. 10-3. Frequency response curve of a tuned RF amplifier. Relative amplitudes are not to scale.

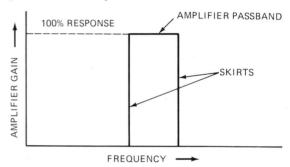

Fig. 10-4. An ideal RF response curve with vertical sides and flat top.

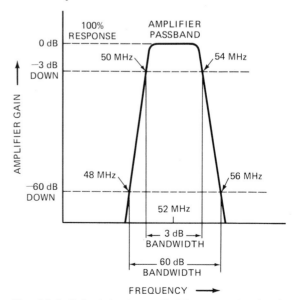

Fig. 10-5. Selectivity is specified by comparing bandwidths on the curve where response is down −3 and −60 dB.

The edge frequencies are at -3 dB down on the curve. This gain is down 29.3 percent, which means 70.7 percent of the maximum gain.

The other two points for measuring bandwidth are at -60 dB down. That value, in dB, is a voltage ratio of $1:1000$.

To calculate the selectivity factor, the formula is

$$\text{Selectivity} = \frac{-60\text{-dB bandwidth}}{-3\text{-dB bandwidth}} \quad \textbf{(10-1)}$$

For example, look at the frequencies marked in Fig. 10-5. At the -3-dB frequencies, the bandwidth is $54 - 50 = 4$ MHz for this curve centered around the f_r of 52 MHz. At the -60-dB frequencies, the bandwidth is $56 - 48 = 8$ MHz. Using the 8 and 4 MHz bandwidths, we have

$$\text{Selectivity} = \frac{8\text{ MHz}}{4\text{ MHz}} = \frac{2}{1}$$

Note that the selectivity factor is expressed as a ratio, here $2:1$. The smaller the ratio the better the selectivity. Then the curve has steep sides because the -60-dB bandwidth is not much more than the -3-dB bandwidth.

RF Gain   Another measure of performance is the ability to provide enough gain for RF signals in the passband. For example, in radio and television receivers, the RF signal from the antenna may have an amplitude of 2 to 5 μV. In order for it to be detected the signal must be amplified to the level of 2 to 5 V, approximately.

The amount of voltage gain needed, then, is 1,000,000. It must be provided by tuned amplifiers. Before detection, the amplifiers for the carrier signal are tuned. After the detector recovers the modulation, the signal can be amplified in untuned stages.

The number of tuned stages needed will depend on how much gain each stage can contribute to the overall gain. Remember that the total gain of cascaded stages is the product of the individual gain values. For instance, if each tuned stage can provide a gain of 100, only three stages will be needed for an overall gain of 1,000,000. The calculations are

$$100 \times 100 \times 100 = 1,000,000$$

When the gain is expressed in dB values, the individual gain figures are added. A voltage gain of 100 is equal to 40 dB. The overall gain for the three stages is then

$$40 + 40 + 40 = 120 \text{ dB}$$

RF Transistors   As the signal frequency for a transistor amplifier is increased, the amount of output eventually falls off because the amplification ability of the transistor drops. This is due to two factors: transit time and internal capacitance. It takes time for the charge carriers from the emitter to diffuse through the base to the collector junction. If that *transit time* is comparable with the period of the signal frequency, the gain will be reduced. Another effect is caused by capacitance of the collector junction. As the signal frequency goes up, the reactance of the internal capacitance goes down. The reactance shunts the collector output circuit with a low impedance when the frequency is high enough. The low output impedance causes low gain.

Because of those factors, RF transistors are designed to have a thin base to reduce diffusion time and a small junction area for minimum capacitance. Still, any transistor has some limit at which its gain drops to unity, or 1, meaning there is no amplification.

Gain-Bandwidth Product   The gain or beta (β) and the frequency limits are indicated by the gain-bandwidth product f_T. The formula for f_T is

$$f_T = \beta \times \text{operating frequency} \quad \textbf{(10-2)}$$

As an example, the 2N5179 transistor has f_T of 1000 MHz. That specification means that the transistor has an effective β of 10 at 100 MHz, because the product 10×100 MHz is 1000 MHz. The transistor could have an effective β of 20 at 500 MHz.

Test Point Questions 10-1
(Answers on Page 225)

a. Does the slope of the skirts on an RF response curve determine selectivity or sensitivity?
b. An RF amplifier has a gain of 10 at 100 percent response. What is the gain at the edge frequencies -3 dB down?

10-2
TYPICAL RF AMPLIFER STAGE

In Fig. 10-6a the RF input signal is coupled to the base of Q1. This transistor is a silicon NPN in the CE circuit. C_1 is the input coupling capacitor. The amplified output is taken from the collector, which has a tuned circuit with L_A and C_A. The output signal is coupled by C_5 to the next stage.

In the base-emitter circuit a combination of fixed bias at the base and self-bias at the emitter is used for bias stabilization. R_3 produces the emitter bias, and C_3 is the emitter bypass for RF signal frequencies. The R_1R_2 voltage divider supplies the required forward voltage at the base. This fixed bias has positive polarity from the collector supply V_{CC}. Note that C_2 bypasses RF signal frequencies around the dc supply voltage.

It is important to realize that the collector itself is not bypassed. The collector must provide the amplified output signal.

Equivalent Circuit for the Collector Load In Fig. 10-6b, the circuit is illustrated only for the ac signal, without any dc voltages. The main idea here is that the side of L_A connected to V_{CC} is grounded by the RF bypass C_2. Therefore, L_A is effectively in parallel with C_A for the RF signal. This resonant circuit is the collector load impedance for the amplified output signal from the tuned RF amplifier.

R_4 represents the input resistance of the next stage, typically between base and emitter in the CE circuit. Because of the low reactance of coupling capacitor C_5, the high side of R_4 is effectively connected to C_A for the RF signal. On the low side, both C_A and R_4 are grounded. They are in parallel, therefore, with R_4 shunting the tuned circuit of C_A and L_A. The shunt resistance of R_4 determines the impedance and bandwidth of the resonant circuit.

Resonant Response Assume the inductance of L_A in Fig. 10-6 is 200 μH and the capacitance of C_A is 126 pF. Then we can calculate the resonant frequency of the tuned circuit:

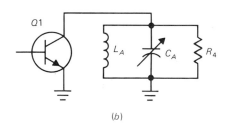

Fig. 10-6. Typical RF amplifier stage. (a) Schematic diagram. (b) Equivalent circuit showing the parallel-resonant circuit for the collector load impedance.

$$f_r = \frac{1}{2\pi\sqrt{LC}}$$

$$= \frac{1}{2\pi\sqrt{200 \times 10^{-6} \times 126 \times 10^{-12}}}$$

$$= \frac{0.159 \times 10^9}{\sqrt{200 \times 126}} = 0.001 \times 10^9$$

$$f_r = 1 \text{ MHz}$$

Remember that the impedance of a parallel-resonant circuit is maximum at f_r. Also, the voltage gain of a CE amplifier is higher with higher values for the collector load impedance. Therefore, the tuned RF amplifier has maximum gain at the resonant frequency. This amplifier selects RF signals at and near the frequency of 1 MHz. The bandwidth depends on the Q of the tuned circuit.

Transistor Capacitances Although the values are low, in the picofarad range, the reactance can be low enough in RF amplifiers to affect the gain and the tuned circuits. Typical examples for a junction transistor, as a CE amplifier, are as follows: the C_{in} is $C_{BE} = 10$ pF; the C_{out} is $C_{CE} = 5$ pF; the C_{CB} is 8 pF between collector and base.

Low reactance for C_{in} reduces the signal input. Also, low reactance for C_{out} decreases the output load impedance , which reduces the gain. Finally, low reactance for C_{CB} allows internal feedback of the amplified output signal back to the input.

Miller Effect In addition, amplification of the ac signal increases the dynamic value of input capacitance. The reason is that the inverted output signal can feed back to C_{in} and reduce its voltage. This action, which is equivalent to increasing C_{in}, is called the *Miller effect*. For the same C values considered before and a voltage gain of 12, the dynamic value of input capacitance is

$$C_{in} + (C_{CB} \times A_V), \text{ approximately}$$

or,

$$10 + (8 \times 12) = 106 \text{ pF}$$

Test Point Questions 10-2
(Answers on Page 225)

Refer to Fig. 10-6.

a. When C_A is increased, will the resonant frequency increase or decrease?
b. When V_E is 3 V and R_3 is 1 kΩ, what is the value of I_E?
c. Would the emitter bypass C_3 be 1600 pF or 1600 μF?
d. When V_E is 3 V, how much voltage across R_1 is required for 0.6 V forward bias?

10-3
SINGLE-TUNED CIRCUIT RESPONSE

The two main characteristics of an *LC* tuned circuit are its resonant frequency and its Q. The f_r determines what signal frequencies can be amplified. The Q determines bandwidth.

Q, or Quality Factor In general, Q is determined by the ratio of stored or reactive energy to dissipated or resistive energy. The reactive energy of a coil is determined by X_L. The coil's dissipated energy is determined by the series resistance r_S of the coil winding. For a coil, therefore,

$$Q = \frac{X_L}{r_S} \qquad (10\text{-}3)$$

As an example, assume a coil has X_L of 1000 Ω, at a specific frequency, and a total resistance of 5 Ω for r_S. Then the Q of the coil is 1000/5 = 200. There are no units for Q because the ohms in the ratio cancel.

Capacitors also have resistive losses, mainly from heating effects in the dielectric. As a general rule, however, the capacitors used in tuned RF amplifiers have a much greater Q than the coils. Therefore, the Q of a tuned circuit usually depends on the Q of the coil. The lowest Q for a circuit determines the quality factor, just as the weakest link determines the strength of a chain.

Figure 10-7 shows how to calculate Q for two equivalent examples. In Fig. 10-7a, the series-resonant circuit has a coil with X_L of 1000 Ω and an r_S of 5 Ω. The Q then is 1000/5 = 200 of the coil. This value is also the Q of the tuned circuit.

In Fig. 10-7b, the equivalent parallel-response circuit is shown. This case is important in regard to tuned RF amplifiers because the collector load impedance is generally a parallel-resonant circuit. The same L and C are used as in Fig. 10-7a, but for parallel resonance it is more useful to consider Q in terms of the parallel resistance R_P shunting the tuned circuit. Therefore, the series resistance of the coil r_S is converted to the equivalent shunt resistance R_P. The conversion formula is

$$R_P = Q^2 \times r_S \qquad \textbf{(10-4)}$$

In this example,

$$R_P = 200^2 \times 5 \ \Omega$$
$$= 200,000 \ \Omega$$

To calculate the Q with the parallel R_P instead of the series r_S, the formula is

$$Q = \frac{R_P}{X_L} \qquad \textbf{(10-5)}$$

Note that here X_L is in the denominator to be divided into the higher value for R_P. In this example,

$$Q = \frac{200,000}{1000} = 200$$

The Q of this tuned circuit is 200 calculated either from the series r_S of the coil or its equivalent shunt R_P. This conversion applies for high-Q circuits with Q values of 10 or more.

Q and Bandwidth (BW) In Fig. 10-8 two impedance curves for parallel-resonant circuits are compared. Both have the same resonant frequency but different values of Q. Note that the high-Q curve has a higher impedance at f_r. However, this

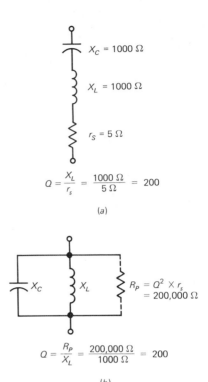

$$Q = \frac{X_L}{r_s} = \frac{1000 \ \Omega}{5 \ \Omega} = 200$$

(a)

$$Q = \frac{R_P}{X_L} = \frac{200,000 \ \Omega}{1000 \ \Omega} = 200$$

(b)

Fig. 10-7. Calculating Q for series and parallel circuits. (a) Series-resonant circuit with a series r_S equal to the resistance of the coil. (b) Parallel-resonant circuit with an equivalent shunt damping resistance R_P.

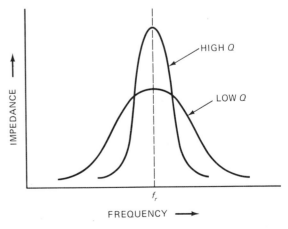

Fig. 10-8. Impedance curves for parallel resonance with high Q and low Q.

curve is narrower, which means a smaller bandwidth. The lower-Q curve has more bandwidth but less gain.

High impedance for the amplifier load means high gain. Also, a narrow response curve allows good selectivity. However, additional factors may make it desirable to have more bandwidth for a tuned RF amplifier. First, a modulated RF signal consists of a band of frequencies, produced by modulation, around the center carrier frequency. The bandwidth must be great enough to include all the frequencies in order to preserve the modulation in the signal. Another factor is that too much gain may make the RF amplifier unstable and have a tendency to break into oscillations.

The bandwidth of a single tuned circuit can be calculated from the formula

$$BW = \frac{f_r}{Q} \qquad \textbf{(10-6)}$$

For the example in Fig. 10-9, f_r is 1 MHz and the Q is 50. Then

$$BW = \frac{1 \times 10^6}{50}$$

$$= 0.02 \times 10^6 = 20 \times 10^3$$

$$BW = 20 \text{ kHz}$$

Bandwidth, like f_r, is in units of frequency. The value of 20 kHz is the bandwidth at the -3 dB, or half-power points on the response curve. That is, the BW is between the edge frequencies with 70.7 percent response.

Test Point Questions 10-3
(Answers on Page 225)

a. X_L is 800 Ω and $r_S = 16$ Ω. Calculate the Q.
b. Q is 50 at an f_r of 500 kHz. Calculate the bandwidth.

10-4
SHUNT DAMPING RESISTOR

Figure 10-10a shows the same resonant circuit as in Fig. 10-7 but with a 15-kΩ damping resistor R_D. The resistance is not the equivalent R_P of the coil; it is an actual carbon resistor connected in parallel. The purpose of R_D is to lower the Q and increase the bandwidth of the resonant LC circuit.

The reason why shunt resistance lowers the Q is that the parallel path is a branch for resistive current. Lower values of R_D result in less Q and more bandwidth. There is more resistive branch

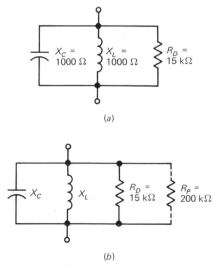

(a)

(b)

Fig. 10-10. Shunt damping resistor R_D connected across an LC-tuned circuit. (a) Actual circuit with $R_D = 15$ kΩ. (b) Equivalent R_P of coil in parallel with R_D. The coil is the same as shown in Fig. 10-7.

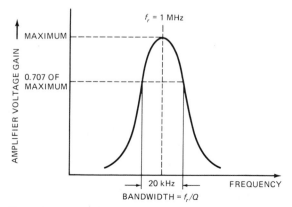

Fig. 10-9. Example of half-power bandwidth (BW) on response curve for a resonant circuit with Q = 50.

current relative to the reactive current in the L and C branches. In other words, a lower value for the shunt R_D means more damping.

Damping the Transient Response *Damping* basically refers to the inhibiting effect resistance has on the transient response of the tuned circuit. A transient is a rapid change in voltage or current. The amplitude is high for a short time, and then it falls off.

Any transient, such as an abrupt burst of signal energy, can set the tuned circuit into oscillation. The LC circuit oscillates at its natural resonant frequency, which is f_r. The oscillation is usually started by a self-induced voltage across L. It should be noted that the oscillations can also be caused by stray feedback.

A tuned circuit with high Q oscillates for a longer period of time before resistive losses in the coil and circuit damp the oscillations. A damping resistance increases these resistive losses and decreases the time it takes for the oscillations to decay to zero. For that reason, a tuned circuit is often damped to improve its transient response. A better transient response means the circuit cannot break into oscillations so easily. This damping effect is the same as lowering the Q and increasing the bandwidth for the resonant response curve of the tuned circuit.

The damping R can be in either series or parallel with L, but a shunt damping resistor, such as R_D in Fig. 10-10, is generally used. The reason is that the parallel resistance does not alter the symmetry of the frequency-response curve.

Loaded Q_L with Damping We can calculate the Q in the circuit of Fig. 10-10 to see how the shunt damping resistor R_D loads down the tuned circuit for a lower Q and more bandwidth. Remember that, for parallel R, the Q is calculated as R/X_L. In Fig. 10-10a, the Q is equal to R_D/X_L, or $15,000/1000 = 15$. This Q is with R_D as a shunt load. Compare this with Fig. 10-7, where the same circuit has an unloaded Q of 200.

The shunt R_D for damping is also called a *swamping resistor*. The name indicates that the effect of

high Q in the tuned circuit itself is overcome by the lowered Q with shunt damping. Specifically, the equivalent shunt resistance of the coil, which is relatively high, is reduced to the value of the parallel R_D because it is much lower.

The effect is shown in Fig. 10-10b with 200 kΩ for R_P of the coil. However, the equivalent parallel R of 15 kΩ for R_D and 200 kΩ for R_P is practically 15 kΩ because it is so much smaller. The total R_T of two resistances in parallel is determined mainly by the smaller R. In other words, the high R_P of a high-Q coil is swamped by the lower value of R_D in parallel.

Furthermore, the maximum Z of a parallel-tuned circuit at resonance becomes equal to R_D because of its swamping effect. Then the gain of a tuned amplifier is determined by the amount of R_D.

In summary, the shunt damping R_D is used for the purpose of lowering the Q of a single-tuned circuit for more bandwidth. R_D can also be used on either side of a double-tuned circuit or on both sides. When R_D is low enough on one side of a double-tuned circuit, the swamping effect results in the same frequency response as in a single-tuned circuit.

Test Point Questions 10-4
(Answers on Page 225)

In Fig. 10-10, let R_D be 20 kΩ instead of 15 kΩ.

a. Calculate Q for the tuned circuit shown in Fig. 10-10a.
b. Calculate the total resistance for R_D and R_P in parallel, Fig. 10-10b.
c. Which R_D provides a broader bandwidth, 20 or 15 kΩ?

10-5
SINGLE-TUNED TRANSFORMER COUPLING

In Fig. 10-11, Q1 is an RF amplifier in cascade with another RF amplifier, Q2. Both are NPN transistors in the CE circuit. The interstage transformer T_1 is used to couple the amplified RF signal from the collector output of Q1 to the base input

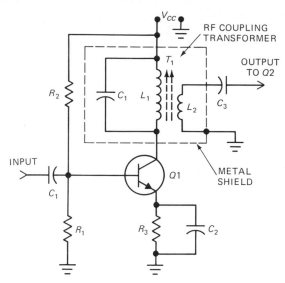

Fig. 10-11. Single-tuned transformer-coupled RF amplifier.

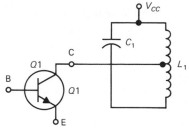

Fig. 10-12. Tapped coil L_1 in the collector circuit is used for impedance matching.

circuit of $Q2$. The primary winding L_1 is tuned by C_1 to provide a parallel-resonant circuit for the collector load impedance. L_1 couples the RF signal into L_2 by mutual induction. No capacitor is needed for the coupling, but C_3 blocks the dc bias voltage on $Q2$ to prevent a short circuit to ground through the L_2 winding.

Note that T_1 is enclosed in a metal shield, as indicated by the broken lines. The shield is often used with RF transformers. Its purpose is to prevent pickup of interference by L_1 or L_2 from external signals and to prevent the transformer itself from radiating RF signals to other parts of the circuit. Radiation is always a problem whenever a component has current flow at radio frequencies. The shield isolates the transformer from RF interference in or out.

Alignment   Tuned transformers in RF amplifiers are usually adjustable so the circuit can be tuned to amplify the desired RF signal frequencies. A common method is the adjustable core, which is indicated by the dashed arrows to the right of L_1 in Fig. 10-11. A threaded screw is attached to a ferrite slug to move the core in and out of the coil and thereby adjust the coil's inductance. The res-

onant frequency of the tuned circuit can then be adjusted by varying the L.

The process of tuning successive amplifiers to the frequency of the desired signal is called *alignment* of the tuned circuits. A special plastic alignment tool is used to prevent stray capacitance from detuning the circuits.

Impedance Match of Transformer T_1 in Fig. 10-11 has a step-down turns ratio from L_1 to L_2. The purpose is to match the low impedance of the base input circuit of $Q2$ to the higher impedance of the collector output circuit of $Q1$.

The maximum-power-transfer theorem states that most power will be delivered to a load when the impedance of the load is the same as that of the signal source. In Fig. 10-11, the collector circuit of $Q1$ is the signal source and the base circuit of $Q2$ is the load. When the impedances are not equal, a transformer can be used to correct the mismatch.

Tapped Coil in Collector Circuit   Another technique for impedance matching is shown in Fig. 10-12. In that circuit, the coil L_1 for the tuned circuit is tapped down for the collector connection. In that way, more L and less C can be used for a high Q and Z in the resonant circuit. However, the collector is at a point of lower impedance. If the collector were connected across the full coil, the circuit would be loaded down too much. In effect, the tap allows the collector to be matched to the higher impedance of the parallel-resonant circuit. The result is good selectivity in the tuned circuit but with the required impedance values.

a. In Fig. 10-11, is Q1 a CE, CB, or CC circuit?
b. In Fig. 10-11, which symbol indicates the secondary of the RF coupling transformer?
c. In Fig. 10-11, which component forms the tuned circuit with the primary of T_1?

10-6
DOUBLE-TUNED TRANSFORMERS

The single-tuned response curve does not have enough bandwidth for many applications. As an example, Fig. 10-13 shows the desired frequency response for the output from the chroma bandpass amplifier in a color television receiver. This amplifier provides the required gain and selectivity for the 3.58-MHz modulated chroma signal before it is demodulated. More chroma signal means more color in the reproduced picture.

Note that the response curve has steep skirts and a 1-MHz bandwidth at the -3-dB edge frequencies. This passband of ±0.5 MHz is wide compared with an f_r of 3.58 MHz at the center. The double hump on the response curve is a characteristic of a double-tuned circuit transformer that has close coupling to increase the bandwidth.

The circuit for a chroma bandpass amplifier is shown in Fig. 10-14. Bandpass transformer T_1 is double-tuned; both primary and secondary are resonant at the same frequency. Two arrows for the adjustable core indicate the alignment adjustments. The left arrow is for tuning L_P. It is usually at the top of the shield cover. The secondary can be tuned from the bottom.

Coefficient of Coupling The shape of the frequency-response curve of a double-tuned transformer depends on the value of the mutual inductance L_M between L_P and L_S. Specifically, the coupling coefficient is

$$k = \frac{L_M}{\sqrt{L_P L_S}} \qquad (10\text{-}7)$$

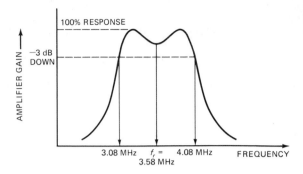

Fig. 10-13. Example of double-tuned response curve for the 3.58-MHz chroma signal in color television receiver.

where k has no units because it is a ratio of inductance values. Typical values of k in RF transformers are 0.01 to 0.05.

The primary and secondary windings are usually wound on the same insulated form. When they are close to each other, the mutual inductance and k are relatively great. The condition is often called *tight coupling*, which means that L_P and L_S are close. When the windings are more separated, a lower L_M results in a lower k. That condition is called *loose coupling*.

Critical Coupling k_C The important case of critical coupling occurs when the coefficient of coupling has the following value:

$$\text{Critical } k_C = \frac{1}{\sqrt{Q_P Q_S}} \qquad (10\text{-}8)$$

A double-tuned transformer with critical coupling provides a response curve that approaches the ideal. The top is flat for uniform gain over a broad range of frequencies. Also, the skirt selectivity is sharper than that of a single-tuned circuit.

In Fig. 10-15 three response curves for different values of k are compared. Actually, the comparison is with respect to the critical coupling curve a with k_C of 0.025. For loose coupling with a smaller k, the response would be like that of a single-tuned circuit. Increasing k broadens the top of the curve

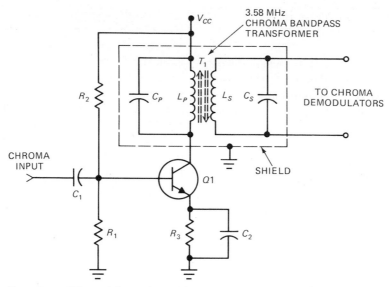

Fig. 10-14. RF amplifier with double-tuned transformer T_1. The circuit is a chroma bandpass amplifier in a color television receiver.

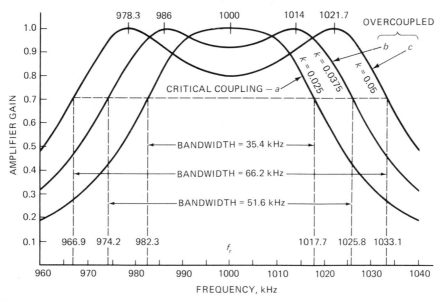

Fig. 10-15. Response curves for different values of coupling coefficient k. Bandwidth increases as k increases.

and makes the sides sharper up to the point of critical coupling.

When k is greater than the critical value, the transformer is said to be *over-coupled*. The tight coupling increases the bandwidth more, but the curve has a double-hump appearance as in the curves of Fig. 10-15b and c. The chroma response curve in Fig. 10-13 also has a slight double hump. However, too much overcoupling can produce an excessive dip at the resonant frequency.

In summary, critical coupling k_C increases the bandwidth and sharpens the skirts without a double hump on the curve. Loose coupling, which is less than k_C, has less bandwidth. Coupling more than k_C is tight coupling or overcoupling for more bandwidth, but with a double hump.

Test Point Questions 10-6
(Answers on Page 225)

a. What is the -3-dB bandwidth for the curve in Fig. 10-13?
b. What is the value of k_C for the curve with critical coupling shown in Fig. 10-15?
c. Which allows more bandwidth, a single-tuned response or a double-tuned response with critical coupling?

10-7
STAGGER TUNING

In stagger tuning two or more single-tuned stages are used and each resonant circuit is staggered at a different frequency. The advantages of stagger tuning are increased bandwidth and good skirt selectivity.

When all cascaded stages are resonant at the same frequency, the combination is said to have *synchronous tuning*. Cascading increases gain, but synchronous tuning reduces the overall bandwidth sharply. The reason is the reduced response for frequencies just a little off the f_r. The reduction factor is multiplied by the effect of the cascaded gain.

In Fig. 10-16 the circuit is a two-stage RF amplifier in which an IC unit is used for each stage and the frequencies for the tuned circuits are staggered. In stage 1 the T_1 primary is tuned to the resonant frequency f_{r_1}. Also, in stage 2 the T_2 primary is tuned to f_{r_2}. That resonant frequency is a little above f_{r_1} as shown by the response curves in Fig. 10-17.

In Fig. 10-17, the smaller curve at the left for f_{r_1} is a typical single-tuned response. The curve at the right is similar but for the higher resonant frequency f_{r_2}. However, the overall response for both stages is flatter and wider at the top. Its center frequency f_C is midway between f_{r_1} and f_{r_2}.

The overall response curve combines the gain values for the response curves of both stage 1 and stage 2. For frequencies in the passband where one curve is low, the other is high. At f_C exactly, both curves have the same medium value of gain. As a result, the response for both stages has a wide passband of frequencies with uniform gain.

When there are two stages, the combination is

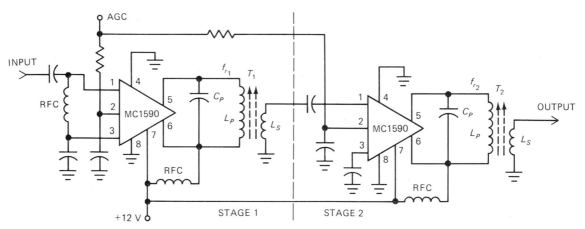

Fig. 10-16. Integrated circuit (IC) amplifiers for two tuned stages with resonant frequencies staggered at f_{r_1} and f_{r_2}. The AGC connection is for automatic gain control.

a staggered pair. The two resonant frequencies are symmetrical below and above f_C. Both tuned circuits have the same Q. For a staggered triple, one stage tuned to f_C is combined with a staggered pair. The Q values and resonant frequencies are designed to provide uniform gain through the passband.

Test Point Questions 10-7
(Answers on Page 225)

a. Does synchronous tuning with cascaded stages increase or decrease the overall bandwidth?

b. In Fig. 10-16, give the letter designations for the transformers used for the two single-tuned stages.

c. In Fig. 10-17, with f_{r_1} = 41 MHz and the critical coupling f_C = 43 MHz, what is f_{r_2}?

10-8
WAVE TRAPS

A resonant circuit that is used to reject an undesired frequency is called a *wave trap* or *trap circuit*. The rejection or attenuation is accomplished by a sharp reduction in amplifier gain at the trap frequency. Three examples of trap circuits are shown in Fig. 10-18. In all cases, the wave trap is a resonant LC circuit tuned to the rejection frequency.

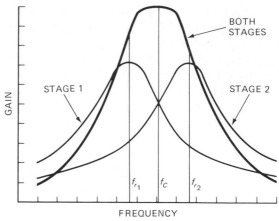

Fig. 10-17. Gain-frequency response curves for the stagger-tuned pair of amplifiers in Fig. 10-16.

However, the method of reducing gain depends on how the trap is connected.

Parallel-Resonant Trap In the trap method of Fig. 10-18a, L_2 tunes with C_2 to form a parallel-resonant circuit. Both L_2 and C_2 form the wave trap. The trap itself, however, is in series with the signal path to the next amplifier stage.

Parallel resonance provides a high impedance at the resonant frequency. When the trap is in series, though, the high impedance produces maximum voltage across the L_2C_2 combination at the rejection frequency with minimum voltage for in-

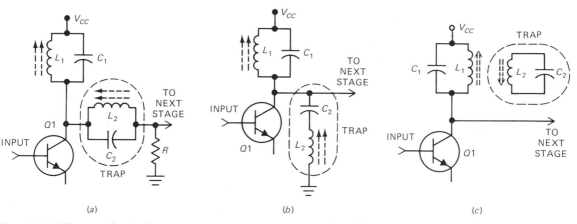

Fig. 10-18. Three methods of using a trap circuit to reject an undesired frequency. (a) Parallel-resonant circuit in series with the signal path to the next stage. (b) Series-resonant circuit in shunt with the collector load impedance. (c) Absorption trap with L_2 inductively coupled to L_1.

put to the next stage. The trap itself has maximum voltage, but that voltage is not connected across the next input circuit because neither side of the trap circuit is grounded.

The trap circuit is in series with the input resistance of the following amplifier to form a voltage divider from the collector of Q1 to chassis ground. More voltage across the trap means less voltage for the other part of the divider.

It should be noted that frequencies in the passband of the amplifier have their normal gain. The tuned circuit with L_1 and C_1 provides the required collector load impedance. Only the frequency to be attenuated is rejected by the wave trap.

Series-Resonant Trap

The same results can be obtained with the series-resonant wave trap shown in Fig. 10-18b. There L_2 and C_2 form a series-resonant circuit. Series resonance provides a very low impedance at resonance. However, this trap circuit is connected in shunt with the collector output circuit of Q1. The rejection frequency is shorted to ground, therefore, by the low impedance of the series-resonant trap.

Absorption Trap

In the trap method of Fig. 10-18c, L_2 in the trap circuit is inductively coupled to L_1 in the collector tuned circuit. There is no direct connection, but the two tuned circuits are coupled by transformer action. When the trap is resonant at its rejection frequency, maximum current flows in the L_2C_2 circuit. The current absorbs energy from the collector circuit. The effect is to load down the L_1C_1 tuned circuit, which lowers the circuit's Q. As a result, the amplifier has low output impedance and little gain at the trap frequency.

An Application of Wave Traps

The frequency response curve in Fig. 10-19 shows how trap circuits are used to reject undesired frequencies. This curve is for the picture IF amplifier in a television receiver. Trap circuits are used to prevent sound signal from interfering with the picture signal. The effect of audio voltage coupled to the picture tube

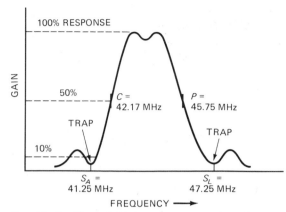

Fig. 10-19. Typical IF response curve of a television receiver showing the effect of wave traps in rejecting 47.25 and 41.25 MHz.

is to produce dark and light bars across the screen that vary with the sound signal.

The frequencies marked in Fig. 10-19 can be summarized as follows:

S_A 41.25 MHz for the associated sound signal. This is the sound for the channel tuned in.
C 42.17 MHz for the color subcarrier signal.
P 45.75 MHz for the picture carrier signal.
S_L 47.25 MHz for the lower adjacent sound signal. This is the sound for the channel below the station tuned in.

All these are intermediate frequencies in the picture IF amplifier of the television receiver.

The trap at 41.25 MHz is adjusted to reduce the associated sound signal in the picture IF amplifier in order to prevent sound bars in the picture. The trap at 47.25 MHz reduces this response to prevent interference from the lower adjacent channel. Trap circuits are tuned by feeding in the signal at the rejection frequency and adjusting for a sharp dip in the output.

Test Point Questions 10-8
(Answers on Page 225)

a. In which figure, 10-18a or 10-18b, does the trap act as a short circuit at the rejection frequency?
b. In Fig. 10-18c, is L_1C_1 or L_2C_2 tuned to the rejection frequency?

10-9
NEUTRALIZATION OF RF AMPLIFIERS

Transistors are *bilateral devices*, which means that signal can feed in two directions. One direction is from input to output for amplification; the other is from output to input as internal feedback. The amount of feedback depends on the internal properties of the transistor and the type of circuit for the amplifier. Those factors are especially important for RF amplifiers because even a low internal capacitance can have reactance low enough to allow feedback at radio frequencies.

A serious problem with feedback is amplifier instability. The amplified signal in the output that is fed back may have the correct phase shift, at some frequency, to arrive at the input with the same phase as the input signal. That would produce positive feedback, which is used for oscillator circuits. The circuit then becomes an oscillator, instead of being an RF amplifier. When that happens, the amplifier is said to be unstable because it can break into oscillations.

Internal Feedback   Figure 10-20 shows the path for feedback inside a junction transistor. C_{BC} is the capacitance of the junction between collector and base. Typical values are 5 to 20 pF. Although low, that capacitance can feed part of the amplified RF signal from the collector output back to the base input. R_{BC} represents the resistive component of the feedback path.

The two components C_{BC} and R_{BC} form an *RC* phase-shift network. At some frequency the phase shift, combined with the phase shift in the collector output circuit, can produce feedback that is in phase with the base input signal. The amplifier then can oscillate at that frequency.

External Neutralizing Capacitor   Another feedback path can be established in the RF amplifier to neutralize the effect of internal feedback. Usually, a *neutralizing capacitor* that has a value approximately the same as that of the internal capacitance is used. An example is C_2 in the RF amplifier circuit shown in Fig. 10-21. The neu-

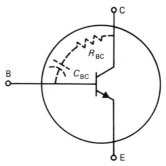

Fig. 10-20. Internal feedback path in a junction transistor.

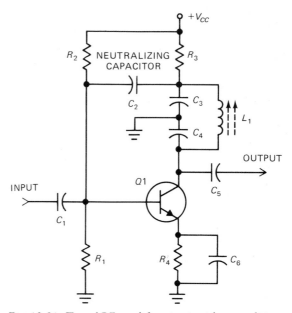

Fig. 10-21. Tuned RF amplifier circuit with neutralizing capacitor C_2.

tralizing signal fed back through C_2 must be about the same amplitude as the internal feedback, but out of phase with it. Then the internal feedback is canceled. The result is stable amplifier operation because of the neutralized circuit.

Neutralized RF Amplifier   In the circuit of Fig. 10-21, C_2 is the neutralizing capacitor. It feeds part of the amplified collector output signal back to the base. However, the polarity of the neutralizing signal must be opposite that of the amplified output signal. The reason is that the amplified signal feeds back through the internal capacitance. Remember that the neutralizing signal cancels the internal feedback.

In order to provide a neutralizing signal, the collector tuned circuit uses the two capacitors C_3 and C_4 to resonate with L_1. The junction of C_3 and C_4 is grounded. Then the ungrounded side of C_3 at the top and C_4 at the bottom have signals 180° out of phase. The signal voltage across C_3 is the feedback for the neutralizing capacitor C_2. Furthermore, R_3 is used to isolate the tuned circuit from V_{CC} so that the top of C_3 can have a signal voltage without any bypassing from the power supply.

The same results can be produced with a tapped coil for L_1, instead of the capacitive voltage divider, to supply the neutralizing signal. In either case, though, note that the amplified RF signal is available at the collector terminal for the output to the next stage.

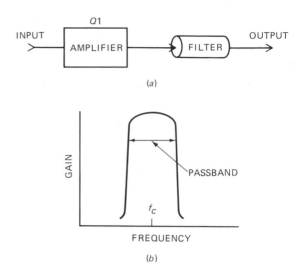

(a)

(b)

Fig. 10-22. (a) Application of a mechanical or crystal filter for a bandpass amplifier Q1. (b) Response curve shows the filter passes only the desired band of frequencies centered around f_C.

Test Point Questions 10-9
(Answers on Page 225)

a. Does neutralization make the stability of an RF amplifier better or worse?
b. In Fig. 10-21, which is the neutralizing capacitor?

10-10
MECHANICAL AND CRYSTAL FILTERS

Mechanical and crystal filters are often used as bandpass filters instead of *LC* resonant circuits. Actually, they are mechanical resonators. The advantages are higher Q and sharper skirt selectivity compared with electrical resonance. Furthermore, mechanical filters can be used for audio frequencies, at which resonant *LC* circuits are not practical.

In crystal filters a quartz disk is used as the mechanical resonator. Ceramic filters are similar; a ceramic material such as lead titanate is used instead of quartz. Both types are applications of the piezoelectric effect, which converts the voltage

variations across a crystal into mechanical vibrations, or vice versa.

The magnetostrictive effect also is used in mechanical filters. In magnetostriction, variations in magnetic field strength produce a mechanical twist in a thin rod.

Both crystal and mechanical filters are examples of electromechanical *transducers*. They convert electric energy to mechanical energy, or vice versa.

Figure 10-22 is an illustration of how a mechanical or crystal filter is used for a bandpass amplifier circuit. The input can have a wide range of frequencies. However, the filter in the output circuit takes the place of a parallel-resonant *LC* circuit for high impedance. Only the frequencies in the passband of the filter have appreciable gain, therefore. The response curve (Fig. 10-22b) shows high gain for the passband; it has a relatively flat top around the center at f_C. The sharp slope of the skirts at the sides means that any frequencies close to the passband, but not in it, are rejected. There is little gain for frequencies either below or above the passband.

Structure of a Mechanical Filter In Fig. 10-23 the signal to be filtered is applied to the transducer

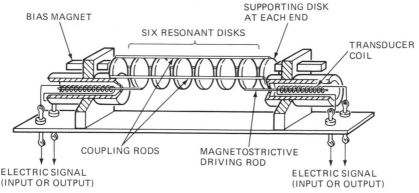

Fig. 10-23. Internal structure of a mechanical filter.

coil at one end. Either end of the filter can be used for input, with the output being at the other side. The coil is in a steady magnetic field produced by a permanent magnet for bias. Signal current in the coil produces a magnetic field, which varies with the signal. The interaction of the two fields causes motion in the coil, which now acts as a transducer. The motion of the coil is transferred to the magnetostrictive driving rod, which is attached to a resonant disk. There are six disks resonant at different frequencies in the passband; they are coupled to each other by thin rods.

Each disk has a vibrating frequency that is fixed by its physical dimensions. Signals that are different from the natural resonant frequencies of the disks do not transfer through the filter. The disks are very active, though, at signal frequencies within the passband. As a result, those signal variations are coupled through the filter to the second transducer coil at the opposite end. There the mechanical variations are converted back into current variations as the coil moves back and forth in the field of the bias magnet.

Another mechanical filter is shown in Fig. 10-24. In this type the *flexure mode* is used: the coupling bar flexes or bends in its mechanical motion. The input signal is applied to a ceramic crystal which acts as the transducer at one end. Because of the piezoelectric effect, the crystal vibrates. The vibrations are transferred to the metal bar on which the crystal is mounted. The bar flexes along its

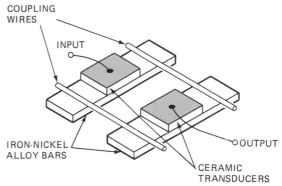

Fig. 10-24. Structure of a flexure-mode mechanical filter.

length. The natural resonant frequency of the variations depends on the length and thickness of the bar. Again, only signal frequencies in the passband of the filter can produce appreciable flexing.

The coupling wires between the two bars make one bar vibrate when the other does. For the output, the ceramic crystal uses the piezoelectric effect to convert the mechanical variations back into signal voltage.

Crystal Filters By itself, a quartz or ceramic crystal has very high Q and narrow bandwidth compared with a resonant LC circuit. However, a crystal filter uses two or more crystals to achieve the required bandwidth with excellent skirt selectivity.

In Fig. 10-25 is shown a typical arrangement with two crystals X_1 and X_2. This connection is called a *half-lattice*. In general, a lattice network has cross-connections. The center-tapped coil L is used to provide a phase reversal of the two crystals at opposite ends. With this arrangement, it is possible to balance out the shunt capacitance C_1 or C_2 of each crystal. The crystals contribute to the overall frequency response by acting as series-resonant circuits. However, the complete filter circuit provides high impedance by parallel resonance with the inductance of L.

The half-lattice circuit becomes the basic building block for more complex filters. For example, a filter with six crystals for a bandpass RF amplifier is shown in Fig. 10-26. Three half-lattice sections are used in the filter. The single center-tapped coil L_2 serves the first two sections, and in the third section L_3 is used for balancing out the shunt capacitances of the crystals. The variable coils L_1 at the input and L_4 at the output are set for the desired bandpass frequencies within the range of the crystal filter.

The number of crystals used is often referred to as the number of *poles* in the filter. In Fig. 10-26, a six-pole filter is used. With more poles in the filter, it is possible to have sharper skirt selectivity.

Monolithic Crystal Filter The filter shown in Fig. 10-26 is built up from discrete, or separate, components. The entire assembly is sealed in a metal can. However, the size is relatively large compared with the method of forming the components on a single, monolithic quartz wafer as shown in Fig. 10-27. Two or more pairs of electrodes are deposited on opposite sides of the wafer. Each area between a top and bottom electrode becomes a resonator. The coupling between resonator sections depends on the distance separating the sections.

The monolithic crystal filters produce results similar to those with discrete crystal filters. The advantages are that monolithic filters are smaller and cost less. A four-pole monolithic crystal filter is small enough to be packaged in a TO-5 case.

Filter Applications Mechanical filters can be used for audio and radio frequencies up to 800 kHz, approximately. The crystal and ceramic filters are

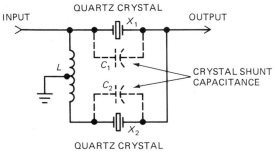

Fig. 10-25. Half-lattice circuit for a crystal filter.

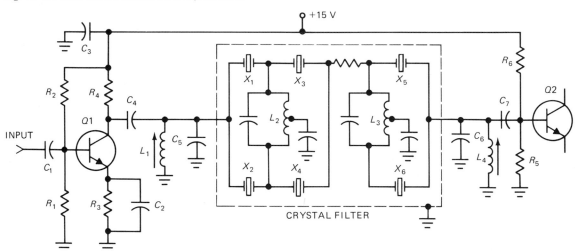

Fig. 10-26. Circuit of an RF amplifier using a crystal filter.

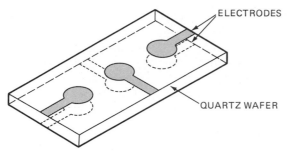

Fig. 10-27. Construction of a monolithic crystal filter.

for radio frequencies, generally from 800 kHz up to about 70 MHz. Ceramic filters have lower Q than quartz filters, but they cost less. In all these filters, the width of the passband can be from 0.1 to 10 percent of the center frequency.

The filter applications are in bandpass amplifiers requiring high Q with sharp skirt selectivity. It should be noted, though, that the filter passband cannot be varied to achieve tuning over a range of center frequencies. One application is in the IF amplifier of a radio receiver, in which the intermediate frequencies of the desired signal remain the same for all stations tuned in. Another application is in a sideband filter at the transmitter when single-sideband transmission is used.

Test Point Questions 10-10
(Answers on Page 225)

Answer True or False.

a. A crystal filter has sharper skirt selectivity than a resonant LC circuit.
b. A transducer can convert between mechanical and electric energy.
c. A monolithic crystal filter is smaller than a filter with discrete components.

10-11
RANDOM NOISE

All materials generate some noise because of the random motion of electrons. Electric circuits produce even more noise as the heating effect of the current adds thermal energy to the random action. In an amplifier circuit the random voltage

or current, called *noise*, can have enough amplitude to interfere with the desired signal. Especially in RF amplifiers with very weak input signal, it becomes very important that the amplifier circuit itself generate very little noise. Otherwise, the signal is masked by the effect of the noise. Thermal noise is also called *Johnson noise*.

Random noise is also called *white noise* because it produces snow on the screen of a television picture tube. In a radio the noise produces a continuous rushing sound often described as hissing or frying.

Random noise is different from the pulse-type noise produced by sparking in electrical equipment. The pulse interference is intermittent, and the pulses can be reduced in amplitude by noise limiter circuits. However, random noise has a continuous frequency spectrum of constant voltage and current amplitudes. This noise cannot be filtered out of the signal. The only solution is to have more signal to start with.

Actually, the random noise sets a limit on the weakest signal the amplifier can handle and still be useful. When the signal is not much more than the noise level, all the gain in the world cannot help, because the noise is amplified with the signal. In such a case a television picture would be masked by snow. Also, voice and music would be lost in background noise. This factor is the reason why the first RF stage should be a low-noise amplifier.

In addition to thermal noise, transistors generate random noise because of random flow of the internal charge carriers. The FET has less noise than NPN or PNP bipolar types.

This important comparison is specified as the *signal-to-noise* (S/N) ratio. The minimum S/N ratio for acceptable results is at least 10:1 in voltage and 100:1 in power, or 20 dB.

Test Point Questions 10-11
(Answers on Page 225)

Answer True or False.

a. Random noise can produce snow in a television picture.
b. The S/N ratio of a signal of 80 μV with 2 μV of noise is 40, or 32 dB.

SUMMARY

1. RF amplifiers are generally tuned by resonant circuits to amplify a band of radio frequencies. The high impedance for parallel resonance allows high amplifier gain for frequencies at and near resonance.

2. The RF response curve is a plot of relative gain of the amplifier at different frequencies, as shown in Fig. 10-3. High gain means good sensitivity for amplifying weak signals.

3. The bandwidth is usually measured between the frequencies at which the gain is 70.7 percent of 100 percent response, or -3 dB down. Bandwidth can be calculated as f_r/Q.

4. Selectivity is a measure of how sharp the skirts are at the edges of the response curve to reject undesired signal frequencies. The selectivity factor can be specified as a ratio of the bandwidth at -60 dB to the bandwidth at -3 dB.

5. For an RF amplifier with a single-tuned circuit, the response is centered around the resonant frequency at which $f_r = 1/(2\pi \sqrt{LC})$.

6. A single-tuned circuit has $Q = X_L/r_S$, where r_S is the internal series resistance of the coil. With a shunt damping resistor, $Q = R_D/X_L$, where R_D is the parallel resistance across L and C.

7. A higher Q means higher impedance at resonance with more amplifier gain and a narrower bandwidth for sharp selectivity.

8. When amplifiers are in cascade, the overall gain is the product of the individual values in the stages. That effect reduces the overall bandwidth for cascaded RF amplifiers tuned to the same resonant frequency. Cascaded amplifiers tuned to the same frequency are said to have synchronous tuning.

9. In stagger tuning, separate stages are resonant at different frequencies. The purpose is to increase the bandwidth of the cascaded amplifier stages.

10. Double-tuned transformers are often used for more bandwidth and sharper selectivity compared with a single-tuned circuit.

11. The double-tuned response depends on the coupling between primary L_P and secondary L_S. In general, the coupling coefficient $k = L_M/\sqrt{L_P L_S}$. The critical coupling is $k_C = 1/\sqrt{Q_P Q_S}$. The bandwidth is maximum without double peaks at critical coupling. Less than k_C is loose coupling; more than k_C is tight coupling or overcoupling.

12. A wave trap is a resonant circuit tuned to reject an undesired signal by reducing the gain at the trap frequency.

13. In a neutralizing circuit an external feedback capacitor is used to cancel the internal feedback of the transistor. The purpose of neutralization is to stabilize an RF amplifier to prevent oscillations.

14. Random noise is generated in all components by thermal energy. The effects are snow in a television picture and background noise in the sound.

15. An RF amplifier with a weak signal must have low noise for a good signal-to-noise ratio.

16. Mechanical and crystal filters instead of LC resonant circuits can be used in amplifiers requiring a wide bandpass with high Q and good skirt selectivity.

SELF-EXAMINATION
(Answers at back of book)

Choose (a), (b), (c), or (d).

1. The gain of an RF amplifier with a tuned LC circuit for the collector is maximum at the resonant frequency because of (a) series resonance, (b) parallel resonance, (c) low Q, (d) high emitter resistance.
2. The Q of a single-tuned LC circuit is lower when its (a) series r_S is reduced, (b) shunt R_D is reduced, (c) shunt R_D is increased, (d) capacitance is reduced.
3. Snow in a television picture is a result of (a) high Q in the tuned circuits, (b) excessive gain, (c) random noise in the signal, (d) insufficient wave traps.
4. The response curve of a double-tuned circuit shows two peaks when the transformer has (a) loose coupling, (b) critical coupling, (c) tight coupling, (d) excessive damping.
5. Which of the following methods cannot be used to increase the bandwidth for cascaded RF amplifier stages? (a) A shunt damping resistor across each tuned circuit; (b) stagger tuning; (c) double-tuned transformers with tighter coupling; (d) wave traps in each stage.
6. Which of the following circuits has the greatest bandwidth? (a) f_r is 50 MHz with Q of 50; (b) f_r is 455 kHz with Q of 100; (c) f_r is 1 MHz with Q of 100; (d) f_r is 1 MHz with Q of 10.
7. Refer to the wave trap of Fig. 10-18a. The desired frequency is 42.5 MHz, and the interfering frequency is 41.25 MHz. Therefore, (a) L_1C_1 is tuned to 41.25 MHz, (b) L_2C_2 is tuned to 42.5 MHz, (c) L_1C_1 is tuned to 82.5 MHz, (d) L_2C_2 is tuned to 41.25 MHz.
8. In Fig. 10-6, L_A is 120 μH and C_A is 30 pF. The RF amplifier is tuned to what frequency? (a) 0.52 MHz; (b) 1.2 MHz; (c) 2.65 MHz; (d) 10.7 MHz.
9. The purpose of using a tapped coil for L_1 in Fig. 10-12 is to get (a) increased bandwidth, (b) impedance matching, (c) the effect of a wave trap, (d) forward bias on Q1.
10. When both the primary and secondary resonant circuits of a double-tuned transformer have a Q of 50, the value of critical coupling k_C is (a) 50, (b) 0.5, (c) 1, (d) $\frac{1}{50}$, or 0.2.

ESSAY QUESTIONS

1. What is the difference between sensitivity and selectivity?
2. Define the following: **a.** half-power bandwidth, **b.** selectivity factor, **c.** cascaded gain, **d.** S/N ratio, **e.** damping.
3. **a.** Give the formula for calculating the coefficient of coupling k for a double-tuned transformer. **b.** Give the formula for critical coupling k_C.
4. Draw the schematic diagram of a single-tuned CE amplifier resonant at 14.8 MHz. Use an NPN transistor, and show dc voltages.

5. Draw the schematic diagram of a double-tuned CE amplifier with both primary and secondary resonant at 14.8 MHz.
6. Compare the following three frequency response curves: **a.** single-tuned, **b.** double-tuned with critical coupling, **c.** double-tuned with tight coupling.
7. What is the advantage of stagger tuning with cascaded RF amplifier stages?
8. **a.** What is the purpose of a wave trap? **b.** Show two types of wave-trap circuits.
9. An RF amplifier is tuned to 1.5 MHz. The tuning capacitance is then increased from 20 to 180 pF, or nine times. Why will the new resonant frequency be reduced to 0.5 MHz?
10. Describe two differences between damping with series resistance and damping with shunt resistance.
11. What is the difference between a bandpass filter and a wave trap?
12. Give two types of bandpass filters that do not use *LC* circuits.
13. **a.** What is the advantage of high Q and narrow bandwidth in an RF amplifier? **b.** Why is more bandwidth necessary in some cases? **c.** Give two methods of increasing bandwidth.
14. What is meant by the gain-bandwidth product of an RF transistor?
15. What is meant by the alignment of tuned circuits?
16. What is the purpose of neutralization in an RF amplifier?
17. Define the following: **a.** piezoelectric effect, **b.** transducer, **c.** magnetostriction.
18. Compare the construction of and applications for mechanical and crystal filters.
19. Why are random, thermal, and white noise almost impossible to filter out of the desired signal?
20. Why is it important that the first stage in a radio receiver have a good signal-to-noise ratio?

PROBLEMS
(Answers to odd-numbered problems at back of book)

1. An RF amplifier has a frequency response curve with the following characteristics:
   Down 1 dB at 1.4 and 1.5 MHz
   Down 3 dB at 1.3 and 1.6 MHz
   Down 60 dB at 0.9 and 2.0 MHz
   a. What is the standard bandwidth? **b.** Calculate the selectivity factor.
2. Three RF amplifiers in cascade have a voltage gain of 50 for each stage. What is the overall voltage gain?
3. Two RF amplifiers are in cascade; one has a gain of 20 dB and the other 40 dB. What is the overall gain in **a.** dB and **b.** voltage?
4. Refer to the emitter resistor R_3 in Fig. 10-11. If I_E is 5 mA and R_3 is 470 Ω, calculate V_E.
5. An RF amplifier is tuned with a parallel *LC* circuit consisting of a 3.3-μH coil and 76-pF capacitor. **a.** Calculate the resonant frequency f_r. **b.** Calculate the Q of the tuned circuit at f_r when r_S of the coil is 2 Ω. **c.** The input to the next

stage represents a shunt load of 10 kΩ. Calculate the loaded Q of the tuned circuit. **d.** Calculate the half-power bandwidth of this tuned amplifier for the loaded Q.

6. Refer to the emitter bypass capacitor C_2 in the amplifier of Fig. 10-11. What value of C_2 is needed to bypass the 470 Ω of R_2 at 10 MHz? (*Hint:* X_C of C_2 is one-tenth of R_2.)

7. Calculate the coupling coefficient k of an RF transformer with the following values: L_P of 400 μH, L_S of 200 μH, and L_M of 30 μH.

SPECIAL QUESTIONS

1. Why is tuning so important in a radio or television receiver?
2. Give one or more examples of random noise preventing good reception.
3. Describe briefly two applications of crystal and ceramic transducers other than bandpass filters.

ANSWERS TO TEST POINT QUESTIONS

10-1	**a.** Selectivity		10-5	**a.** CE		10-8	**a.** Fig. 10-18b	
	b. 7.07			**b.** L_2			**b.** L_2C_2	
10-2	**a.** Decrease			**c.** C_1		10-9	**a.** Better	
	b. 3 mA		10-6	**a.** 1 MHz			**b.** C_2	
	c. 1600 pF			**b.** 0.025		10-10	**a.** T	
	d. 3.6 V			**c.** Double-tuned			**b.** T	
10-3	**a.** 50		10-7	**a.** Decrease			**c.** T	
	b. 10 kHz			**b.** T_1 and T_2		10-11	**a.** T	
10-4	**a.** 20			**c.** 45 MHz			**b.** T	
	b. 20 kΩ							
	c. 15 kΩ							

Chapter 11

Oscillator Circuits

An oscillator uses a transistor or vacuum tube in a circuit to generate an ac output. The oscillator circuit is basically an amplifier, but feedback from its output to its input enables the oscillator to sustain the output without the need for a signal from a preceding stage. The most common oscillators are of the tuned RF type with inductance L and capacitance C. They produce a sine-wave output at the resonant frequency of the LC circuit as shown in Fig. 11-1.

Every transmitter needs an oscillator to generate the RF carrier waves. Also, an oscillator is used in the frequency converter of superhetrodyne receivers for the IF signal. As another example, in digital circuits the multivibrator clock generator is an oscillator that produces pulses at a specific frequency. Finally, a signal generator is essentially an oscillator used to supply test signals. Either audio or radio frequencies can be produced by oscillator circuits with sine-wave or nonsinusoidal waveforms. More details are given in the following topics:

11-1
OSCILLATOR REQUIREMENTS

Any device that repeats two opposite actions at a regular rate is an oscillator. The oscillating values alternate periodically. One cycle includes the time for both alternations. A perfect example of mechanical oscillation is a pendulum swinging back and forth. The corresponding electrical example is a repetition of alternations in voltage or current around a middle value, specifically the type with sine-wave variations in amplitude as in Fig. 11-1. Such oscillations can be produced by an LC circuit at its natural resonant frequency when electric energy is supplied.

How an *LC* Circuit Oscillates   In the circuit of Fig. 11-2a, the battery supplies energy to the capacitance C in parallel with the inductance L. Under the conditions shown no oscillations are produced. The battery voltage simply charges the capacitor to the dc level V_C and produces direct current I_L in the coil. However, energy is stored in the electric field of the capacitor and the magnetic field of the inductor.

In Fig. 11-2b the switch is opened to disconnect the battery voltage. As a result, the action of the LC circuit can be considered by itself. Without the battery voltage, I_L and V_C drop toward zero.

When I_L starts to decay, its changing magnetic field produces a self-induced voltage V_L. Now the

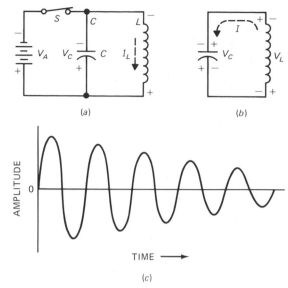

(a)   (b)

(c)

Fig. 11-2. How an LC circuit oscillates at its natural resonant frequency. (a) When the switch is closed the battery charges C and produces current in L. (b) When the switch is opened the battery voltage is removed from the circuit. Now C and L interchange energy. (c) Sine waves of oscillating V and I.

coil is a voltage source. However, the polarity of V_L tends to keep I_L in the same direction, which allows C to discharge. In fact, V_L will charge C in the opposite polarity. When the value of V_C becomes more than the value of the decaying V_L, the capacitor supplies discharge current I_C. The current is in the direction opposite from that of the original I_L.

The final result is that the inductance and capacitance interchange energy to produce alternations in V and I at the natural resonant frequency of the LC circuit, as shown in Fig. 11-2c. Whenever I_L goes through zero, it is changing to induce V_L. Whenever V_C goes through zero, it is changing to produce I_C. The waveform is a continuous sine wave because the V and I values cannot change abruptly.

The amplitude of the oscillations in V and I becomes smaller as energy is dissipated in the resistance of the circuit. This waveform is a damped sine wave decaying toward zero as shown in Fig. 11-2c.

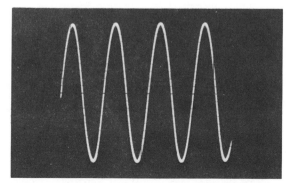

Fig. 11-1. Sine-wave output from a tuned RF oscillator as viewed on an oscilloscope screen. Four cycles are shown.

The action of an *LC* circuit in producing sine-wave oscillations from its stored energy is called *ringing*. Actually, a coil in any circuit can produce ringing with its stray capacitance in parallel. A sharp drop of *I* is often followed by sine-wave ringing, especially with a high-Q coil.

The *LC* combination is also called a *tank circuit* because it stores the energy for oscillations. The ability of the tank circuit to produce sine waves from either a dc input or pulses is called *flywheel effect*.

In a practical oscillator circuit, a transistor, IC unit, or tube is used for amplification. Then the oscillator can provide feedback from the output circuit to the input to sustain oscillations. Energy for the continuous oscillations comes from the dc power supply for the amplifier. Actually, the circuit converts the dc supply voltage to an ac output from the oscillator.

This type of oscillator with a tuned *LC* circuit is used to generate a sine-wave output at radio frequencies up to about 300 MHz. At audio frequencies, the values of *L* and *C* needed would be too large. Then *RC* feedback oscillator circuits (Sec. 11-10) can be used. At frequencies above 300 MHz, UHF and microwave oscillators are used, as described in Chap. 14.

Oscillator Frequency In a tuned RF oscillator, the output is essentially the resonant frequency of the *LC* circuit, where

$$f_r = \frac{1}{2\pi\sqrt{LC}} \qquad (11\text{-}1)$$

For practical values in RF circuits, *L* is in microhenrys and *C* is in picofarads. The resulting f_r is in gigahertz (10^9 Hz). As an example, with *L* of 80 μH and *C* of 20 pF,

$$f_r = \frac{1}{2\pi\sqrt{80 \times 20}} = \frac{1}{2\pi\sqrt{1600}}$$

$$= \frac{1}{2\pi \times 40} = \frac{1}{80\pi}$$

$$f_r = 0.004 \text{ GHz, or } 4 \text{ MHz}$$

For good stability, the tuned circuit should have a high Q to prevent any shift in frequency. Furthermore, the *L* and *C* must be constant with changes in temperature so that the oscillator frequency does not drift. Keeping the dc supply voltage constant also improves the frequency stability.

Relaxation Oscillators A different type of oscillator uses transistors, an IC unit, or tubes in a circuit in which feedback makes the amplifier alternate between conduction and cutoff at a regular rate. The output voltage then oscillates between the extremes of high and low values. This oscillating waveform is either a square wave or a train of rectangular pulses.

One complete cycle of oscillations includes the conduction, or ON, time and the OFF time. Such a circuit is a relaxation oscillator, because the cutoff time is a relaxed state. Relaxation oscillators are used as pulse generators for either audio or radio frequencies up to about 30 MHz.

The two basic types of pulse generator circuits are the blocking oscillator (BO) and multivibrator (MV). A blocking oscillator uses one amplifier stage with a transformer for feedback. A multivibrator uses two amplifier stages. The output of each stage drives the input of the other stage for feedback. Multivibrator circuits are explained in detail in Sec. 11-12.

Oscillator Waveforms Four examples of oscillator waveform are shown in Fig. 11-3. In Fig. 11-3a the sine wave is from a tuned *LC* oscillator. In Fig. 11-3b the square-wave output is from a symmetrical pulse oscillator with equal ON and OFF times. Such a circuit is a square-wave generator.

The rectangular wave in Fig. 11-3c is actually an unsymmetrical square wave. Either the ON or OFF time is longer than the other. Then the periods for high and low voltage levels are not equal.

The sawtooth waveform in Fig. 11-3d can be derived from the rectangular waveform in Fig. 11-3c. The method is to use the rectangular voltage for charging and discharging a capacitor C through a series resistance R. When the applied voltage is high, it charges C slowly through a high R to

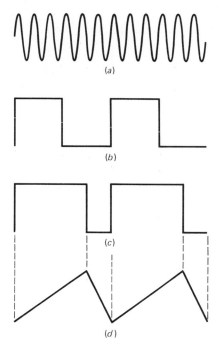

Fig. 11-3. Oscillator output waveforms (a) Sine wave. (b) Square wave. (c) Rectangular wave. (d) Sawtooth or ramp wave.

produce the linear rise or ramp. When the charging voltage drops, C discharges fast through a low R. The frequency of the resultant sawtooth voltage is the same as that of the rectangular waveform. Such a circuit is called a *sawtooth generator*. The sawtooth waveform is also called a *ramp voltage*.

Test Point Questions 11-1
(Answers on Page 251)

a. In Fig. 11-2, what two components form the oscillating circuit?

b. When L and C are increased, does f_r increase or decrease ?

11-2
TICKLER FEEDBACK OSCILLATOR

The operation of the tickler feedback oscillator circuit is illustrated in Fig. 11-4. The amplifier circuit of Fig. 11-4a becomes the oscillator of Fig.

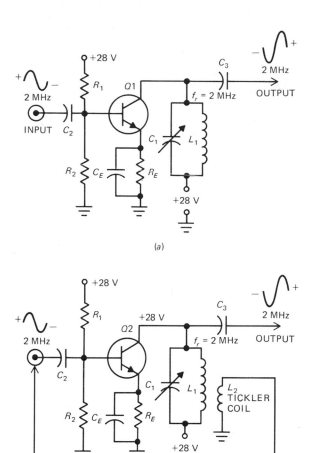

Fig. 11-4. Converting a tuned RF amplifier stage to an oscillator circuit with positive feedback. (a) Amplifier with a 2-MHz signal for input and output. (b) Oscillator circuit with a tickler coil for feedback from output to input. No external signal is needed for the 2-MHz output.

11-4b because positive feedback is added. Feedback is provided by the small inductance L_2, which is called a *tickler coil*.

In the amplifier of Fig. 11-4a, transistor Q1 is an NPN silicon type connected in the common-emitter (CE) circuit. The collector output is tuned by L_1C_1 to a radio frequency of 2 MHz. The input signal is applied to the base by C_2. The amplified output from the collector is coupled by C_3 to the next stage. Note that the amplified output is inverted by the CE amplifier.

Positive Feedback What makes the circuit shown in Fig. 11-4b an oscillator is the positive feedback from tickler coil L_2. Note that L_2 is transformer-coupled to L_1 in the tuned circuit. As a result, part of the 2-MHz output from the collector is coupled into L_2.

The voltage across L_2 can be either in phase with the voltage across L_1 or out of phase by 180^0. That depends on how the tickler coil L_2 is wound and which end is grounded. What the circuit needs is a phase inversion. Then the two phase inversions make the feedback in phase with the applied input signal. One inversion is in the collector circuit of the CE amplifier, and the other inversion is in the transformer coupling between L_1 and L_2.

The feedback is an ac voltage that has positive and negative polarities. Describing the feedback as positive means that its phase aids the input variations. The result is regeneration, because the oscillator amplifies its own feedback.

Names for Oscillators Oscillator circuits are named either for the inventor or to describe the method of feedback. In general, a circuit with a tickler feedback coil, as shown in Fig. 11-4, is called an Armstrong oscillator. In the Hartley circuit, the oscillator coil is tapped for feedback. In the Colpitts oscillator, the feedback is provided by a capacitive voltage divider.

Furthermore, different methods of connecting the tickler feedback coil and the LC tank circuit can be used. An example of the tuned-base, collector-feedback oscillator is shown in Fig. 11-5. In that circuit the tickler coil L_2 has the collector current I_C. Any change in I_C induces voltage into L_1. Therefore, the tuned circuit is excited by the feedback to produce oscillations.

Variable-Frequency Oscillator In the circuits of Figs. 11-4 and 11-5 the tuning capacitor C_1 is varied to change the oscillator frequency. A variable inductance can be used instead to vary the frequency. Such circuits are called *variable-fre-*

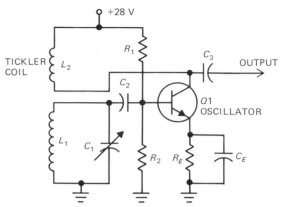

Fig. 11-5. Tuned-base, collector feedback circuit. The tickler feedback coil is L_2.

quency oscillators (VFO). Crystal-controlled oscillators are examples of oscillators in which the frequency is fixed by the crystal.

Oscillator Output The ac output of the oscillator is usually coupled to another stage for amplification. Either capacitive coupling can be used or inductive coupling with a transformer.

The oscillator is driving two circuits: its own feedback path and the input to the next stage. Both circuits are in parallel with the oscillator output. The feedback keeps the circuit oscillating. The output path allows the oscillations to be used in the next circuit. The power for both parallel circuits constitutes the load on the oscillator. The loading reduces the effective Q of the oscillator-tuned circuit.

Testing for Oscillator Output The fact that the oscillator is actually generating ac output can be checked in several ways. An oscilloscope shows the ac voltage in the oscillator circuit or at the input to the next stage. As another method, when the oscillator has enough power output, it can light a small neon bulb or fluorescent lamp placed near the LC tank circuit.

Probably the most convenient test is to measure the bias on the oscillator with a dc voltmeter. In oscillator circuits, the feedback signal usually is rectified to provide signal bias. The correct dc bias shows that the oscillator is oscillating because it produces feedback. Use a high-impedance meter to prevent loading down the oscillator.

Test Point Questions 11-2
(Answers on Page 251)

Refer to the tuned RF feedback oscillator shown in Fig. 11-4b.

a. Which component is the tickler coil?
b. Is the feedback regenerative or degenerative?
c. Which component varies the oscillator frequency?

11-3
HARTLEY OSCILLATOR CIRCUIT

The mark of a Hartley oscillator is a tuned LC circuit with a tapped coil for inductive feedback, instead of a separate tickler coil. See Fig. 11-6. C_1 and L_1 form the tuned circuit. The tap at point G on L_1 is used to supply the collector voltage. L_2 in this line is an RF choke.

Point G at the tap is effectively at ground potential for ac signal because of the bypass capacitor C_4. The oscillator output is taken from the collector. This ac voltage is V_{AG}, from point A on L_1 to the grounded tap at G. At the opposite side of the tap, the ac feedback is V_{BG} coupled by C_2 to the base. The feedback is positive because it is out of phase by 180° with V_{AG}. As a result of the

positive feedback, the circuit is an oscillator that generates an ac output at the resonant frequency of the LC circuit.

Consider the dc voltages now. The V_C is at 28 V because the dc resistance of the RF coils L_1 and L_2 is negligible. The emitter has self-bias of 1 V from R_E with C_E for bias stabilization. Forward voltage at the base is provided by the R_1R_2 divider from the supply of $+28$ V. The net V_{BE} is $1.4 - 1.0 = 0.4$ V. The bias of 0.4 V is less than the cutoff value of 0.5 V, but the positive peaks of feedback voltage drive the base positive to drive Q1 into conduction. The oscillator operates class C.

The functions of all the components in Fig. 11-6 can be summarized as follows:

L_1 Inductance for tuned circuit, tapped for feedback.

C_1 Capacitance for tuned circuit, variable for tuning.

L_2 RF choke to keep the oscillator signal out of power supply.

C_4 RF bypass to ground the tap on L_1 for ac.

C_2 Couples the ac feedback to the base while blocking the dc supply voltage.

R_2 Load resistor for the base. RC coupling with C_2 for the signal. Dc divider with R_1 for base bias. RC filter with C_2 for the rectified signal bias.

R_1 Provides forward dc bias with R_2 for the base.

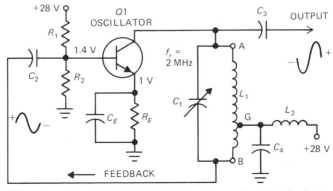

Fig. 11-6. Hartley oscillator circuit with a tapped coil L_1 for feedback.

Phase Inversion with a Tapped Coil The reason why the tap on L_1 provides positive feedback is illustrated in Fig. 11-7. Just L_1 alone is shown with the coil in two parts L_A and L_B so you can see the polarities of induced voltage. Consider electron flow into point A. This current flows through the turns of L_A between points A and G to the V^+ supply. The turns of L_B are not in that path. However, all the turns are on one coil. Therefore, L_B is transformer-coupled to L_A.

In the ac variations, assume that I is increasing. Then, by Lenz' law, the self-induced voltage V_{AG} is negative at point A to oppose the increase of I. Furthermore, this induced voltage is negative for the entire coil. All the turns are wound in one direction, and they are in the same magnetic field.

Point A is the negative end of the induced voltage compared with any turns farther down on the coil. The highest voltage is across all the turns from A to B. If we look at the induced voltage from B, the point is positive compared with any turns farther up on the coil. At any point tapped between A and B, therefore, the two points have opposite polarities with respect to the tap.

With ac variations in I, ac voltages are induced. Still, V_{AG} and V_{BG} are out of phase by 180°. When one is maximum negative, the other is maximum positive. Since the tap at G is grounded, V_{AG} and V_{BG} are ac voltages in opposite polarity with respect to chassis ground.

In the Hartley circuit of Fig. 11-6, V_{BG} provides the positive feedback required for input to the base. In general, the tap provides feedback voltage of about one-third the total voltage across the coil.

The amplification by Q1 has one phase inversion, and the feedback from the tapped coil has another phase inversion. As a result, the positive feedback results in regeneration for the oscillator circuit.

Series Feed and Shunt Feed The method of feed refers to how the V^+ voltage is fed to the output electrode. In the Hartley circuit of Fig. 11-6, series feed is used because the LC tuned circuit is in series with the 28-V supply voltage. A disadvantage of series feed is that direct current flows

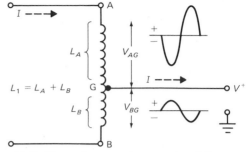

Fig. 11-7. The voltage in the tapped coil L_B is out of phase by 180° with the voltage in L_A.

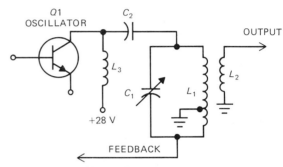

Fig. 11-8. Hartley oscillator circuit with shunt feed of V^+ through L_2 to the collector.

through L_1 in the tuned circuit. Also, V^+ is applied to the tuning capacitor C_1.

A Hartley oscillator circuit with shunt feed is shown in Fig. 11-8. Note that the 28 V for V_C is supplied through L_2. The choke is needed as a load impedance for the collector circuit. If the V^+ supply were connected directly to the collector, it would short the ac signal across the tuned circuit. Also, the coupling capacitor C_2 must be used to prevent dc voltage from being shorted through the grounded coil L_1. The two extra components are necessary with shunt feed, but there is no dc voltage or current in the tuned circuit.

Note that C_2 connects the tank circuit to the collector for its ac load impedance. The oscillator output is inductively coupled by L_1 to L_2 for the next stage.

Electron-Coupled Oscillator (ECO) Figure 11-9 shows a Hartley circuit with a pentode tube. This

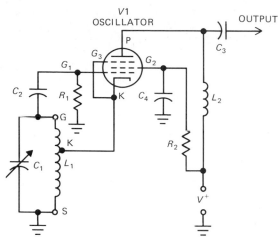

Fig. 11-9. Electron-coupled oscillator circuit using pentode tube. See text for explanation of the circuit.

oscillator is said to be electron-coupled to the output. The grids G_1 and G_2 are used with the cathode to form a triode oscillator. However, oscillator voltage at the control grid varies the electron flow for the plate current. As a result, the oscillator output is in the plate circuit by electron coupling to the tuned circuit L_1 and C_1 at the control grid. The RF choke L_2 is the plate load impedance. The oscillator output is coupled by C_3 to the next stage. The advantage of electron coupling is that the oscillator is isolated from the output load. The isolation allows better frequency stability.

The ECO has shunt feed because there is no dc supply voltage in the tuned circuit. Actually, the screen grid G_2 serves as the anode for a triode oscillator. Since the bypass C_4 grounds the ac signal, the screen grid is effectively connected to the low side of the L_1C_1 tuned circuit at point S. The screen-grid resistor R_2 just provides the desired value of dc voltage for G_2.

In the tuned circuit the tap on L_1 divides the circuit into two ac voltages of opposite phase. V_{GK} is the feedback voltage for the control grid. C_2 is the coupling capacitor, which also provides grid-leak bias with R_1.

Now look at the oscillator voltage V_{SK} at the opposite side of the tap on L_1. We would usually consider chassis ground to be the reference for sig-

nal voltage. However, that applies only when the cathode is grounded. Here the cathode has signal voltage from point K to ground. Therefore, the cathode is the reference. The turns from S to K provide the return path to the cathode for signal current from G_2. The electron flow from S to K induces a voltage across L_1. The V_{GK} is out of phase by 180° with V_{SK}, and the result is positive feedback for the oscillator.

The oscillator circuit of Fig. 11-9 illustrates another feature with the rotor of the tuning capacitor C_1 grounded. This method eliminates the effect of body capacitance in tuning the oscillator. Since the rotor is grounded, the body capacitance to ground does not change the capacitance of C_1. Otherwise, the added capacitance when the rotor is touched can affect the oscillator frequency.

Test Point Questions 11-3
(Answers on Page 251)

a. In the Hartley oscillator circuit of Fig. 11-6, is positive feedback provided by the tap on L_1, L_2, or C_1?
b. In Fig. 11-7, is the polarity of V_{AG} the same as or opposite that of V_{BG}?
c. Does the oscillator circuit of Fig. 11-8 have series or shunt feed for V^+?

11-4
COLPITTS OSCILLATOR CIRCUIT

A unique feature of the Colpitts oscillator is a capacitive voltage divider for feedback, as shown in Fig. 11-10. C_A and C_B form the series divider across the tuning coil L_1 in the collector circuit. The oscillator voltage across C_B is positive feedback to the base.

The junction between C_A and C_B is grounded. As a result, the capacitive divider is equivalent to a tapped coil for the oscillator signal. The ac voltages V_{CA} and V_{CB} then have opposite polarities to ground. The positive feedback of V_{CB} is coupled through C_2, which blocks the dc collector voltage from the base circuit.

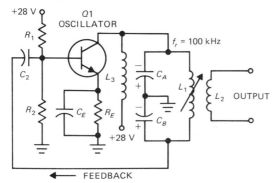

Fig. 11-10. Colpitts oscillator circuit with a capacitive voltage divider for feedback.

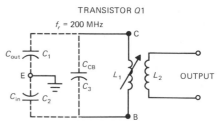

Fig. 11-11. Tuned circuit for ultra-audion oscillator. Only the transistor internal capacitances are used.

The oscillator output is inductively coupled to the next stage by L_2. Shunt feed is used for the collector voltage through L_3. L_3 is an RF choke used to prevent shorting the oscillator signal through the power supply.

Since the capacitance of the resonant LC circuit is divided in the Colpitts oscillator, tuning is generally done by varying L_1 as shown in Fig. 11-10. Either the inductance is adjusted or different coils can be switched in. Otherwise, C_A and C_B would have to be ganged for tuning. Varying only one C would change the amount of feedback voltage.

The Colpitts oscillator is used either for low radio frequencies, such as the 100 kHz of Fig. 11-10, or for the VHF band up to 300 MHz. At low frequencies, a tuning capacitor may be too large. In the VHF band, it may be too difficult to tap the very small coil necessary for the Hartley circuit.

Ultra-audion Oscillator The ultra-audion oscillator is a modified Colpitts circuit, but only the internal capacitances of the transistors are used. Figure 11-11 shows the equivalent tuned circuit. Note that C_3, C_2, and C_1 are internal capacitances shown here for a junction transistor. The dashed lines indicate that they are not physical capacitors wired into the circuit. The ultra-audion oscillator is used for the VHF band above 30 MHz, where the small internal capacitances alone provide enough C for the tuned circuit.

The divider with C_1 and C_2 of Fig. 11-11 corresponds to the divider with C_A and C_B of Fig. 11-10. C_1 is the Q1 output capacitance between collector and emitter. C_2 is the input capacitance between base and emitter. C_3 across L_1 is the transistor capacitance between base and collector; it corresponds to C_{GP} in a triode tube. These internal capacitance values are typically 2 to 10 pF in a small RF transistor.

Test Point Questions 11-4
(Answer on Page 251)

a. In the Colpitts circuit of Fig. 11-10, is the voltage across C_A or C_B used for positive feedback?
b. Which component is varied to tune the oscillator of Fig. 11-10?

11-5
TUNED-PLATE TUNED-GRID (TPTG) OSCILLATOR

As shown in the tuned-plate tuned-grid (TPTG) oscillator circuit of Fig. 11-12, a triode tube is used with two LC tuned circuits, one for the plate and the other for the control grid. Feedback from the plate is coupled through the internal grid-plate capacitance C_{GP}. The circuit can also be used with transistors.

In the TPTG oscillator, both tuned circuits are slightly inductive at the oscillator frequency. Then the inductive reactance resonates with the reactance of C_{GP}. The tank with the higher Q has a greater effect on the frequency. Usually, C_2 in the plate circuit is adjusted to set the oscillator frequency and C_1 varies the amount of grid excita-

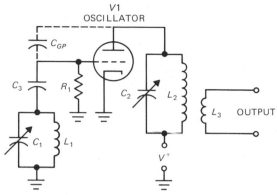

Fig. 11-12. Tuned-plate tuned-grid (TPTG) oscillator circuit using a triode tube. Feedback is through the internal grid-plate capacitance C_{GP}.

tion. The oscillator output is coupled by L_3 to the next stage.

Consider the dc voltages. The plate circuit has series feed for the V^+ voltage. The grid has signal bias with the R_1C_3 grid-leak combination. The negative bias is rectified oscillator signal provided by the grid excitation. The presence of the negative grid-leak bias, measured with a dc voltmeter, shows that the oscillator is working.

Test Point Questions 11-5
(Answers on Page 251)

a. In the circuit of Fig. 11-12, is the feedback coupled by C_{GP}, C_1, or C_2?
b. Is grid-leak bias a positive or negative dc voltage?

11-6
CRYSTAL OSCILLATORS

A quartz crystal is often used in an oscillator when it is necessary to have accurate frequency control. At its resonant frequency, the crystal is an electromechanical oscillator that is equivalent to an LC tuned circuit but with a much higher Q. Figure 11-13 shows a crystal in its housing. Note the exact frequency specification of 3579.545 kHz. The resonant frequency of a crystal is usually between 0.5 and 30 MHz. In an oscillator circuit, the crystal

Fig. 11-13. Crystal in its housing. Height is 1 in.

takes the place of the LC circuit in determining the frequency.

Crystal oscillators are generally used in mobile equipment, such as CB radios, for both the transmitter and receiver. Broadcast transmitters must use crystal control to provide the exact carrier frequency with good stability and very little tolerance. The frequency drift of a crystal is less than 1 Hz per 10^6 Hz. For test equipment, signal generators often have a crystal oscillator for calibrating the variable-frequency oscillator.

Piezoelectric Effect When a crystal is compressed, expanded, or twisted it generates a low voltage output through the piezoeletric effect. The reverse action also takes place, that is, voltage input can distort the crystal mechanically. As a result, the crystal can be excited to oscillate at a specific resonant frequency which depends on its dimensions. The thinner the crystal the higher the frequency of oscillations.

Crystal Cuts A piece of quartz looks like frosted glass. The raw crystal has a hexagonal shape, as shown in Fig. 11-14. Thin pieces from the raw crystal are sliced and then polished for mounting in a crystal holder. Typical dimensions are 0.5 to 1.0 in [12.7 to 25.4 mm] for the length or width. The thickness may be 0.3 in [7.6 mm], or less.

The cuts are named according to how they are sliced with respect to the axes of the crystal. In Fig. 11-14 note that the vertical line down the center to the point at the bottom is the Z axis. Any line through two opposite points on the hexagon is an X axis. Any line through opposite faces with a 90° angle to the face is a Y axis.

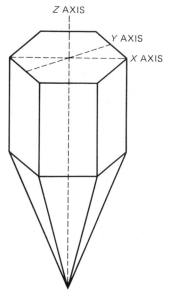

Fig. 11-14. The axes used for different crystal cuts.

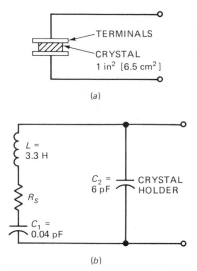

Fig. 11-15. Equivalent circuit of crystal unit. (*a*) Crystal in its holder. (*b*) Equivalent resonant circuit. Values for 500 kHz, approximately.

When a slice is parallel to the Z axis and its faces are perpendicular to the X axis, the plate is an X cut. The Y cut has its faces perpendicular to the Y axis. However, slightly different angles can be used for cuts with different names, such as AT, BT, CT, and GT, in order to obtain the desired characteristics for frequency and temperature. These special cuts usually depend on the torsion or shear of the crystal instead of compression. The GT cut has the lowest temperature coefficient, meaning that the frequency does not change with temperature. The AT and BT cuts are used at the highest frequencies.

A quartz crystal is very fragile, even in its holder. It should not be dropped. Also, too much excitation voltage can fracture it.

Crystal Equivalent Circuit The vibrating crystal in its holder shown in Fig. 11-15*a* is equivalent to the resonant LC circuit shown in Fig. 11-15*b*. L corresponds to the mass of the crystal; C_1 is the crystal's compliance or ability to change mechanically; R_s is the electrical equivalent of mechanical friction. Note the very high L/C ratio for series

resonance with the 3.3-H value of L and the 0.04-pF value of C_1. R_S is low compared with the reactance. As a result, Q values of 10,000 to 50,000 are typical. C_2 is the output capacitance of the crystal holder.

Crystal-Controlled Oscillator Circuit The circuit of Fig. 11-16 is similar to the TPTG oscillator of Fig. 11-12, except that a transistor is used in-

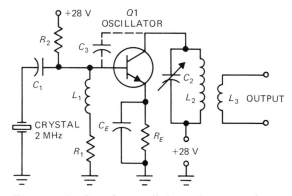

Fig. 11-16. Crystal-controlled oscillator, similar to TPTG circuit of Fig. 11-12.

stead of a tube and a 2-MHz crystal is used to determine the oscillator frequency. Feedback is through C_3, the internal capacitance between collector and base.

In the base circuit the R_1R_2 voltage divider supplies forward voltage from $+28$ V. The capacitor C_1 is used to keep the dc base voltage off the crystal. However, C_1 may be omitted because the crystal holder itself is a capacitor. The RF choke L_1 provides high impedance for the crystal output to the base.

In the emitter circuit, R_E, with its bypass C_E, is used for bias stabilization. In the collector, the LC tuned circuit provides oscillator output inductively coupled by L_3 to the next stage.

Pierce Crystal Oscillator   The Pierce crystal-controlled oscillator circuit shown in Fig. 11-17 is used often because it requires very few components and has good frequency stability. In a Pierce oscillator, one resonant circuit is connected between the output and input electrodes. Here the crystal across the collector and base determines the oscillator frequency. C_2 and C_3 form a capacitive voltage divider for feedback in a modified Colpitts circuit. The ac voltage across C_3 is positive feedback to the base.

In the base circuit the R_1R_2 divider supplies forward voltage from the $+28$ V. Bias stabilization is provided by the R_EC_E combination in the emitter circuit.

In the collector circuit the RF choke L_1 is used for shunt feed from the $+28$ V of the power supply. C_1 couples the oscillator output to the next stage. Also, C_1 blocks the 28 V from the crystal.

Crystal Oscillator Frequency   A crystal has a specific frequency, generally 0.5 to 30 MHz, but additional circuits can be used with crystal-controlled oscillators to obtain different frequencies. For higher values, a frequency multiplier circuit is used. Doublers and triplers are common. These amplifiers use LC circuits tuned to a harmonic of the crystal oscillator frequency. As an example, an oscillator output of 15 MHz can be raised to

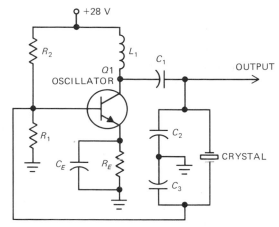

Fig. 11-17. Pierce crystal-controlled oscillator circuit.

45 MHz by a tripler circuit. Multipliers are used when it is necessary to have a frequency higher than the natural frequency of the crystal.

For a lower value, a frequency divider can be used. Digital counters are used to divide the crystal oscillator frequency by almost any value. As an example, an oscillator output of 1000 kHz can be divided by a factor of 100 down to 10 kHz.

Test Point Questions 11-6
(Answers on Page 251)

a. Is a typical resonant frequency for a crystal oscillator 3275 Hz or 3.275 MHz?
b. Is a typical Q for a crystal 25 or 25,000?
c. Is the Pierce crystal oscillator of Fig. 11-17 similar to a Colpitts or a Hartley circuit?

11-7
VOLTAGE-CONTROLLED
OSCILLATOR (VCO)

The voltage-controlled oscillator (VCO) circuit is used for electronic tuning of the oscillator frequency. The method requires the use of a semiconductor capacitive diode, also called a *varicap* or *varactor*. Its capacitance varies with the amount of reverse voltage. When there is a capacitive diode across L in the tuned circuit of the oscillator, its frequency is varied by controlling the dc voltage across the diode.

Varactor Diodes A PN junction with reverse voltage is actually a capacitor. The P and N electrodes are the two conductor plates on the sides of the depletion zone at the junction. This zone

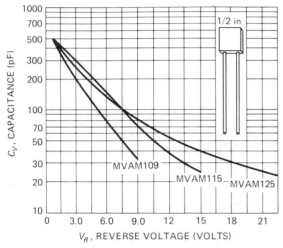

Fig. 11-18. Varactor, or voltage-controlled capacitive diode characteristic curves of three commercial varactors. Curves show that capacitance C_V decreases with increases in the reverse voltage. The insert shows a typical varactor.

serves as an insulator because it has no free charges. C may be 80 pF or more with a reverse voltage of 6 V, as typical values. The reverse voltage can be negative at the anode of the diode or positive at the cathode.

Most importantly, C changes with the amount of reverse voltage. The graph in Fig. 11-18 shows how C_V decreases as the reverse voltage increases. The reason is that more reverse voltage widens the depletion area. The effect is equivalent to more insulation distance between the plates, which reduces the capacitance.

VCO Circuit In Fig. 11-19, transistor $Q1$ is a Hartley oscillator. The tuned circuit consists of the tapped coil L_1 across the capacitive diodes D_1 and D_2. Both cathodes have a positive dc control voltage for the reverse voltage to vary C_V. This capacitance controls the oscillator frequency. Two diodes are used in series. The purpose is to balance out the effect of oscillator voltage on the diodes.

The oscillator output from the source electrode of $Q1$ is directly coupled to the gate of $Q2$. Its output is from the source in the source-follower circuit, which corresponds to an emitter follower. The purpose is to use $Q2$ as a buffer stage, which isolates the oscillator output of $Q1$ from the load connected to $Q2$. The advantage is better fre-

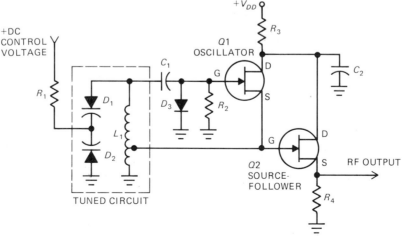

Fig. 11-19. Circuit for a voltage-controlled oscillator (VCO). $D1$ and $D2$ are capacitive diodes used to control the oscillator frequency. $Q1$ and $Q2$ are field-effect transistors.

quency stability. Both $Q1$ and $Q2$ are the N-channel junction field-effect transistor (JFET) type.

The functions of all the components in Fig. 11-19 can be summarized as follows:

$Q1$	Hartley oscillator transistor
D_1 and D_2	Capacitive diodes to control the oscillator frequency
L_1	Oscillator coil
C_1	Coupling capacitor
R_2	Dc return for the gate of $Q1$
D_3	Rectifies signal for signal bias at the gate of $Q1$
C_2	RF bypass for the drain electrodes.
R_3	Isolates drain of $Q1$ from the dc supply voltage, for the ac signal. Provides the drain voltage for $Q1$ and $Q2$.
V_{DD}	Dc supply voltage for drain of $Q1$ and $Q2$
$Q2$	Source-follower transistor
R_4	Output load resistance for the source electrode of $Q2$

The VCO has many applications because of its ability to control the oscillator frequency with a dc voltage. As an example, the push-button tuning on television receivers switches in the dc control voltage to set the frequency for each channel. The application is called *electronic tuning*, although the dc control voltage is adjusted manually. However, the dc voltage can also be adjusted automatically by an electronic control circuit. The method is to use the phase-locked loop circuit that is described in the next section.

Test Point Questions 11-7
(Answers on Page 251)

a. Does Fig. 11-19 show a Hartley or a Colpitts oscillator?
b. Will a higher positive dc control voltage in the circuit of Fig. 11-19 make the oscillator frequency higher or lower?

11-8
PHASE-LOCKED LOOP (PLL)

The purpose of the phase-locked loop (PLL) circuit is to make a variable-frequency oscillator lock in at the frequency and phase angle of a standard frequency used as the reference. Then the oscillator has the same accuracy as the standard.

The method is to use a phase detector to compare the two frequencies. Any difference of phase results in an error signal that indicates how much the oscillator differs from the standard. The error signal is a dc control voltage that can be used to correct the oscillator frequency.

The phase detector, or comparator, uses two diodes in a balanced rectifier circuit. It has two input signals. The amount of rectified dc output depends on the difference in phase between the two input frequencies. Essentially, the circuit is very similar to the FM detector used for FM signals. (FM detectors are explained in Chap. 17.) However, the phase detector is designed to compare the phase angles within one cycle.

A voltage controlled oscillator is perfect for use with a PLL circuit because the varactor uses a dc control voltage to determine the oscillator frequency and the PLL circuit produces a dc control voltage. A typical arrangement is shown in Fig. 11-20. The four main parts of the circuit are:

1. *Phase detector or comparator.* The two inputs are signals from the VCO with a frequency f_O and a separate oscillator with the standard frequency f_S used as the reference. The output is an error signal that indicates whether f_O is the same as f_S or has a different phase. The detector can have opposite polarities of dc output voltage for lagging and leading phase angles between the two input signals.

2. *Low-pass filter.* This *RC* circuit removes the ac signal variations of the two oscillators from the rectified dc output of the phase detector. Input to the filter is the dc error signal with ac ripple, but the output is a filtered dc control voltage for the dc amplifier.

3. *Dc amplifier.* The dc amplifier increases the amount of dc control voltage for better control.

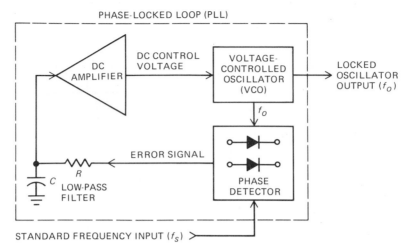

PHASE-LOCKED LOOP (PLL)

DC AMPLIFIER

DC CONTROL VOLTAGE

VOLTAGE-CONTROLLED OSCILLATOR (VCO)

LOCKED OSCILLATOR OUTPUT (f_O)

f_O

ERROR SIGNAL

R

C LOW-PASS FILTER

PHASE DETECTOR

STANDARD FREQUENCY INPUT (f_S)

Fig. 11-20. Basic parts of phase-locked loop or PLL circuit. The purpose is to lock in the frequency of the VCO to a reference frequency.

The amplifier output provides the desired dc level for the control voltage in the polarity needed for the varactor in the VCO.

4. *Voltage-controlled oscillator.* The VCO uses a varactor to set the oscillator frequency, as in the circuit of Fig. 11-19. Input from the dc amplifier keeps the VCO locked into the frequency and phase of the standard oscillator.

The PLL circuit has three operating modes for the VCO. They are free-running, capture, and locked-in, or tracking. If f_O is too far from f_S, the PLL circuit cannot lock in the oscillator. Without lock-in, the VCO is a *free-running oscillator.* Also, the free-running frequency will be produced without any dc control voltage if the PLL circuit is disconnected. When the VCO frequency is within the range of the PLL circuit, however, it produces dc control voltage that reduces the difference between the two frequencies. Once the control voltage starts to change the VCO frequency, the oscillator is in the *capture state.* When f_O is exactly the same as f_S, the VCO is in the *lock-in state.* The PLL will hold the VCO locked in as long as the dc control voltage is applied.

There are many applications for the PLL circuit. In general, when the circuit is used to control the frequency of an oscillator, it is called *automatic*

frequency control (AFC). That function in the RF tuner of television receivers is also called *automatic fine tuning* (AFT). In the horizontal synchronizing circuits, the application is *horizontal AFC.*

One of the most important uses is with a crystal-controlled oscillator as the standard source for a reference. Then the PLL circuit makes an oscillator without a crystal have the same frequency stability as the crystal oscillator. An example of that application is explained in the next section.

Test Point Questions 11-8
(Answers on Page 251)

Refer to Fig. 11-20.

a. Are two diodes used for the phase detector or the dc amplifier?
b. Which stage has a varactor?
c. Label the frequencies of the two input signals to the phase detector.

11-9
FREQUENCY SYNTHESIS

A crystal-controlled oscillator has excellent frequency stability, but the crystal has only one output frequency. A variable-frequency oscillator, on

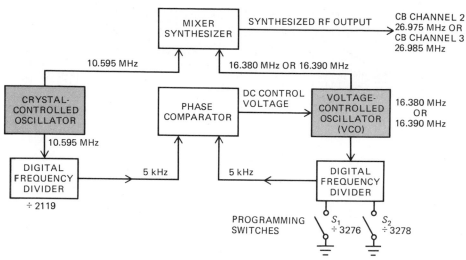

Fig. 11-21. An example of frequency synthesis for transmitter outputs at 26.975 and 26.985 MHz. These are two channel frequencies for citizens band (CB) radio.

the other hand, can produce many output frequencies, but it does not have the accuracy of a crystal. In many applications, especially in citizens' band (CB) radio, it is necessary to have many channel frequencies with crystal-controlled accuracy. Instead of having a separate crystal for each channel, however, frequency synthesis is used to provide multiple channel frequencies from one crystal. Synthesis means putting together, or mixing, two frequencies to provide the desired output.

An example of frequency synthesis is shown in Fig. 11-21 for the specific case of two CB channels with one crystal. (The frequencies of all 40 CB channels are listed in Appendix B.) Actually, all the transmitter frequencies can be synthesized from the same crystal. The crystal accuracy for the output frequency results from having the VCO locked into the crystal oscillator by means of the PLL.

The crystal oscillator provides the frequency standard. In this example, the oscillator operates at 10.595 MHz. However, the output is divided down by the fixed ratio of 1:2119 to provide 5 kHz as the reference frequency for the PLL circuit.

The VCO operates at different frequencies. In this example, the frequency can be either 16.380 or 16.390 MHz. However, the VCO output is di-

vided down by the ratio needed to produce 5 kHz for the PLL.

The 5-kHz signal from the VCO and the 5-kHz reference signal from the crystal oscillator are compared in the phase detector of the PLL. As a result, the dc control voltage locks in the VCO. Its frequency can change only if the standard reference changes.

Each frequency divider is a group of digital flip-flop circuits. For the crystal oscillator output of 10.595 MHz, or 10,595 kHz, the divisor is the same 2119 all the time to produce divided output of 5 kHz. The calculations are

$$\frac{10{,}595 \text{ kHz}}{2119} = 5 \text{ kHz}$$

The frequency division on the VCO is changed according to the frequency desired in the output in order to get 5 kHz for the PLL. In Fig. 11-21, S_1 is closed for the divisor of 3276 with the VCO at 16.380 MHz, or 16,380 kHz. Then

$$\frac{16{,}380 \text{ kHz}}{3276} = 5 \text{ kHz}$$

For the other VCO frequency of 16.390 MHz, or 16,390 kHz, switch S_2 is closed for the division by 3278. Then

$$\frac{16,390 \text{ kHz}}{3278} = 5 \text{ kHz}$$

The mixer or synthesizer combines the direct signals from the VCO and crystal oscillator to produce the sum frequency, which is the synthesized output. These examples are

$$10.595 + 16.380 = 26.975 \text{ MHz}$$

This is the transmitter frequency for CB channel 2. For the transmitter frequency on CB channel 3,

$$10.595 + 16.390 = 26.985 \text{ MHz}$$

These are the two synthesized output frequencies. Additional frequencies can be produced by changing the division on the VCO. The accuracy of each synthesized frequency is crystal-controlled because one component is from the crystal oscillator and the other component, from the VCO, is locked into the crystal oscillator by the PLL circuit.

Test Point Questions 11-9
(Answers on Page 251)

Refer to Fig. 11-21.

a. Is the input to the phase comparator 5 kHz or 10.595 MHz?
b. Do the programming switches change the frequency division on the VCO or the crystal oscillator?
c. When the VCO is operating at 16.400 MHz, what divisor is needed to get 5 kHz?

11-10
RC FEEDBACK OSCILLATORS

At audio frequencies, the LC values for a tuned circuit would be too large. However, RC networks can be used to provide the phase shift for positive feedback. Then the frequency of oscillations depends on the RC values.

The feedback path in the oscillator circuit of Fig. 11-22 includes three RC circuits: R_1C_1, R_2C_2, and R_3C_3. Assume they all have the same values of R and C. At one specific frequency at which they each have a phase shift of 60°, therefore, the feedback is shifted by $3 \times 60° = 180°$. Since the amplifier Q1 has a phase inversion of 180°, the total phase shift of 360° provides positive feedback to make the circuit oscillate. The oscillator output is sine waves at a frequency of 20 Hz to 200 kHz, approximately. The frequency of the oscillator is varied by changing the RC values in the feedback circuit.

In the circuit of Fig. 11-22, C_4 is the coupling capacitor for the oscillator output. R_L is the output load resistance for the collector circuit, returning to the $+28$-V supply. Self-bias for stabilization is provided by R_EC_E in the emitter circuit. The required forward bias for the base is supplied by the R_4R_3 voltage divider from $+28$ V.

Test Point Questions 11-10
(Answers on Page 251)

a. Is a typical frequency for an RC feedback oscillator 1 kHz or 10 MHz?
b. In Fig. 11-22, is C_3, C_4, or C_E in the feedback circuit?

11-11
WIEN-BRIDGE OSCILLATOR WITH AN OPERATIONAL AMPLIFIER

The Wien-bridge oscillator is an RC feedback oscillator, but the circuit uses inverting and noninverting amplification for negative and positive feedback. Those functions can be provided by an operational amplifier (μA) IC unit as shown in Fig. 11-23a. The amplifier output is fed back to the $-$ and $+$ input terminals.

The path with R_2 and R_1 provides negative feedback to the $-$ terminal. The purpose is to reduce amplitude distortion.

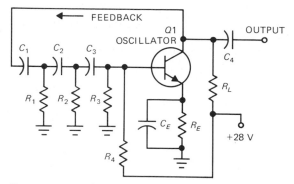

Fig. 11-22. RC feedback oscillator circuit.

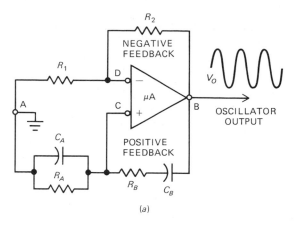

(a)

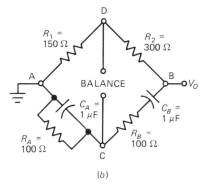

(b)

Fig. 11-23. Wien-bridge oscillator using an operational amplifier. (a) Circuit. DC supply voltages are not shown. (b) Wien bridge alone. Values shown are for balance at 1592 Hz. See text for calculations.

Positive feedback is applied through the path with $R_A C_A$ and $R_B C_B$ to the $+$ terminal. This feedback produces oscillations.

The circuit oscillates only at the frequency at which the bridge is balanced. Then the positive feedback balances out the negative feedback.

For the Wien bridge of Fig. 11-23b the following relative values are shown: $R_2 = 2R_1$, $R_A = R_B$, and $C_A = C_B$. Also, R_A and R_B are about two-thirds the value of R_1. C_A and C_B have values that can have X_C equal to R_A or R_B at the operating frequency. Then the bridge is balanced and the circuit oscillates at the frequency

$$ f = \frac{1}{2\pi(R_B C_B)} \qquad \textbf{(11-2)} $$

where f is in hertz, R is in ohms, and C is in farads. For the values in Fig. 11-23

$$ f = \frac{1}{2\pi(100 \times 1 \times 10^{-6})} = \frac{10^4}{2\pi} $$
$$ f = 1592 \text{ Hz} $$

At that frequency the bridge is balanced because the X_C of C_A or C_B is equal to R_A or R_B. Then the impedance Z_{BC} between points B and C with the series capacitor is equal to twice Z_{CA} with the parallel capacitor. The ratio is 1.4/0.707, approximately. This proportion is the same as the resistive branch with R_2 twice R_1.

At the oscillating frequency, the positive feedback voltage across $R_A C_A$ just equals or barely exceeds the negative feedback voltage across R_1. At any other frequency, the negative feedback is greater which prevents oscillations.

The oscillator frequency is generally varied by rotating C_A and C_B together on a common shaft. Because of the good sine-wave output it produces the Wien-bridge oscillator is often used for audio signal generators with a frequency range of 20 Hz to 200 kHz. Different values for R_A and R_B can be switched in to change the frequency range. Also R_1 is usually a temperature-sensitive resistance to stabilize the output.

Answer True or False.

a. The Wien-bridge oscillator does not require any *LC*-tuned circuit.
b. In the op amp of Fig. 11-23*a*, the V_O output is inverted from the + input terminal.

11-12
MULTIVIBRATORS

The basic MV circuit shown in Fig. 11-24 is a relaxation oscillator for generating square-wave pulses. Two amplifier stages are used. The output of Q1 drives the input of Q2, and the output of Q2 is fed back to Q1. Since each CE amplifier inverts the polarity of the signal, the two stages provide positive feedback, which results in oscillations. The MV oscillator has the practical feature that it does not need a transformer for feedback.

The oscillations are in the ON-OFF conditions for each amplifier. When one stage conducts, it cuts off the other. When the OFF stage starts to conduct, that action cuts off the stage that was on. The circuit is considered a relaxation oscillator because one stage is always resting while it is cut off. Any circuit that alternates between conduction and cutoff at a regular rate can be considered a relaxation oscillator.

The rate at which Q1 and Q2 of Fig. 11-24*a* are cut off determines the oscillator frequency. One cycle includes the time for both stages. As an example, when each stage is cut off for 0.5 ms, the total cutoff period T is 2×0.5 ms = 1 ms. The MV frequency is then 1000 Hz. That would be the frequency of the square-wave signal output in Fig. 11-24*b*.

The multivibrator is a very popular circuit with many applications. It can be used as a simple, economical audio oscillator for a tone generator. In digital logic circuits, it is the clock generator that provides timing pulses. Also, it is used as a sawtooth generator with an *RC* waveshaping circuit. A useful feature is that it can easily be synchronized with external trigger pulses to make it lock in at the frequency of the synchronizing signal.

Collector-Coupled MV Circuit In the collector-coupled MV circuit of Fig. 11-24*a* the collector of Q1 is coupled through C_2 to the base of Q2. Also, the collector of Q2 feeds back to the base of Q1 through C_1. The two stages are *cross-coupled*. Each stage has a collector load R_{L_1} or R_{L_2}. Forward bias for the base is provided by R_3 and R_4 from the dc supply of 5 V. In the bias network R_1 is in the discharge path of C_1, and R_2 is in the discharge path of C_2.

Initially, one stage conducts slightly more than the other even when the amplifier circuits are sym-

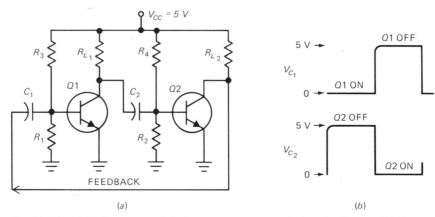

Fig. 11-24. (*a*) Collector-coupled, free-running symmetrical multivibrator. (*b*) Output voltage at collector of Q1 or Q2.

metrical. Any slight difference is amplified by both stages because of the positive feedback. It should be noted that more I_C means less V_C, which provides negative going voltage drive to the next stage.

Assume $Q1$ is conducting more than $Q2$ at the start. The feedback makes $Q1$ conduct more current, and there is less current in $Q2$. Very quickly, then, $Q1$ conducts maximum current to cut off $Q2$. How long $Q2$ remains off depends on the R_2C_2 time constant in the coupling for its base input circuit.

Then $Q2$ starts to conduct. This change is amplified to produce maximum current in $Q2$, which cuts off $Q1$. As a result, the stages alternate between conduction and cutoff.

The collector voltage V_C provides a square-wave output voltage as shown in Fig. 11-24b. The variations are between 5 and 0 V. The 5 V for V_C equals the dc supply voltage when the stage is cut off, without any I_C. The zero value of V_C results when the maximum I_C produces a 5-V drop across R_L.

Remember that a drop in $+V_C$ is a negative change. When V_C goes from 5 to 0 V, the base circuit of the next stage has a drive of -5 V.

How long the base signal is at -5 V depends on the time for the coupling capacitor to discharge with a lower value of V_C. As C_1 or C_2 discharges, the voltage across R_1 or R_2 decreases from -5 V toward zero. When the negative drive at the base drops enough, the stage can start conducting. The stage that conducts can then cut off the other.

When the cutoff time is the same for each stage, the multivibrator is symmetrical and a square-wave output is produced. An unsymmetrical MV provides unequal pulse widths. The output voltage can be taken from the collector of either $Q1$ or $Q2$ with inverted polarities.

Emitter-Coupled MV Circuit For the circuit of the emitter-coupled MV see Fig. 11-25. The collector of $Q1$ drives the base of $Q2$ through the coupling circuit with C_C and R_2. However, $Q2$ feeds back to $Q1$ only through the common R_E.

$Q2$ is cut off by a negative base signal when $Q1$ conducts. In the opposite case, $Q1$ is cut off by a

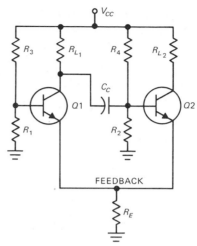

Fig. 11-25. Emitter-coupled multivibrator circuit.

positive emitter voltage across R_E when $Q2$ conducts. At that time, $Q2$ cannot cut itself off because R_2 has a positive base drive from $Q1$.

The emitter-coupled circuit is automatically unsymmetrical with $Q2$ off for more time than $Q1$. The reason is that $Q2$ has an RC coupling circuit for negative base drive but $Q1$ has not.

Free-Running or Astable MV The circuits shown in Figs. 11-24 and 11-25 are free-running, which means that the MV oscillates by itself without the need for any external signal to change states. The free-running MV is astable, because it is not stable when either stage is cut off. The circuit naturally changes states at the frequency of oscillations.

Monostable or One-Shot MV The monostable or one-shot MV circuit has one stable state with one stage conducting. An input pulse is needed to trigger the OFF stage into conduction. Then the MV goes through one cycle of changes and back to its original conditions. This type of MV circuit has reverse bias at the base to hold off conduction until a trigger pulse arrives.

Bistable MV or Flip-Flop The bistable MV or flip-flop also needs trigger pulses, but it has two stable states for either conducting stage. When an input pulse turns on the OFF stage, the conditions reverse and stay that way. Another trigger pulse is needed to put the MV back to its original conditions. The name *flip-flop* describes this idea of switching states one way and then the opposite way.

Eccles-Jordan MV   The Eccles-Jordan MV circuit is a collector-coupled, bistable multivibrator. It is the basic flip-flop commonly used for digital logic circuits.

Schmitt Trigger   The Schmitt trigger circuit is an emitter-coupled, bistable MV. It is often used to change a sine-wave input to a square-wave output. Then the square waves can provide sharp trigger pulses for controlling another circuit.

Test Point Questions 11-12
(Answers on Page 251)

a. In the circuit of Fig. 11-24, does $Q2$ feed back to $Q1$ through C_2, R_2, or C_1?

b. In the circuit of Fig. 11-25, does $Q2$ feed back to $Q1$ through C_C, R_2, or R_E?

c. In a free-running MV, each stage is cut off for $1 \ \mu s$. What is the oscillator frequency?

11-13
SIGNAL GENERATORS

The signal generator is an important application of oscillators. Actually, an RF signal generator is a miniature transmitter to supply an RF output at practically any frequency. An AF signal generator provides an audio signal. Signal generators are often used in servicing radio receivers, television receivers, and audio amplifiers. The three most common types are as follows.

1. *AM RF signal generator.* This generator supplies RF signals at any one frequency in its range (Fig. 11-26). The RF output is available with or without modulation by an internal audio signal. In addition, the audio signal is supplied at separate output terminals. The audio frequency is usually 400 or 1000 Hz.

2. *FM RF signal generator.* A generator of this type is generally called a *sweep generator* because its frequency-modulated output sweeps through a range of radio frequencies. It is used for alignment of receivers, with an oscilloscope in the output, to obtain a visual response curve for tuned amplifiers. This method can be used for AM or FM receivers.

3. *Audio signal generator.* The audio signal generator provides any frequency in the range from 20 Hz at the low end, usually, to more than 100 kHz at the high end. The waveform may be a sine wave or a square wave. When additional waveforms are available, such as a sawtooth and a triangle, the unit is called a *waveform generator* or *function generator.*

RF generators supply sine-wave signals. Maximum signal output is about 0.5 V or less, with 50-Ω shielded coaxial output. Audio signal generators usually have a maximum output of about 2 V.

A Typical RF Signal Generator   Refer to Fig. 11-26. The range switch at the right selects one of five bands A to E for the output frequency of the RF signal generator. The lowest band, A, is 0.31 to 1.1 MHz, or 310 to 1100 kHz. The highest band, E, is 32 to 110 MHz. Within each band the dial pointer at the left is set for a specific frequency that is indicated on the scales. RF output is taken from the terminals at the right; the RF level is adjusted with the amplitude control next to the terminals.

Audio output is taken from the terminals at the left. Its amplitude control sets the audio level. The audio section is turned on for either audio output alone from the audio output terminals or modulated RF signal from the RF output terminals. The audio frequency here is 1000 Hz. Some generators have provision for varying the audio frequency.

Signal Injection   The use of the signal injection technique to localize a defective stage of an AM

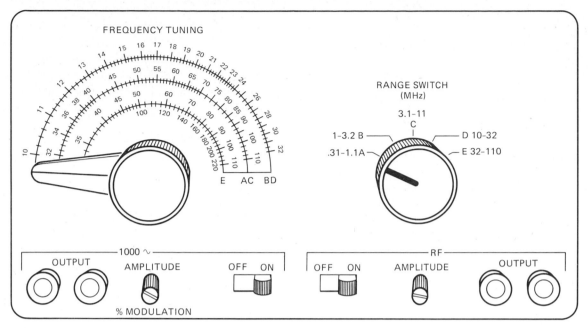

Fig. 11-26. Typical AM, RF signal generator. RF bands are A, B, C, D, and E from 310 kHz to 110 MHz. Width is 11 in. (*Heath Co.*)

radio is illustrated in Fig. 11-27. The general idea is to couple the signal input to each stage and note if it produces an output. The output of the radio shown in Fig. 11-27 is the loudspeaker which produces sound. You start at the loudspeaker and work back to the front. In that way you know which stages are good for testing the others.

Three types of signals are needed from the generator:

1. Audio signal alone for the audio section.
2. Modulated RF signal at 455 kHz for the IF section at 455 kHz. In the receiver, the audio modulation is detected to produce audio output. The 455 kHz is a standard IF value for AM radios.
3. Modulated signal for the RF section at any frequency between 535 and 1605 kHz. Be sure the receiver tuning dial is set to the same frequency as the generator.

As an example of localizing a defective stage, suppose the trouble is lack of an audio output. Feed

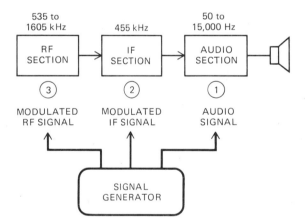

Fig. 11-27. Use of RF signal generator for signal injection in AM radio to localize a defective stage.

audio signal to the collector and base electrodes of each audio amplifier. If the audio tone is heard, the stages are good. Now change the signal to modulated RF output at 455 kHz and check the IF section. Assume there are three IF stages. Assume also that input to the third stage produces an audio output but input to the second stage does not. Therefore, the second IF stage is defective, because it cannot amplify its input signal. The trouble can be the transistor itself, the dc supply voltages, or the transistor's ac coupling circuit to the next stage.

Test Point Questions 11-13
(Answers on Page 251)

a. In Fig. 11-26, which band is for 455 kHz?
b. Is the left or right pair of output terminals used when checking an audio amplifier?

SUMMARY

1. An oscillator circuit is basically an amplifier with positive feedback from output to input.
2. Positive feedback has the same polarity as an input signal that would be applied for amplification.
3. An LC circuit oscillates at its natural resonant frequency $f_r = 1/(2\pi\sqrt{LC})$.
4. In a tickler feedback oscillator, the positive feedback is obtained by a tickler coil inductively coupled to the LC tuned circuit.
5. The Hartley oscillator circuit uses a tapped coil in the LC tuned circuit for inductive feedback instead of the tickler coil.
6. The Colpitts oscillator circuit uses a capacitive voltage divider for feedback.
7. The ultra-audion oscillator is a modified Colpitts circuit using the interelectrode capacitances.
8. The tuned-plate tuned-grid oscillator uses two tuned circuits with feedback through C_{GP}.
9. Crystal-controlled oscillators are used for the best frequency stability. The crystal is equivalent to a tuned circuit with very high Q.
10. The voltage-controlled oscillator (VCO) has a capacitive diode in the tuned circuit to control the oscillator frequency. C varies with the amount of reverse voltage across the diode.
11. The phase-locked loop (PLL) uses a phase detector to compare one input frequency with a reference standard. Dc control voltage from the detector can make a VCO lock in at the same frequency and phase as a crystal oscillator.
12. A circuit for frequency synthesis uses a PLL with a VCO and frequency dividers to provide different output frequencies from one crystal-controlled oscillator. See Fig. 11-21.
13. RC feedback oscillators are used as audio oscillators. The frequency depends on the values of R and C.
14. The Wien-bridge oscillator uses RC feedback in a balanced-bridge circuit. It provides sine-wave audio output with excellent waveform.
15. The multivibrator (MV) circuit uses two amplifier stages, and the output of one stage drives the input of the other. The oscillations are in the ON-OFF states for each stage. Two basic circuits are the collector-coupled MV and the emitter-coupled MV.
16. Signal generators are test equipment useful for checking receivers. The generator supplies both RF and AF signals.

SELF-EXAMINATION

Choose (a), (b), (c), or (d).

1. The two main parts of a tank circuit are (a) R and C, (b) L and C, (c) R and L, (d) R and an NPN transistor.
2. When L is doubled and C is halved, the f_r is (a) double, (b) one-half, (c) one-quarter, (d) the same.
3. Positive feedback is the same as (a) regeneration, (b) degeneration, (c) negative feedback, (d) frequency synthesis.
4. The oscillator that uses a tapped coil in the LC tuned circuit is the (a) Pierce, (b) Colpitts, (c) Hartley, (d) Armstrong.
5. The circuit that uses a capacitive voltage divider to provide feedback is the (a) Colpitts, (b) Hartley, (c) Armstrong, (d) multivibrator.
6. RF feedback oscillators are usually tuned by varying the (a) bias, (b) supply voltage, (c) L or C, (d) load impedance.
7. The oscillator with the best frequency stability and accuracy is the (a) Hartley, (b) Colpitts, (c) tickler feedback, (d) crystal-controlled oscillator.
8. A typical frequency for an RC feedback oscillator is (a) 1 kHz, (b) 100 MHz, (c) 1000 MHz, (d) 1 GHz.
9. When a 6-MHz crystal is followed by a frequency tripler, the output will be (a) 6 MHz, (b) 2 MHz, (c) 18 MHz, (d) 54 MHz.
10. Which of the following statements is true of the Hartley oscillator shown in Fig. 11-6? (a) Positive feedback is coupled to the emitter; (b) C_2 is the tuning capacitor; (c) L_1 is used as an RF choke; (d) series feed is used for V^+.
11. The varactor in a voltage-controlled oscillator needs (a) audio signal below 1000 Hz, (b) reverse dc control voltage, (c) forward bias, (d) parallel capacitance more than 5 μF.
12. The phase comparator in a PLL circuit is used to provide (a) one-half the crystal oscillator frequency, (b) RF output with audio modulation, (c) dc control voltage, (d) double the crystal oscillator signal.
13. In Fig. 11-21 one of the synthesized output frequencies is (a) 5 kHz, (b) 10.595 MHz, (c) 16.380 MHz, (d) 26.975 MHz.
14. In Fig. 11-21 the frequency division for the crystal oscillator is (a) 1/2119, (b) 1/3276, (c) 1/3278, (d) 1/3280.
15. Which of the following applies to the multivibrator shown in Fig. 11-24? (a) Schmitt trigger; (b) bistable flip-flop; (c) free-running, (d) emitter-coupled.
16. Which of the following applies to the signal generator shown in Fig. 11-26? (a) Function generator; (b) FM sweep generator; (c) square-wave generator; (d) modulated or unmodulated RF output.

ESSAY QUESTIONS

1. Name five types of tuned RF feedback oscillator circuits.
2. In a tuned RF feedback oscillator, which part of the circuit **a.** generates oscillations, **b.** provides feedback, **c.** supplies dc power?

3. Define the following: ringing, damped oscillations, tank circuit, flywheel effect, frequency multiplier.
4. Compare the uses for positive feedback and negative feedback.
5. Compare the effect of regeneration and degeneration.
6. Name two types of relaxation oscillator circuits.
7. Draw the circuit diagram of a Hartley oscillator with series feed.
8. Draw the circuit diagram of a Colpitts oscillator with shunt feed.
9. What is the advantage of crystal control for an oscillator?
10. Name two types of cuts for quartz crystals in oscillator circuits.
11. What is meant by frequency synthesis?
12. Compare the connections for series feed and shunt feed in a Hartley oscillator.
13. What is the difference between the Hartley and Colpitts oscillator circuits?
14. Compare the method of feedback for the Hartley oscillator and TPTG circuit.
15. Give two ways to test whether an oscillator is generating an ac output.
16. What is meant by the abbreviations VFO, VCO, PLL, ECO, and TPTG for oscillator circuits?
17. Give the input and output voltages for a PLL circuit?
18. Explain briefly how the frequency of the VCO shown in Fig. 11-19 is changed.
19. Explain briefly how 26.975 MHz is synthesized for the RF output in the circuit of Fig. 11-21.
20. Which is better for audio frequencies, the TPTG or the Wien-bridge oscillator? Explain why.
21. Which six components determine the frequency of the RC feedback oscillator shown in Fig. 11-22?
22. Which two components of the Wien-bridge oscillator shown in Fig. 11-23 determine the amount of negative feedback?
23. Compare the circuits of a collector-coupled and an emitter-coupled multivibrator.
24. Give the uses of three types of signal generators.

PROBLEMS
(Answers to odd-number problems at back of book)

1. Calculate the frequency of a tuned RF feedback oscillator with 240 μH for L and 180 pF for C.
2. If the C in Problem 1 is reduced to 90 pF, what will be the new oscillator frequency?
3. What value of L is needed with a C of 20 pF for an oscillator frequency of 110.7 MHz?
4. For the C of 20 pF what value of L is needed for the frequency of 55.35 MHz?
5. For the Wien-bridge circuit shown in Fig. 11-23b, calculate X_C for the 1 μF of C_A or C_B at the oscillator frequency of 1592 Hz.
6. Show the calculations for synthesizing 26.965 MHz for the RF output in the transmitter of Fig. 11-21.

SPECIAL QUESTIONS

1. Give two applications of audio oscillators and two of tuned RF oscillators.
2. Name two types of oscillators for the microwave band.
3. Explain briefly how division by 2119 can be obtained with digital dividers.

ANSWERS TO TEST POINT QUESTIONS

11-1 a. L and C
      b. Decrease
11-2 a. L_2
      b. Regenerative
      c. C_1
11-3 a. L_1
      b. Opposite
      c. Shunt
11-4 a. C_B
      b. L_1

11-5 a. C_{GP}
      b. Negative
11-6 a. 3.275 MHz
      b. 25,000
      c. Colpitts
11-7 a. Hartley
      b. Higher
11-8 a. Phase detector
      b. VCO
      c. f_O and f_S

11-9   a. 5 kHz
       b. VCO
       c. 3280
11-10 a. 1 kHz
       b. C_3
11-11 a. T
       b. F
11-12 a. C_1
       b. R_E
       c. 0.5 MHz
11-13 a. Band A
       b. Left

Chapter 12
Modulation and Transmitters

The transmitter sends out an RF signal to the receiver. Two features of the transmitter are appreciable RF power output and a specific frequency of the transmitted RF signal. The more power in the radiated signal, the greater is the transmission distance. Furthermore, the frequency of the RF signal is in an assigned band. Transmitters are regulated by the Federal Communications Commission (FCC).

The heart of any transmitter is a tuned RF oscillator stage that generates the RF carrier wave. In addition, though, transmitters usually include amplifiers for the oscillator output and circuits for modulating the RF wave. Modulation is the process of varying the RF output in step with a lower-frequency signal such as an audio signal. The most common modulating methods are amplitude modulation (AM) and frequency modulation (FM). More details are explained in the following topics:

12-1
TRANSMITTER REQUIREMENTS

Figure 12-1 shows a small transmitter used for radio telephone communication. The unit shown is designed to broadcast over the amateur radio band from 16 to 54 MHz at a rated power of 100 W.

The block diagram in Fig. 12-2 illustrates the stages in an AM transmitter with audio modulation. Starting at the top, left, the oscillator stage generates the RF carrier wave. Typical circuits are the Hartley, Colpitts, and crystal-controlled oscillators.

Fig. 12-1. A 100-W transmitter for radiotelephone operation in the amateur bands from 16 to 54 MHz. (*Heath Co.*)

The next stage in the diagram is an RF amplifier for the oscillator output. This amplifier has two functions. First, it provides enough signal to drive the power amplifier. Also, the driver separates the oscillator from the power amplifier to reduce loading on the oscillator circuit and improve frequency stability.

The power amplifier provides the amount of antenna current needed for the desired power output. More antenna current means that a stronger electromagnetic wave is radiated. Therefore, the signal can be transmitted over a longer distance. In addition, the power amplifier of Fig. 12-2 is modulated by the audio signal. The amplitude variations of the modulated RF signal are present in the output of the power amplifier and the antenna circuit.

A simpler transmitter can have just the master oscillator and power amplifier (MOPA). An MOPA transmitter is generally keyed on and off for radiotelegraphy.

The power supply is needed to provide dc voltage to all the stages.

RF Carrier Frequencies   The transmitted carrier signal must be a radio frequency in order to have efficient radiation by the antenna. Audio frequen-

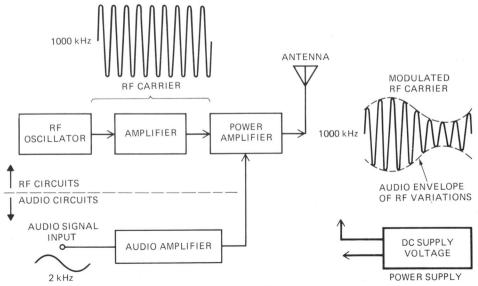

Fig. 12-2. Block diagram of AM transmitter with audio modulation.

Table 12-1
Major Radio-Frequency Bands

Name	Abbreviation	Frequencies	Notes
Very low frequency	VLF	Below 30 kHz	Radio location equipment
Low frequency	LF	30 to 300 kHz	Maritime radio navigation
Medium frequency	MF	300 kHz to 3 MHz	Includes AM radio broadcast band
High frequency	HF	3 to 30 MHz	Includes citizens' band (CB) radio
Very high frequency	VHF	30 to 300 MHz	Includes FM radio broadcast band and television VHF channels 2 to 13
Ultra high frequency	UHF	300 MHz to 3 GHz*	Includes television UHF channels 14 to 83
Super high frequency	SHF	3 to 30 GHz*	Satellite communications
Extremely high frequency	EHF	30 to 300 GHz*	Satellite communications

*1 GHz = 1×10^9 Hz. Microwaves are from 0.2 to 100 GHz.

cies cannot be radiated to any great extent. By means of the modulation, then, the carrier provides the transmission of the RF wave while the modulating signal has the desired information.

In general, radio frequencies extend from 20 kHz up to many gigahertz. (One gigahertz equals 10^9 Hz.) Specifically, RF transmission is generally considered in the frequency bands listed in Table 12-1. These are the major divisions in the RF spectrum. Note that each band is higher by a multiple of 10. As an example, the VHF band at 30 to 300 MHz is 10 times higher than the HF band at 3 to 30 MHz.

Frequency Tolerance The frequency tolerance specifies the required stability of the carrier frequency. In the AM radio broadcast band, the tolerance is ±20 Hz. As an example, that tolerance means that a carrier frequency of 600 kHz cannot exceed 600,020 Hz nor drop below 599,980 Hz. The calculations for 600 kHz (600,000 Hz) are

$$600,000 + 20 = 600,020 \text{ Hz}$$

and

$$600,000 - 20 = 599,980 \text{ Hz}$$

In the citizens' band (CB) class D service around 27 MHz, the tolerance is ±0.005 percent. With a carrier frequency at 27 MHz,

$$
\begin{aligned}
\text{Tolerance} &= 27 \times 10^6 \times 0.005 \text{ percent} \\
&= 27,000,000 \times 0.00005 \\
&= 1350 \text{ Hz}
\end{aligned}
$$

The transmitted carrier cannot be above or below the assigned frequency by more than 1350 Hz.

For FM and TV stations in the VHF band of 30 to 300 MHz, the frequency tolerance is ±1 kHz. As an example, consider an FM radio station with an assigned carrier frequency of 96 MHz, which is 96,000 kHz. The average frequency cannot be more than 96,001 kHz, or 96.001 MHz, nor less than 95,999 kHz, or 95.999 MHz.

Test Point Questions 12-1
(Answers on Page 286)

a. In Fig. 12-2, which stage generates the RF carrier?

b. In Fig. 12-2, what is the frequency of the audio modulation?

c. Give the frequencies in the VHF radio band.

12-2
PRINCIPLES OF MODULATION

The process of modulation can be defined as varying the RF carrier wave in accordance with the intelligence or information in a lower-frequency signal. During modulation, the basic sine waves in the RF carrier are converted into a complex waveform. The waveform is still an RF signal, but it contains the variations of the modulating signal. The RF variations allow transmission over long distances, and the modulation has the desired intelligence.

The Baseband Signal   The lower-frequency signal that modulates the RF carrier is called the *baseband signal,* a general term that includes all types of modulating information. Three examples are:

1. Audio-signal modulation for voice or music information.
2. Video-signal modulation for picture information.
3. Pulse data for numerical information.

Referring back to Fig. 12-2, the 2 kHz signal is an example of modulation using audio frequencies. In general, the AF range for baseband signals can be 300 to 3000 Hz for voice communications, 50 to 5000 Hz for AM radio broadcasting, and 50 to 15,000 Hz for FM radio broadcasting.

As another example of baseband modulation, the video signal for the picture information in television has frequencies from 30 Hz to 4 MHz, approximately. Note that here the video signal frequencies can be as high as 4 MHz, which itself is a radio frequency. However, the frequencies of the picture carrier signal in the VHF and UHF bands are still much higher than the baseband frequencies.

The frequencies in the baseband signal must always be much lower than the RF carrier frequency being modulated. The reason is that, at any one instant on the baseband signal, the time must be long enough to include many cycles of the RF carrier wave. Otherwise, the individual cycles in the carrier wave would not accurately follow the variations in the baseband modulation. The result would be a distorted output signal.

Types of Modulation   The two most common methods of having the baseband signal vary the

RF carrier wave are amplitude modulation (AM) and frequency modulation (FM). In AM, the instantaneous amplitude of the RF carrier varies with the modulating signal. An example of AM is shown by the waveform for the modulated RF carrier in Fig. 12-2. To recover the modulation, the AM detector circuit in the receiver must extract the amplitude variations of the RF carrier signal.

In FM the varying amplitudes of the modulating signal are made to change the instantaneous frequency of the RF carrier wave. In effect, the frequency of the RF carrier is *wobbled* around its center value, or resting frequency. However, the average frequency is still the assigned carrier frequency. In order to recover the modulation, the FM detector circuit in the receiver must convert the frequency variations into corresponding amplitude variations and then extract the modulating signal.

Still another important method is *pulse modulation;* the RF carrier wave is transmitted in pulses. The position of the pulses, or the pulse width, can be made to vary in accordance with the modulating information.

AM and FM are the two methods used in commercial radio and television broadcasting. In the standard radio broadcast band of 535 to 1605 kHz AM is used. The FM radio broadcast band is 88 to 108 MHz. In television, the video modulation produces an AM picture carrier signal. However, the associated sound in television is an FM signal.

In addition, AM and FM are commonly used in two-way radio systems. In the CB radio service AM is used.

Test Point Questions 12-2
(Answers on Page 286)

a. Is the baseband frequency higher or lower than the RF carrier?
b. In the standard radio broadcast band is AM or FM used?
c. Is AM or FM used for the picture carrier signal in television?

12-3
AMPLITUDE MODULATION

The block diagram in Fig. 12-2 illustrates a typical AM radio telephone transmitter. *Radiotelephone* indicates voice communications, but, in general, it means all types of audio information including music. For modulation of the transmitter, note that the final RF power amplifier has two input signals. One is the carrier at 1000 kHz from the RF section; the other is the audio-modulating voltage at 2 kHz from the audio section of the transmitter. The output of the audio amplifier is connected to the RF amplifier to modulate its output.

Audio Section The audio output of some devices, for example, a microphone, is quite weak. Typical values are in millivolts. To have enough voltage for modulation, the audio signal must be amplified. For voice, the first audio stage is called the *speech amplifier*. The last stage is the audio modulator. That stage is a power amplifier to feed the modulation to the RF section.

Plate Modulation The plate modulation method for AM is illustrated in Fig. 12-3. In this case modulation takes place in the plate circuit of the final RF power amplifier. The voltage values for each cycle of audio modulation are in Table 12-2.

Note that the audio voltage across the secondary of the modulation transformer T_2 is in series with the V^+ supply and the RF amplifier plate circuit. Therefore, the audio-modulating voltage varies the plate voltage of the RF amplifier at the audio rate.

When the audio voltage is at zero, the RF amplifier plate voltage is 600 V, as it is without any modulation. However, when the audio-modulating voltage is at $+600$ V, the plate supply for the RF amplifier is effectively 1200 V. That value is equal to 600 V for V^+ from the dc power supply plus 600 V from T_2. Then the amount of RF output is double the unmodulated carrier level. For the opposite case, when the audio-modulating voltage is at -600 V, the plate supply for the RF amplifier is $600 - 600 = 0$ V. Then the RF output is zero.

As a result, the varying amplitudes of the RF carrier wave provide an outline that corresponds to the audio-modulating voltage. The outline shown in dashed lines at the top and bottom of the waveform is called the *modulation envelope*. Both the positive and negative peaks of the AM carrier are symmetrical around the center axis of the carrier, because both envelopes have exactly the same variations.

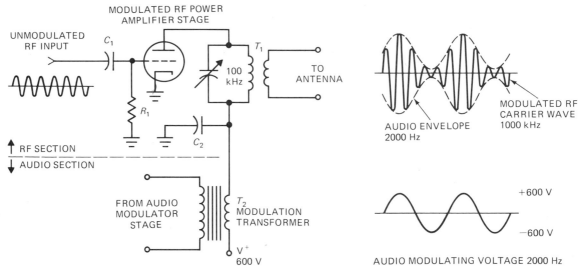

Fig. 12-3. Circuit for plate modulation of a triode tube used as the final RF amplifier. The 1000-kHz RF carrier is amplitude-modulated by the 2-kHz audio signal. See voltages in Table 12-2.

Table 12-2
Modulation Values for Fig. 12-3

Audio Voltage V_A	Plate Supply V^+	RF Amplifier Plate Voltage $(V_A + V^+)$, V	Modulated RF Signal Amplitude
0	600	600	Carrier level
+600	600	1200	Double carrier level
0	600	600	Carrier level
−600	600	0	Zero
0	600	600	Carrier level

It should be noted that the envelope really exists not as an audio signal in itself, but only as amplitude variations of the RF carrier. In fact, the top and bottom envelopes vary in opposite polarities. In terms of audio voltage, the two opposite envelopes cannot be used at the same time to provide the audio information. That is why the modulated carrier must be rectified in order to detect the audio modulation. The detector can rectify either the positive or negative half-cycles of the RF carrier. Then the amplitude variations in either the positive or negative envelope can provide audio signal output.

Percent of Modulation The percent of modulation specifies the variation in the AM signal compared with the unmodulated carrier. It is also called the modulation factor m. The formula for calculating the percent of modulation is

$$m, \% = \frac{max - min}{max + min} \times 100 \quad \textbf{(12-1)}$$

The max and min values are voltages or currents on the modulated carrier. An example of 100 percent modulation is illustrated in Fig. 12-4a. There the max value is 8 A and the min value is 0 A for the modulated current on the antenna. Then

$$m = \frac{8 - 0}{8 + 0} \times 100$$

$$= \frac{8}{8} \times 100$$

$$m = 1 \times 100 = 100 \text{ percent modulation}$$

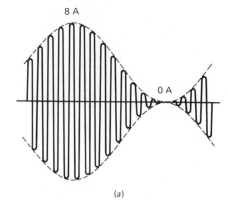

(a)

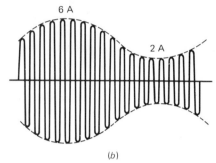

(b)

Fig. 12-4. Examples of the percent of modulation, or modulation factor m. (a) 100 percent. (b) 50 percent.

Note that, for 100 percent modulation, the RF signal varies up to double the unmodulated carrier level and down to zero.

As another example, 50 percent modulation is shown in Fig. 12-4b. The max and min values of antenna current are 6 and 2 A, respectively. Then

$$m = \frac{6 - 2}{6 + 2} \times 100$$

$$= \frac{4}{8} \times 100$$

$$m = 0.5 \times 100 = 50 \text{ percent modulation}$$

Overmodulation It is important to have a high percent of modulation at the transmitter. That allows the most audio signal to be recovered by the detector circuit in the receiver. The modulation is generally maintained close to 100 percent but not at that level, because complex audio waveforms can exceed the 100 percent level. More than 100 percent is called *overmodulation*. In the example of Fig. 12-5, note that the overmodulated carrier signal is actually off for part of the time. The result is two big problems. First, the modulating signal is badly distorted. Also, the effect of turning the carrier signal on and off is to generate new frequencies that interfere with nearby channels. The generation of interfering frequencies outside the assigned channel is called *splatter*.

Oscilloscope Modulation Patterns   The oscilloscope can be used to check the modulation. By connecting the modulated RF signal to the vertical input terminals and using the internal horizontal sweep of the oscilloscope, the signal waveforms shown in Figs. 12-4 and 12-5 can be observed. However, a more accurate analysis can be made by using the audio-modulating voltage for the oscilloscope horizontal input, instead of the internal sweep, as shown in Fig. 12-6.

The trapezoidal patterns in Fig. 12-6 present a visual display of the maximum and minimum values in the modulated wave. The height marked *A* is the maximum, and *B* is the minimum. Therefore, the percent of modulation can be calculated as $(A - B)/(A + B)$ corresponding to formula 12-1. The examples in Fig. 12-6a and 12-6b are for 33⅓ and 100 percent modulation. In Fig. 12-6c the modulation is more than 100 percent because the overmodulation results in zero carrier level over part of the modulation cycle. Values more than 100 percent cannot be calculated by the formula.

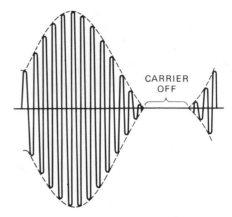

Fig. 12-5. Example of overmodulation.

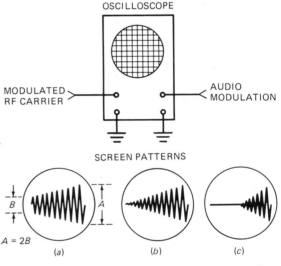

Fig. 12-6. Trapezoidal modulation patterns. Oscilloscope connections are shown at top. See text for calculations. (*a*) 33⅓ percent modulation. (*b*) 100 percent modulation. (*c*) Overmodulation.

Test Point Questions 12-3
(Answers on Page 286)

a. In Fig. 12-3, what is the RF amplifier plate voltage when the audio-modulating voltage is +600 V?

b. In Fig. 12-3, what is the frequency of the modulated carrier wave?

c. The maximum *I* or *V* on the modulated carrier wave is 5 units, and the minimum is 3 units. Calculate the percent of modulation.

12-4
HIGH- AND LOW-LEVEL MODULATION

The terms high- and low-level modulation refer to the RF power level at which the baseband signal modulates the output. *High-level modulation* means that the modulating signal is applied when the RF power is the rated output to the antenna. An example is the circuit shown in Fig. 12-3, where the

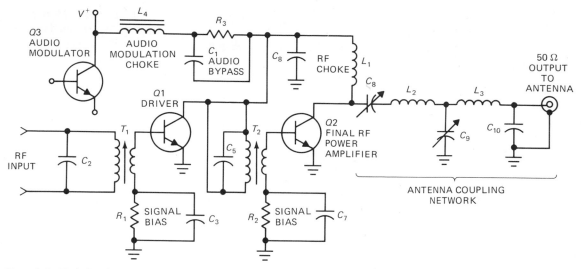

Fig. 12-7. High-level modulation in a transistor collector circuit. Both $Q2$ and $Q1$ are modulated.

audio signal modulates the plate circuit of the final RF power amplifier. Another example, of *collector modulation* with transistors, is shown in Fig. 12-7.

Collector modulation in junction transistors corresponds to plate modulation in tubes. However, both the power amplifier and its driver stage usually are collector-modulated. In Fig. 12-7, note that the collector circuit of $Q2$ and $Q1$ has audio-modulating voltage applied from $Q3$. The reason is that modulating only the collector of the output transistor will not produce 100 percent modulation. During the part of the modulation cycle when the collector voltage is low, saturation in the transistor does not allow the RF output to follow the modulation. Also, part of the RF drive feeds through to the antenna. Therefore, it is necessary to modulate the collector circuit of both the driver stage and the final RF amplifier.

Notice that the collector of the modulated RF stage $Q2$ is in series with the audio choke L_4 for its dc supply voltage. The series path is through RF choke L_1, the dc voltage-dropping resistor R_3, the audio choke L_4, to V^+. As a result, the audio voltage across the choke can modulate the RF collector voltage without the need for an audio-modulation transformer. Such a circuit with series con-

nections for V^+ with an audio-modulating choke is called *Heising modulation*. The same method is often used for plate modulation with tubes.

High-level modulation is efficient and has good linearity hence little distortion, but it requires appreciable audio power. Specifically, for 100 percent modulation the required audio power is one-half the RF power. For example, with an RF carrier output of 1 kW, the audio power needed for 100 percent modulation is 0.5 kW.

Low-Level Modulation   In low-level modulation the modulating signal is applied in the RF circuits in which the RF power level is much less than the output to the antenna. The advantage is that much less modulating power is needed. With AM, though, there is a problem in preserving the modulation envelope in the amplifiers that follow the modulating stage. Class A or AB operation must be used rather than class C with its high efficiency. In general, the lower efficiency resulting from not using class C operation cancels any advantage derived from using less audio power. It should be noted that these limitations do not apply to FM, because FM does not produce an envelope of amplitude variations.

Answer True or False.

a. Both Fig. 12-3 and Fig. 12-7 show high-level modulation.
b. In Fig. 12-7, both $Q1$ and $Q2$ are modulated by audio voltage.
c. Class A operation has the highest efficiency.
d. The advantage of low-level modulation is that less power is needed for the baseband signal.

12-5
AM TRANSMITTERS

Figure 12-8 is a block diagram of a typical AM transmitter. The first stage at the left is the oscillator that provides a stable source of RF energy to generate the carrier frequency. Next is the buffer stage to amplify the RF output from the oscillator. The main function of this amplifier is to buffer, or protect, the oscillator from the rest of the transmitter RF circuits. Otherwise, changes in tuning or loading conditions could affect the stability of the oscillator frequency. The buffer stage is generally an emitter follower. This circuit provides good isolation because of its high input impedance, which minimizes any loading effect on the oscillator.

The next stage in the RF chain is the *intermediate power amplifier* (IPA). It provides the power gain needed to drive the final power amplifier. The IPA can also be considered a driver stage. More than one IPA stage may be used for the required power. The IPA generally operates class C for maximum efficiency. That operation is permissible because the RF carrier is not modulated here.

The final RF amplifier provides the power output needed for the antenna circuit. Again class C operation is used for best efficiency. High-level modulation is shown in Fig. 12-8 with the audio signal applied in the output circuit of the RF power amplifier. Either plate or collector modulation can be used with tubes or transistors. In general, transistors are available for RF power output up to about 100 W. Vacuum tubes are used for transmitters with power output in the kilowatt range.

In the audio section for the modulating signal, the first stage is an amplifier for the microphone signal. The next stage also is an audio amplifier, but it has the additional function of compressing the peak amplitudes in the audio signal to allow a higher average percent of modulation. Otherwise, the average modulation level would have to be lowered in order to prevent overmodulation on peak voltages.

The last stage in the audio section is a power amplifier. It provides the amount of audio signal needed to modulate the final RF power amplifier.

Transmitter Tuning   Tuning is essentially a question of setting all the tuned circuits for resonance at the required frequency. In a transmitter the tuning is critical because it allows maximum power output for the best transmission. Furthermore, cor-

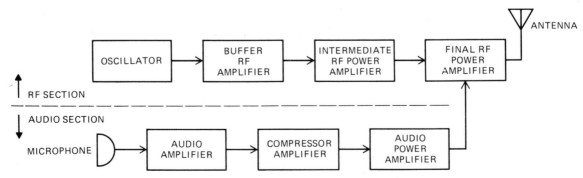

Fig. 12-8. Block diagram of typical AM transmitter in which high-level modulation is used.

rect tuning provides maximum efficiency in the operation of the RF amplifiers.

To be more specific, let us assume the transmitter RF carrier output is at 500 kHz. Then the oscillator, buffer, IPA, and final RF amplifier are tuned to that frequency. All the amplifier stages are tuned for maximum output at the oscillator frequency which is 500 kHz. Consider the buffer amplifier first. It is tuned for a maximum 500-kHz signal into the IPA. Then the IPA is tuned for a maximum 500-kHz signal into the final RF amplifier. As the last step, the output of the power amplifier is tuned for the antenna circuit.

When an RF amplifier is tuned, a dc meter can be used to indicate maximum RF signal output. Measuring with dc is much easier than reading RF voltage or current. The two dc methods are:

1. The RF output of one stage is tuned until there is maximum dc signal bias at the input to the next stage.
2. The RF output is tuned until minimum dc plate current or collector current flows in the same stage.

Both methods are illustrated in Fig. 12-9, but only one is used at a time. Remember that signal bias is rectified signal input for either tubes or transistors. We assume there is enough drive to rectify the signal input. These stages usually operate class C.

Why the dc collector current drops when the LC circuit is tuned to resonance (Fig. 12-9) may not seem so obvious. The reason is that, when the output circuit is tuned to the frequency of the input signal, the amplifier has maximum ac output voltage. As a result, the collector voltage varies between high and low values. Off resonance, the collector voltage remains at a high value. When the collector circuit is tuned, then, the dc milliammeter shows a sharp dip to a minimum that indicates maximum ac signal.

Antenna-Matching Network The output circuit of the final RF amplifier has additional requirements because the output signal must be coupled to the antenna. Figure 12-10 shows a circuit for matching the high impedance of the output of the final RF power amplifier to the low impedance of the antenna circuit. Inductor L_1 and capacitors C_2 and C_3 form a π network. C_2 is used to tune the amplifier by adjusting for minimum plate current. C_3 is used to tune the antenna circuit for the required amount of loading on the amplifier. The dc blocking capacitor C_1 is needed to prevent a

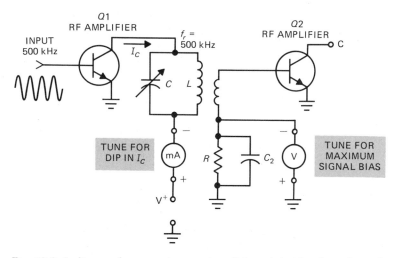

Fig. 12-9. Indicators for transmitter tuning. C is varied either for a dip in I_C or maximum signal bias in the next stage.

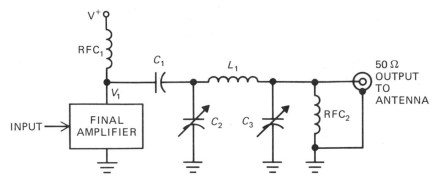

Fig. 12-10. Use of a π-matching network for coupling the transmitter output to the antenna.

short circuit of the dc supply voltage to ground through the choke RFC_2. This choke provides a dc path to ground for the antenna circuit. The other choke, RFC_1, isolates the antenna circuit from the low impedance of the power supply.

The idea of loading the power amplifier is to have the correct amount of power on the antenna. That value can be indicated by an RF meter in the antenna circuit or a dc meter for plate current in the amplifier. In Fig. 12-10, C_3 on the output side of the π network varies the loading by changing the impedance in the antenna circuit. That factor determines how much of the amplifier output is coupled to the antenna. C_2 on the input side of the π network tunes the amplifier plate circuit for resonance. For the example of a tube as the final RF power amplifier, the tuning and loading procedure requires the following steps:

1. Tune C_2 for minimum dc plate current.
2. Tune C_3 for either a specified value of dc plate current or RF antenna current.
3. Retune C_2 for minimum dc plate current.
4. Readjust C_3 and C_2 to obtain the required loading for the desired amount of output, but with C_2 tuned for minimum plate current.

The final step must be to tune C_2 for minimum dc plate current. Otherwise, the amplifier will not operate at maximum efficiency. In a transmitter, efficiency is of extreme importance because of the high power involved.

Test Point Questions 12-5
(Answers on Page 287)

Answer True or False.

a. A buffer amplifier can use the emitter-follower circuit.
b. RF power amplifiers of the unmodulated carrier signal usually operate class C.
c. In Fig. 12-9, the RF amplifier $Q1$ can be tuned for maximum I_C.
d. In Fig. 12-10, the antenna loading capacitor C_3 is usually adjusted for minimum output.

12-6
SIDEBAND FREQUENCIES

In an AM signal, the changes in amplitude caused by modulation actually generate new frequencies slightly higher and lower than the carrier frequency. The components close to the carrier are the side frequencies.

As an example, a 1000-kHz carrier with 2-kHz audio modulation is shown in Fig. 12-11. In Fig. 12-11a, the usual modulated waveform of the AM signal is shown as a *time-domain graph*. The display is a plot of amplitude variations on the vertical axis with respect to time on the horizontal axis. You can see this waveform can be viewed on an oscilloscope by connecting the AM signal to the vertical input terminals while using the internal horizontal sweep. The internal sweep provides horizontal deflection that is linear with respect to time. As a result, the horizontal axis of the display is a linear time base with equal displacements for equal periods of time. By setting the horizontal sweep frequency for the audio-modulating frequency the audio-modulation envelope can be seen.

The *frequency domain graph* in Fig. 12-11*b* has the horizontal axis in units of frequency. The dis-

play shows the RF side frequencies generated by the modulation but without the amplitude variations of the carrier. The frequency domain graph cannot be shown by ordinary oscilloscopes. However, a special type called the *spectrum analyzer* displays the frequency components. A modern analyzer is shown in Fig. 12-12.

Calculating the Side Frequencies Each RF side frequency differs from the RF carrier frequency by the value of the audio-modulating frequency. To calculate the side frequencies:

$$f_{US} = f_c + f_a$$
$$f_{LS} = f_c - f_a$$

(12-2)

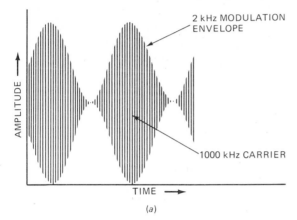

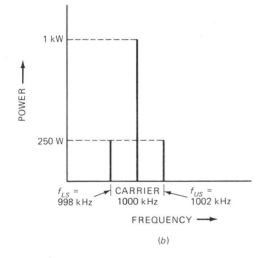

Fig. 12-11. Two ways to view an AM signal, shown for 100 percent modulation. (*a*) Time-domain graph with the envelope of audio modulation. (*b*) Frequency-domain graph with RF sideband frequencies that differ from the carrier frequency by the frequency of the audio modulation.

Fig. 12-12. Spectrum analyzer used to display sideband frequencies on oscilloscope screen. (*Hewlett Packard Co.*)

Chapter 12/Modulation and Transmitters **263**

where f_c is the carrier frequency and f_a is the audio-modulating frequency. For the example in Fig. 12-11b, the upper side frequency is

$$f_{US} = 1000 + 2 = 1000 \text{ kHz}$$

and the lower side frequency is

$$f_{LS} = 1000 - 2 = 998 \text{ kHz}$$

The modulation always generates both upper and lower side frequencies. Each is an RF value close to the carrier frequency. In fact, the side-frequency components can be called *side carriers*.

Modulation Energy in the Sidebands For 100 percent modulation, each side-frequency component has one-fourth of the power, or one-half of the voltage and current, compared with the carrier power without modulation. The two side frequencies combined have one-half the unmodulated carrier power, or the AM signal has 50 percent more power than the same carrier without modulation. The audio modulator supplies the added power.

What Is an AM Signal? The question of whether the AM signal is a carrier with varying amplitudes or a constant-amplitude carrier with upper and lower side frequencies can be answered by saying that the two concepts add up to the same thing. The carrier with its side frequencies is shown in Fig. 12-13a, and the amplitude variations in the AM carrier are shown in Fig. 12-13b. If the three graphs in Fig. 12-13a were combined to an exact scale, the result would be the AM waveform in Fig. 12-13b. When the amplitudes of the side frequencies are added to the carrier with the same polarity, the carrier amplitude increases. When the side frequencies cancel part of the carrier amplitudes because of opposite polarities, the carrier amplitude decreases. The constant carrier with its side frequencies (Fig. 12-13a) can be considered as one package that is equivalent to the single AM waveform in Fig. 12-13b.

We can consider audio-modulating information either as the amplitude variations of the carrier or as the varying energy of the side carriers. The proof is that if the AM signal were to be passed through an extremely sharp filter that had a bandwidth too narrow to include the side frequencies, the output would be the carrier without any modulation. That is why tuned RF amplifiers for an AM carrier signal must have enough bandwidth to include the side frequencies.

The RF side carriers should not be compared with the audio-modulation envelope in an AM signal. The envelope is an audio frequency, but the side carriers are radio frequencies. In our example, the modulation envelope is at 2000 Hz, or

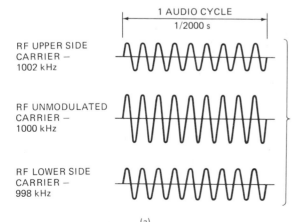

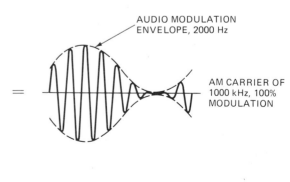

(a)   (b)

Fig. 12-13. The carrier plus the upper and lower side frequencies shown in (a) correspond to the AM waveform shown in (b).

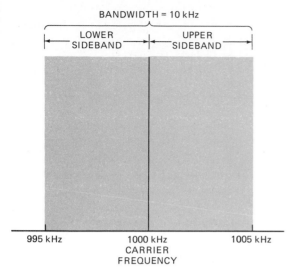

Fig. 12-14. Sidebands for an AM signal with bandwidth of ±5 or 10 kHz.

2 kHz, and the side frequencies are 1002 and 998 kHz.

Another question is how the side frequencies are produced. Remember that any frequency represents a change of voltage or current amplitude in a specific period of time. When the carrier amplitude is varied by modulation, the time rate of change is altered, or the slope of the voltage and current values in the RF carrier is changed. When the carrier amplitude is increased, its slope increases, which corresponds to a higher frequency. For the opposite effect, when the carrier amplitude is decreased, lower frequencies are generated. The carrier still has its original frequency at the times when the modulating voltage is zero.

The fact that different side frequencies are produced in the AM signal does not constitute frequency modulation. There are important differences in FM, as explained in Sec. 12-8. The main principle of FM is that the carrier frequency varies with the amount of audio-modulating voltage, not the audio frequency.

Bandwidth of an AM Signal So far, only one pair of side frequencies for one modulating frequency has been considered. When there is a band

of frequencies as in a typical audio signal, such as voice or music, the side frequencies are grouped in sidebands. These are shown in Fig. 12-14 for an AF band of 0 to 5000 Hz. The carrier is at 1 MHz, or 1000 kHz.

Each audio frequency produces a pair of RF side frequencies. All the higher frequencies are in the upper sideband. That band includes the frequencies from 1000 to 1005 kHz. Also, all the lower frequencies are in the lower sideband of 995 to 1000 kHz. The result is a bandwidth of ±5 kHz, or 10 kHz.

Test Point Questions 12-6
(Answers on Page 287)

a. Is a modulation envelope shown in Fig. 12-13a or Fig. 12-13b?
b. Are the RF side carrier frequencies shown in Fig. 12-13a or Fig. 12-13b?
c. Calculate the upper and lower side frequencies for 3-kHz audio modulation with a 27-MHz carrier frequency.
d. Calculate the total bandwidth needed for an AM signal at 55.25 MHz with 0.5-MHz video modulation.

12-7
SINGLE-SIDEBAND TRANSMISSION

It is not necessary to transmit both sidebands, since each sideband has the same modulating information. The advantage is that only one-half the bandwidth is needed for the AM carrier signal compared with double-sideband transmission. It should be noted that the modulation always produces double sidebands, but one can be filtered out before transmission. Either the upper or lower sideband may be used.

The disadvantage is that the modulation factor is reduced when only one sideband is used. Figure 12-15 shows how the carrier with one sideband corresponds to 50 percent modulation compared with 100 percent with two sidebands.

Note that the envelope of the AM carrier with the single-sideband in Fig. 12-15 still corresponds

Fig. 12-15. Modulation with only one sideband plus the carrier. Either the upper or lower sideband can be used. The result is equivalent to one-half the modulation with both sidebands.

to the 2-kHz audio modulation. Furthermore, the AM carrier signal still has the envelope top and bottom. One RF side frequency is filtered out, but not the audio envelope.

Carrier Suppression Another great improvement in transmitter efficiency is made possible by reducing the carrier power. The reason why that can be done is that all the modulating information of the AM signal is in the sidebands. Without the sidebands, the carrier itself does not have any intelligence.

If a signal is without the carrier, the carrier can be reinserted at the receiver. The frequency and phase of the reinserted carrier must be exactly the same as at the transmitter; otherwise, the detected signal will be distorted.

There are several methods of transmission that are often used with reduced carrier to save power and with a single sideband to save bandwidth. They are for such communications applications as military radio, mobile radio, citizen's radio, and amateur radio. Several examples follow.

SSBRC Transmission The abbreviation SSBRC stands for single sideband with reduced carrier power. The presence of some carrier power makes the receiver tuning easier.

SSBSC Transmission The abbreviation SSBSC stands for single sideband with suppressed carrier, which means no carrier. When there is no carrier, the carrier must be reinserted at the receiver. A pilot signal may be transmitted to indicate carrier frequency and phase for the receiver demodulation.

DSSC Transmission The abbreviation DSSC stands for double sidebands with suppressed carrier.

Vestigial Sideband Transmission The vestigial sideband method is applied to transmission of the picture carrier amplitude-modulated with a video signal for television broadcasting. Full carrier power is used. In effect, all the frequency components in the upper sideband are transmitted, but only a vestige, or part, of the lower sideband is transmitted. The television receiver compensates for the vestigial sideband transmission by providing more gain for the video-modulating frequencies transmitted with only one sideband.

Balanced Modulator The balanced modulator circuit can be used at the transmitter to suppress the carrier. Figure 12-16 shows a balanced modulator in which four diodes are used in a Wheatstone bridge. There are two pairs of input connections. Audio signal input at the left is applied by the AF modulating transformer T_1. That voltage is connected across terminals AA of the bridge. At the bottom of the diagram, the RF carrier input is applied across the terminals of the bridge at RR.

The output from RF transformer T_2 consists of the RF sidebands produced by audio modulation but without the carrier frequency itself. Although the primary of T_2 goes to the audio input terminals AA, the RF coupling capacitors C_1 and C_2 block the audio component. As a result, terminals SS have only the RF sideband output.

When there is no audio input or the audio-modulating voltage is at zero, the bridge balances out the RF carrier. The carrier signal itself, then, cannot produce any output.

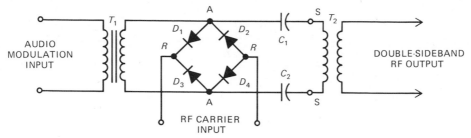

Fig. 12-16. Balanced modulator in which diode bridge circuit is used to suppress the carrier.

Now let an audio-modulating voltage be applied. The audio voltage makes one terminal A more positive and the other A more negative. Then the diodes are unbalanced. That effect allows the modulated RF signal to be produced for the output.

It should be noted that both upper and lower sidebands are produced in the output. The result is a double-sideband, suppressed-carrier signal.

Sideband Filter Although double sidebands can be transmitted, usually one sideband is filtered out of a suppressed-carrier signal. By using a mechanical or crystal filter with sharp selectivity, one sideband can be eliminated without affecting the other. The frequency-response curve for a sideband filter is illustrated in Fig. 12-17. The numerical values are for a CB radio with the carrier at 27 MHz and 3 kHz for the audio modulation. The upper sideband extends up to 27.003 MHz and the lower sideband down to 26.997 MHz. The side carriers are shown with higher amplitudes for the midfrequency components because the midfrequency components have stronger audio modulation.

In Fig. 12-17 the upper sideband is transmitted and the lower sideband is filtered out. Those effects can be reversed, however, to transmit the lower instead of the upper sideband.

Single-Sideband Transmitter Figure 12-18 is a block diagram of a single sideband transmitter. The balanced modulator has an output of both upper and lower sidebands without the carrier. However, the filters marked USB and LSB can

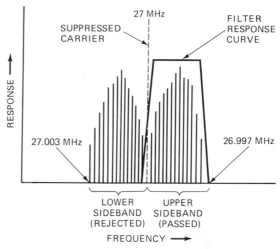

Fig. 12-17. Response of a sideband filter used to remove one of the two sidebands.

filter out either one of the sidebands. The single-sideband output is then applied to a mixer stage. Also coupled into the mixer is the desired frequency from a synthesizer. Then the signal from the mixer can be heterodyned to the desired transmit frequency. In that way any number of output channels can be produced from a single set of sideband filters. Because of the expense, it would not be practical to provide a separate set of filters for every channel.

It should be noted that class C operation cannot be used for all the stages following the sideband filter because the RF signal has the audio modulation. That factor limits efficiency. However, the advantage of not transmitting the carrier and eliminating one sideband more than makes up for the disadvantage. For example, a 100-W single-side-

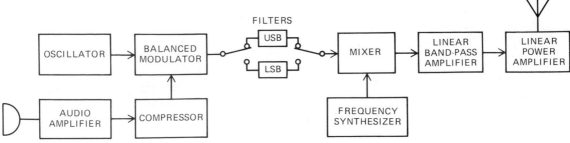

Fig. 12-18. Block diagram of a single-sideband transmitter.

band transmitter with a suppressed carrier is as effective for communications as a conventional 800-W AM transmitter.

Test Point Questions 12-7
(Answers on Page 287)

Answer True or False.

a. A single-sideband signal has one-half the bandwidth of a double-sideband signal.
b. A sideband filter is designed for a band of audio frequencies.
c. A balanced modulator produces a DSSC signal.
d. The carrier is suppressed in a SSBSC signal.

12-8
FREQUENCY MODULATION

The difference between frequency and amplitude modulation is illustrated in Fig. 12-19. The AM signal (Fig. 12-19a) has varying amplitudes but the constant carrier frequency of 100 kHz. The FM signal (Fig. 12-19b) has constant amplitudes but varying frequencies above and below the 100-kHz center frequency for the carrier. Those changes in the frequency of the RF carrier are produced by the audio-modulating voltage. The amount of change, or the *frequency deviation*, increases with an increase in the modulating signal voltage.

The audio frequency does *not* determine the RF carrier deviation. In this example, the frequency deviation of the RF carrier is 30 kHz. As a com-

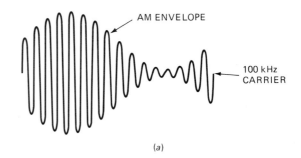

(a)

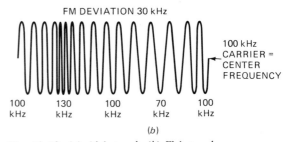

(b)

Fig. 12-19. (a) AM signal. (b) FM signal.

parison of FM and AM, the main features of the two methods of modulation are listed in Table 12-3.

Center Frequency The center frequency is the frequency of the transmitted RF carrier without modulation, or when the modulating voltage is at its zero value. In Fig. 12-19b the center frequency is 100 kHz. The center frequency is also called the *rest frequency*.

Frequency Deviation The frequency deviation is the amount of change from the center frequency.

Table 12-3
Comparison of FM and AM Signals

FM	AM
Carrier amplitude is constant	Carrier amplitude varies with modulation
Carrier frequency varies with modulation	Carrier frequency is constant
Modulating voltage *amplitude* determines RF carrier *frequency*	Modulating voltage *amplitude* determines RF carrier *amplitude*
Modulating frequency is the rate of frequency deviations in the RF carrier	Modulating frequency is the rate of amplitude changes in the RF carrier

In Fig. 12-19*b* it is 30 kHz. The amount of frequency deviation depends on the amplitude of the audio-modulating voltage. The peak audio voltage produces the peak frequency deviation.

Frequency Swing The frequency swing is the total deviation above and below center frequency. The swing in Fig. 12-19*b* is ±30 kHz, or a total of 60 kHz.

Advantages of FM Frequency modulation provides a transmission system that can be practically free from noise. Most types of noise produce AM variations in the signal. Examples are lightning discharge, noise from sparking in electrical equipment, and random noise in electronic components; all are variations in voltage and current amplitudes. In FM, however, the desired signal is in the variations of frequency in the carrier signal. To recover the desired audio signal without noise at the receiver, two steps are necessary:

1. An FM detector circuit that responds to the changes in carrier frequency is used to recover the original audio signal.
2. A limiting circuit is used to remove amplitude variations in the FM signal. Any amplitude changes in the RF carrier can only represent noise and interference.

Both requirements are easily met in FM receiver circuits.

Another advantage of frequency modulation is transmitter efficiency. All RF stages can operate class C because the carrier signal does not have any modulation envelope.

Circuit for Frequency Modulation In Fig. 12-20, $Q1$ is an RF oscillator used as the carrier generator; it is an N-channel JFET. The Hartley circuit is used here with C_3 and the tapped coil L_1 for the tuned circuit. However, the capacitive diode D_1 is effectively in parallel to control the frequency. An audio input voltage is applied to D_1 to modulate the frequency of the transmitted carrier. The audio voltage varies C of the varactor diode D_1. These changes in C control the frequency of the oscillator.

Circuit for Phase Modulation The phase modulation method is generally used with a crystal-

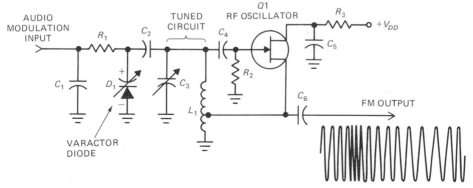

Fig. 12-20. A frequency-modulated oscillator circuit. Q1 is an N-channel junction field-effect transistor (JFET).

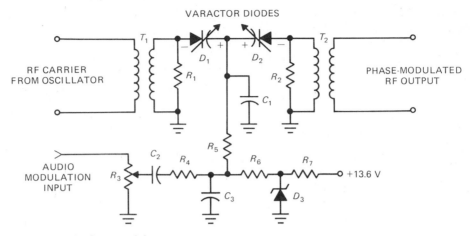

Fig. 12-21. A phase modulator circuit.

controlled oscillator to shift the phase angle of the RF output in step with the audio modulation. However, any change in phase is equivalent to a change in frequency. The result is *equivalent FM* or *indirect FM*. Phase modulation is used with crystal oscillators because the frequency cannot deviate very much.

In the circuit of Fig. 12-21 the carrier output from a separate RF oscillator (not shown) is coupled by T_1 to the capacitive diodes D_1 and D_2. The audio input voltage shifts the bias above and below its average dc value. As a result, the capacitance of D_1 and D_2 changes. The effect causes a shift in tuning of the secondary of RF transformer T_1 and the primary of T_2. Then the phase angle of the RF carrier is shifted from its average value. The result is phase modulation.

Comparison of FM and PM   There are two important differences between phase modulation (PM) and the direct FM system. In phase modulation, the equivalent FM has more frequency deviation at higher audio frequencies. The reason is that a change in phase angle at a faster rate at higher audio frequencies corresponds to a greater frequency change. However, the audio-modulating signal can be *predistorted* to correct for the excessive swing. In the circuit of Fig. 12-21, R_4 and C_3 form the predistortion network to reduce

the modulating voltage for high audio frequencies. The result with this correction is the same signal for transmission as in direct FM. An FM receiver operates the same way with an FM signal produced by either PM or direct FM. Note that resistor R_3 can be adjusted to set the level of audio modulation.

Another difference is that phase modulation can produce only small changes compared with direct FM. However, the frequency deviation can be increased, if necessary, by using frequency-multiplier circuits. For example, a frequency-doubler or -tripler stage multiplies the carrier frequency and its frequency deviation by a factor of 2 or 3.

Sidebands in an FM Signal   Figure 12-22 shows the spectrum of an FM signal with its sideband frequencies. The example is of a 1-MHz carrier modulated with a 5-kHz audio tone. Note that six pairs of side frequencies are produced above and below the carrier frequency, compared with only one pair in AM. Specifically, the first upper side frequency in this example is

1000 kHz + 5 kHz = 1005 kHz, or 1.005 MHz

The additional side frequencies in the upper sideband are 1010, 1015, 1020, 1025, and 1030 kHz. Also, the first lower side frequency is

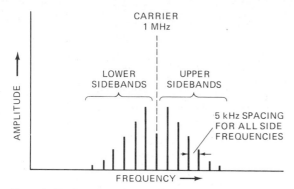

Fig. 12-22. Frequency spectrum for an FM signal with multiple sidebands. The carrier frequency is 1 MHz. The audio-modulating frequency is 5 kHz.

$$1000 \text{ kHz} - 5 \text{ kHz} = 995 \text{ kHz, or } 0.995 \text{ MHz}$$

The additional side frequencies in the lower sideband are 990, 985, 980, 975, and 970 kHz.

An FM signal inherently uses more bandwidth than an AM signal with the same modulation because of the additional sideband frequencies. It should be noted that the number of pairs of side frequencies in an FM signal can be more or less than the example of six here. The number of side frequencies produced above and below the carrier depends on the modulation index.

Modulation Index The formula is

$$M = \frac{\Delta f_c}{f_a} \qquad\qquad (12\text{-}3)$$

where M is the modulation index, Δf_c is the amount of frequency deviation in the RF carrier produced by the modulating voltage, and f_a is the audio-modulating frequency. As an example, assume a frequency deviation of 20 kHz in the RF carrier produced by an audio-modulating frequency of 5 kHz. Then

$$M = \frac{\Delta f_c}{f_a} = \frac{20 \text{ kHz}}{5 \text{ kHz}} = 4$$

There are no units for M because it is a ratio of frequencies.

Percent of Modulation In FM, the percent of modulation is an arbitrary value in terms of the maximum frequency deviation allowed by the FCC. For commercial FM broadcast stations, 100 percent modulation is defined as 75 kHz deviation. Then a frequency deviation of 15 kHz, for example, corresponds to 20 percent modulation. The calculations are

$$\frac{15}{75} = \frac{1}{5} = 0.2 = 20 \text{ percent}$$

For the sound transmitter of television broadcast stations, the deviation of 25 kHz is defined as 100 percent modulation. Less frequency deviation is used in order to conserve space in the television channel.

With FM radio communications equipment for voice the maximum frequency deviation is generally 5 kHz. Such a system can be considered as *narrow-band* FM.

Keep in mind that the percent of modulation varies with the intensity of the audio voltage. When audio signals are weak, the modulating voltage is small and there is little frequency deviation of the carrier and a low percent of modulation. When signals are louder, the audio voltage is greater and it produces more frequency deviation and a higher percent of modulation. The deviation produced by the loudest audio signal should be the amount defined as 100 percent modulation.

Deviation Ratio The deviation ratio compares the maximum amount of frequency deviation at 100 percent modulation with the highest audio-modulating frequency. For the commercial FM radio broadcast band, for example, those values are 75 kHz/15 kHz = 5 for the deviation ratio.

Preemphasis and Deemphasis Preemphasis refers to boosting the relative amplitudes of the modulating voltage for higher audio frequencies from 2 to approximately 15 kHz. Deemphasis means attentuating those frequencies by the amount by which they were boosted. However, the preemphasis is done at the transmitter and the deemphasis is done in the receiver. The purpose is to improve the signal-to-noise ratio for FM reception. A time constant of 75 μs is specified in the RC or L/R network for preemphasis and deemphasis.

Test Point Questions 12-8
(Answers on Page 287)

a. Which system is free from noise, FM or AM?
b. Is the carrier deviation determined by the modulating voltage or the frequency?
c. Is phase modulation direct or indirect FM?
d. Calculate the percent of modulation for 45-kHz deviation in the FM broadcast band.
e. How much deviation is 100 percent modulation of the FM sound signal in television broadcasting?

12-9
MULTIPLEXING

The term *multiplexing* is applied to the process of placing more than one channel of information on a single carrier. In stereo broadcasting, left and right audio signals are multiplexed on the RF carrier wave of the broadcast station. In color television, the chroma video signal for color is multiplexed with the monochrome video signal for black and white.

A common multiplexing technique is to use a *subcarrier,* which has a lower frequency than the main carrier. In stereo broadcasting, one audio signal modulates the subcarrier. The other audio signal modulates the main carrier, which is also modulated by the subcarrier. The transmitted RF carrier then has two channels of information. The receiver can separate the subcarrier signal by using resonant circuits tuned to the subcarrier frequency.

Another multiplex method is to use FM for one signal and AM for the other and have both modulate one RF carrier wave. The receiver can separate one signal from the other because of the different modulation. The technique of different modulation methods can also be used for a subcarrier signal and the main carrier.

There are three important applications of multiplexed signals in broadcasting.

1. *Stereo sound in the commercial FM radio broadcast band of 88 to 108 MHz.* Practically all FM radio stations broadcast their programs in stereo.

   In this system, the left and right audio signals are encoded as sum (L + R) and difference (L − R) signals. The (L + R) signal is transmitted as the modulation on the main carrier. Also, multiplexed with this modulation is a 38-kHz subcarrier modulated by (L − R). The 38-kHz modulation is the stereo signal.

   The purpose of the encoding of the left and right channels is to transmit (L + R) as a normal audio signal for receivers not equipped for stereo. More details of the encoding at the transmitter and decoding of the multiplexed stereo signal are explained in Chap. 17, on FM receivers.

2. *Stereo sound in the standard AM radio broadcast band of 535 to 1605 kHz.* This is one of a number of systems under study for approval by the FCC. In this method of multiplexing one audio channel uses amplitude modulation while the other uses phase modulation, both on the same RF carrier.

3. *Color television.* Practically all TV stations broadcast their programs in color, although the programs are viewed in black and white on a monochrome receiver. The station transmits a chroma signal on a 3.58-MHz color subcarrier, which modulates the main carrier. That carrier also has the modulation information for black-and-white details in the picture. At the television receiver, the 3.58-MHz subcarrier is used to provide a color video signal. For the picture tube, the color signal is combined with the monochrome signal to reproduce the image with full color superimposed on the black-and-white details. The color picture tube has a fluorescent screen that emits red, green, and blue light. By mixing these three colors, all colors, including white, can be reproduced on the TV screen.

It should be noted that all of the multiplex systems for broadcasting have the feature of *compatibility*, which means that a normal signal is transmitted for reception by receivers that do not have provision for decoding the multiplexed signal. A monophonic FM receiver for only one audio channel can use the transmitted audio signal for normal sound output, but not in stereo. A monochrome TV receiver can use the transmitted video signal for a normal picture, but not in color.

12-10
FREQUENCY MULTIPLIERS

Transmitters in the VHF and UHF bands have an output frequency that is often too high for the master oscillator stage. Either the oscillator circuit is not practical or the stability may not be good enough at such a high frequency. Crystal oscillators generally operate below 30 MHz, approximately.

The solution is to use frequency multiplication for the output of the master oscillator, as shown in Fig. 12-23. Doublers and triplers are common. Such circuits are RF amplifiers, but the output is tuned to a harmonic multiple of the input frequency. Then the maximum amplified output is produced at an exact multiple frequency of the input signal. A frequency multiplier stage operates class C, so that the distortion can produce harmonic frequencies.

The oscillator shown in Fig. 12-23 operates at 12.5 MHz. The output is tripled to 3 x 12.5 = 37.5 MHz. This signal then is doubled to 2 x 37.5 = 75 MHz. Finally, the frequency is doubled again to $2 \times 75 = 150$ MHz for a transmitter output at the assigned carrier frequency. Overall, the oscillator frequency of 12.5 MHz is multiplied by $3 \times 2 \times 2 = 12$ to produce 12×12.5 MHz, or 150 MHz, for the carrier signal output radiated from the transmitting antenna.

Multiplying the Deviation in an FM Signal
Frequency multiplication can be used in either AM or FM transmitters. In an AM transmitter, though, the oscillator frequency must be multiplied before the modulation. When an FM signal is involved, however, frequency multiplication of the modulated carrier has the additional advantage of multiplying the amount of frequency deviation. In general, frequency multiplication is used in FM transmitters because such transmitters operate in the VHF band.

As an example, suppose that the 12.5-MHz oscillator output in Fig. 12-23 is frequency-modulated with a maximum deviation of 6 kHz. The input is tripled in the next stage. The tripled output frequency is 37.5 MHz for the carrier, and the amount of deviation also is tripled for $3 \times 6 = 18$ kHz. The reason is that all input frequencies are tripled, because the third-harmonic components are at or near the resonant frequency of the output circuit. The output circuit is tuned to the third harmonic of the input.

For the total multiplication of $3 \times 2 \times 2 = 12$, the final amount of frequency deviation is 6 kHz $\times$ 12, or 72 kHz. This is the total deviation in the transmitted FM signal.

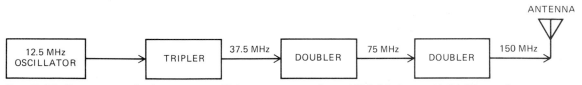

Fig. 12-23. Frequency multiplication in an FM transmitter produces 150 MHz from a 12.5 MHz oscillator.

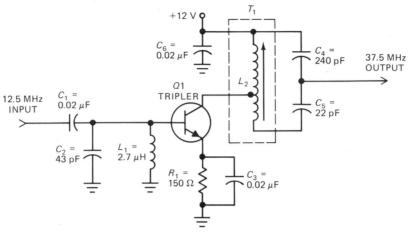

Fig. 12-24. A frequency tripler circuit.

Frequency multiplication is usually necessary with a phase-modulated signal to increase the amount of equivalent FM. The original frequency deviation is very small in PM to avoid distortion in the modulated signal.

Frequency Multiplier Circuit Figure 12-24 shows the circuit for the tripler stage in the transmitter of Fig. 12-23. The transistor amplifier Q1 operates class C; signal bias in the base circuit is produced by the driver in the preceding oscillator stage. Base current in Q1 charges coupling capacitor C_1 for the bias. Although the bias has reverse polarity, the positive peaks of the input signal drive Q1 into conduction. The pulses of current in class C operation provide nonlinear distortion. That effect is desirable for a frequency multiplier, because the distortion generates harmonic frequencies. The second or third harmonic is generally used because the amplitudes of the higher harmonic frequencies are greatly reduced.

Test Point Questions 12-10
(Answers on Page 287)

a. A frequency doubler has an FM signal input at 13 MHz with a deviation of 5 kHz. What is the output frequency for the carrier?

b. In question a, what is the frequency deviation of the frequency doubler's output signal?

c. Should a frequency multiplier stage operate class A or class C?

12-11
PULSE MODULATION

A pulse transmitter is turned on and off at a regular rate to transmit pulses of RF energy. The ratio of time on to time off is the *duty cycle*. Usually, the time on is less than the time off for a low duty cycle. That feature allows a significant savings in power compared with constant-carrier types of transmission. Specifically, a pulse-modulated transmitter produces high values of peak power for short periods of time. These peak power levels can be much greater than the average power-output rating.

Without any modulation, all the transmitted pulses have the same amplitude, width, and spacing. However, the modulating intelligence can be used to control one characteristic of the pulses. Common methods are pulse-amplitude modulation (PAM), pulse-width modulation (PWM), pulse-frequency modulation (PFM), and pulse-coded modulation (PCM). Their waveforms are shown in Fig. 12-25.

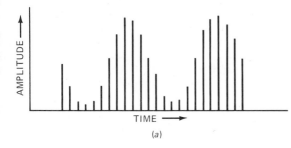

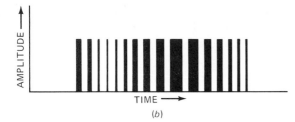

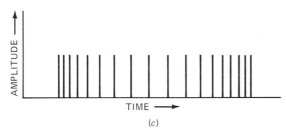

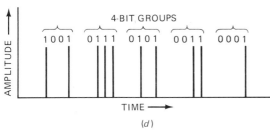

Fig. 12-25. Forms of pulse modulation. (a) Pulse-amplitude modulation (PAM). (b) Pulse-width modulation (PWM). (c) Pulse-frequency modulation (PFM). (d) Pulse-coded modulation (PCM).

Pulse-Coded Modulation Pulse-coded modulation is shown in Fig. 12-25d. Different methods can be used, but the example of Fig. 12-25d illustrates the important case of coding with the binary digits 0 and 1. (See Chap. 8 for an explanation of binary numbers.) The presence of the pulse, that is the transmitter is on, represents binary num-

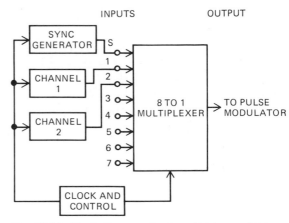

Fig. 12-26. Block diagram of a time-division multiplexing system.

ber 1. The absence of a pulse (the transmitter is off) represents binary 0. Each value of 0 or 1 is one bit of information.

In Fig. 12-25d, a group of four pulses is used to represent words of four bits each. As examples, the 4-bit words here are

$$1001 = \text{decimal } 9$$
$$0111 = \text{decimal } 7$$
$$0101 = \text{decimal } 5$$
$$0011 = \text{decimal } 3$$
$$0001 = \text{decimal } 1$$

Note how the train of pulses represents the binary digits. The first pulse in the group of four at the left is on, which represents binary 1. Also, the last pulse is on. The space between is for two missing pulses. Each space represents a binary 0. The 4-bit group therefore indicates 1001. In that way, the pulses transmit information using the binary words. Furthermore, binary coding can be used to represent all types of analog information, such as an audio signal. Converters are used to change between analog and digital information.

Time-Division Multiplexing Figure 12-26 is a block diagram of a system that can multiplex up to seven different modulating signals on a single pulsed carrier. The 8-to-1 multiplexer shown at the right is the electronic equivalent of a rotary switch. It has inputs for seven modulation channels but only a single output. The clock and control circuits shown at the bottom step the multiplexer through its different input channels.

The input at the top, marked S, is a synchronizing signal from the sync generator. The input allows a synchronizing signal to be transmitted first. After synchronization is achieved, each of the seven modulating channels of information has access to the pulse modulator one at a time. The 8-to-1 multiplexer connects each input, in turn, to the modulator at the output. After the last channel has been transmitted, the sync code is once again sent, and the process repeats itself. Although the information is sent in sequential order, fast rates of transfer are possible because of the time-division multiplexer's high speed digital circuits.

Applications of Pulse Modulation Pulse transmission has the disadvantage of requiring a very wide frequency band because of the rectangular waveshape of the pulses. Waveforms with fast rise and fall times have frequency components much higher than the pulse repetition rate. The reason is harmonic frequencies in the pulse waveforms.

Pulse transmission is often used in the microwave band, where the frequency spectrum can accommodate wide-band signals. One of the first uses of pulse transmission was in radar equipment. Pulse systems are also used for radio navigation, automatic landing equipment, data communications, telemetry, satellite communications, and many other applications.

An important advantage of pulse modulation, besides power efficiency, is an enhanced signal-to-noise ratio. Except in PAM systems, all the output pulses are at full transmitter power for a strong signal. Any interfering noise pulses can be rejected by limiter circuits in the receiver. If there should be any missing pulses, the error can be detected by logic and timing circuits. Special signals called *parity bits* or *parity words* are injected to detect errors. All those advantages combined make pulse

transmission the best of all types to use when signal-to-noise ratio is the deciding factor.

12-12
KEYING METHODS

In the early days of radio, intelligence was conveyed by turning the transmitter on and off in a pattern or code called *keying*. The Morse code, which was commonly used, is a system of dots and dashes in the transmission. A dot, called *dit*, is a short unit pulse. The longer dash, called *dah*, is three unit pulses long. The space between is one unit pulse. Each letter, digit, or symbol is coded as a group of dots and dashes. For instance, the letter *a* is dit-dah. The space between letters is three unit pulses. Between words or groups of symbols, seven unit pulses are used.

Transmission by keying methods is *radiotelegraph* communication; voice transmission is *radiotelephone*. For radiotelegraph with an unmodulated carrier wave (CW), the transmission is interrupted CW. When the keyed carrier has 1000-Hz tone modulation, the transmission is modulated CW, or MCW.

Today, CW and MCW transmissions are still used by radio amateurs and in some military applications. The advantage is simple equipment at the transmitter and narrow bandwidth at the receiver for a good signal-to-noise ratio. Sometimes a CW transmission can get through to the receiver under conditions so very poor that radio communication by other methods would be impossible.

The basic idea in radiotelegraphy is to key the

transmitter on and off. The key serves as a switch. One method is to open the cathode return circuit in a vacuum-tube amplifier. That circuit is *cathode-keying*. In a transistor amplifier the corresponding method is to open the emitter return circuit. Another technique is to apply a negative blocking bias to the control grid of a tube. Then the key is used to short out the blocking bias and the stage turns on. That method is called *blocked-grid keying*. The corresponding method with transistors is to remove a reverse bias at the base electrode.

Key Clicks Whatever the keying method, it must not act too abruptly. If the rise and fall times are too sharp at the start and end of the keyed waveform, key clicks will result. The clicks are broadband interference signals that can appear many kilohertz from the desired signal frequency. The effect is unnecessary interference with other radio communications.

Click filters are used to lengthen the rise and fall times of the keyed signal. Figure 12-27 shows the envelope for the keyed transmissions. Too much filtering must be avoided because it makes the break between characters too soft, and therefore difficult to decipher. The click filter itself is an *RC* network with the proper time constant.

Circuit of a Keyed Transmitter A schematic diagram of the keyed transmitter is shown in Fig. 12-28. Note that the dc supply voltage is applied to the oscillator transistor $Q1$ and final power amplifier $Q3$ without any keying in those circuits. In general, it is preferable not to key the oscillator stage. Keying could produce a frequency shift, which causes an FM "chirp" in the signal.

The final power amplifier operates class C. It does not conduct without an input signal. Therefore, keying is not necessary for this stage.

Keying for the transmitter shown in Fig. 12-28 is done in the buffer stage $Q2$ by means of the keying transistor $Q4$. That transistor is in the path of the dc supply voltage for the buffer amplifier. The keyer $Q4$ must conduct in order to connect $Q2$ to the 12-V supply. $Q4$ is off because the base circuit is open unless the key is closed. Closing the

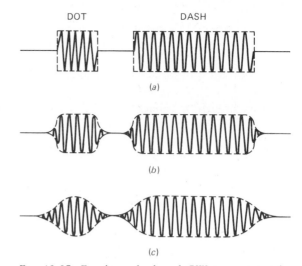

Fig. 12-27. Envelopes for keyed CW transmission by using the dots and dashes of Morse Code. (*a*) Wave with key clicks. (*b*) Wave with filtering—no clicks. (*c*) Excessive filtering—transmission blurred and indistinct.

key allows $Q4$ to conduct. Then $Q2$ conducts to drive the final power amplifier for the transmitter output to the antenna.

Two *RC* circuits are used for click filters to shape the rising and falling edges of the keyed envelope. When the key is closed, R_6 in the collector circuit of $Q4$ limits the charging of C_6. The time is at the rising edge of the signal pulse. When the key is opened, C_7 will supply some base current through R_7 for $Q4$ to smooth the falling edge.

Frequency-Shift Keying The frequency-shift keying technique does not turn the transmitter on and off. Instead, the keying shifts the frequency of the carrier from one value to another. The two values are called *mark* and *space*. The amount of frequency shift can be wide or narrow depending on the keying speed and the application. The shift is easy to accomplish. All that is required is a control voltage and a capacitive diode in the oscillator circuit. As the control voltage shifts back and forth for mark and space, the oscillator will shift back and forth between the two frequencies.

Frequency-shift keying is used in the radioteletype. In the Teletype machine a mechanical or

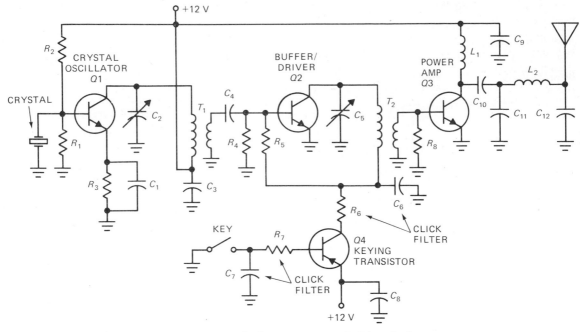

Fig. 12-28. Solid-state CW transmitter. Note the keying circuit with Q4 at the bottom.

electronic means is used to translate characters into a defined series of mark and space signals. Upon reception, the mark and space signals cause the message to be printed automatically.

An important difference between CW and Teletype is in the codes used. The International Morse Code is used in CW, and either the Baudot or ASCII code is used in the Teletype. The Baudot is an older code for Teletype machines. It is based on the *baud unit*, which is one dot per second for binary signals. The newer code is ASCII, for American Standard Code for Information Interchange.

Test Point Questions 12-12
(Answers on Page 287)

Answer True or False.

a. Blocked-grid keying is similar to frequency-shift keying.
b. The RF stage keyed in Fig. 12-28 is Q2.

12-13
PARASITIC OSCILLATIONS

Parasitic oscillations are undesired oscillations that are not related to the frequency of the master oscillator. The parasitics are produced in the RF amplifier stages or even the oscillator itself. The cause of parasitic oscillations is stray capacitive and inductive coupling between components. Transmitter circuits are susceptible to parasitics because the high power involved means that a small amount of stray coupling still provides appreciable feedback. The parasitics occur with either transistor amplifiers or vacuum tubes.

It should be noted that parasitic oscillations can occur even when an RF amplifier is neutralized to compensate for internal feedback. The neutralization eliminates oscillations at the signal frequency, but parasitics can occur at any frequency.

Low-Frequency Parasitics An example of a transistor amplifier is shown in Fig. 12-29. The function of C_1 is as a coupling capacitor for the signal input. However, C_1 also forms a tuned circuit with the choke L_1. Another combination is the bypass capacitor C_2 with the choke L_2. If the two combinations nearly match in resonant frequency, parasitic oscillations will result. The frequency of the parasitics is low, compared with the signal frequency, because coupling or bypass capacitors and chokes are relatively large. Furthermore, the transistor gain increases at lower frequencies, which makes parasitic oscillations even more likely.

There are several ways to prevent low-frequency parasitics. The Q of the parasitic tuned circuits can be lowered. For instance, a resistor might be added in series with the coil L_1 (Fig. 12-29). There is not much current in the base circuit, so the resistive losses are tolerable. Another technique is to use coils and capacitors with values that cannot match the resonant frequency of one parasitic tank circuit to any other. In addition, the use of low-Q decoupling chokes, such as ferrite beads, in the connecting leads helps to reduce parasitics. Finally, it is a good idea to use resistors, instead of choke coils, when possible.

High-Frequency Parasitics   Figure 12-30 shows an example of high-frequency parasitics. The parasitic tuned circuit of a vacuum-tube amplifier is in heavy lines. Each connecting lead itself is a small stray inductance. That, combined with capacitance, results in a parasitic tuned circuit at a much higher frequency than the signal frequency.

The method of curing such high-frequency parasitics is to spoil the Q of the parasitic tank circuit. An example is to form a parasitic suppressor network consisting of several turns of heavy wire in shunt with a noninductive resistor of about 50 Ω. The suppressor combination is connected in series with the plate lead or cathode lead of the amplifier.

Detection of Parasitics   The parasitic oscillations often cause obvious symptoms. In severe cases, components in the transmitter may arc, overheat, or burn out. Transmitter tuning may be

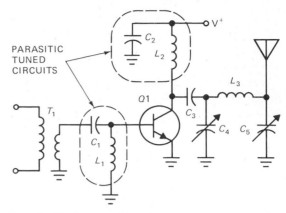

Fig. 12-29. Sources of low-frequency parasitic oscillations in a transistor amplifier.

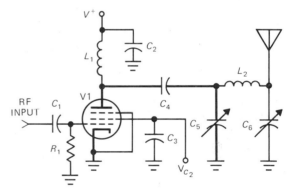

Fig. 12-30. Very high frequency vacuum tube amplifier. The parasitic circuit is shown in heavy lines.

erratic, and interference with nearby receivers is often noticed.

A spectrum analyzer is very helpful for detecting and identifying the frequency of the parasitic. Wavemeters of the absorption type also can be used. An older and simpler technique is to use a neon lamp. The drive to the amplifier stage is disabled, but power is on, and the lamp is held near the components to find parasitic oscillations. The technique is not applicable in many cases, however, because safety interlocks often turn the power off to avoid dangerous voltages inside the transmitter.

a. Can coupling and bypass capacitors be a source of low-frequency or high-frequency parasitics?
b. Can the inductance of connecting leads be a source of low-frequency or high-frequency parasitics?

12-14
FCC ALLOCATIONS

Every radio transmitter has to have a certain portion of the frequency spectrum reserved for its operation to avoid interfering with other radio services. The frequency allocations by the Federal Communications Commission (FCC) are in Appendix B.

Furthermore, frequencies for all countries are allocated by the World Administrative Radio Conference (WARC). Even the frequencies for satellite communications and the orbital positions of the satellites are regulated.

A radio station's band of frequencies is its *channel*. In many cases, a channel is used by more than one station because there are not enough channels for exclusive use. The frequency spectrum is very crowded with radio services.

Low-powered stations can serve a small community by providing signals over a relatively short distance, and other stations in distant cities may use the same channel without interference. Time restrictions are often placed on low-powered stations. The reason is that ionospheric conditions change and stations will interfere with each other if they operate on a 24-hour (h) schedule. For the same reason, some stations are required to reduce power in the late afternoon or evening.

Standard AM Radio Broadcast Band The standard AM radio broadcast frequencies set by the FCC are 535 to 1605 kHz, which may be extended to 520 to 1640 kHz, as recommended by the WARC. The channels are 10 kHz apart. For instance, 510 and 520 kHz are assigned channels for two AM radio stations, but not in the same city. Tolerance for the assigned carrier frequency is ± 20 Hz.

Incidentally, K and W are assigned as the first letters for broadcast stations in the United States. The call letters of stations east of the Mississippi begin with W, as in WNBC, New York, and those to the west begin with K, as in KTVV, San Francisco. The exceptions are a few stations that started operation before the system was established in 1912, notably KDKA, Pittsburgh.

Commercial FM Radio Broadcast Band The commercial FM radio broadcast frequencies assigned by the FCC are 88 to 108 MHz. Each channel has a bandwidth of 0.2 MHz, or 200 kHz. The channel includes a 25-kHz guard band at each end with ± 75 kHz as maximum deviation for the FM signal. Stations are assigned frequencies such as 96.3 and 96.5 MHz. However, stations in the same city are widely separated in frequency. Tolerance for the assigned carrier frequency is ± 2 kHz.

Commercial Television Broadcast Stations Each commercial television broadcast channel is 6 MHz wide to include the AM picture signal and FM sound signal, which are spaced 4.5 MHz apart. The 3.58-MHz color subcarrier signal is multiplexed with the black-and-white video modulation on the main picture carrier. There are 12 VHF channels 2 to 13 and the UHF channels 14 to 83, as listed in Appendix B. For example, channel 2 is 54 to 60 MHz. The picture carrier is at 55.25 MHz and the sound carrier at 55.75 MHz, with 4.5 MHz between the two. All TV stations have the same distribution of frequencies in the 6-MHz channel, whether in the VHF or UHF band.

Short-Wave Radio The short-wave stations for international broadcasting operate in the 3- to 30-MHz band allocated for the purpose.

Amateur Radio Bands The amateur or "ham" radio bands are designated by the FCC and are based on international recommendations by the WARC. Harmonic intervals are generally so assigned that one antenna can serve for multiple frequencies. These bands are listed in the *ARRL Handbook.*

Citizens' Band Radio In the popular citizens' band are forty 10-kHz channels from 26.965 to 27.405 MHz. All the CB channels are listed in Appendix B. Class D service is assigned to the citizens' band, which means maximum power output of 4 W. No operator license is required for mobile CB equipment. Amplitude modulation, with either double- or single-sideband transmission, is used for voice telephony.

Standard Time and Frequency Broadcasts The National Bureau of Standards operates radio stations WWV in Fort Collins, Colorado, and WWVH in Hawaii to broadcast precise time and frequency standards that can be used by tuning to those stations. The assigned carrier frequencies and power output are listed in Table 12-4. The carrier itself is a standard frequency that can be used to calibrate receivers and signal generators. Also, the exact time is broadcast at regular intervals.

Emergency Broadcast System The emergency broadcast system (EBS) is composed of AM, FM, and TV broadcast stations, with additional organizations, which are to transmit information to the public should it be necessary in an emergency. To be ready, the stations broadcast test signals every 1 to 3 months. The signals are audio tones at 853 and 960 Hz. They are attention signals. Muted receivers at the stations would be automatically turned on to call attention to an emergency. Then the stations would be in communication to broad-

Table 12-4
Standard Time and Frequency Stations

WWV, MHz	WWVH, MHz
2.5	5
5	10
10	15
15	
20	
25	

cast emergency information in an organized system. The EBS system replaces the older CONELRAD system for emergency broadcasts at 640 and 1240 kHz.

Test Point Questions 12-14
(Answers on Page 287)

Give the frequencies for the following:

a. AM radio broadcoat band
b. FM radio broadcast band
c. Television channel 13
d. CB channel 9

12-15
TYPES OF EMISSION

Emission refers to the type of signal transmitted. Designations are assigned by the FCC to different modulation and transmission characteristics. The letter symbol A is for amplitude modulation; F is for frequency or phase modulation; and P is for pulse modulation. Also, the numbers 0 to 9 are used with the letters for different signals. For example, A2 is modulated CW telegraphy, but A3 is for telephony. Radiotelegraphy uses keying codes, and radiotelephony is for voice communications. All the symbols for the different types of emission are listed in Tables 12-5 to 12-7.

For the AM emission, Table 12-5, note that A1 is for radiotelegraphy with CW transmission and A2 is for MCW. The keyed CW signal is also called *interrupted CW,* or *ICW.*

The AM picture signal for television is a special type of emission A5C because of the video modulation and vestigial sidebands. A4 or A4A emission is used for facsimile, which is a system for reproducing graphic material at a remote location.

In the FM emissions list in Table 12-6, note that F3 is the symbol for the signal in the FM radio broadcast band and F5 is the symbol for the FM sound signal in television. The difference is that in F3 the maximum deviation is 75 kHz for 100 percent modulation, but in F5 the maximum is 25 kHz.

Test Point Questions 12-15
(Answers on Page 287)

Give the symbols for the following types of emission:

a. Amplitude-modulated voice telephony
b. Tone-modulated keyed telegraphy
c. Frequency-modulated voice telephony with 75-kHz deviation
d. AM picture carrier signal for television
e. Pulse-coded modulation for telephony

Table 12-5
Types of Emission for Amplitude Modulation

Symbol	Transmission of Characteristics
A0	Continuous carrier; no modulation and no keying (CW)
A1	Telegraphy with keying (ICW)
A2	Telegraphy with tone or tones (MCW)
A3	Telephony for voice, with double sidebands
A3A	Telephony (single sideband, reduced carrier)
A3J	Telephony (single sideband, suppressed carrier)
A3B	Telephony (independent sidebands)
A4	Facsimile
A4A	Facsimile (single sideband, reduced carrier)
A5C	Television (vestigial sideband)
A7A	Telegraphy (multichannel SSB reduced carrier)
A9B	Special (cases not covered by the above)

Table 12-6
Types of Emission for Frequency Modulation

Symbol	Transmission of Characteristics
F1	Telegraphy with frequency-shift keying
F2	Telegraphy with tone modulation
F3	Telephony for FM broadcasting with 75-kHz deviation
F4	Facsimile
F5	Television sound with 25-kHz deviation
F6	Telegraphy (four-frequency diplex)
F9	Special (cases not covered by the above)

Table 12-7
Types of Emission for Pulse Modulation

Symbol	Transmission Characteristics
P0	Radar (pulsed carrier)
P1D	Telegraphy (pulsed carrier with or without tones)
P2D	Telegraphy (pulse-amplitude modulation)
P2E	Telegraphy (pulse-frequency or width modulation)
P2F	Telegraphy (pulse-phase or position modulation)
P3D	Telephony (pulse-amplitude modulation)
P3E	Telephony (pulse-frequency or width modulation)
P3F	Telephony (pulse-phase or position modulation)
P3G	Telephony (pulse-coded modulation)
P9	Special (cases not covered by the above)

SUMMARY

1. The RF section of a radio transmitter generally consists of a master oscillator, the final power amplifier to feed the antenna, and intermediate power amplifiers, or buffer stages.
2. A modulated signal combines the RF carrier wave with a lower-frequency signal that has the desired information. The modulating intelligence is the baseband signal. Typical are audio signals for voice, video signal for picture, and pulses for digital information.

3. The three main types of modulation are amplitude modulation (AM), frequency modulation (FM), and pulse modulation (PM).

4. For 100 percent modulation in AM, the modulation envelope has a peak value double the unmodulated carrier level.

5. High-level modulation means that the modulating signal is applied where the RF power level is the rated output to the antenna.

6. Amplitude modulation results in RF sideband frequencies above and below the carrier by an amount equal to the modulating frequency. The modulating information is in the sidebands. Both sidebands may be used for double-sideband transmission, or only one with single-sideband transmission for less bandwidth.

7. The RF carrier itself can be suppressed or reduced for less transmitter power. A balanced modulator is a circuit used to suppress the carrier itself while passing the sideband frequencies.

8. The RF amplifiers in transmitters generally operate class C for maximum efficiency. However, any stage amplifying an AM wave must operate class A to preserve the modulation envelope.

9. The main advantage of frequency modulation, compared with AM, is that FM is practically free from noise.

10. In FM the modulating *voltage* produces proportional changes in carrier *frequency*. The average or middle value of the carrier is center, or rest, frequency. The change from the center is the frequency deviation.

11. In FM, 100 percent deviation is defined as the maximum deviation of 75 kHz in the FM radio band or 25 kHz in the FM sound signal in television.

12. A frequency multiplier stage is a class C amplifier in which the output circuit is tuned to a harmonic of the input signal. Frequency doublers or triplers are typical. With an FM signal, the carrier and its frequency deviation are multiplied.

13. A subcarrier has a lower frequency than the main carrier. The purpose is usually to have the subcarrier modulate the main carrier. This technique is used to multiplex two different signals on one carrier wave.

14. Pulse modulation allows high transmitter efficiency and good signal-to-noise ratio, but it requires appreciable bandwidth. The main methods are pulse-amplitude modulation (PAM), pulse-width modulation (PWM), pulse-frequency modulation (PFM), and pulse-coded modulation (PCM).

15. Two common methods of keying the transmitter on and off for radiotelegraph transmission are cathode keying and blocked-grid keying.

16. In frequency-shift keying, the carrier frequency is changed from one value to another instead of turning the transmitter on and off.

17. Parasitics are undesired oscillations in RF amplifiers at a frequency that is not related to the signal frequency. The parasitic oscillations waste power and may cause erratic tuning.

18. Some common types of emission are A1 for keyed CW, A2 for MCW in radiotelegraphy, and A3 for voice telephony with amplitude modulation.

SELF-EXAMINATION
(Answers at back of book)

Choose (a), (b), (c), or (d).

1. In which of the following frequency bands are the standard AM radio broadcast stations? (a) MF, (b) HF, (c) VHF, (d) UHF.
2. The commercial FM radio broadcast band is (a) 535 to 1605 kHz, (b) 27 to 29 kHz, (c) 88 to 108 MHz, (d) 300 to 3000 MHz.
3. The frequency tolerance for the RF carrier in the standard AM radio broadcast band is (a) zero, (b) ± 20 Hz, (c) ± 1000 Hz, (d) ± 25 kHz.
4. Which of the following is *not* a baseband signal for modulation? (a) Audio signal; (b) video signal; (c) RF carrier, (d) binary-coded pulses.
5. In an AM signal, the peak carrier amplitude is double the unmodulated level. The percent modulation is (a) 20, (b) 50, (c) 100, (d) 200.
6. For an FM broadcast station, the maximum deviation produced by audio modulation is 45 kHz. That percent modulation is (a) 10, (b) 45, (c) 60, (d) 100.
7. The final power amplifier in an FM transmitter usually operates class (a) A, (b) B, (c) C, (d) D.
8. A transistor RF power amplifier can be tuned for (a) minimum I_C in the next stage, (b) zero signal in the next stage, (c) minimum I_C in the same stage, (d) maximum I_C in the same stage.
9. A 900-kHz carrier is amplitude-modulated with a 4000-Hz audio tone. The lower and upper side frequencies are (a) 450 and 1800 kHz, (b) 800 and 1000 kHz, (c) 896 and 904 kHz, (d) 4000 and 8000 kHz.
10. The purpose of a balanced modulator circuit is to eliminate the (a) carrier, (b) upper sideband, (c) lower sideband, (d) baseband signal.
11. A frequency multipler circuit (a) operates class A, (b) is tuned to a harmonic of the input signal, (c) needs parasitic oscillations, (d) is usually pulse-modulated.
12. The type of emission for the standard AM radio broadcast band is (a) A5C, (b) A0, (c) A3, (d) F3.

ESSAY QUESTIONS

1. Name the two stages in the MOPA transmitter for radiotelegraph operation.
2. Name two more stages needed for the MOPA transmitter for radiotelephony.
3. Define the following: baseband signal, subcarrier, multiplexing.
4. What applications would you think of for the following frequencies: 38 and 710 kHz; 3.58; 4.5; 27.065; 96.3; and 211.25 MHz?
5. Name three types of modulation.
6. What is the difference between high- and low-level modulation? Give two examples of high-level modulation.
7. Explain the difference between the frequency of the modulation envelope in AM and the sideband frequencies.

8. What are the main advantages of FM over AM?
9. Give two differences between FM and AM.
10. Why is phase modulation also called indirect FM?
11. Define center frequency, deviation, swing, and modulation index of an FM signal.
12. What is the maximum frequency deviation of the loudest audio signal in the FM radio band? Of the FM sound signal in television?
13. What is the function of a frequency multiplier stage? Give an FM signal example. Why does the stage operate class C?
14. Give the function of each of the following: master oscillator, buffer amplifier, IPA, final power amplifier.
15. What is the purpose of using a balanced modulator?
16. What do the following abbreviations mean with respect to the carrier and its sidebands in radio transmission: SSBRC, SSBSC, DSSC? Give the emission type designations for SSBRC and SSBSC.
17. Explain briefly why an RF amplifier can be tuned for minimum dc collector current to provide maximum ac signal drive to the next stage.
18. Why is the connection of a transmitter to an antenna so important?
19. Define preemphasis and deemphasis for FM. Give the purpose of this technique. What is the specified time constant?
20. Give four methods of pulse modulation.
21. Describe briefly two methods of keying the transmitter on and off for radio-telegraphy. How do those methods differ from frequency-shift keying?
22. Why does neutralization of an RF amplifier have little effect in reducing parasitic oscillations?
23. Give the frequencies of the HF, VHF, and UHF bands.
24. Give the frequencies of the AM radio broadcast band, FM radio broadcast band, citizens' radio band (CB), and the UHF television channels 14 to 83.
25. Give the functions of $Q1$, $Q2$, $Q3$, and $Q4$ in the circuit for a keyed transmitter shown in Fig. 12-28. Which components are used for the two click filters?
26. Give the emission designation for each of the following signals: **a.** AM radio, **b.** FM radio, **c.** AM picture signal, **d.** FM sound signal in television, **e.** voice telephony in citizens' band with double sidebands and normal carrier amplitude.

PROBLEMS
(Answers to odd-numbered problems at back of book)

1. In an AM signal the peak antenna current is 12 A and the minimum is 3 A. Calculate the percent of modulation.
2. A 5-kW AM transmitter is 50 percent modulated. What is the sideband power?
3. In a 5-kW AM transmitter high-level modulation is used. What is the value of the audio signal power needed for 100 percent modulation?
4. An AM transmitter uses audio-modulation frequencies up to 4000 Hz. What bandwidth is needed for the AM signal with double sidebands?

5. A radio transmitter is rated as having a frequency accuracy of ± 0.2 parts per million. Calculate the maximum frequency error at 150 MHz.
6. Calculate the upper and lower side frequencies of an AM signal at 900 kHz with 5000-Hz audio modulation. What is the frequency of the modulation envelope?
7. Calculate the percent modulation of a signal in the FM broadcast band at 92 MHz with 20 kHz frequency deviation.
8. Calculate the percent modulation of the FM sound signal with 20-kHz deviation in television broadcasting.
9. An FM signal has a frequency deviation of 20 kHz produced by 2-V audio signal at 4000 Hz. **a** Calculate the modulation index. **b** If there are eight pairs of sidebands in the FM signal, what is the required bandwidth?
10. An oscillator at 4.2 MHz is followed by two frequency doublers and two triplers. Calculate the output frequency of the transmitter.
11. An FM broadcast station is assigned to operate at 91.9 MHz. What is the highest legal carrier frequency of this transmitter?
12. A 1-MHz carrier is amplitude-modulated by a 5-kHz audio signal. **a.** Calculate the upper and lower side frequencies, in megahertz. **b.** What bandwidth is needed for the AM signal, in kilohertz?

SPECIAL QUESTIONS

1. List in separate groups as many radio services as you can for AM, FM, and pulse modulation.
2. Give three or more applications for a radio transmitter.
3. Explain the functions of a radio transmitter and radio receiver. What factors determine the capacity for good communications between transmitter and receiver?
4. Compare AM and FM signals in terms of the sideband frequencies produced by audio modulation.
5. Compare audio and video signals in terms of the modulation frequencies and the resultant sidebands.
6. Give the frequencies for two TV stations, FM radio stations, and AM radio stations in your area.

ANSWERS TO TEST POINT QUESTIONS

12-1   **a.** Oscillator   12-3   **a.** 1200 V
        b. 2000 Hz   **b.** 1000 kHz
        c. 30 to 300 MHz   **c.** 25 percent
12-2   **a.** Lower   12-4   **a.** T
        b. AM   **b.** T
        c. AM   **c.** F
                                        d. T

12-5 a. T
      b. T
      c. F
      d. F
12-6 a. Fig. 12-13*b*
      b. Fig. 12-13*a*
      c. 27.003 and 26.997 MHz
      d. 1 MHz
12-7 a. T
      b. F
      c. T
      d. T
12-8 a. FM
      b. Voltage
      c. Indirect
      d. 60 percent
      e. 25 kHz
12-9 a. 38 kHz
      b. 3.58 MHz
      c. (L + R) and (L − R)

12-10 a. 26 MHz
      b. 10 kHz
      c. Class C
12-11 a. F
      b. T
      c. F
12-12 a. T
      b. F
12-13 a. Low-frequency
      b. High-frequency
12-14 a. 535 to 1605 kHz
      b. 88 to 108 MHz
      c. 210 to 216 MHz
      d. 27.065 MHz
12-15 a. A3
      b. A2
      c. F3
      d. A5C
      e. P3G

Chapter 13

Antennas and Transmission Lines

An antenna is just a metal conductor, usually either a length of wire or hollow tubing. A conductor must be used for the antenna so that current can flow in it. At the transmitter the antenna current produces electromagnetic radio waves. The wave consists of varying electric and magnetic fields that move out into space from the antenna. At the receiver, radio waves induce a current in the antenna. This current is the input signal to the receiver.

Both the transmitter and the receiver antennas have essentially the same requirements but opposite functions. In a CB radio, as an example, the same antenna is used for transmitting and receiving.

The transmission line is the connecting link to the antenna. Usually the line consists of a pair of wire conductors with constant spacing. The function is to conduct current without any radiation. More details of radio waves, antennas, and lines are given in the following topics:

13-1
ELECTROMAGNETIC RADIO WAVES

Antennas are conductors designed to radiate electromagnetic waves or to receive radiated waves that are present in the air or space. Antennas are made in a wide range of sizes and shapes to serve particular applications. Two examples of familiar antennas are shown in Fig. 13–1.

When there is current in it, the antenna always has an associated magnetic field in the space around it. When the intensity of the magnetic field changes or the field itself is moved, an induced voltage is generated. The voltage always has an associated electric field. The result is two varying fields, one with magnetic flux and the other with electric lines of force.

Actually, the fields in space are more important than the conductors. Any changing magnetic field will generate an electric field. Also, any changing electric field will generate a magnetic field. The two fields are illustrated in Fig. 13-2 with the direction of propagation through space. The electric field is labeled E; the magnetic field is labeled with the letter H, which is the symbol for magnetic field intensity. Both fields vary in strength while moving in the direction of propagation P. Visualize the arrows in three dimensions. The fields E and H are perpendicular to each other, and both are at right angles to the direction of wave motion. The electric and magnetic fields form an electro-

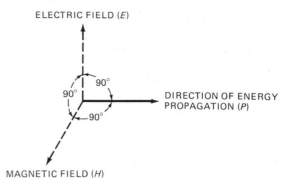

Fig. 13-2. Components of an electromagnetic radio wave and its propagation in space.

magnetic wave propagating itself through space. The energy in the wave motion is divided equally between the electric and magnetic components.

Velocity of Radio Waves   In general, an electromagnetic wave is a form of radiation that transmits energy through space. Light radiation, heat radiation, x-rays, and radio waves are examples of electromagnetic waves. All electromagnetic waves are propagated through space with the velocity of light, which has the symbol c. The velocity is

$$c = 186,000 \text{ miles per second (mi/s)}$$

$$c = 300,000,000, \text{ or } 3 \times 10^8, \text{ m/s}$$

$$c = 3 \times 10^{10} \text{ cm/s}$$

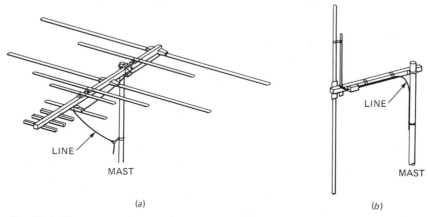

Fig. 13-1. Two common types of antennas. (a) Television receiver antenna. (b) CB radio antenna used for either transmitting or receiving.

The velocity can be considered practically the same in air or a vacuum.

In radio waves the frequency of variation in the intensity of E and H is the same as the frequency of variation in antenna current that produces the electromagnetic wave. In other words, the frequency of the field is the frequency of the source that generates the field.

Wavelength of Radio Waves We can now consider the length of the electromagnetic wave traveling in space. Its symbol is the Greek letter lambda, λ. One wavelength is the distance between two points along the direction of propagation, where E and H repeat their values of intensity.

The wavelength λ, frequency f, and velocity of light c are related to each other by the formula

$$\lambda = \frac{c}{f} \qquad (13-1)$$

If as an example, a radio station is broadcasting at 1 MHz, the wavelength of the electromagnetic wave can be calculated as

$$\lambda = \frac{300{,}000{,}000 \text{ m/s}}{1{,}000{,}000 \text{ Hz}}$$
$$\lambda = 300 \text{ m}$$

The 300 m is approximately 325 yards, or 975 ft.

The higher the frequency the shorter the wavelength. Frequencies in the citizens' radio band around 27 MHz have shorter wave lengths than the AM radio broadcast band of 535 to 1605 kHz.

Test Point Questions 13-1
(Answers on Page 320)

a. Name the two field components of an electromagnetic wave.

b. Is the wavelength shorter or longer at higher frequencies?

c. What is the velocity of electromagnetic radio waves, in meters per second?

13-2
PRINCIPLES OF RADIATION

The electric and magnetic lines of force make the electromagnetic field move out from the antenna into the space around it. The radiation mechanism is illustrated by the example in Fig. 13-3. In Fig. 13-3a the capacitor plates for C_A represent the antenna conductors in Fig. 13-3b, which are opened up to radiate the field into space. The capacitance has voltage variations associated with the electric field. L_A represents the inductance of the antenna conductors. The current variations produce the magnetic field.

Essentially, the radiation is a matter of having voltage and current variations with a frequency

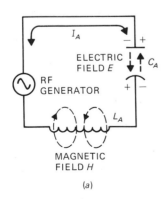
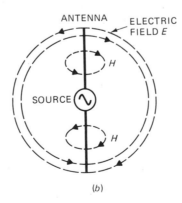
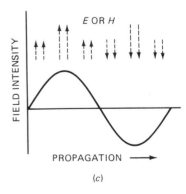

Fig. 13-3. Radiation of an electromagnetic wave into space. (*a*) Equivalent circuit of the antenna with inductance L_A and capacitance C_A. (*b*) Antenna conductors extend the fields E and H into the space around the antenna. (*c*) How the field intensity of the radiated wave varies with its propagation.

high enough to allow the energy in the electromagnetic field to move out into space. In fact, that condition specifies what is meant by radio frequencies. Above 30 kHz, approximately, radio frequencies are high enough to be radiated efficiently from an antenna.

Consider the equivalent circuit shown in Fig. 13-3a. The RF generator is an ac source used to charge the capacitor C_A. The inductance L_A provides series resonance at the frequency of the ac source so that the maximum amount of alternating current is delivered to the antenna circuit.

The capacitor C_A is charged whenever its voltage is less than the source voltage. That charge produces one polarity of voltage across C_A. When the source voltage decreases, C_A is able to discharge. Then the source voltage charges C_A in the opposite polarity on the next half-cycle. Therefore, the ac voltage across C_A has a varying electric field in the dielectric; it is indicated by the dashed lines for E. The frequency of the variations is the same as the source frequency.

The inductance L_A has the series current. Its associated magnetic field is indicated by the dashed lines for H. The frequency of the variations of I_A and H also are the same as the source frequency.

So far, we have just a series-resonant circuit. In Fig. 13-3b, however, the capacitor plates are opened up in the form of the two conductors of the antenna. The purpose is to make the field lines for E and H extend into space around the antenna.

The two conductors produce a capacitive effect especially at their ends, where charge can accumulate. The dielectric of the capacitor is the space around the antenna. That space has the electric lines of force shown by the dashed lines for E. Their direction is up and down, like that of the antenna. Also, the current in the antenna conductors produces the magnetic field indicated by H. The magnetic lines of force are in a plane perpendicular to that of the electric field. It should be noted that although the antenna is shown vertical here, it can be horizontal instead.

When the rate of change is very fast and at a great enough distance from the antenna, the field lines cannot collapse into the antenna conductor before the current and voltage produce an expanding field again. As a result, part of the field energy is repelled away from the antenna.

As the lines of force in a field move into space, they can generate their own fields. A magnetic field in motion generates an electric field. That action corresponds to voltage induced in a coil that is cut by magnetic lines of force. Also, an electric field in motion generates a magnetic field. That action corresponds to the charge or discharge current produced by a capacitor when its voltage changes.

The Radio Wave in Space Figure 13-3c illustrates how the electric field E or magnetic field H goes through cycles of variations as it is propagated outward from the antenna conductors. It should be noted that E and H are 90° out of phase with each other in time. Also, their field lines are at right angles to each other. Finally, propagation is in the direction of motion of the field lines. If we visualize three planes in space, the direction of propagation is at right angles to both E or H.

As a result, an electromagnetic field is radiated into space. The radio wave has left the antenna to be propagated in all directions away from the antenna conductor. Energy is radiated in that way from all types of antennas. Any conductor with current can produce some radiation. Practically, though, only radio frequencies above 30 kHz are effective for transmitting electromagnetic waves. The higher the frequency the more efficient the radiation. At frequencies below 30 kHz the antenna would have to be much too long. In addition, the variations in field intensity are not fast enough to create significant radiation from the antenna.

Radiation and Induction Fields In addition to the energy radiated from the antenna, there are local fields associated with the current and voltage in any circuit. To distinguish between the radiated energy and the local fields near the conductor, one is called the *radiation field* and the other the *induction field*. Beyond a distance of a few wavelengths, however, the induction field need not be considered, because its intensity falls off rapidly.

The radiation field is the only one considered for radio waves, because it can be propagated over long distances.

Interference with Radio Waves Any radiation picked up by the receiving antenna which is not the desired signal can be considered interference. The interference can be from a radiation field, induction field, or both. As an example, if you live near a busy highway, passing cars can produce interference in radio and television receivers. The energy that reaches the receiver antenna is the radiation field of the ignition system. Close to the car, the local induction field of the ignition coil also can produce interference. Ignition interference produces short, horizontal black flashes in the television picture and crackling noise in the sound. Additional examples of interference are radiation produced by sparking motors and from fluorescent lights.

Another phenomenon interferes with good reception by reducing the strength of the signal at the receiver. Any metal conductor can actually serve as an antenna. When metal structures are near the receiving antenna, they reduce the field intensity. The metal acts as a short circuit for the electromagnetic wave. That is why it is difficult to have good reception of radio signals inside a metal enclosure or near a large metal structure. Insulating materials, however, do not affect the electromagnetic radio wave.

Test Point Questions 13-2
(Answers on Page 320)

a. Which is better for radiation from an antenna, a high or low frequency?
b. Is the local field close to a conductor the induction or the radiation field?
c. A magnetic field in motion generates E or H?

13-3
ANTENNA CHARACTERISTICS

The choice of an antenna type depends mainly on the operating frequencies. The higher the frequency, the shorter is the required length of the antenna conductor.

An antenna can be resonant or nonresonant. A resonant antenna has a specific length for a particular frequency. Especially in the VHF band of 30 to 300 MHz, a half-wave is a practical length. The television antenna shown in Fig. 13-1a is an example of a half-wave dipole with a length about 8 ft end to end. Such an antenna is used for the television and FM radio bands.

In general, the larger the antenna the more energy it can radiate for a transmitter or intercept for a receiver. The resonant length of an antenna, however, provides the most efficient operation.

Grounded and Ungrounded Antennas An example of an antenna that operates independently of earth ground is illustrated by the two poles, or *dipole*, shown in Fig. 13-3. Each pole has signal current with a symmetrical source at the center. Earth ground is not necessary because the antenna capacitance C_A is between the two poles.

The grounded antenna shown in Fig. 13-4 depends on earth ground as one side of the antenna circuit for radiating the electromagnetic wave. The antenna wire has insulators for isolation from the metal mast that supports the antenna structure. Antenna current is supplied by the lead-in wire at one end. This antenna is the *inverted-L* type.

The return side of the feed, however, is connected to earth ground. Then the antenna has signal voltage with respect to earth and the radiated wave depends on the ground characteristics.

Distributed Inductance L_A and Capacitance C_A An antenna looks only like a length of wire or a rod, but electrically it has inductance and capacitance distributed over its length. That is illustrated in Fig. 13-5. In Fig. 13-5a current in the antenna wire must flow through a small equivalent inductance L_A. The L corresponds to a small coil. The coil has lumped inductance concentrated in a small area. The antenna has distributed inductance from one end of the conductor to the opposite end.

As shown in Fig. 13-5b, the antenna wire has distributed capacitance C_A to earth ground. C_A is

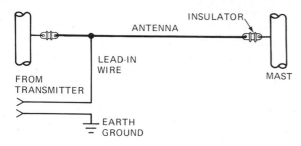

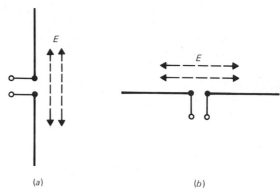

Fig. 13-4. Example of an antenna using earth ground for one side of the antenna circuit.

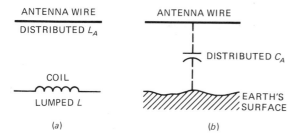

Fig. 13-5. Distributed inductance and capacitance of an antenna. (a) Inductance L_A of the antenna wire. (b) Capacitance C_A of the antenna to earth ground.

Fig. 13-6. Antenna polarization in terms of electric field E: (a) vertical; (b) Horizontal.

indicated at one point in the figure, but the entire wire length has shunt capacitance that produces the total C_A. In the dipole antenna, C_A is the capacitance from one pole to the other. In a grounded antenna, C_A is the capacitance to ground.

Resonant Antennas   The values of L_A and C_A can be used to make the antenna act like a series-resonant circuit. The advantage is maximum antenna current at the resonant frequency. There is no coil and there is no physical capacitor, but the electrical characteristics of the antenna correspond to those of an *LC* circuit.

The physical length of the antenna, with respect to the wavelength of the radio wave, determines the resonant frequency. Generally, the length of a half-wave at the operating frequency is used for a resonant antenna in the VHF band of 30 to 300 MHz. At those frequencies, a half-wave is about 1.5 to 15 ft [0.46 to 4.6 m]. Those lengths are practical for an ungrounded half-wave dipole antenna that can be supported on a mast suffi-

ciently high to be independent of earth ground. At lower frequencies, a quarter-wave resonant antenna is used. It is usually a grounded antenna, and ground takes the place of one pole.

Polarization   The antenna conductor can be mounted either vertically or horizontally. In either case, the electric field E has lines of force in the same direction as the antenna. See Fig. 13-6 for the case of a dipole antenna.

The direction of polarization of a radio wave is defined as the direction of the electric field E. As a result, a vertical antenna and its transmitted radio wave are vertically polarized, as shown in Fig. 13-6a. In Fig. 13-6b the polarization is horizontal for a horizontal antenna. The transmitter and receiver antennas should have the same direction of polarization for maximum signal pickup.

Horizontal polarization is generally used at frequencies in the VHF band of 30 to 300 MHz. The reason is that most noise interference in that band is vertically polarized. Therefore, horizontally polarized antennas should be less susceptible to noise pickup. In television broadcasting horizontal polarization is used, which is why antennas for TV reception are horizontal.

A new method for the VHF band combines horizontal and vertical polarization, which results in *circular polarization*. The advantage is that the receiving antenna can be either vertical or horizontal for good signal pickup. That feature is important with indoor antennas. At the present time, circular polarization is used for the transmitted signals in the FM radio broadcast band. It is also being considered as a standard of transmission in television broadcasting.

Microvolts per Meter   The microvolts per meter unit is generally used to indicate the field strength of the transmitted radio wave in space. One meter is approximately 40 in, or a little more than a yard. As an example, assume that a half-wave dipole antenna one meter long can pick up 300 μV of signal. The field strength then is 300 μV/m. The height of the receiving antenna is standardized, usually at 30 ft for the VHF band. Also, the polarization of the antenna is the same as that of the transmitted radio wave.

Note that the same field produces more antenna signal in a longer antenna at lower frequencies, assuming both antennas are resonant. For instance, an antenna with a length of 2 m would pick up 600 μV of signal at the same field strength of 300 μV/m.

Test Point Questions 13-3
(Answers on Page 320)

Answer True or False.

a. A half-wave dipole is a resonant antenna.
b. A horizontal antenna transmits radio signal with horizontal polarization.
c. The field strength of 200 μV/m will produce 200 μV of signal in an antenna 2 m long.

13-4
HOW THE ANTENNA PROVIDES SIGNAL FOR THE RECEIVER

The electromagnetic field is radiated from the transmitting antenna in all directions. Neither the air nor any insulating materials have any effect on propagation of the radio signal. As the electromagnetic field cuts across the metal conductor of the receiving antenna, however, a very low voltage is induced. That antenna signal is generally of the order of microvolts. An amplitude of 1 to 5 mV, or 1000 to 5000 μV, is a strong antenna signal.

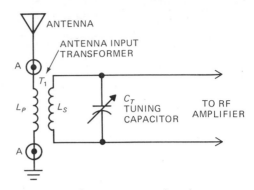

Fig. 13-7. How the antenna supplies the input signal to a receiver.

It should be noted that although the electromagnetic field induces a signal in the receiving antenna, the radio wave continues to be propagated in space as though the antenna were not there. The amount of energy extracted by the receiving antenna is so small that it has practically no effect on the field. All receivers in the service area around the transmitter can receive the radiated signal.

The antenna input circuit of a receiver is illustrated in Fig. 13-7. Note the triangle symbol used for any type of antenna. The voltage induced in the antenna produces signal current in the primary L_P of T_1, which is the antenna input transformer. The terminals marked A on the back of the receiver are for the antenna connections. By transformer action, then, the signal is coupled to the secondary L_S. The capacitor C_T tunes the secondary for the desired carrier frequency.

A receiver antenna generally will pick up all signals in the operating band of the receiver, but its tuned circuits select the desired station. The selected RF signal is then increased to the required strength by the amplifier circuits.

It is important to realize that any receiver needs an antenna to pick up the radio signal. Some receivers have built-in antennas that are not obvious, but there must be an antenna to feed signal to the first RF amplifier.

In a receiver that is well-shielded, such as an auto radio, there will be practically no signal unless the antenna is connected. Even a small piece of wire 1 or 2 ft [30.5 to 61 cm] long for an external receiver antenna will make a big difference in reception compared with no antenna at all. If you touch the antenna terminals on a receiver, without an antenna, your body capacitance can provide radio signal.

Test Point Questions 13-4
(Answers on Page 320)

a. Is the typical antenna signal at the receiver 120 μV or 5 V?
b. Can the transmitted radio wave produce a signal in a conductor or an insulator?

13-5
HALF-WAVE DIPOLE ANTENNA

On the basis of the velocity of light of an electromagnetic radio wave in space, the formula for calculating a half-wavelength can be derived as $L = 492/f$, where f is in megahertz and the length of L is in feet converted from meters or miles. However, the resonant length of a half-wave dipole conductor is slightly less than a half-wavelength in free space. The reason is that the antenna has capacitance that alters the current distribution at the ends. This *end effect* requires foreshortening of the conductor for the resonant length. For a half-wave dipole,

$$L = \frac{468}{f} \qquad (13\text{--}2)$$

where L is in feet and f is in megahertz.

As an example, at 100 MHz the half-wave dipole length is

$$L = \frac{468}{100} = 4.68 \text{ ft } [1.43 \text{ m}]$$

Each pole is a quarter-wave or 2.34 ft [71.3 cm] for this example as shown in Fig. 13-8a. The insulating space between the two poles is negligible. The dipole can be mounted either horizontally or vertically. A half-wave dipole is also called a *Hertz antenna*.

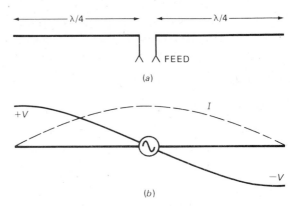

Fig. 13-8. Half-wave dipole antenna. (*a*) Each pole of the antenna is a quarter wavelength long. (*b*) Current *I* and voltage *V* distribution along the conductors.

Current and Voltage Distribution   When the dipole antenna is fed by the transmitter, electrons travel along the conductor to the open ends. This motion of electrons is the antenna current *I* in Fig. 13-8*b*. When the electrons reach the ends of the conductor, electric charge is accumulated. That charge provides the voltage *V* at the ends. The charge at each end provides a potential for moving the electrons in the opposite direction, thereby reversing the direction of current. As a result, the current is zero at the ends and two currents of equal amplitude flow in opposite directions.

Farther back from the ends of the conductor, the outgoing and returning currents are not equal because the charges causing the *I* have been supplied to the antenna at different times of the RF cycle. Maximum current is at the center, where the reflected current can add to the original current from the ac source.

The ends of the antenna are points of maximum voltage. The waveform for the voltage distribution along the antenna shows $+V$ and $-V$ at the two ends because the two poles have opposite polarities from the ac source.

The current and voltage distribution illustrate why the physical length makes the antenna resonant at a specific frequency. The electron charges in the antenna conductor must travel from the center out to the ends and back to the center in the time of one half-cycle of the ac source. Then the current and voltage values on the antenna are maximum. At other frequencies, partial cancellation of the incident and returning electrons reduces the amount of antenna current or voltage.

Standing Waves The waveforms for the values of I and V along the antenna shown in Fig. 13-8b are known as *standing waves*. Points of minimum I or V are *nodes*; the maximum points are *loops*. In a half-wave antenna, successive loops or nodes for V and I are a half-wavelength apart.

It is important to realize that the standing wave is not a picture of the current or voltage. Actually, each point on the conductor has a continuously changing voltage and current that varies with respect to time at the frequency of the ac source. However, the peak amplitude of the ac variations is different along the length of the antenna. That distribution of peak values does not change along the antenna length, though, and the result is a stationary graph or standing wave of I and V values from one end to the other.

Standing Wave Ratio The *standing wave ratio* (SWR) compares the voltage at a maximum point to that at a minimum point. For instance, when a dipole antenna has V at the ends 10 times more that at the center, the SWR is 10. A resonant antenna has a high SWR, which indicates high values for V and I in the standing waves. Then the efficiency is high for a transmitting antenna that radiates or a receiving antenna that picks up signal. However, a nonresonant transmission line that feeds the antenna should have an SWR of 1. That is the lowest possible SWR, and it indicates there are no standing waves.

Radiation Resistance The fact that an antenna has specific values of V and I means that it has a definite impedance. If the antenna is resonant, the feed point is resistive. That R is called the *radiation resistance* of either a transmitting or a receiving antenna.

The radiation resistance of a half-wave dipole is 72 Ω at the center. Farther out, the antenna has a reactive impedance with higher values. The ends have the highest impedance; typical values are up to several thousand ohms.

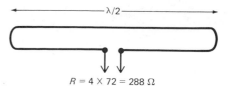

Fig. 13-9. Folded dipole antenna.

Folded Dipole Antenna As shown in Fig. 13-9, the folded dipole antenna is constructed of two half-wave conductors joined at the ends but with one conductor open at the center where it connects to the transmission line. The spacing between the two conductors is small compared with a half-wavelength.

More conductor is used than in a straight dipole, but the bottom part is folded back in quarter-wave sections so that the folded dipole still occupies a half-wave in space. The antenna characteristics are essentially the same as those of a straight dipole, except that R is $4 \times 72 = 288\ \Omega$ for a folded dipole. That value is convenient for matching to a 300-Ω transmission line.

Polar Directivity Pattern The direction in which an antenna transmits or receives best is shown by a graph as in Fig. 13-10. The pattern is that of a half-wave dipole mounted horizontally. However, the response of a vertical antenna is equal in all directions in the horizontal plane.

The graph shows signal strength in polar coordinates for magnitude and direction. The angle indicates direction, and the length of the radial arm is the amount of signal voltage. The pattern of a transmitting antenna shows in which direction the antenna radiates the most signal. The pattern of a receiving antenna shows the direction from which the most signal is induced. Rotating the antenna for the best signal is called *orientation*.

A half-wave dipole at its fundamental resonant frequency has the double lobes shown in Fig. 13-10 for a figure-8 directivity pattern. The antenna receives best from the broadside direction, front and back, with little signal off the ends. At the transmitter, multiple crossed dipoles can be used for radiation in all directions.

Harmonic Antennas An antenna can be used at harmonic frequencies of the fundamental half-wave resonance. However, the directional pattern shifts at odd multiples of the resonant frequency, as shown in Fig. 13-11. The half-wave dipole of Fig. 13-11a has the normal figure-8 pattern. The same antenna length for double the frequency, or 2f (Fig. 13-11b), acts as a full-wave dipole. It still has the same general response but with wider lobes as compared with a half-wave dipole for 2f. Also, there is more signal for a full-wave antenna than for a half-wave antenna because more conductor is used. Finally, at the third harmonic frequency, or 3f, the directional pattern splits into the side lobes shown in Fig. 13-11c. Now the antenna receives or transmits little in the broadside direction. The best results are at an angle of 37°.

V-Dipole Antenna In some cases the two poles are angled in the form of a V to take advantage of the directional response at the third harmonic frequency. Then a receiving antenna can be used to cover a 3:1 range of frequencies for resonance. Dipole antennas in the form of a V are often used for television receiving antennas.

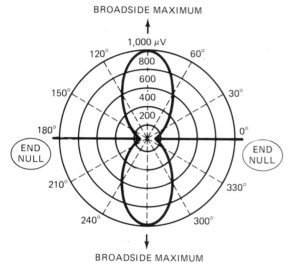

Fig. 13-10. Polar directivity pattern for a half-wave dipole antenna.

Test Point Questions 13-5
(Answers on Page 320)

a. Calculate the length of a half-wave dipole at 200 MHz.
b. What is the radiation resistance of a center-fed half-wave dipole?
c. Has a half-wave dipole maximum response in the broadside direction or off the ends?

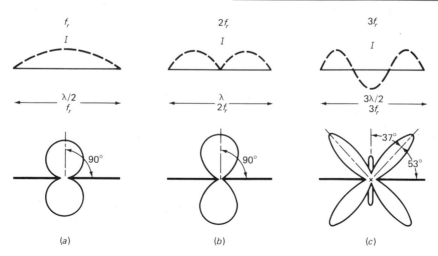

Fig. 13-11. Directional patterns for a center-fed dipole at the fundamental and harmonic frequencies. (a) Figure-8 pattern of fundamental half-wave resonance. (b) Full-wave operation at the second harmonic. (c) Lobes split to change the directional pattern at the third harmonic, where the antenna length is 3λ/2.

13-6
QUARTER-WAVE GROUNDED ANTENNA

When the antenna in Fig. 13-12 is connected to earth ground, or even close, the ground becomes part of the radiating system. Here the vertical conductor produces a vertical electrostatic field that extends through ground. The earth acts as a conducting surface, or mirror, for the radiation. As a result, the λ/4 grounded antenna shown in Fig. 13-12 has a λ/4 image within the ground. Then the antenna voltage and current distributions become similar to those of a vertical half-wave dipole in free space.

The quarter-wave grounded type, called a *Marconi antenna,* is generally used at frequencies below 30 MHz. Since λ/4 is one-half the length of λ/2, the required quarter-wavelength is easier to obtain for lower frequencies, which need longer antennas.

The quarter-wave antenna is operated at frequencies of λ/4 or odd multiples such as 3λ/4 or 5λ/4. Then the feed point at the grounded end is a point of high current and low voltage. Its radiation resistance is approximately 36 Ω. That value is convenient for feeding with 50-Ω coaxial cable.

Directional Patterns The vertical antenna transmits a radio wave with vertical polarization. The directional pattern for transmitting or receiving includes all angles in a circle in a horizontal plane around the vertical antenna; see Fig. 13-13a. In Fig. 13-13b directivity in the vertical plane is shown. The two half-lobes shown represent part of the figure-8 pattern of a dipole. The vertical antenna and its mirror image on the ground correspond to a dipole. However, the lower half of each lobe is canceled by reflected waves from the ground. The loss of radiation actually increases the signal strength above ground in the horizontal and upward direction.

The radiation has two components: *ground waves* and *sky waves.* A ground wave is radiated out from the antenna in the horizontal direction along the surface of the earth. The sky wave has a sharper vertical angle; it radiates energy up into the sky. Some of that energy may be lost, but reflection

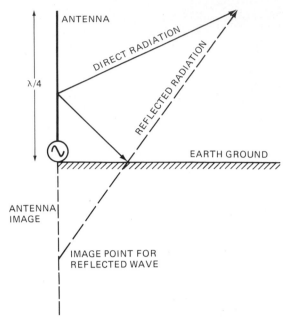

Fig. 13-12. How a quarter-wave grounded antenna produces radiation.

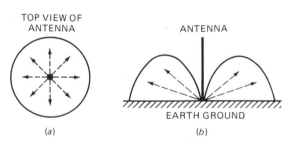

Fig. 13-13. Directional patterns for a vertical quarter-wave grounded antenna. Dashed arrows indicate field lines. (a) Horizontal plane in all directions around the antenna conductor. (b) Field lines in the vertical plane.

from the ionized layer of air above the earth makes it possible to transmit over very long distances.

Applications for grounded quarter-wave antennas include AM radio broadcasting in the 535 to 1605-kHz band, CB radio at about 27 MHz, and shortwave mobile radio communications at about 3 to 30 MHz.

Antenna Counterpoise When the earth does not have good conductivity, it may be necessary to install an artificial ground with a network of metal conductors imbedded in the earth for a grounded antenna. The metal structure that takes the place of earth ground is called a *counterpoise*. The surface of the metallic counterpoise should be at least equal to that of the antenna, and preferably much larger.

When the metal counterpoise is not actually in the earth, the conductor is called a *ground plane*. The structure usually consists of rods or wires extending radially out from the antenna base. As a result, a vertical antenna can be mounted high up to be independent of earth ground.

The action of a counterpoise illustrates how radio transmission and reception can be accomplished when there is no earth ground. In an airplane, for instance, the metal frame and skin act as a counterpoise that serves as ground for all the radio equipment. Similarly, the metal chassis of an automobile, is ground for the antenna circuit. As a final example, one pole of an ungrounded dipole antenna can be considered the counterpoise for the other pole.

Ground-Plane Antenna In Fig. 13-14 the λ/4 vertical antenna is shown mounted on a plate attached to the supporting mast. The antenna itself is insulated from the plate and its metal conductors as the ground plane. With feed from 50-Ω coaxial cable, the inside conductor is connected to the

antenna and the outside shield is connected to the plate and ground plane. The construction can be used for mounting at least a quarter-wavelength above the earth. This type is called a *monopole* or *whip* antenna. It is used for frequencies of 3 to 30 MHz, at which a quarter-wave is not too long for the whip.

Test Point Questions 13-6
(Answers on Page 320)

Answer True or False.

a. A grounded λ/4 antenna is one-half the length of a λ/2 dipole.

b. A vertical antenna transmits in all directions in a circle in the horizontal plane.

c. A grounded vertical antenna transmits ground waves and sky waves.

d. A counterpoise must be a good insulator.

13-7
ANTENNA LOADING

When the antenna is not long enough for resonance, a small inductance can be added in series with the antenna; see Fig. 13-15a. The inductance is called a *loading coil* because it increases antenna load current by making the antenna resonant. A loading coil makes the antenna electrically longer by adding the inductance that would be there if the antenna were the right length. The reason for using an antenna shorter than neccessary is that the required physical length may not be practical for mounting.

Another way to make an antenna electrically longer is to use capacitive top loading. A metal cylinder, disk, or spoked wheel at the top end of an antenna conductor adds shunt capacitance to ground (Fig. 13-15b). The effect corresponds to a longer antenna because more C requires a longer time to charge the antenna capacitance.

It is also possible to use a series capacitor to make the antenna electrically shorter if that effect should be needed. Inserting series capacitance de-

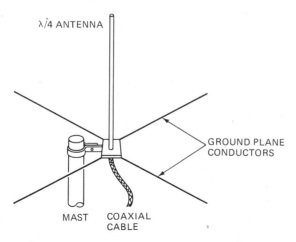

Fig. 13-14. Construction of ground-plane antenna.

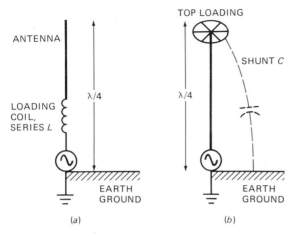

Fig. 13-15. Antenna loading to increase electrical length. (*a*) Loading coil used to increase series inductance *L*, (*b*) Top loading to increase shunt capacitance *C*.

creases the overall antenna capacitance, which reduces the electrical length of the antenna.

In summary, the three methods of antenna loading are:

1. Increasing series inductance by adding a loading coil makes the antenna electrically longer.
2. Increasing shunt capacitance by top loading with a metal conductor makes the antenna electrically longer.
3. Adding series capacitance makes the antenna electrically shorter.

The antenna loading coil of a whip antenna is usually mounted in the base, which is called *base loading*. That method is commonly used for CB antennas. At 27 MHz, a quarter-wave is approximately 8½ ft. That length is a little too much for a practical whip antenna on an automobile, but 5 to 6 ft can be used with base loading.

Test Point Questions 13-7
(Answers on Page 320)

Which of the following can make the antenna electrically longer? (1) Series capacitor; (2) series loading coil; (3) capacitive top loading; (4) vertical polarization.

13-8
ANTENNA ARRAYS

An array is a group of antenna conductors with a common transmission line. The purpose is to increase the antenna signal, by having more conductors for the antenna. Arrays also improve the directional response of the antenna. Antenna arrays can use half-wave dipoles or quarter-wave grounded antennas. Also, the array can be horizontal or vertical.

Antenna Gain   The greater the gain of a transmitting antenna the stronger the radiated field. A high-gain antenna for a receiver provides more signal pickup.

The antenna gain of an array is measured against the signal strength of a standard dipole antenna. Actually, an antenna cannot amplify, but the effect of the array is to have more signal than the reference dipole. The reason for antenna gain is simply the use of more conductors with the correct phasing between them.

As an example, an array with two dipoles generally can provide about twice the signal voltage of one dipole. Then the voltage gain is a little under 2, or about 5 dB.

Front-to-Back Ratio   The array is often designed to increase the antenna gain at the front and have less signal from the back. That difference in directional response is the *front-to-back* ratio of the antenna. When that design is used, it is important that the antenna be mounted to face in the desired direction.

Test Point Questions 13-8
(Answers on Page 320)

a. An antenna gain of four times in voltage is how many dB?
b. An antenna receives twice as much signal from the front as from the back. What is the front-to-back ratio, in dB?

13-9
PARASITIC ARRAYS

A parasitic element in an array is a conductor near the antenna but not connected to it. The antenna, which has the connections to the transmission line, is the *driven element*. As shown in Fig. 13-16, a parasitic conductor in back of the antenna is called a *reflector*. When the parasitic conductor is in front, it is a *director*, as shown in Fig. 13-17. The driven element and the parasitic element are actually coupled to each other by the electromagnetic field in space. A straight dipole antenna or a folded dipole is generally the driven element.

Dipole with Reflector The parasitic reflector of Fig. 13-16 is 0.2λ behind the dipole to reinforce signals received from the front. Also, the reflector is about 5 percent longer than the dipole. The added length compensates for the spacing.

Antenna operation depends on the radiated signal from the parasitic reflector. From the front direction, the radio signal produces current in the reflector 90° later than in the dipole. The 90° lag results from the reflector spacing and length. The current in the reflector then radiates signal to the driven element. The radiated signal takes 90° of the cycle to reach the dipole. Meanwhile, the signal at the dipole has varied through 180° of the cycle. The radiated signal and the direct signal at the dipole then are in the same phase. As a result, they combine to provide more antenna signal for the transmission line.

For a signal arriving from the back, however, the radiated signal from the reflector is 90° out of phase with the antenna current in the dipole. That combination results in less antenna signal than the additive signals from the front.

The gain of a dipole with a reflector is 5 dB with a front-to-back ratio of 3 dB, approximately. The impedance of the antenna is about one-half that of the driven element alone.

Dipole with Director In Fig. 13-17 the parasitic director is 0.12λ in front of the dipole and is about

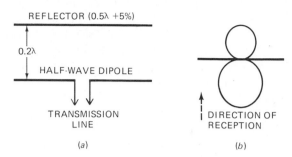

Fig. 13-16. Dipole with a reflector in back of the receiving antenna.

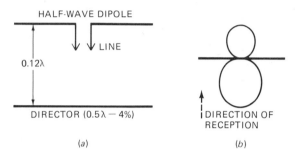

Fig. 13-17. Dipole with a director in front of the receiving antenna.

4 percent shorter. Compared with a reflector, the director is shorter and is mounted at the front instead of the back. However, the results are approximately the same as with a reflector. The dipole with a director still receives more signal from the front than the back.

In many cases, a director is combined with a dipole and reflector. Such an array has an antenna gain of approximately 7 dB.

Yagi Antenna The Yagi array has a dipole with one reflector and two or more directors. Its advantages are high gain, good front-to-back ratio, and very narrow directional response. However, its resonant response is very sharp, which means that it cannot be used for a broad range of frequencies.

UHF Dipole with Corner Reflector For the UHF band of 300 to 3000 MHz, a half-wave dipole is quite small. At 1000 MHz, for example, a half-wave is 468/1000 = 0.468 ft [14.3 cm], which is only 5.6 in. Relatively large reflectors can be used, therefore, to provide high antenna gain. As shown in Fig. 13-18, a conducting sheet is used as a corner reflector behind the UHF dipole. Note that the width of the reflector is 2 to 3λ. The reflector can be either solid metal or a grid of wires.

The dipole is mounted along the line bisecting the 90° corner; it is insulated from the reflector at the back. Maximum transmission or reception is along that line to or from the front.

For the values in Fig. 13-18, the antenna gain is about 10 dB. The antenna impedance is approximately 72 Ω.

The reflector can also be constructed in a parabolic shape. When the parabolic reflector is at least 10 times the length of the dipole, the antenna acts as a point source of energy for the reflector. Then the array has very sharp directivity with a beam width of 2 to 3°, which is similar to a focused beam of light.

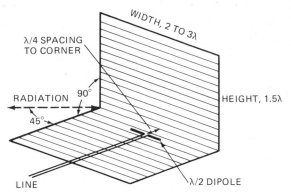

Fig. 13-18. UHF dipole with a corner reflector.

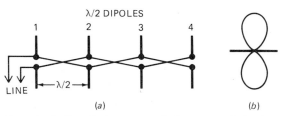

Fig. 13-19. Broadside array with four vertical dipoles. (a) Array with transposed line connections. (b) Response in horizontal plane broadside to the array.

Test Point Questions 13-9
(Answers on Page 320)

Answer True or False.

a. A parasitic reflector is in back of the dipole, but a director is in front.
b. The Yagi antenna uses a reflector and directors.
c. A reflector cannot be used with a folded dipole.

13-10
DRIVEN ARRAYS

In a driven array, all the antennas are connected to the transmission line. How the antenna signals combine depends on their spacing and the phasing of the transmission line feed. The results are high gain with more antenna conductors and sharper directional response. Either λ/2 or λ/4 antennas can be used, and they can be mounted vertically or horizontally.

The directional response is specified with respect to the plane of the array. Try to visualize all the antennas embedded in one sheet or broad surface

for the conductors. When the antenna is vertical, the sheet or plane of the array is vertical; when the antenna is horizontal, the plane of the array is horizontal. In a broadside array, the directional response is broadside, or perpendicular to the plane of the array. In an end-fire array, the directional response is in the plane of the array. Maximum gain is off the ends of the array in the line of the successive antennas.

Broadside Array Figure 13-19 shows four vertical half-wave dipoles spaced a half-wave apart. Imagine that the dipoles are up and down in front of you. The directional gain is then broadside in the horizontal plane; it is toward you and away in the opposite direction.

For a transmitting antenna, the operation of this array is as follows. The space between antennas is a half-wave. Also, the length of the line is a half-wave between antennas. However, the line connections are crisscrossed or transposed. The transposing changes the phase of the antenna driving signal by 180°. Also, the length of the line for a half-wave corresponds to 180° in phase. All the antennas have feed in the same phase, because of the antenna spacing and transposed lines.

Consider a point in space in front of you, along the midpoint of the array. At that point, the electromagnetic fields of all the antennas can combine because the distance to the radiators is approximately the same. Off to the right or left, though, there is cancellation of the fields from each of the antennas for distances that differ by a half-wavelength. The overall directional pattern is shown in Fig. 13-19b.

End-Fire Array   In Fig. 13-20 two half-wave dipoles are shown mounted horizontally one behind the other by $\lambda/4$. Both are driven antennas connected to the transmission line. No parasitic elements are needed because the quarter-wave spacing makes the array unidirectional. The antenna that has the line feed at point X is the back of the array. Both antenna gain and front-to-back ratio are approximately 3 dB.

Operation of the array for reception can be analyzed as follows. From the front, antenna 1 intercepts signal $\lambda/4$ sooner than antenna 2. However, the $\lambda/4$ line delivers the signal at point X in the same phase as the signal in antenna 2. The signals add, then, for reception from the front.

A signal from the back is intercepted by antenna 1 a quarter-cycle later than by antenna 2. Furthermore, the additional $\lambda/4$ line delivers the signal from antenna 1 to point X 180° out of phase with the signal on antenna 2. The two signals cancel, and the result is minimum reception from the back.

The array shown in Fig. 13-20 is end-fire, because maximum transmission or reception is off the ends of the plane for the two antennas in the horizontal direction. The array is unidirectional because of the quarter-wave spacing between antennas.

Vertical and Horizontal Stacking   *Stacking* just means mounting two or more antennas in the same direction. The purpose is to increase antenna gain and sharpen directivity. In an end-fire array such as that of Fig. 13-20 the antennas can be considered as stacked horizontally one behind the other. More than two antennas can be used in that way. Also, antenna arrays can be stacked.

A common stacking technique is mounting horizontal end-fire arrays one above the other on a vertical mast. Each array in the stack is an *antenna bay*. More gain results; horizontal directivity is sharper; and there is less gain in the vertical direction.

The general method of vertical stacking is to use two or three bays a half-wave apart. All are connected to a common transmission line. The connections must be symmetrical so that all the arrays will deliver a signal to the feed point at the same time. When there are two bays, the feed has $\lambda/4$ lines to both antennas. When there are three bays, the feed is at the center antenna and there are $\lambda/2$ lines to the other two bays. The connections between the antenna bays are called *phasing rods*.

Collinear Array   In the collinear array the horizontal antenna elements are placed end-to-end as shown in Fig. 13-21. The two $\lambda/2$ dipoles on both sides of the feed line are joined by $\lambda/4$ phasing rods or lines. Those connections can actually be folded $\lambda/2$ sections of line. Current in the phasing sections is delayed by $2 \times 90° = 180°$ between the antenna elements. Therefore, each of the $\lambda/2$ dipoles has current of the same phase.

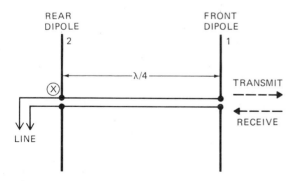

Fig. 13-20. Two horizontal dipoles mounted one behind the other in an end-fire array.

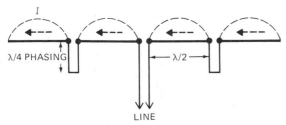

Fig. 13-21. Collinear array with half-wave dipole antennas.

Consider a point in space broadside to the line of the antennas, at the middle, and in the front or back. This point is equidistant from opposite ends of the array. Then the radiated fields of a transmitting antenna combine. If the antenna is for receiving, the induced antenna currents add for the line feed. Off the ends of the array, however, the dipoles have a canceling effect because the signals are out of phase. As a result, the directional response is maximum broadside to the line of the array but minimum off the ends.

Test Point Questions 13-10
(Answers on Page 320)

a. How many parasitic elements are used in a driven array?
b. Are three horizontal $\lambda/2$ dipoles stacked vertically with feed at the center called a broadside or an end-fire array?
c. An array supplies four times the signal voltage of a half-wave dipole. What is the antenna gain, in dB?

13-11
LONG-WIRE ANTENNAS

The name *long-wire antenna* is used for antennas that are much more than a half-wavelength. They are generally 2 to 6λ, in length. As the antenna is made longer in terms of half-waves, the current distribution in the conductor increases the directivity along the line of the antenna wire itself. The directional response is an extreme example of the side lobes produced by harmonic resonances on a dipole antenna.

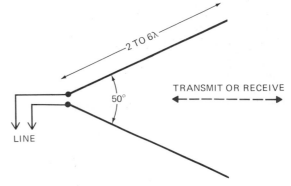

Fig. 13-22. V antenna is an example of a long-wire antenna. This antenna is bidirectional.

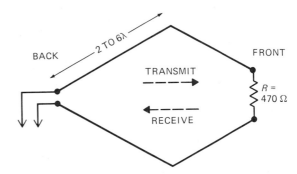

Fig. 13-23. Rhombic antenna consisting of two V antennas. This antenna is unidirectional because of R at the front.

V Antenna The response of a long wire by itself is in the direction of the conductor. When two long wires are combined in the form of a V, as in Fig. 13-22, the lobes of the directional patterns for both conductors reinforce along a line through the middle. Therefore, the V antenna transmits or receives best along the line bisecting the angle of the V. The greater the electrical length of the two legs, the smaller the included angle should be for maximum gain. For the example shown in Fig. 13-22, with legs of 4λ, the gain is about 7 dB.

Rhombic Antenna The rhombic antenna consists of two V antennas, as shown in Fig. 13-23. To make it unidirectional, the rhombic antenna can be terminated with a 470-Ω resistor as an approximate match to a 300-Ω line. That end then becomes the front of the antenna.

For the example shown in Fig. 13-23, with legs of 4λ, the gain is about 10 dB. The rhombic antenna has the advantages of uniform impedance over a wide frequency range of 3:1, high gain, and sharp directivity.

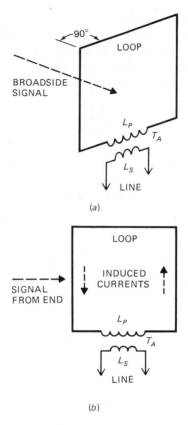

(a)

(b)

Fig. 13-24. Loop antenna. (a) Minimum signal, or null, is broadside to the loop. (b) Maximum signal pickup is from either end of the loop.

Test Point Questions 13-11
(Answers on Page 321)

Answer True or False.

a. A long-wire antenna cannot have a length greater than 0.75λ, approximately.
b. The V and rhombic antennas have maximum gain along a line bisecting the V.

13-12
LOOP ANTENNAS

The loop antenna is a coil of wire that serves as a very directional receiving antenna. The loop shown in Fig. 13-24 is square, but it can be circular. Maximum signal is received off the ends of the loop and there is a sharp null for minimum signal in the broadside direction. When used with direction-finding equipment, a vertical loop is rotated to find the null, which indicates where the signal is coming from.

In Fig. 13-24a, the vertical loop is shown broadside to the signal for a null. When a signal is vertically polarized, antenna current is induced mainly in the vertical conductors. The vertical conductors are connected by the horizontal conductors to opposite ends of the antenna input transformer T_A. From the broadside direction, the vertical conductors have approximately the same signal. The reason is that both conductors intercept the radio wave at the same time. However, the two conductors feed opposite ends of T_A. The currents are canceled in the transformer, therefore, and the result is practically no signal for the receiver.

When signal is received off the ends of the loop, as shown in Fig. 13-24b, the currents in the vertical conductors are not in the same phase because of their spacing. Actually, any signal components of opposite phase in the vertical conductors allow T_A to provide input signal for the receiver. The reason is that opposite phases for currents in opposite directions in the primary L_P can produce signal current in L_S. More turns are generally used in the loop to increase the amount of antenna signal.

Sense Antenna The sense antenna is a separate vertical antenna used with the loop for direction-finding equipment. It permits the broadside responses from the front and back of the loop to be distinguished from each other.

Adcock Antenna The Adcock antenna array consists of two vertical antennas with transformer coupling that cancels in-phase signals. The effect is similar to that of a loop antenna. The Adcock antenna, then, can be used for direction-finding equipment. However, the sensitivity is limited because the antenna is essentially a one-turn loop.

Test Point Questions 13-12
(Answers on Page 321)

a. Does the null of a loop antenna occur with a broadside signal or with a signal off the ends?
b. Is a sense antenna a vertical conductor or a loop?

13-13
PROPAGATION OF RADIO WAVES

The forward travel of electromagnetic radio waves is called *propagation*, because the electric and magnetic fields generate their own wave motion for spreading out in all directions. There are several ways for the transmitted radio wave to reach the receiving antenna. For convenience, they can be classified as follows:

1. *Ground wave.* This radio signal is transmitted close to the earth and in it.
2. *Surface wave.* This radio signal is transmitted just above the earth's surface, mainly in a direct line to the receiving antenna.
3. *Sky wave.* This radio signal is transmitted upward into the sky. Part of the energy is lost, but some of the sky wave returns to the earth from the ionosphere by reflection and refraction.

Those three types of radio waves are illustrated at the antenna in Fig. 13-25. Note the charged layers in the ionosphere, high above the earth.

In general, high frequencies above 30 MHz are propagated mainly by surface waves for line-of-sight transmission to the receiving antenna. Typical transmission distances are up to 75 mi.

The greatest transmission distance is obtained with sky waves. When there is enough power for the radio signal, reflection from the ionosphere can allow reception more than 1000 mi away. Long-distance communications is often abbreviated as DX.

The Ionosphere Ultraviolet light and other radiation from the sun produce ionization in the upper part of the earth's atmosphere. The charged particles include free electrons and ions in a rarefied gas. Most of the effect on radio waves is from the electron density in the upper atmosphere at a height of 70 to 225 mi [112.6 to 362 km]. This

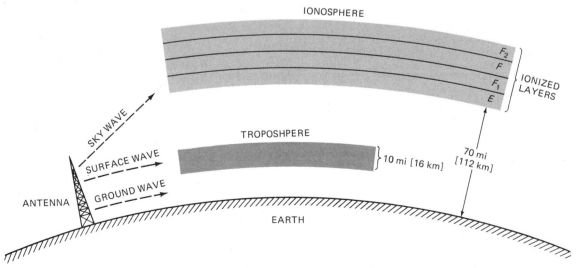

Fig. 13-25. Propagation of radio waves in the space around the earth. Distances are not to scale.

charged region is called the *ionosphere*, and it is also known as the *Kennelly-Heaviside* layer.

As indicated in Fig. 13-25, the ionosphere can be considered in layers labeled E, F_1, F, and F_2 according to their charge density. The E layer, closest to earth, has a constant height of about 70 mi [112.6 km]. However, the charge density can vary with the time of day; it depends on the amount of ultraviolet rays from the sun. Maximum charge occurs at noon and minimum during the dark hours of the morning.

At higher altitudes, the F layer is about 185 mi [298 km] above the earth. However, that area of charge exists as a single layer only at night. During the day, the upper and lower sections break into the separate layers F_1 and F_2. The F_2 layer has the greatest degree of charge density because it is closest to the sun.

The ionosphere is important for long-distance communications below 30 MHz, approximately, because the radio waves can be either reflected or refracted to be returned to earth. At higher frequencies, the radio waves continue through the ionosphere into space until all the energy is dissipated. The dependence of the charge density of the ionosphere on the sun is the reason such radio communications can vary with the time of day and season of the year.

Reflection and Refraction of Radio Waves Reflection from the ionosphere means that the waves are turned back at the edge of the charged layer. The phenomenon is similar to the reflection of light from a mirror. Refraction means that the wave enters the charged layer and then is bent away to return at an angle smaller than that of reflected wave. Those effects are illustrated in Fig. 13-26.

Both reflection and refraction occur whenever a wave enters a medium with new characteristics. The result is a discontinuity in the propagation. Radio waves entering the ionosphere encounter a greater charge density. The change in electrical characteristics alters the direction of the electrical and magnetic fields of the radio wave. When there is a slight decrease of velocity in the charged me-

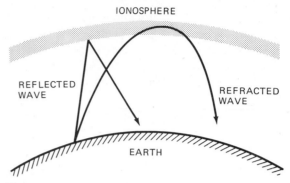

Fig. 13-26. Reflection and refraction of radio waves by the ionosphere.

dium, the wavefront is tilted away from a 90° angle of incidence. The amount of reflection or refraction in the ionosphere depends on the ionospheric charge density and the frequency of the radio wave.

The difference in the two effects is that reflection is from the edge of an ionosphere layer, but a refracted wave actually travels within the layer. Frequencies that are too high to be reflected can still be bent back by refraction.

Critical Frequencies of the Ionosphere Consider radio waves sent upward at 90° toward the ionosphere. At high frequencies the waves will just go through and continue into space. At lower frequencies, part of the energy can be reflected or refracted back to earth. The highest frequency that can be bent back from a specific layer of the ionosphere is called the *critical frequency*. All lower frequencies are also returned to earth. It should be noted that higher frequencies can be reflected, but the angle of incidence must be less than 90°.

Zone of Silence In Fig. 13-27 are shown the results of propagation when the ground wave or surface wave is combined with the sky wave returned from the ionosphere. However, unless the sky wave reaches the earth just where the ground wave ends, there will be an area where no signals can be received either way. The distance from the end of the ground wave to the point where the sky wave first comes back to earth is *the zone of silence*. The remedy is to change the angle of radiation.

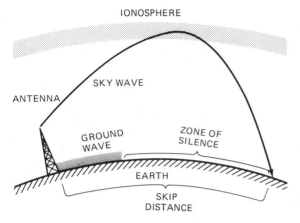

Fig. 13-27. Skip distance and the zone of silence with sky waves.

Skip Distance *The skip distance* is the total distance from the transmitting antenna to the point where the sky wave returns to earth. As an example, in Fig. 13-27 assume that the ground wave extends for 40 mi [64.4 km] but the sky wave skips over that area and is reflected back to earth 300 mi [482.7 km] away. The 300 mi, then, is the skip distance.

It is possible for the sky wave to be reflected from the earth and travel back to the ionosphere again and thereby result in multiple reflections. Energy is lost, but signals have been known to hop their way around the earth in that way. It takes approximately 0.14 s for a radio signal to travel once around the earth, assuming a circular surface path. Naturally, a path resulting from multiple reflection from the ionosphere would take longer.

Troposphere The *troposphere* is about 10 mi [16.09 km] thick immediately above the surface of the earth. Refer again to Fig. 13-25. The troposphere is the medium for transmission of surface waves.

Unlike the ionosphere, the troposphere has no layers of charged particles. However, there may be discontinuities at the boundaries of air masses that have different moisture content. Then reflections can be produced by the troposphere for radio waves above 30 MHz, approximately.

Scatter Propagation *Scatter* refers to a disordered change in direction and polarization of the radio wave. A nonuniform medium, including rain drops, snow, sleet, or fog, causes the scattering. Actually, the scatter propagation allows reception in areas that would otherwise have no signal. One example is ionospheric scatter, which provides signals in locations that are in the zone of silence for sky waves. Also, tropospheric scatter results in weak but reliable signals several hundred miles beyond the line-of-sight transmission distance. Tropospheric scatter propagation is used for radio communications in the VHF, UHF, and SHF bands.

Test Point Questions 13-13
(Answers on Page 321)

Answer True or False.

a. Sky waves are reflected from the ionosphere.
b. The skip distance is equal to the zone of silence.
c. Surface waves are propagated through the troposphere.

13-14
LINE-OF-SIGHT TRANSMISSION

In the VHF band of 30 to 300 MHz and at higher frequencies, there is little signal from the ionosphere. Also, the ground-wave propagation is limited because of absorption losses at high frequencies. In the VHF range, therefore, the RF signal is transmitted by surface waves directly to the receiving antenna. That propagation in a direct path is line-of-sight transmission. It applies to antennas in the FM radio broadcast band of 88 to 108 MHz and television broadcasting in the VHF and UHF bands.

In line-of-sight transmission, the height of the antenna is very important for good reception. The purpose is to overcome the shadowing effect of the curvature of the earth (Fig. 13-28). That is why the transmitting antennas for broadcasting are located at the highest available points. In large urban areas this is generally the top of a skyscraper. The

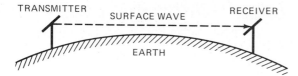

Fig. 13-28. Line-of-sight transmission.

distance to the horizon is then as great as possible. The receiving antenna also should be high.

The distance to the horizon can be calculated as

$$D = 1.41\sqrt{H} \qquad (13\text{-}3)$$

where D is in miles and the antenna height H is in feet. As an example, for a transmitting antenna at 900 ft [274.3 m]

$$D = 1.41 \sqrt{900}$$

$$= 1.41 \times 30$$

$$D = 42.3 \text{ mi [68.1 km]}$$

The distance to the horizon from the receiving antenna also is taken into account. Assume that value is 14.1 mi [22.7 km] for an antenna height of 100 ft [30.5 m]. Then the total line-of-sight distance is 42.3 + 14.1 = 56.4 mi [90.8 km].

The line-of-sight transmission of frequencies in the VHF and UHF bands is similar to transmission of visible light. A disadvantage of line-of-sight transmission is radio shadows. As a result of these shadows, there is little or no signal behind an obstruction. Mountains, forests, and tall buildings attenuate the signal. Even an airplane in the sky can reflect signals in the VHF and UHF bands because their wavelength is comparable in size with the size of the obstruction.

An advantage of line-of-sight transmission in the VHF and UHF bands is the absence of atmospheric static. The static is usually associated with sky waves from the ionosphere. However, the VHF band is subject to man-made interference from electrical equipment.

Test Point Questions 13-14
(Answers on Page 321)

a. To increase the distance for line-of-sight transmission, should the antenna be mounted higher or lower?

b. Is a typical distance for line-of-sight transmission 30 or 300 mi [48.27 or 482.7 km]?

c. Is a typical frequency for line-of-sight transmission 1.2 or 120 MHz?

13-15
TYPES OF TRANSMISSION LINES

A transmission line is just a pair of wire conductors with constant spacing to prevent radiation. It should be noted that antenna conductors are opened outward so that they can radiate an electromagnetic field into space. The function of a transmission line, however, is only to conduct current from one point to another. Typical lines are shown in Fig. 13-29.

Three important requirements of the line are (1) minimum losses, (2) no reflections of signal on the line, and (3) no stray radiation or pickup of the signal by the line itself. Line losses attenuate the signal because of I^2R dissipation in the conductors. To prevent reflections, the line is terminated in its characteristic impedance. More details of this requirement are explained in Sec. 13-16, which explains what is meant by the characteristic impedance of a transmission line.

Balanced Lines   A line is balanced when each of the two conductors has the same capacitance to ground. The balanced line is connected to opposite ends of a transformer that is center-tapped to ground. For reception, any in-phase signal currents caused by stray pickup are canceled because of the balanced antenna input circuit. In feeding a transmitting antenna, the opposite-phase currents in the line have fields that cancel. Examples of balance are the open-wire line (Fig. 13-29a) and twin lead (Fig. 13-29b).

Shielded Line   Shielded line is completely enclosed with a metal sheath or braid that is grounded to serve as a shield for the inner conductor, as shown in Fig. 13-29c. This shielded line is unbalanced, but some types have two conductors. Either way, the shield prevents the inner conductor from receiving or radiating any signal. This function belongs to the antenna.

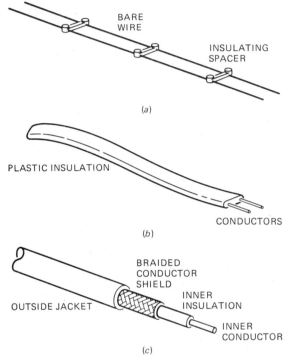

Fig. 13-29. Three main types of transmission lines. (a) Open-wire line. (b) Twin lead. (c) Coaxial cable.

Comparison of Lines   Table 13-1 compares four types of transmission lines. The open-wire line is usually made of bare wire with insulating spacers. It has the lowest losses and relatively high impedance. Twin lead is similar but easier to handle, because the conductors are imbedded in a plastic ribbon. The 300-Ω twin lead is common because television receivers and FM radios have an input impedance of 300 Ω. Both the open-wire line and twin lead are balanced but not shielded. Both have much lower line losses than the shielded coaxial cable.

However, coaxial cable is used very often because it is strong and is little affected by the weather. In many cases, the shielding is necessary to prevent the line from picking up interference. Coaxial cable generally has a characteristic impedence of 50 or 75 Ω.

The coaxial cable is unbalanced, but a balancing transformer, called a *balun*, can be used. This unit converts, either way, between balanced and unbalanced circuits.

Velocity Factor   In Table 13-1, V is a ratio of the velocity of an electromagnetic wave traveling through the dielectric of the transmission line to the velocity of the wave in free space. V is always less than 1 because the velocity is reduced by the dielectric. V is used in calculating the wavelength of a signal on the line. Then the length of the line must be decreased by the factor of V.

Table 13-1
Common Types of Transmission Lines

Type	Characteristic Impedance Z_0, Ω	Attenuation at 100 MHz, dB/100 ft	Capacitance per foot, pF	Velocity factor V†
Open-wire line	300 to 600	0.2	1	0.98
Twin lead	300*	1.2	6	0.8
Coaxial cable,‡ RG 59 U	75 Ω	3.7	21	0.66
Coaxial cable, RG 58 U	50 Ω	5.5	28	0.66

*Twin lead is also available in Z_0 of 150 and 75 Ω with closer spacing.
†V is the ratio of velocity in a dielectric to velocity in free space.
‡RG 59 U cable is a little wider than RG 58 U.

Answer True or False.

a. A typical impedance for coaxial cable is 300 Ω.
b. The coaxial cable in Fig. 13-29 is shielded but not balanced.
c. Twin lead has higher Z_0 but lower losses than coaxial cable.

13-16
CHARACTERISTIC IMPEDANCE Z_0

When the length of a transmission line is comparable to the wavelength of the signal frequency, the distributed inductance and capacitance of the line become important. The reason is that the line is long enough to have standing waves of voltage and current. Then the line could radiate instead of just deliver energy to or from the antenna.

The constant spacing between conductors results in a constant value of L and C per unit length of the line. L is the inductance affecting the current in the conductors; C is the capacitance affecting the voltage across the line. Those factors provide a specific impedance that is characteristic of any length of the line. The characteristic impedance can be calculated as

$$Z_0 = \sqrt{\frac{L}{C}} \quad \Omega \qquad (13\text{–}4)$$

where L is in henrys and C in farads for any unit length that is the same for both L and C. As an example, assume a line with an L of 0.54 μH for the two conductors and a C of 6 pF each for a 1-ft length of line. Then

$$Z_0 = \sqrt{\frac{0.54 \times 10^{-6}}{6 \times 10^{-12}}}$$
$$= \sqrt{0.09 \times 10^6}$$
$$Z_0 = 300 \ \Omega$$

This characteristic impedance of 300 Ω is the usual Z_0 for the twin lead generally used with television receivers to match the 300-Ω impedance at the antenna input terminals. The Z_0 of a line is also called the *surge impedance*.

The characteristic impedance is the same for any length of line. In other words, 2 or 200 ft [0.61 or 61 m] of line has the same Z_0 because the characteristic L and C apply per unit of length. However, a longer line has more resistive losses, because resistance increases with length of the conductor.

It should be noted that Z_0 is an ac characteristic, which cannot be measured with an ohmmeter. You can use an ohmmeter to measure the low resistance of the conductors, but only to see if the line is open.

The Infinite Line The reason Z_0 is important is that it is the value of the resistance to use at the end of the line to make the line function as though it were infinitely long. Then the line cannot have any standing waves because there is no end to produce reflected waves of V and I.

The idea of an infinite line is not as unusual as it may seem. All we mean is that current from a source at one end is delivered to a load resistance at the other end without any waves reflected back from the load to the source. For the example in Fig. 13-30, the 50-ft [15.2-m] length of 300-Ω transmission line is terminated with a 300-Ω load resistance across the end terminals. Because of the impedance match of line and load, maximum power is delivered to the load and there are no reflections.

The line can be considered infinitely long because it has no discontinuity. All along its length,

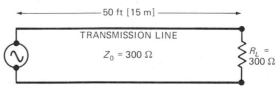

Fig. 13-30. Terminating a line in its characteristic impedance makes the line appear infinitely long.

the Z_0 is 300 Ω across the line for its characteristic L and C. At the end terminals, the 300-Ω R across the line is the same as the Z_0 at all other points. The line thinks it is infinitely long because there is no end to the 300-Ω impedance between the two conductors.

Nonresonant, or Flat, Line The term *nonresonant*, or *flat*, *line* describes a transmission line terminated in its Z_0. The term *flat* means the line does not have standing waves. The length of such a line is not critical. As an example, assume that 300-Ω twin lead is connected to the 300-Ω antenna input terminals on a television receiver. The line can be cut at any convenient length because there are no standing waves. However, more length still produces slightly more resistance losses. Also, any extra length should not be coiled because of the inductive effect.

A practical test of whether the line is nonresonant or has standing waves is to hold the line to add hand capacitance. With a flat line, there may be slightly less signal. With a resonant line, there will be a big difference. There will be more or less signal as you touch the line at different points along its length.

A resonant line has standing waves with different values of signal current and voltage along it. Its length is critical because a connection at different points on it provides different amounts of signal. Also, the hand capacitance across the line alters the standing-wave pattern. A flat line does not have standing waves, though, which allows the conductors simply to deliver power to the load.

Z_0 of Open-Wire Line In terms of physical construction, the characteristic impedance depends on the size and spacing of the conductors. The Z_0 of parallel conductors with air as the dielectric between them is

$$Z_0 = 276 \times \log \frac{s}{r} \qquad (13\text{-}5)$$

where Z_0 is in ohms, s is the distance between centers, r is the radius of each conductor and r and

s are in the same units. As an example, consider an open-wire line using No. 18 gage wire. This wire has a radius of 0.02 in [0.51 mm]. If the spacing is 1 in [2.54 cm],

$$Z_0 = 276 \times \log \frac{1}{0.02}$$
$$= 276 \times \log 50$$

From a calculator or a log table

$$\log 50 = 1.7$$

Then

$$Z_0 = 276 \times 1.7$$
$$Z_0 = 469 \ \Omega$$

The Z_0 of parallel conductors embedded in plastic is lower because plastic increases the capacitance between conductors.

Z_0 of Coaxial Cable The Z_0 of a coaxial line with air insulation between the inner and outer conductors is

$$Z_0 = 138 \times \log \frac{d_o}{d_i} \qquad (13\text{-}6)$$

As an example, assume the inner conductor wire to have a diameter d_i of 0.08 in [2 mm] and the diameter d_o of the outer conductor is 0.25 in [6.35 mm].

$$Z_0 = 138 \times \log \frac{0.25}{0.08}$$
$$= 138 \times \log 3.125$$
$$Z_0 = 68 \ \Omega \quad (\text{approx})$$

When the insulation between conductors is plastic, the Z_0 is lower because the insulation increases the capacitance between conductors.

Answer True or False.

a. Z_0 is an ac characteristic that is the same for any length of line.
b. The characteristic impedance can be measured approximately with a sensitive ohmmeter.
c. A line terminated in its Z_0 is a flat, or nonresonant line.

13-17
RESONANT LINE SECTIONS

When a transmission line is *not* terminated in its characteristic impedance, reflections are produced from its end. The result is standing waves of I and V, as on an antenna. A particular length of $\lambda/4$ or $\lambda/2$ can then serve as a resonant circuit. The four main examples, of $\lambda/4$ and $\lambda/2$ either open or shorted at the end, are shown in Fig. 13-31. A line section is also called a *stub*.

Since the action of a resonant stub depends on having standing waves produced by reflections, the end is purposely not terminated in Z_0. Instead, the end of the line is left open or the two conductors are shorted. Both methods represent the extreme means for making the termination of the line different from its characteristic impedance.

Open End In general, an open end must be a point of high V and low I. There is no place for electrons to go, except to charge the ends of the conductors. Therefore, I is low. Also, V is high as the open end accumulates charges. With high V and low I, an open end must be a point of high impedance, because the V/I ratio is high.

Shorted End In general, a shorted end must be a point of low V and high I. Charge cannot accumulate in the continuous conductor at the shorted end. With low V and high I, a shorted end has low Z because the V/I ratio is low.

Shorted $\lambda/4$ Stub In Fig. 13-31a the standing-wave pattern results in low V, high I, and low Z at the shorted end. A quarter-wave back, however, the maximum values become minimum, and vice versa. To the ac source at the input end, the $\lambda/4$ stub looks like a high impedance where the line is connected to the source. That effect of a high input impedance at a particular frequency makes the shorted $\lambda/4$ stub correspond to a parallel-resonant tuned circuit.

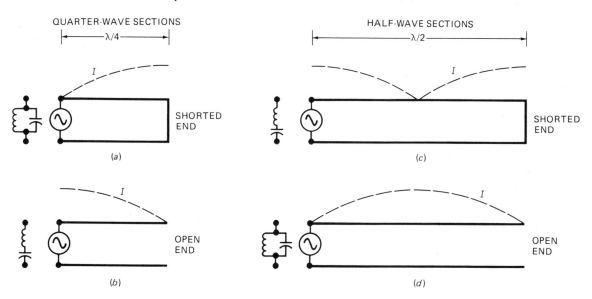

Fig. 13-31. Transmission-line sections, or stubs, as resonant circuits. (*a*) $\lambda/4$ stub shorted at the end. (*b*) Open $\lambda/4$ stub. (*c*) Shorted $\lambda/2$ section. (*d*) Open $\lambda/2$ section.

Open λ/4 Stub In Fig. 13-31b the conditions are opposite those in Fig. 13-31a. The open end has high V, low I, and high Z. A quarter-wave back, however, the ac source sees low V, high I, and low Z. That effect corresponds to a series-resonant tuned circuit.

λ/2 Stubs The λ/2 lengths of resonant line sections repeat the same end values of V, I, and Z a half-wave back at the source. As shown in Fig. 13-31c, a shorted end with low Z also provides low Z at the source. A shorted λ/2 section corresponds to a series-resonant circuit, therefore. It is like an open λ/4 section.

In Fig. 13-31d an open end for the λ/2 section provides high Z at the end and at the source. An open λ/2 section corresponds to a parallel-resonant circuit, therefore. It is like a shorted λ/4 section.

Uses for Resonant Stubs Quarter-wave sections are generally used because they are shorter. Typical applications are:

1. Resonant line, instead of an LC tank circuit, for the VHF and UHF bands where the L and C values would be too small. Specifically, a shorted λ/4 section serves as a parallel-resonant circuit for an amplifier or oscillator.
2. Wave trap to reject an undesired signal. Specifically, an open stub at the antenna input terminals on the receiver will short out an undesired signal frequency that makes the line λ/4 long.
3. Phasing elements for connections to the dipoles in an antenna array. Specifically, a length of λ/4 shifts the phase by 90°, and λ/2 shifts the phase by 180°.

In addition, resonant lines are used for wavelength and frequency measurements.

Lecher Line In the Lecher line application, open-wire with a length of λ/2 to 5λ is used. An adjustable shorting bar is moved along the line to vary the standing-wave pattern. Points of maximum I or V can be indicated by a light bulb, an RF current meter in the shorting bar, or an RF voltmeter. The distance between two maximum or minimum points is a half-wavelength.

Slotted Line In the slotted-line construction, a tubular coaxial line with a slot cut along the outer conductor is used. For wavelength and frequency measurements, an RF voltmeter with its probe is moved along the slot to indicate points of maximum and minimum V.

Test Point Questions 13-17
(Answers on Page 321)

a. For a parallel-resonant circuit, is a λ/4 stub open or shorted at the ends?
b. Is the phase angle corresponding to λ/4 in a standing-wave pattern 90 or 180°?

13-18
FEEDING AND MATCHING THE ANTENNA

The method of feed refers to how the line is connected to the antenna. When the impedance of the line matches the antenna Z, the result is maximum transfer of power. Also, there are no standing waves on the line and no reflections. The antenna, though, should have maximum standing waves. It should be noted that when one side of the line becomes open, the other conductor can serve as an antenna.

Remember that the center of a λ/2 dipole has high I and low Z. The Z of a straight dipole is 72 Ω; a folded dipole is 300 Ω. The low Z also applies to the grounded side of a λ/4 antenna, which is typically 36 Ω. An open end on the antenna, however, has low I and high Z.

Center feed means the line is connected to the middle of a half-wave dipole. With *end feed*, the line is at an open end of the antenna. *Current feed* means the line is connected to a point of high I. With *voltage feed*, the line is at a point of high V.

A half-wave dipole with center feed automatically is current-fed. Also, a quarter-wave antenna is current-fed at the grounded side. The grounded

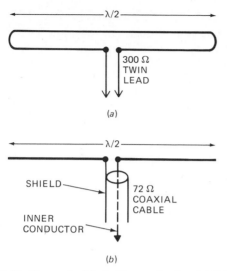

Fig. 13-32. Center feed for half-wave dipoles. (*a*) Folded dipole with a 300-Ω twin lead. (*b*) Straight dipole with a 72-Ω coaxial cable.

$\lambda/4$ antennas are generally fed with coaxial cable, since neither is balanced.

Resonant Feeder Line The resonant feeder line has a critical length, usually $\lambda/4$ or an odd multiple. It is used to obtain high impedance for the end feed of an antenna, because $\lambda/4$ on a standing-wave pattern inverts between low and high Z. However, since a resonant feeder has standing waves, the high values of V and I at the loop points cause considerable line losses.

Nonresonant Feeder Line A nonresonant feeder line is terminated in its characteristic impedance Z_0. A nonresonant or flat line can be any length. A match at only one end still allows the line to be flat, without standing waves. Transmission lines are available only in a limited range for Z_0. Common values are 50 or 72 Ω for coaxial cable, 300 Ω for twin-lead, and 600 Ω for an open-wire line. A line can be considered flat when its SWR is 1.5 or less.

Center Feed for Half-Wave Dipole Two examples of center feed for half-wave dipoles with

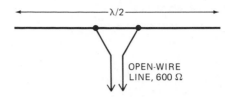

Fig. 13-33. Delta match for feeding a dipole with an open-wire line. Z_0 of the line 600 Ω.

nonresonant feeder lines are shown in Fig. 13-32. In Fig. 13-32*a* the folded dipole has Z of 300 Ω at the center, which matches the 300-Ω twin lead. In Fig. 13-32*b* the straight dipole has 72-Ω Z, the same as the coaxial cable. Note that the inner conductor of the cable is connected to one pole and the outer shield to the other pole. That method is unbalanced, but it can be used. If desired, a balancing unit or balun can be connected between the coaxial line and antenna.

Delta Match Shown in Fig. 13-33 is an open-wire line feeding a half-wave straight dipole. The antenna Z is 72 Ω, and the Z_0 of the line is 600 Ω. In order to match the antenna, the section of line near the antenna is fanned out to points that have higher Z than the center. That method is called a *delta match.* Note that the antenna conductor need not be split in the middle, because the line is not connected to the center anyway.

End Feed with $\lambda/4$ Matching Stub In Fig. 13-34, C_1 and C_2 tune the antenna coupling transformer T_1 for series resonance. This circuit has low Z. The end of the $\lambda/2$ dipole has high Z. Therefore, the quarter-wave stub is used to match the impedances. In general, $\lambda/4$ or any odd multiple inverts from low Z at one end to high Z at the other end.

Shunt and Series Feed The comparison of shunt and series feed for a grounded quarter-wave antenna is shown in Fig. 13-35. The method of Fig. 13-35*a* is shunt feed because the antenna has its own ground in parallel with the coupling circuit. The variable capacitor C adjusts for different connection points on the antenna. The method

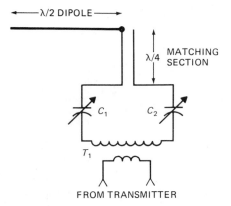

Fig. 13-34. End feed for a dipole with a λ/4 matching stub to provide high Z.

of Fig. 13-35*b* is series feed because the antenna is grounded through the coupling circuit. This π-type network provides an impedance match between the antenna and line.

Test Point Questions 13-18
(Answers on Page 321)

a. Does the center of a half-wave dipole have current or voltage feed?

b. Is the circuit in Fig. 13-34 current or voltage-fed?

c. Is a delta match shown for the antenna in Fig. 13-33 or 13-34?

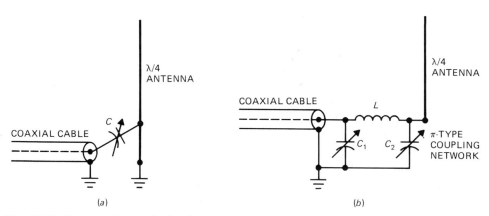

Fig. 13-35. Shunt and series feed with a grounded λ/4 antenna. (a) Shunt connection with C to vary coupling. (b) Series feed. L, C₁, and C₂ form a π-type antenna-coupling network.

SUMMARY

1. A radio wave consists of electric and magnetic fields propagated through space at the velocity of light, which is 3×10^8 m/s.
2. The wavelength of a radio wave is $\lambda = c/f$, where c is 3×10^8 m/s and f is the frequency. Higher-frequency waves have shorter wavelengths.
3. An antenna is a conductor with voltage and current that can radiate an electromagnetic field at radio frequencies.
4. A resonant antenna has a physical length that allows maximum I and V on the conductor at the operating frequency.

5. The half-wave dipole is the basic type of resonant antenna. The half-wave-length is $L = 468/f$, where f is in megahertz and L is in feet. The half-wave dipole is a Hertz antenna.

6. The quarter-wave antenna generally has one $\lambda/4$ pole with earth ground instead of the missing quarter-wave. The quarter-wave type is a Marconi antenna.

7. An antenna counterpoise is a metal structure that substitutes for earth ground.

8. A horizontal antenna provides a radio wave with horizontal polarization, defined as the direction of the electric field. A vertical antenna has vertical polarization.

9. Standing waves indicate repetitive values of V and I on the antenna conductor. The standing-wave ratio (SWR) is the ratio of V or I at a peak or loop to the value at a null or node.

10. Radiation resistance of an antenna is the ratio of V to I at one point on the conductor. R is 72 Ω at the center of a straight dipole, 300 Ω for a folded dipole, and 36 Ω at the grounded end of a $\lambda/4$ antenna.

11. The directional response of a half-wave dipole is a figure-8 pattern with maximum response broadside and minimum off the ends.

12. An antenna loading coil is a series inductance used to make the antenna electrically longer.

13. An array is a combination of two or more antenna conductors connected to one transmission line.

14. Antenna gain is the ratio of the antenna array signal to the signal for a standard half-wave dipole.

15. Parasitic elements are antenna conductors not directly connected to the transmission line. A parasitic reflector is mounted in back of the dipole, and a director is mounted in front of the dipole.

16. A broadside array has antenna gain in the direction perpendicular to the plane of the conductors. An end-fire array has gain off the ends of the antenna plane.

17. A long-wire antenna has a length much greater than $\lambda/2$. Typical lengths are 2λ to 6λ. Common types are the V and rhombic antennas. The directional response is along the line bisecting the V.

18. A loop antenna has directional response with a sharp null broadside to the conductor. The feature is applied to radio direction-finding.

19. The ionosphere is a charged region of the atmosphere about 70 to 225 mi [112.6 to 362 km] above the earth. Radio waves below 30 MHz, approximately, can be reflected from the ionosphere and result in sky waves.

20. The troposphere is about 10 mi [16.09 km] thick; it is just above the surface of the earth.

21. Radio waves are propagated by sky waves from the ionosphere, surface waves through the troposphere, and groundwaves in or close to the earth.

22. Skip distance is the length of the path for a sky wave from the transmitter to the point at which the wave returns to earth.

23. The zone of silence is the area without radio signal between the point the ground waves end and sky waves are reflected back to earth.

24. Line-of-sight transmission means that the surface wave, like light waves, is propagated in a straight line to the horizon. It is a characteristic of radio signals in the UHF and VHF bands.

25. A transmission line consists of a pair of conductors. The constant spacing provides a characteristic impedance Z_0, which is independent of the length of the line.

26. The three main types of transmission lines are coaxial cable with Z_0 of 50 or 72 Ω, twin lead with Z_0 of 300 Ω, and open-wire line with Z_0 of 400 to 600 Ω.

27. Terminating a transmission line with a resistance equal to Z_0 is an impedance match that makes the line equivalent to an infinite line. The result is maximum transfer of power from line to load, no reflections, and no standing waves on the line.

28. A resonant line section or stub is a $\lambda/4$ or $\lambda/2$ length either open or shorted at the end. The $\lambda/4$ stub provides an impedance transformation between high and low Z at opposite ends.

29. A flat line is nonresonant because it is terminated in its Z_0. As a result, its length is not critical.

30. A balancing unit, or balun, can be used at either end of a line to convert between 72-Ω unbalanced Z and 300-Ω balanced Z.

SELF-EXAMINATION

Match the numbered statements at the left with the lettered statements at the right.

1. Velocity of light	(a) $\lambda/4$ stub
2. Sky wave	(b) Flat line
3. Half-wave dipole at 100 MHz	(c) Parasitic element
4. Antenna feed	(d) 30 to 300 MHz
5. Loop antenna	(e) Longer electrical length
6. Folded dipole	(f) Horizontal polarization
7. Reflector	(g) 50 Ω
8. Phase shift of 90°	(h) Ionosphere
9. SWR of 1	(i) Horizon distance
10. Vertical stacking	(j) 4.68 ft
11. VHF band	(k) 3×10^8 m/s
12. $\lambda/4$ grounded antenna	(l) Counterpoise
13. Horizontal antenna	(m) Broadside array
14. Coaxial cable	(n) Long-wire antenna
15. Line-of-sight transmission	(o) Sharp broadside null
16. Artificial ground	(p) Delta match
17. Antenna loading coil	(q) Marconi antenna
18. Rhombic antenna	(r) 300 Ω

ESSAY QUESTIONS

1. Give the velocity of radio waves, in meters per second and centimeters per second.
2. Whigh frequency is best for radio transmission: 40 Hz, 40 kHz, or 40 MHz? Briefly explain why.
3. What is the difference between the radiated field and the induction field?
4. What is the function of the transmission line for a transmitter and a receiver?
5. Give a typical value for an antenna signal at the receiver.
6. Define the following: ionosphere, troposphere, sky wave, ground wave, skip distance, and zone of silence.
7. Compare the directional characteristics of a vertical loop antenna and a horizontal half-wave dipole.
8. Compare the folded dipole and straight dipole antennas.
9. What is meant by a harmonic antenna?
10. Define **a.** antenna polarization and **b.** radiation resistance.
11. Give two differences between the Hertz and Marconi antennas.
12. What is the function of a counterpoise?
13. Give two methods of making an antenna electrically longer.
14. What is the difference between a parasitic array and a driven array?
15. What is the difference between broadside and end-fire arrays?
16. Give the meanings of the following abbreviations: Z_0, SWR, μV/m.
17. What is meant by a flat transmission line?
18. Define antenna gain and front-to-back ratio.
19. Why is a transmission line terminated in its characteristic impedance?
20. What is meant by an infinite line?
21. Give two characteristics of a resonant $\lambda/4$ stub.
22. What is meant by **a.** parasitic reflector, **b.** director, **c.** Yagi array?
23. Show two examples of long-wire antennas.
24. Describe briefly three types of transmission lines.
25. Define the characteristic impedance of a line.
26. Show the standing wave pattern of V and I on **a.** half-wave dipole and **b.** grounded quarter-wave antennas.
27. Compare current feed and voltage feed for an antenna.
28. Compare the use of a flat line and a resonant line for feeding an antenna.
29. What is the purpose of a delta match?
30. Describe briefly how to check a transmission line to see if it is open.

PROBLEMS
(Answers to odd-numbered problems at back of book)

1. Calculate the wavelength, in meters, of radio waves in free space at the following frequencies: **a.** 1000 kHz in the AM radio band, **b.** 14.5 MHz in mobile radio service, **c.** 100 MHz in the FM radio band, **d.** 800 MHz in the UHF television band.

2. Convert the wavelength of 3 m to feet.
3. Calculate the length of a half-wave dipole, in feet, at the following frequencies: **a.** 59 MHz in the band for VHF television channel 2, **b.** 98 MHz in the FM band, **c.** 177 MHz in the band for VHF television channel 7, **d.** 600 MHz in the band for UHF television channel 35.
4. At what frequency will a dipole antenna 7.83 ft [2.39 m] long be operating at the third harmonic of its half-wave resonance?
5. The standing-wave pattern on an antenna has 6.2 A at a loop and 0.4 A at a node. Calculate the SWR.
6. An antenna 4 m long is in a field with a strength of 180 μV/m. What is the value of the signal produced in the antenna?
7. An antenna array has 2400 μV of signal compared with 600 μV for a standard reference dipole. Calculate the antenna gain, in dB.
8. Calculate the horizon distance, in miles, from an antenna at a height 500 ft above the earth.
9. Calculate the characteristic impedance of an open-wire line that has ¾-in spacing between conductors with a radius of 0.02 in.
10. Calculate the characteristic impedance of a line that has L of 0.34 μH and C of 1.8 pF for each foot of line.

SPECIAL QUESTIONS

1. Name one or more types of antennas that you have seen used with receivers or transmitters.
2. Repeat Question 1 but for transmission lines.
3. Describe briefly one or more types of antennas not illustrated in this chapter.

ANSWERS TO TEST POINT QUESTIONS

13-1	**a.**	E and H	13-6	**a.**	T
	b.	Shorter		**b.**	T
	c.	3×10^8 m/s		**c.**	T
13-2	**a.**	High		**d.**	F
	b.	Induction	13-7		(2) and (3)
	c.	E	13-8	**a.**	12 dB
13-3	**a.**	T		**b.**	6 dB
	b.	T	13-9	**a.**	T
	c.	F		**b.**	T
13-4	**a.**	120 μV		**c.**	F
	b.	Conductor	13-10	**a.**	None
13-5	**a.**	2.34 ft		**b.**	Broadside
	b.	72 Ω		**c.**	12 dB
	c.	Broadside			

13-11 a. F
b. T
13-12 a. Broadside signal
b. Vertical conductor
13-13 a. T
b. F
c. T
13-14 a. Higher
b. 30 mi
c. 120 MHz

13-15 a. F
b. T
c. T
13-16 a. T
b. F
c. T
13-17 a. Shorted
b. 90°
13-18 a. Current
b. Voltage-fed
c. Fig. 13-33

Chapter 14
Microwaves

The term *microwaves* refers to high frequencies (above 300 MHz) and short wavelengths (less than 1 m). At those frequencies and wavelengths the components depend on the varying electromagnetic field rather than on the current in a wire conductor or the voltage across two points. Instead of resonant LC circuits and conventional wire conductors, therefore, resonant cavities and waveguides are often used at microwave frequencies. More details are given in the following topics:

14-1
MICROWAVE BANDS

The microwave frequencies span three major bands at the highest end of the RF spectrum. They are the ultra-high frequency (UHF) band of 0.3 to 3.0 GHz, the super-high frequency (SHF) band of 3 to 30 GHz, and the extra-high frequency (EHF) band of 30 to 300 GHz.

Note that 1 GHz = 10^9 Hz, or 1 GHz = 1000 MHz. The UHF band can be specified as 300 to 3000 MHz, or 0.3 to 3 GHz.

Such high frequencies are often specified in wavelengths. The wavelengths range from 1 m at 0.3 GHz down to 1 mm as the frequency increases to 3000 GHz. Remember that all radio waves have the velocity of light, 3×10^8 m/s. Also, the wavelength λ equals v/f. For 15 GHz, as an example, the wavelength

$$\lambda = \frac{3 \times 10^8 \text{ m/s}}{15 \times 10^9 \text{ Hz}}$$

$$= 0.2 \times 10^{-1} \text{ m}$$

$$\lambda = 0.02 \text{ m}$$

The 0.3 to 300-GHz band is called *microwaves* because λ is a fraction of a meter.

All in all, the microwave band covers a range of 1000:1 either in f or λ: 0.3 to 300 GHz in frequencies and 1 m to 1 mm in wavelength. For f,

$$f = \frac{300 \text{ GHz}}{0.3 \text{ GHz}} = 1000$$

For λ,

$$\lambda = \frac{1 \text{ m}}{1 \text{ mm}} = 10^3 = 1000$$

The vast majority of frequencies in the radio spectrum are in the microwave range. That means the potential for microwave applications is very great because of the large bandwidth available.

Microwave Letter Designations The letter designations in Table 14-1 are commonly used for specific parts of the microwave band. They do not have any official international standing, but they are convenient for specifying microwave applications. Most commercial microwave equipment is made for specific bands from 1 to 40 GHz. Note that the new letter designations C to K are in ascending order of frequencies. The old S band of 2 to 4 GHz is split into the E and F bands, which are popular microwave frequencies.

Microwave Applications Probably the most common use of microwaves is in the field of telecommunications: the transmission of information, including analog and digital signals, from one point to another. Figure 14-1 shows the antenna tower of a microwave radio relay station for a long-distance telephone network. The 5.925- to 6.425-GHz band is used for microwave relay stations in telecommunications.

The large bandwidth of a microwave signal makes the signal very attractive for such applications. At 6 GHz, or 6000 MHz, as an example, a bandwidth of 2 percent is equal to 120 MHz.

Another important application of microwaves is in radar systems. (The term *radar* is an abbreviation for *radio detection and ranging*.) A radar system uses a transmitter and receiver operating in the gigahertz range. The transmitter sends out an RF wave in a narrow beam from a directional antenna. When the RF wave strikes an object in the radio beam, part of the RF energy is reflected

Table 14-1
Microwave Bands

Old	P	L	S		C		X	Ku	K	Ka
New	C	D	E	F	G	H	I	J		K

0.5	1	2	3	4	6	8	10	20	40

Frequencies, GHz

Fig. 14-1. Antenna tower of a microwave telecommunications relay station. (*Western Electric*)

back. The receiver detects the reflected signal and compares it with the original transmission. To display the information, an oscilloscope screen shows the location of the object with its distance and angle. The antenna rotates to cover different sectors. Radar equipment operates on 8.5 to 9.2 GHz and 13.25 to 13.40 GHz. Radar altimeters that measure height above ground operate at 15.7 to 17.7 GHz. The military uses a radar system, called identify-friend-or foe (IFF), for identifying aircraft. The IFF system operates in the 1030- to 1090-MHz band.

Microwaves are also used for distance-measuring equipment (DME) in air navigation. The DME system indicates miles to the airport. The frequencies used are 962 to 1213 MHz.

Another application is microwave links used in television for relaying signals from a remote location to the studio, or from the studio to the transmitter, and for intercity networks. The studio-to-transmitter microwave links use 947 to 952 MHz.

Some of the industrial applications of microwaves are RF heating, bonding, and gluing. A popular consumer application is the radar oven in which microwaves are used for very fast cooking. Microwave ovens operate at 2.45 GHz.

With the advent of communication via satellites in space, that application of microwaves has become very important. More details of satellite communications are given in Sec. 14-8.

Fiber Optics As the frequency of microwaves is increased, the radiation approaches the short wavelengths of visible light. Actually, in a communications system with fiber optics, visible light is used as the carrier of modulated signals. One application is in long-distance telephone transmission lines. Instead of using radio, cables containing optical fibers are used. They transmit light by internal reflections along the entire length. The advantage is that the losses are very small.

At the transmitting end, a laser beam is modulated to produce pulses of light. At the receiving end, special photodiodes convert the light back into electrical signals.

The term *laser* is an abbreviation for *light amplification by stimulated emission of radiation*. The laser is an oscillator circuit that generates coherent light, which means it has a narrow band of wavelengths. A similar device is the *maser* for microwave amplification. The applications are in the field of *optoelectronics*.

Test Point Questions 14-1
(Answers on Page 341)

a. How many megahertz are there in one gigahertz?

b. Is the λ of microwaves at 100 GHz, 0.3 cm, or 100 m?

c. Which new band in Table 14-1 would be used for TV microwave links?

14-2
WAVEGUIDES

As shown in Fig. 14-2, a waveguide is a hollow metal structure used for coupling microwave energy. Transmission lines such as coaxial cables, twin lead, and open-wire lines are generally not used at the highest microwave frequencies because of excessive losses produced by skin effect in the conductors. The skin effect causes current to flow only on or near the surface of a conductor carrying current.

A waveguide does not have any inner conductor. Energy is propagated by the varying electric and magnetic fields, as in the radiation of electromagnetic radio waves. Waveguides can be rectan-

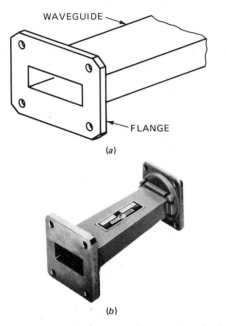

Fig. 14-2. Rectangular waveguide. (a) Detail of end flange. (b) Complete section.

gular (Fig. 14-2), circular, or elliptical in cross section. They are usually made of brass or aluminum to be rustproof. The inside may be silver-plated for minimum losses at higher frequencies. The higher the frequency, the smaller is the required size of the waveguide.

Rectangular Waveguide Figure 14-3 illustrates how a rectangular waveguide can be developed from the idea of an infinite number of quarter-wave shorted stubs on a two-wire transmission line. Air is the dielectric. Transmission of energy in the two-wire line is possible because the stubs have high impedance at a distance of $\lambda/4$ from the shorting bar.

The stubs shown are horizontal. Each point on the line has one $\lambda/4$ stub to the left and another to the right. Two $\lambda/4$ stubs add to equal the width w of the rectangular waveguide. The length of the shorting bars corresponds to the waveguide height h. Every pair of the horizontal $\lambda/4$ stubs is like a window with w equal to $\lambda/2$. The dimension h is not critical, but it must be less than w for a rectangular waveguide. In practical terms, w is the long side and h is the short side.

Consider the waveguide as a hollow conducting tube to restrict the electric and magnetic fields. As shown in Fig. 14-4, propagation of the field energy is from one end to the other within the waveguide. Reflections from the metal walls allow the energy

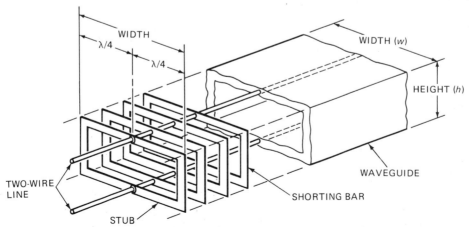

Fig. 14-3. Developing a rectangular waveguide from an infinite number of shorted quarter-wave stubs on a two-wire line.

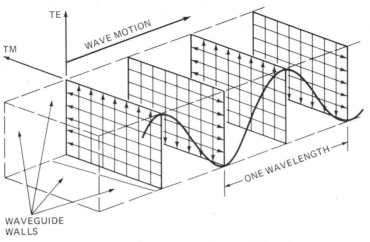

Fig. 14-4. Propagation of the transverse electric (TE) and transverse magnetic (TM) fields inside a waveguide.

to be propagated along the length. Energy can be coupled in or out by inserting a metal probe in the cavity.

Cutoff Frequency   A waveguide acts like a high-pass filter. The low-frequency cutoff is determined by its dimensions. When the frequency is too low for the metal shell to be equivalent to $\lambda/4$ stubs, the field energy is rapidly attenuated along the axis of propagation. At the cutoff frequency and above, however, the waveguide passes energy along its main axis.

As a result, the waveguide serves as a filter that passes only the frequencies above its cutoff value. That characteristic of a waveguide structure is what distinguishes the waveguide from a two-wire line. The line transmits energy down to 0 Hz, or direct current.

Modes of Propagation   The *mode* describes the method of using the electric and magnetic fields for propagation of the microwave energy. Consider a rectangular waveguide. In the transverse electric (TE) mode, the electric field E is transverse to the long axis. *Transverse* means crosswise, in a plane perpendicular to the conductors. That means E is zero in the perpendicular plane along the axis of propagation. There cannot be any electric field in the metal conductor itself. The energy is propagated by magnetic waves.

In the transverse magnetic (TM) mode, the magnetic field H is transverse to the long axis. That means H is zero in the perpendicular plane along the axis of propagation. The energy is propagated by electric waves.

Ordinary transmission lines with two conductors have transverse electric and magnetic fields, and so they are called TEM lines.

Dominant Mode   The dominant mode operation is at the lowest frequency. When the waveguide is rectangular, the dominant mode results from a half-wave for the width, which makes the height less than $\lambda/2$. In TE operation, the symbol is TE_{10} for the dominant mode. The first subscript indicates a half-wave for w and no half-waves for h. As another example, a circular waveguide has a dominant mode TE_{11} or TM_{11} because of the symmetrical cross section.

TEE Junctions   It is sometimes necessary to split the microwave energy in a waveguide system into two or more paths. Then a T, or tee, junction is used, as shown in Fig. 14-5. In Fig. 14-5a the H-plane tee is made with the two branches in the magnetic plane. Note that the junction divides the width w of the waveguide. Two loads connected to the two branches will effectively be in parallel. Therefore, the H-plane type is also called a *shunt tee*.

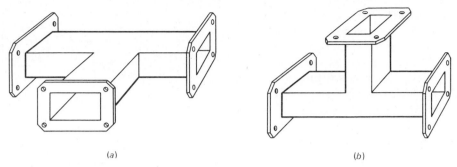

(a)   (b)

Fig. 14-5. Waveguide junctions. (a) H-plane tee. (b) E-plane tee.

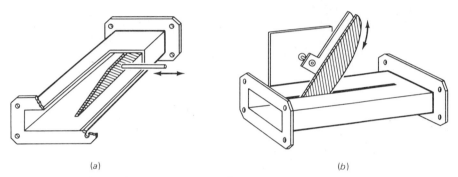

(a)   (b)

Fig. 14-6. Variable attenuators for a waveguide. (a) Lateral-motion type. (b) Flap type.

The waveguide junction shown in Fig. 14-5b is called an *E-plane tee*. Note that the junction is along the height H of the waveguide to provide two branches in the electric plane. That type is also called a *series tee* because two loads connected to the arms are effectively in series.

Waveguide Attenuators Figure 14-6 shows two types of waveguide attenuators. The attenuator itself is a thin dielectric substrate coated with a resistive material. When the attenuator is inserted into the electric field of the waveguide, I^2R losses are produced. The amount of loss depends on the resistance, the attenuator length, and the position of the attenuator in the waveguide. The position can be varied in Fig. 14-6a and 14-6b to control the amount of attenuation.

Waveguide Tuning To tune the waveguide, a metal plate is placed in the hollow structure. A

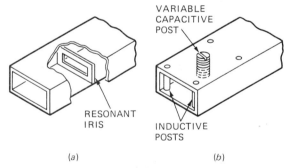

(a)   (b)

Fig. 14-7. Waveguide tuning. (a) Resonant iris. (b) Variable capacitive post.

plate across the small dimension h increases the inductance, and one across the wide dimension w increases the capacitance. Adding a plate across each dimension produces the resonant iris shown in Fig. 14-7a. The iris is also called a *waveguide window*. For variable tuning, screw-type posts can be used, as shown in Fig. 14-7b.

Microwave Horn Antenna The horn antenna is just an open section of a rectangular waveguide; see Fig. 14-8. It is flared out to match the characteristic impedance Z_0 of the guide to the Z_0 of free space. This type is a primary antenna, since it is the source of microwave energy for radiation. A secondary microwave antenna is not a radiator itself, but it has the capacity to focus radiation from a primary antenna. As an example, a parabolic reflector is commonly used as a secondary microwave antenna. At microwave frequencies, the radiation can be focused into a narrow beam, as light can be from a parabolic mirror, for greater intensity and the desired directivity.

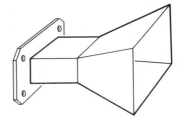

Fig. 14-8. Pyramidal horn for microwave antennas.

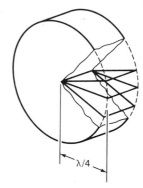

Fig. 14-9. Model of a microwave cavity formed with quarter-wave sections of a transmission line.

Test Point Questions 14-2
(Answers on Page 341)

Answer True or False.

a. Waveguides are used to minimize skin effect at microwave frequencies.
b. A rectangular waveguide serves as a high-pass filter.
c. TE_{10} is the dominant mode for coaxial cable.

14-3
CAVITY RESONATORS

Conventional LC circuits are not generally used at the higher microwave frequencies because the required inductance and capacitance values are too low. However, a shorted quarter-wave transmission line has high impedance at the open end $\lambda/4$ away. The shorted $\lambda/4$ line therefore, corresponds to a parallel-resonant LC circuit, which has high impedance at f_r. That feature suggests that the structure illustrated in Fig. 14-9 can be used for microwaves. If shorted $\lambda/4$ sections are placed closer and closer together, the result will be the continuous circular structure of the resonant cavity shown. The cavity is resonant at the frequency that makes the radius of the circle equal to a quarter-wave. The losses are very low because of little resistance with the large conducting surface. In fact, the Q of a resonant cavity can be as high as 20,000.

A high-Q resonant cavity serves the same purpose as conventional tuned circuits at lower frequencies. It can select one frequency and reject the frequencies above and below, which corresponds to the selectivity of a tuned circuit. That function is as a bandpass filter. Also, resonance can be used to reject one frequency and pass all others as a bandstop filter. The wave trap is an application. Mainly, though, a resonant cavity is used to control the frequency of oscillations when it is used with an active device to form a microwave oscillator circuit.

Figure 14-10 shows a bandpass cavity that can be used to reject frequencies above and below f_r. The cavity acts like a series-resonant circuit to increase the energy at the resonant frequency.

Energy is coupled into the cavity and out by means of small holes, or apertures, in the sides. The amount of coupling is controlled by the size of the aperture. A small aperture provides loose coupling; a larger aperture increases the coupling. Tight coupling loads the cavity—that is, reduces

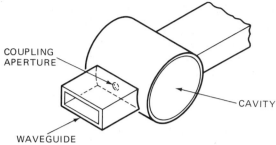

Fig. 14-10. Microwave cavity used as a resonant band-pass element. It rejects frequencies above and below f_r.

the Q and increases the bandwidth—just as in a conventional tuned circuit.

Probe coupling also is popular for transferring energy in or out of the cavity. A probe is simply a length of conductor, such as a wire, inserted into the cavity.

Figure 14-11 shows a bandstop cavity, which acts like a parallel-resonant circuit used as an absorption wave trap to dissipate its energy at the resonant frequency. Note that the cavity can be tuned. As the screw is turned clockwise, a disk moves down. The resultant decrease of internal volume of the cavity reduces the resonant frequency.

Test Point Questions 14-3
(Answers on Page 341)

a. Has a shorted quarter-wave line high or low impedance $\lambda/4$ back?

b. Does a smaller cavity resonate at a higher or lower frequency?

c. Does a large aperture in a cavity provide loose or tight coupling?

14-4
MICROSTRIP

A microstrip transmission line corresponds to a short length of flat coaxial cable, but in printed-circuit form. The technique is commonly used in microwave circuits. Microwave diodes and transistors are generally available in microstrip packages.

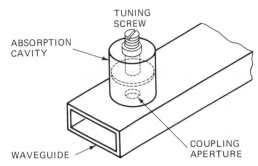

Fig. 14-11. Microwave cavity used as a resonant band-stop element. The function is the same as that of an absorption wave trap.

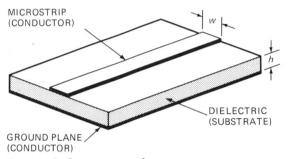

Fig. 14-12. Construction of a microstrip.

Figure 14-12 illustrates the construction of a microstrip. The microstrip consists of a conducting strip printed on one side of a dielectric substrate with a solid ground plane at the bottom. The construction can easily be carried out with photo-etching techniques starting from double-clad (two-sided) printed-circuit board. It is similar to coaxial cable, because both have an inner conductor and an outer conducting plane that is grounded. When the conductor strip is formed between two ground planes, it is called *stripline*.

Three factors determine the characteristic impedance Z_0 of a microstrip. They are thickness of the substrate, the dielectric constant of the substrate, and the width of the conducting strip. Typical values of Z_0 are 10 to 100 Ω.

The microstrip is too lossy for use as a long transmission line; its application is in RF amplifier and oscillator circuits at microwave frequencies. Figure 14-13 shows an RF amplifier for 2.1 GHz. Note the microwave transistor inserted in the mi-

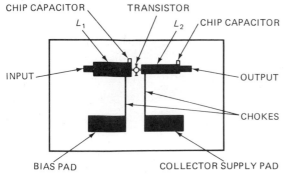

CHIP CAPACITOR   TRANSISTOR

L_1 L_2 CHIP CAPACITOR

INPUT   OUTPUT

CHOKES

BIAS PAD   COLLECTOR SUPPLY PAD

Fig. 14-13. Example of a 2.1-GHz microstrip amplifier, actual size.

crostrip. The ceramic-chip capacitors with L_1 and L_2 are used to match impedances between the line and the transistor. The chips are special capacitors for those high frequencies. Each line-length L_1 or L_2 acts as a quarter-wave matching transformer. The two narrow strips that are vertical in the diagram serve as RF chokes at microwave frequencies. They isolate the transistor collector and base circuits from the dc supply voltage.

Test Point Questions 14-4
(Answers on Page 341)

Answer True or False.

a. Microstrip is similar to a flat coaxial cable but is in printed-circuit form.
b. Microstrip is too lossy for use in microwave oscillator circuits.

14-5
MICROWAVE VACUUM TUBES

The vacuum tube is still used in RF power applications, especially for microwaves. Examples are the magnetron, klystron, and traveling-wave tube (TWT). These tubes are used for high-power microwave amplifiers and oscillators. A magnetron operating in the pulse mode can produce megawatts of peak power for a radar transmitter. The TWT is generally used as a linear amplifier of microwave signals in satellite communications.

Magnetron The structure of a magnetron is shown in Fig. 14-14. The tube is a diode; it has an anode and a heated cathode. The anode is a metal block with machined cavities. The cavity dimensions determine the frequency of oscillation.

A heater wire passes through the cylindrical cathode in the center of the tube to produce thermionic emission of electrons. The anode has positive dc voltage. As a result, electrons from the cathode are attracted to the anode.

The path of the electrons is *cycloidal* because there are two forces acting on the electrons. The forces result from crossed electric and magnetic fields within the tube. In fact, the name *magnetron* comes from the magnets, which can be seen in Fig. 14-14 around the tube. The electric field is a result of the potential difference between anode and cathode provided by the dc supply voltage. The electric field alone tends to produce straight-line motion from cathode to anode. The magnetic field alone produces circular motion. The result of combining the two motions is a cycloidal path for the electrons accelerated toward the anode.

Because of the cycloidal motion, the electrons are alternately accelerating and decelerating. When the electrons are decelerating, they must give up some of their energy as their velocity is reduced. A magnetron is designed to allow the electrons to pump energy into the cavity.

The electric and magnetic fields are adjusted to make the length of the cycloidal loops equal to twice the distance between the cavity openings. Each cavity is a resonator. The spacing makes adjacent cavities have out-of-phase oscillations. The cavity oscillations produce fields that alternately accelerate and decelerate the electrons. As a result, the entire process is regenerative; positive feedback reinforces the oscillations. Probe coupling from one of the cavities provides the means to release the microwave energy from the magnetron tube.

The process of accelerating and decelerating the electrons is called *velocity modulation*. The velocity of the electron stream is alternately increased and decreased; the period is comparable with the total transit time. This is the general principle of microwave tubes.

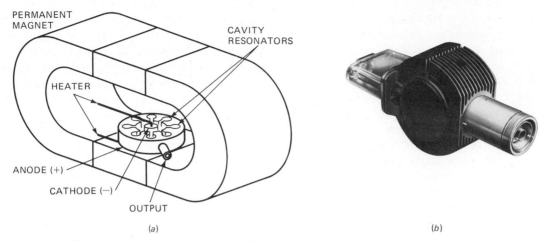

Fig. 14-14. (a) Construction of the magnetron oscillator. Typical frequencies for the resonant cavity are 0.7 to 9.6 GHz. (b) A commercial magnetron. Magnet diameter is 6 in.

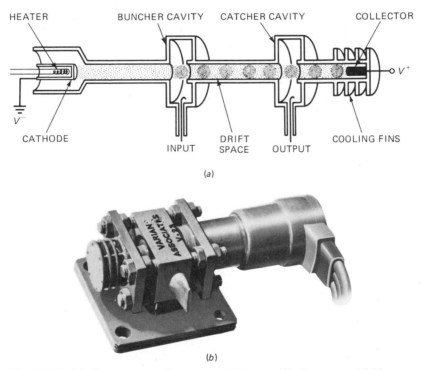

Fig. 14-15. (a) Construction of two-cavity klystron. (b) A commercial klystron. Length is 18 in.

Klystron The structure of the klystron is illustrated in Fig. 14-15. A thermionic cathode at the left releases a stream of electrons attracted to the positive collector electrode at the right end of the tube. The stream passes through two cavities called a *buncher* and a *catcher*, so this type is a two-cavity klystron.

Now let an input signal be applied to the buncher. The ac signal sets up oscillations in the cavity. The cavity oscillations provide reversals within the interior to bunch and expand the electrons. As the velocity-modulated electron stream moves toward the collector, the electrons gain energy because of the acceleration by the positive voltage. Finally, the electron stream moves through the catcher cavity. There the electrons are decelerated; they give up some of their energy to the catcher. As a result, the output probe at the catcher has an amplified version of the input signal applied to the probe at the buncher. As shown here, the klystron is being used as a microwave amplifier. It can be used as an oscillator, though, by feeding some of the amplified output signal at the catcher back to the input at the buncher.

Reflex Klystron   The reflex klystron shown in Fig. 14-16 is used as an oscillator. No external feedback is necessary, because the reflex klystron provides its own internal feedback. Here a single resonant cavity is used instead of the buncher and catcher. Also, a repeller electrode replaces the collector. The repeller is used to turn back the velocity-modulated electrons so that they can give up energy to the cavity. That feature provides the feedback required for oscillations. The reflex klystron oscillator cannot develop the high power of magnetrons, but it has applications in microwave equipment at moderate signal levels.

Traveling-Wave Tube   A problem with magnetrons and klystrons is that they have very high Q because they have resonant cavities. As a result, the bandwidth may be too narrow for some applications. When a bandwidth greater than 10 percent of the resonant frequency is desired, the traveling-wave tube is often used. See Fig. 14-17a for the internal structure of the tube. Thermionic emission is used to produce electrons. The electrons are attracted to the positive collector electrode at the right end of the tube. The signal to be amplified is applied at the waveguide input on the left. Amplified signal is taken from the waveguide output at the right.

The input signal travels along the helix inside the tube. The helix may be thought of as a special coaxial transmission line with large inductance per unit length. The inductance provides the coaxial

circuit with a phase velocity much less than that in free space. In fact, the input signal is slowed down to match the velocity of the electron beam heading toward the collector. As the beam and the input signal move together, their fields interact to produce velocity modulation of the electrons.

Bunching of the beam results as more electrons are decelerated than accelerated. Then there is a net transfer of beam energy to the electromagnetic signal in the helix. That effect allows an amplified version of the input signal to be taken from the output waveguide. The traveling-wave tube, therefore, is a useful microwave amplifier. It can also serve as a microwave oscillator by returning some of the output signal to the input.

Test Point Questions 14-5
(Answers on Page 341)

a. Is the resonant frequency of a magnetron oscillator determined by the cathode or by the cavities?
b. Does velocity modulation of an electron stream require acceleration or deceleration of the electrons, or both acceleration and deceleration?
c. Which microwave amplifier has more bandwidth, the TWT or the klystron?

14-6
MICROWAVE SEMICONDUCTOR DIODES

The first solid-state devices used in the microwave field were point-contact diodes developed for radar systems in the early 1940s. They served as detectors for microwave signals and mixers in frequency conversion. Transistors were not widely applied in microwave equipment until the 1970s. At first, their frequency response was not high enough. Now, however, great advances have made transistors just as useful in high-frequency applications as in lower-frequency ones. Germanium, silicon, gallium arsenide, and indium phosphide are the semiconductor materials used for microwave diodes and transistors.

Point-Contact Diode   The point contact diode consists of a very thin wire in contact with a

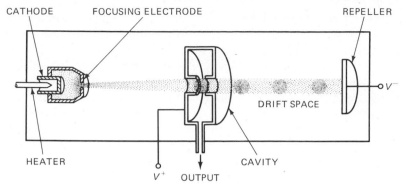

Fig. 14-16. Construction of the reflex klystron.

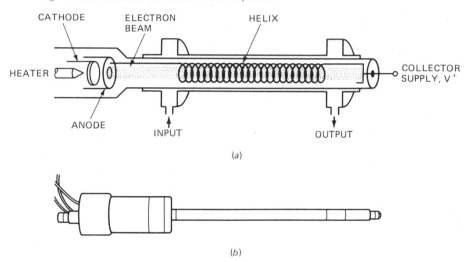

(a)

(b)

Fig. 14-17. (a) Construction of the traveling-wave tube. (b) External appearance of the TWT. Length is 14 in.

semiconductor. The diode is a one-way conductor that serves as a detector. Because of the small contact, its capacitance is low for good performance at high frequencies. The point-contact diode is fragile; it has very low current ratings. It should not be checked with an ohmmeter, since excessive current can damage the contact.

Schottky Diode The Schottky diode replaces the more fragile point-contact diode. It also has a metal-to-semiconductor junction. The metal is the anode, which forms a rectifying junction to the semiconductor. The cathode is just an ohmic connection. Because of its construction, the Schottky diode operates only with electrons as the majority carriers. There is no hole current.

The result is a frequency response that is good enough for microwaves. A diode of the Schottky type is also called a *hot-carrier diode* or *barrier diode.*

The packages for microwave diodes must be small to keep inductance and capacitance as low as possible. As an example, Fig. 14-18 shows a Schottky diode. It is only 2.5 mm in diameter by 1.3 mm. The leads are designed to be soldered into a microstrip circuit. Note the white dot to mark the cathode side.

Tunnel Diodes A small PN junction is used with a very high concentration of doping. The density of impurities makes the depletion zone very thin. When a low voltage is applied, a relatively large

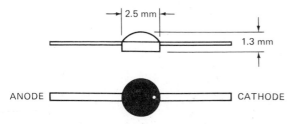

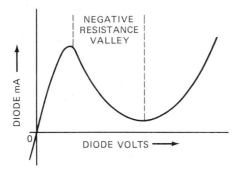

Fig. 14-18. Schottky diode designed to be used in a microstrip circuit.

Fig. 14-19. Volt-ampere characteristic curve of a tunnel diode.

amount of charge carriers can tunnel through the junction easily. The result is the special characteristics illustrated by the tunnel-diode curve in Fig. 14-19. Those features are:

1. The forward current decreases as the forward voltage increases in one section of the curve. The phenomenon is called *negative resistance,* because under normal conditions current increases as voltage increases.
2. There is appreciable reverse current with a small amount of reverse voltage.
3. There is no offset in diode voltage, because current flows with any amount of forward or reverse voltage.

Those features are used for microwave detectors and oscillators. The negative-resistance characteristic is utilized in the diode oscillator. In the application of a *tunnel rectifier,* the diode is used as a *backward detector.* The reverse characteristics provide very good frequency response with low

noise. Also, there is no voltage offset, which allows detection of small signal levels.

Gunn Diode   The Gunn diode is one of a group of microwave diodes in which electromagnetic interactions and transit time within bulk semiconductors are used. It incorporates the transferred-electron effect: electrons move through different areas in the bulk material with different mobilities. The transit time from input to output determines the operating frequency, which is in the gigahertz range. A popular application of Gunn diodes is in microwave oscillators. Typical frequency ranges are 6 to 18 GHz at a power rating of 50 mW for CW transmission. The power rating is much greater for pulse operation.

The bulk semiconductor material is generally indium arsenide or gallium phosphide, which can provide a high-electron-drift velocity for operation at microwave frequencies. Additional types that make use of the bulk effect and transit time in the semiconductor are the IMPATT, TRAPATT, and BARRITT microwave diodes. Those diodes have higher power ratings than the Gunn diode. All of them have the negative-resistance characteristic illustrated in Fig. 14-19, which means they can be used as diode oscillators.

PIN Diode   The abbreviation PIN indicates an intrinsic (I) or undoped layer between P and N semiconductors. The intrinsic layer decreases the capacitance of the diode, which makes it useful for microwaves.

PIN diodes can be used to switch paths for microwave signals. Forward bias turns the diode on so that it can allow a microwave signal to pass. Reverse bias makes the diode have high impedance to block the signal path.

The PIN diode is also used to attenuate microwave signals. Its intrinsic layer gives it a characteristic like a resistor at higher frequencies. Networks of series and parallel PIN diodes used with bias components provide an electronic method of variable signal attenuation. Similar designs are used for amplitude modulation of microwave signals.

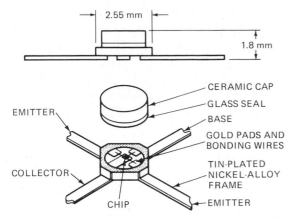

Fig. 14-20. Typical microwave transistor designed to be used in a microstrip circuit. Note double emitter connection.

Test Point Questions 14-6
(Answers on Page 341)

a. Is the advantage of point-contact diodes low capacitance or high power rating?
b. Does the tunnel diode or the Schottky diode have negative resistance?
c. Is the Gunn diode or PIN diode used for microwave oscillators?

14-7
MICROWAVE TRANSISTORS

Generally, two types of microwave transistors are available: the NPN junction transistor made with silicon and the FET. The field-effect transistor for microwaves is called a GASFET because it has gallium arsenide as the semiconductor. GASFET amplifiers are used in the range of 2 to 18 MHz. Output power is in the range of 1 mW to 1 W. It is used as a linear amplifier for AM signals. NPN junction transistors have much higher power ratings, up to 100 W. They are used for both linear amplifiers and class C power amplifiers with CW or FM signals.

The package for a microwave transistor must minimize parasitic shunt capacitance and stray inductance. A typical low-power microwave transistor is shown in Fig. 14-20. The overall dimensions are only 2.55 mm in diameter by 1.8 mm high for the transistor body itself. Note the flat leads. The solder points are next to the transistor, because lead lengths must be at an absolute minimum. The package is designed for mounting in a microstrip circuit.

In microwave transistors for higher power a much larger semiconductor chip is used. However, the resulting parasitic L and C are neutralized in the transistor package. The entire unit is designed to match the transistor to 50-Ω microstrip over a wide range of frequencies.

Test Point Questions 14-7
(Answers on Page 341)

Answer True or False.

a. The GASFET is a field-effect transistor used for microwave amplifiers.
b. Typical power ratings for the GASFET are less than 1 W.

14-8
SATELLITE COMMUNICATIONS

A communications satellite is a combination of rocketry to put the satellite in orbit, microwave electronics for the communications, and solar-energy converters to supply power for the electronic equipment. The first manufactured satellite was launched by the U.S.S.R. in 1957. It was followed by a United States launch in 1958. Then in 1963 the advance in space technology allowed the first synchronous-orbit satellite (SYNCOM) to be launched. Its main function was to relay communications between different points around the earth.

Today, many commercial and government satellites are in circular orbits, in the plane of the earth's equator at a height of about 22,300 mi. In 1981 there were over 55 active communications satellites, and plans for many more. The satellite communications field is becoming crowded.

The rotational period of the communications satellites is approximately 24 h. With occasional minor adjustments the satellites maintain a fixed position relative to the earth, called a *geostationary position.* The term *synchronous* also means that the satellite is in a fixed position in space.

Since the satellite is in a fixed position, earth-station antennas can be mounted in permanent positions and pointed at it. Occasionally, propellant will be released from the satellite to stabilize the antennas along three axes to maintain best performance with the fixed earth-station antennas.

Uplink and Downlink Signals Uplink signals are transmitted from the earth stations. The satellite receives those signals, converts them to different frequencies, and retransmits them back to earth as downlink signals. As an example, Fig. 14-21 shows a simplified block diagram of a 12-channel satellite for television signals. These uplink frequencies are 5.9 to 6.4 GHz.

There are 12 transponders, one for each of the 12 channels. Each channel is 40 MHz wide, because FM is used for the microwave relay signal. However, the satellite signal is converted by the earth station to a standard television signal for broadcasting.

The uplink signals from 5.9 to 6.4 GHz are converted to lower frequencies by the mixer and local oscillator. Each transponder has one channel for the downlink signals to earth. The downlink frequencies are 3.7 to 4.2 GHz.

Note that all 12 channels 40 MHz wide have a total bandwidth of $12 \times 40 = 480$ MHz. The frequency range is only 0.48 GHz, which is less than 1 GHz. The signals use frequency modulation.

The block diagram in Fig. 14-21 does not show the redundant circuits that most satellites have. (Anything redundant is an extra beyond the amount necessary.) However, redundant circuits are used to make satellites more reliable, because the extra circuit can be switched in if there should be a failure. In Fig. 14-21 the input section consisting of the 6-GHz amplifier, mixer, local oscillator, and 4-GHz amplifier will actually be duplicated with identical circuitry. The input section is critical, because a failure here will make all 12 channels inoperative.

Some television satellites pack 24 transponders for 24 channels into the same 0.5-GHz bandwidth.

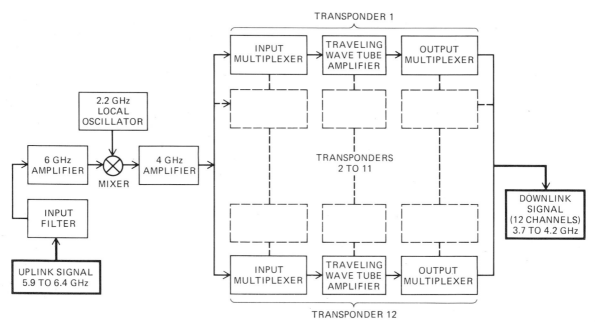

Fig. 14-21. Circuit layout for a satellite with 12 transponders for 12 channels.

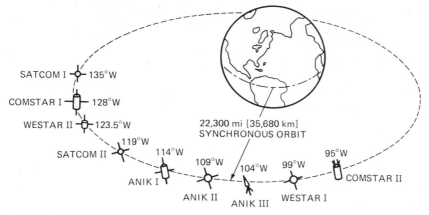

Fig. 14-22. Positions of geostationary, or synchronous, communications satellites.

That is done by vertical polarization of the odd-numbered transponder signals and horizontal polarization of the even-numbered ones. Then odd or even transponder signals can be selected at the earth station by selecting vertical or horizontal antenna polarization.

Satellite Headings All television satellites use the same frequencies, but they transmit in slightly different directions. Figure 14-22 shows several satellites and their angular headings according to the relative position in orbit.

The earth stations can separate the satellite signals by means of very directional receiving antennas. Typical beam widths are 1 to 4°, depending on size, for the antenna directivity.

Satcom I and II are owned by RCA and used for television. Also, Westar I, II, and III owned by Western Union are for television signals. The Anik B satellite is for Canadian television.

The Earth Station A large antenna is necessary to provide sharp directivity and enough gain; see Fig. 14-23. For the downlink signal, a typical satellite transponder might have an output power of 5 W. With an antenna gain of 30 dB at the satellite antenna, the effective radiated power (ERP) for the downlink signal would be $5 \times 1000 = 5000$ W. However, the loss over the long transmission path is more than 196 dB. The signal power reaching the earth station, therefore, is only

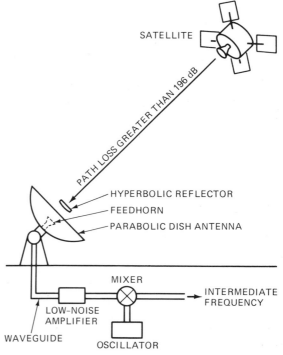

Fig. 14-23. Equipment at a satellite earth terminal.

1.26×10^{-16} W. That value is obtained by calculating a loss of -196 dB from 5000 W.

At the receiver, a parabolic dish antenna with a 6.1 m, or 20 ft diameter, provides nearly 45 dB of gain for the received signal. The antenna gain raises the signal power from 1.26×10^{-16} W up to 3.98×10^{-12} W, which is still low. Therefore, it is important to have a low-noise amplifier (LNA) before the mixer stage in the receiver circuits at the satellite earth terminal (see Fig. 14-23).

Satellite Applications Satellites are used in all worldwide television networks. Domestic satellites also are used for linking cable TV stations. In addition to television, communications satellites serve as relay stations for telephone and digital data transmission. Expanding satellite services include the following:

1. Telemail. Transmission of mail messages by facsimile between post offices.
2. Telemedicine. Consultations, record transfer, and access to large medical diagnostic computers by means of satellites.
3. Video conferencing. Multiple locations can be connected with television signals on demand from a satellite station.
4. Electronic shopping. Viewing and selection of merchandise by TV with computerized billing.
5. Interactive tele-education. Classes in widely separated locations interconnected by television via satellite, with the ability to talk to the instructor.

In the future, communications satellites will probably use the 20 to 30-GHz band. The higher frequencies can provide more bandwidth to handle more communications.

Direct Broadcast Satellites (DBS). This application may become the most important of all. This service has been proposed by Communications Satellite Corporation (Comsat) to transmit or retransmit signals that can be received by the general public. Such a system would bypass our present use of the commerical broadcast channels and cable TV. The DBS service could offer ten or more channels devoted to movies, sports, education, and other programming. Frequencies for DBS are 11.7 to 12.2 GHz, as proposed by the World Administrative Radio Conference (WARC).

Test Point Questions 14-8
(Answers on Page 341)

a. In Fig. 14-21, is 6 MHz used as an uplink or downlink signal?
b. Is the difference in angular headings for satellites usually 5 or 90°?
c. Will the same size antenna dish provide more gain at 6 or at 12 MHz?
d. Is the satellite signal at the earth station generally more or less than 1 mV?

SUMMARY

1. Microwave frequencies range from 0.3 up to 300 GHz, approximately, and the corresponding wavelengths are 1 m down to 1 mm.
2. Microwaves are used for telecommunications, worldwide television, radar, industrial heating, cooking, and research.
3. Waveguides are hollow metal tubes used to propagate microwave energy in the form of electric and magnetic fields. The TE mode indicates a transverse electric field; the TM mode indicates a transverse magnetic field.
4. A waveguide acts as a high-pass filter. It cuts off frequencies below the cutoff value but passes higher frequencies. The smaller the waveguide the higher the cutoff frequency. Waveguides can be tuned with plates, stubs, or screws.
5. A microwave cavity is equivalent to a parallel-resonant LC circuit. The reason is that the cavity can be considered as an infinite number of shorted quarter-wave stubs joined together. The smaller the cavity the higher its resonant

frequency. The resonant cavity is used as a bandpass or bandstop filter and as the tuned circuit for oscillators.

6. Probe and aperture coupling are used to transfer energy in or out of the cavity. A probe is a small metal tip inserted in the field. An aperture is a small opening.

7. Microstrip is a flat coaxial transmission line made as a printed circuit for application in an amplifier or oscillator in which microwave diodes and transistors are used.

8. A magnetron is a vacuum-tube diode with resonant cavities in a strong magnetic field; it is used as a high-power microwave oscillator.

9. A klystron is a microwave amplifier tube in which a cavity provides velocity modulation of the electron beam.

10. Velocity modulation is used in a traveling-wave tube also, but the tube has no cavities. The helix inside provides velocity modulation. The tube is a wideband device compared with the magnetron and klystron.

11. Semiconductor diodes for microwaves include the point-contact, Schottky, tunnel, Gunn, and PIN diodes.

12. Microwave transistors can be the bipolar NPN type with special emitter construction or the gallium arsenide FET, which is called a GASFET. The transistor is usually mounted in a microstrip package.

13. A synchronous, or geostationary, satellite maintains a fixed position relative to the earth's surface.

14. The frequencies used for uplink signals with television satellites are 5.9 to 6.4 GHz; those for downlink signals to earth stations are 3.7 to 4.2 GHz. One satellite can handle as many as 24 channels.

15. The signals from different satellites can be separated by very directional antennas at the receiver with a very narrow beam width. Angular headings differ by 4 or 5°. The received signal is very weak; it requires high antenna gain and a low-noise microwave amplifier.

SELF-EXAMINATION
(Answers at back of book)

Choose (a), (b), (c), or (d).

1. The wavelengths for microwave frequencies are (a) more than 1 m, (b) less than 1 mm, (c) from 1 mm to 1 m, (d) 0.3 to 300 m.

2. The largest application of microwaves is in (a) industrial heating, (b) research, (c) cooking, (d) communications.

3. The angle between electric and magnetic fields in a waveguide is (a) 90°, (b) 0°, (c) 180°, (d) 360°.

4. The dominant mode for rectangular waveguides is (a) TEM, (b) TE_{10}, (c) TE_{11}, (d) TM_{11}.

5. Which of the following is *not* an application of microwave cavities? (a) Bandpass filter; (b) bandstop filter; (c) oscillator frequency control; (d) detector.

6. Microstrip is similar to the (a) rectangular waveguide, (b) circular waveguide, (c) microwave cavity resonator, (d) flat coaxial transmission line.
7. In a magnetron the electrons travel in a cycloidal path because (a) the anode is negative, (b) the cathode is positive, (c) permanent magnets supply a strong field, (d) the cavities are resonant.
8. Which microwave tube uses buncher and catcher cavities? (a) Magnetron; (b) klystron; (c) reflex klystron; (d) traveling-wave tube.
9. Which of the following is used for its negative-resistance characteristic? (a) Point-contact diode; (b) GASFET; (c) Schottky diode; (d) tunnel diode.
10. If all television satellites use the same 3.7- to 4.2-GHz band for downlink signals, how can an earth station select one satellite? (a) High-Q cavity resonators; (b) low-noise parametric amplifiers; (c) narrow-beam receiving antenna; (d) tuning the waveguides.
11. An earth-station receiver has a signal of 4×10^{-12} W across 50 Ω input. What is the signal level, in microvolts? (a) 1.6; (b) 14.14; (c) 328; (d) 1000.
12. A quarter-wave stub shorted at the end has high impedance (a) at the shorted end, (b) $\lambda/4$ back from the short, (c) $\lambda/2$ back from the short, (d) at the center.

ESSAY QUESTIONS

1. Give the frequencies and wavelengths of the microwave band.
2. List the frequencies of the new C and K bands.
3. What frequency is used for microwave ovens?
4. Define laser, maser, and fiber optics.
5. Define TEM, TE, and TM modes.
6. Describe the characteristics of a shorted quarter-wave stub.
7. Why does a rectangular waveguide act as a high-pass filter rather than as a low-pass filter?
8. What determines the resonant frequency of a microwave cavity? What is the function of cavities?
9. Why is it not possible to use conventional L and C for a resonant circuit at microwave frequencies?
10. What is the dominant mode for a rectangular waveguide? For a circular waveguide?
11. Give one method for waveguide tuning.
12. How can energy be coupled into or out of a waveguide?
13. Is a parabolic reflector a primary or secondary microwave antenna? Explain.
14. What is microstrip? How is it used?
15. Name and give typical functions of three types of microwave tubes.
16. Give the three main components of a magnetron.
17. What is the difference between a klystron and a reflex klystron?
18. How does a traveling-wave tube differ from a klystron?
19. What is a GASFET and how is it used?
20. Give two types of microwave diodes and their applications.

21. What is meant by negative resistance?
22. How is a PIN diode constructed?
23. What is a Gunn diode and how is it used?
24. Define a synchronous, or geostationary, satellite orbit.
25. How can an earth station receive signals from different satellites that use the same downlink frequencies?
26. Give typical uplink and downlink frequencies for television satellites.
27. Give three uses of satellite communications.
28. Give two requirements of an earth station for satellites.

PROBLEMS
(Answers to odd-numbered problems at back of book)

1. Calculate the wavelengths in centimeters, corresponding to 6, 12, and 60 GHz.
2. Calculate the frequency, in gigahertz, corresponding to wavelengths of 0.05, 0.025, and 0.005 m.
3. Calculate the length, in centimeters, of a quarter-wave stub, in free space, at 4 GHz. You can ignore velocity factor because of the short length.
4. Calculate the signal level, in microvolts, of 1.26×10^{-10} W across 50 Ω.
5. Calculate the dB loss of a 5000-W signal attenuated to 1.26×10^{-16} W.
6. Give the dB gain of a 1.26×10^{-16} W signal increased to 3.98×10^{-12} W.

SPECIAL QUESTIONS

1. Give three applications of microwaves with which you are familiar.
2. Why is satellite communications especially important for television?
3. Name at least one television program you have seen that must have been transmitted by satellite.
4. Why is waveguide equipment sometimes called "microwave plumbing"?

ANSWERS TO TEST POINT QUESTIONS

14-1 a. 1000
      b. 0.3 cm
      c. C
14-2 a. T
      b. T
      c. F
14-3 a. High
      b. Higher
      c. Tight

14-4 a. T
      b. F
14-5 a. Cavities
      b. Both
      c. TWT
14-6 a. Low capacitance
      b. Tunnel diode
      c. Gunn diode

14-7 a. T
      b. T
14-8 a. Uplink
      b. 5°
      c. 12 MHz
      d. Less

Chapter 15
Receiver Circuits

The receiver input is a modulated RF carrier signal from the transmitter. The output is the desired modulated information, such as an audio signal for a radio or the video signal for a television receiver. Between its input and output, the receiver must (1) select the carrier frequency for the desired station, (2) amplify the modulated carrier signal, (3) detect the signal to recover the modulation, (4) amplify the detector output. In a radio, the audio signal output operates the loudspeaker to reproduce the sound. In a television receiver, the video signal output voltage drives the picture tube.

The antenna signal into the receiver is usually about 0.2 to 2 mV. A diode detector needs about 2 V, however, for minimum distortion. Therefore, the modulated signal must be amplified with a gain of about 1000 to 10,000 to drive the detector.

The superheterodyne circuit is used in just about all receivers; it requires a frequency converter and intermediate frequency section. The main principle is that the frequency-converter circuit shifts the different RF carrier frequencies of all stations to fit the constant IF of the receiver. The standard IF of AM radios is 455 kHz. More details are given in the following topics.

15-1
THE SUPERHETERODYNE RECEIVER

The first block in Fig. 15-1 is an RF amplifier, but that stage may or may not be used. The main requirement is to have the RF signal from the antenna for the mixer section. The mixer combines the RF signal with the output from an RF oscillator. The oscillator is called the *local oscillator* because it is at the receiver, not the transmitter. The oscillator output is not modulated.

Together, the oscillator and mixer stages form the *frequency-converter* section that produces the IF signal. The stages may be separate, or they may be combined in one tube or transistor circuit.

The process of using two waveforms at slightly different frequencies to produce a new frequency is called *heterodyning* or *beating*. Usually, the heterodyning is done to produce a new frequency in the output that is the difference between the two original frequencies.

The term *superheterodyne* applies here because the difference frequency is superaudible, above the AF range. The IF value is still a radio frequency, but it is below the RF band for the receiver.

RF Amplifier   In the circuit of Fig. 15-1 the RF stage is tuned by C_1 in the input and C_2 in the output. Those resonant circuits, shown in block diagram form, are tuned to the carrier frequency of the desired station before the frequency conversion to the IF value. The input is the RF signal from the antenna. The amplified output is coupled into the mixer stage. The RF amplifier tunes through 535 to 1605 kHz in a receiver for the AM radio band.

Frequency Converter   Also coupled into the mixer is the output of the local oscillator. The RF signal shown at 600 kHz beats with the oscillator at 1055 kHz to produce the 455-kHz IF signal. The difference frequency is

$$1055 - 600 = 455 \text{ kHz}$$

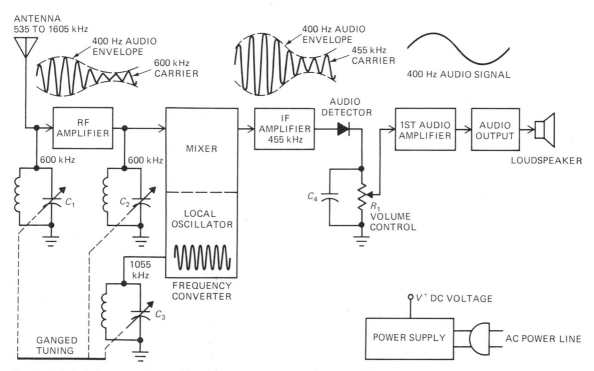

Fig. 15-1. Block diagram of a superheterodyne receiver. Waveshapes and frequencies are shown for the AM broadcast band of 535 to 1605 kHz.

The mixer output circuit is tuned to 455 kHz. The mixer amplifies only the difference-frequency signal because there is no load impedance in the output for any of the input frequencies. Consider the IF section of the receiver as starting with the mixer output circuit.

The IF Section The circuits in the IF section are fixed-tuned for 455 kHz. There may be one, two, three, or more stages tuned to amplify the IF signal.

As shown by the waveforms in Fig. 15-1, the IF signal at 455 kHz is still a modulated signal with the same audio envelope as in the RF signal. No audio signal is produced until the modulated carrier is rectified in the detector stage.

Detector In the detector circuit a diode is used as a half-wave rectifier of the amplitude-modulated IF signal. Only after rectification can the audio variations in amplitude be extracted from the AM signal. The bypass C_4 in the detector output circuit has enough capacitance to filter out 455 kHz but not enough to remove the AF variations. As a result, the detector output has the desired audio-modulating signal.

The detection is very important, but the circuit with just a small semiconductor diode may be hard to find. Sometimes the detector diode for the last IF transformer is in the shielded can housing the transformer.

Audio Section There may be one, two, three, or more audio amplifier stages to provide enough audio signal for the loudspeaker. The manual volume control shown in Fig. 15-1 is a potentiometer used to tap off the desired amount of audio signal voltage. The volume control in the output of the detector is the start of the audio circuits in the receiver.

Incidentally, a *tuner unit* is like a receiver with the RF and IF sections including the detector, but without the audio section. In some applications, a tuner is used with a separate audio amplifier unit. The purpose is usually to have more audio power output than is available in most receivers.

Power Supply Last but not least, the power supply section provides V^+ as the dc supply voltage for all the amplifiers, including the local oscillator. However, the diode detector does not need V^+.

The positive polarity is used for plate and screen voltages on tubes, collector voltage on NPN transistors, and drain voltage for an N-channel FET. In a portable radio, the battery provides the required dc supply voltage for the amplifiers.

Test Point Questions 15-1
(Answers on Page 371)

Refer to Fig. 15-1.

a. Which stage produces IF signal?
b. Which stage rectifies the IF signal?
c. Which capacitor varies the local oscillator frequency?

15-2
HETERODYNING

New frequencies produced by the process of heterodyning are similar to the more familiar example of beats commonly heard with musical sounds. When two tones of nearly the same frequency are produced at the same time, the ear can detect a regular rise and fall in the intensity of the resulting sound. The beat is produced as the two individual sound waves alternately reinforce and then cancel each other.

There is a small, continuously changing phase difference between two waves of slightly different frequencies. After a number of cycles, one wave will be either completely out of or in phase with the other. When the two waves are in phase, they reinforce to produce maximum amplitude of the resultant wave. Later, the two waves will be exactly out of phase and will cancel each other. The rate at which the amplitude of the resultant wave rises and falls is called the *beat frequency*. It is exactly equal to the difference in frequency between the two original waves, since that determines how often the waves can reinforce and cancel each other. When the two frequencies are

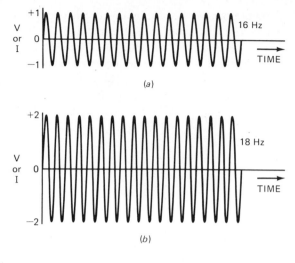

(a)

(b)

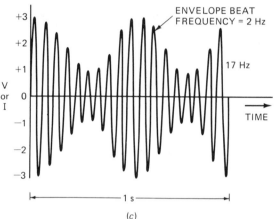

ENVELOPE BEAT
FREQUENCY = 2 Hz

17 Hz

1 s

(c)

Fig. 15-2. How two waveforms produce a beat at the difference frequency. All waveforms are drawn to scale. (a) Wave at 16 Hz. (b) Wave at 18 Hz. (c) Resultant beat consists of the envelope variations at 18 − 16 = 2 Hz.

nearly equal, therefore, we can hear the pulsating intensity of the beat note in addition to the original tones.

Producing the Beat Frequency The same beating action as in sound waves can be produced by combining two voltages or currents. An example is shown in Fig. 15-2. Assume that the wave shown in Fig. 15-2a has a frequency of 16 Hz and the wave shown in Fig. 15-2b has a frequency of 18 Hz. Those two ac voltages may be considered

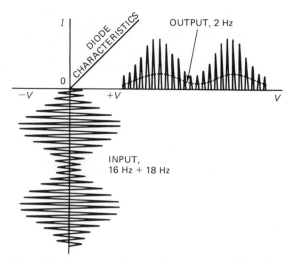

Fig. 15-3. Detecting the envelope shown in Fig. 15-2 c to obtain the beat frequency in the rectified output. Rectification is illustrated with an ideal diode characteristic.

in series with each other in a circuit in which the two waves can combine to produce the resultant wave shown in Fig. 15-2c.

When the voltages are in phase, as at the starting point at the left in the illustration, they add to produce a larger resultant amplitude. A little later, the higher-frequency wave will be one half-cycle ahead. Then the two waves will cancel because they are out of phase.

Consequently, the amplitude of the resultant wave shown in Fig. 15-2c varies at the difference frequency, which is 2 Hz in this example. That is the beat frequency. Low frequencies are illustrated so the waveforms can be shown to scale.

Detecting the Beat The voltage variations of the envelope are both positive and negative in the same amount at the same time. Therefore, the average value of the resultant wave is zero. In order to produce the beat frequency, the resultant wave must be rectified.

Rectification by a diode detector is illustrated in Fig. 15-3. The resultant wave, equal to the sum of the 16- and 18-Hz components, is applied to the diode as signal voltage. Diode current can flow only on the positive half-cycles.

Note that the average value of the rectified output, indicated by the dashed wave in the illustration, varies at the beat rate. The peak values of the output also vary at the difference frequency. That means the output circuit can have current or voltage variations at the beat frequency.

As a result, a new frequency equal to the beat or difference frequency is present in the rectified output, although the beat was just the envelope of the input wave. The undesired frequencies are filtered out, leaving just the beat-frequency signal as the output voltage.

The conclusion is, therefore, that in addition to the mixing of two frequencies, detection is necessary to produce the beat frequency. That is why the frequency-converter or mixer stage in the superheterodyne receiver can be called the *first detector*. It is not necessary to use a diode, though, since nonlinear operation of a transistor or tube amplifier as the mixer can produce the beat frequency.

Zero Beat The zero beat is another application of the heterodyne principle. Two voltages of the same frequency are heterodyned to produce zero output when the frequencies are exactly the same and out of phase by 180°. The usefulness of the zero beat is that it indicates when an unknown frequency to be checked is the same as a known frequency used as the reference.

If the unknown frequency is close to the standard frequency, the beat note will be a low-pitched audio signal equal to the slight difference in frequencies. As the frequencies come closer together, the beat note frequency becomes lower and lower. Finally, there is a null with no output at zero beat when the frequencies are exactly equal.

Producing the Intermediate Frequency The rectified current in the first detector of the superheterodyne receiver includes many frequencies. Those with the greatest amplitude are the signal and oscillator input frequencies, the difference frequency, and the sum frequency. In addition, there can be other frequencies that are combinations of the fundamentals and harmonics of the signal and oscillator frequencies.

As an example, consider a receiver with an IF of 455 kHz that is tuned to a station at 1000 kHz. The oscillator frequency is $1000 + 455 = 1455$ kHz. The main frequencies in the mixer stage are:

Signal frequency	1000 kHz
Oscillator frequency	1455 kHz
Sum frequency	2455 kHz
Difference frequency	455 kHz

Of these, the difference or beat frequency at 455 kHz is the desired one.

The difference is generally used as the IF because it is the lowest frequency. That allows more selectivity and gain with increased stability in the IF section of the receiver.

Why does the mixer supply only 455 kHz? It is just a case of tuning the output circuit to the intermediate frequency. Then there is practically no load impedance, and therefore no gain, for any frequencies except 455 kHz.

Test Point Questions 15-2
(Answers on Page 371)

a. Calculate the beat frequency between 1500 and 1955 kHz.
b. What is the beat frequency between two waveforms at 2.76 MHz if they are 180° out of phase?

15-3
FREQUENCY-CONVERTER CIRCUITS

The frequency-converter section combines the mixer and local oscillator. The Hartley, Colpitts, or tickler-feedback oscillator circuit is generally used. The inputs to the mixer are an RF signal and oscillator voltage. As a result of heterodyning, the output of the mixer stage is the IF signal.

Separate oscillator and mixer stages with tubes or transistors can be used. However, the oscillator and mixer circuits are often combined in one stage. If transistors are used, the base can be used for one input and the emitter for the other. The IF output is taken from the collector.

Oscillator Injection The term *oscillator injection* describes the process of coupling the local oscil-

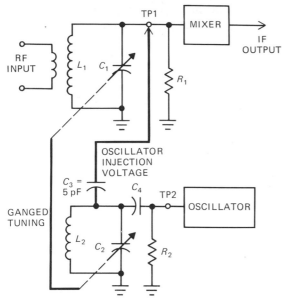

Fig. 15-4. Injection of an oscillator voltage into a separate mixer stage. Test point 1 (TP1) is used to measure dc bias at the mixer produced by the injection voltage. Test point 2 (TP2) is used to measure the oscillator bias.

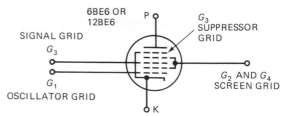

Fig. 15-5. Pentagrid converter tube with the functions of the electrodes. This tube functions as a triode oscillator and pentode mixer.

lator output into the mixer. A typical example is shown in Fig. 15-4; it has separate oscillator and mixer stages. The low value of 5 pF for C_3 indicates loose coupling. Actually, stray capacitance can provide the coupling at high frequencies. Inductive coupling, with an extra winding coupled to L_2 in the oscillator tuned circuit, can be used instead.

There must be enough oscillator voltage coupled into the mixer to produce the heterodyning action. When tubes are used, that value is about 2 to 10 V. When transistors are used, about 1 V is enough oscillator injection voltage.

Too much oscillator voltage is not desirable. First, close coupling may detune the RF and oscillator circuits. Also, the oscillator signal may then be coupled into the antenna circuit, where it will radiate to produce interference in nearby equipment. A superheterodyne receiver can act as an interfering transmitter because of radiation by the local oscillator.

The oscillator injection can be checked by dc voltage measurements. Measure the dc bias on the oscillator. Because that dc voltage is produced by

feedback, the presence of the correct bias shows that the oscillator is operating. Also, the oscillator injection voltage affects the bias at the mixer. Check for the correct dc bias at the electrode that has the oscillator injection voltage. If the receiver does not have oscillator injection for the frequency converter, no stations can be received at all.

Pentagrid Converter The pentagrid tube is common in older AM radios. As shown in Fig. 15-5, there are five grids so that the tube can serve as a combined oscillator and mixer. Actually, the tube is constructed with two control grids, one for the mixer and the other for the oscillator.

In the triode oscillator circuit, G_1 is the oscillator control grid with the common cathode and the screen grid serves as the oscillator anode. The RF signal input is at G_3, which is the signal grid. The plate current varies with both the changes in the G_1 oscillations and the G_3 signal. The two frequencies are heterodyned to produce an IF output in the plate circuit.

Conversion Gain The conversion gain ratio compares IF signal voltage out of the mixer with the RF signal input. Typical values of conversion gain are 10 to 20.

Conversion Transconductance Conversion transconductance is the ratio of IF signal current in the mixer output to RF signal voltage input. The g_m values for the mixer are about one-third the transconductance as a class A amplifier. The conversion transconductance is lower because oscillator injection voltage produces mixer bias near cutoff in order to obtain rectification of the IF beat.

Answer True or False.

a. Typical oscillator injection voltage is 1 V.
b. One tube or transistor can be used as an oscillator-mixer.
c. The IF conversion gain is usually less than 1.

15-4
COMBINED MIXER AND OSCILLATOR CIRCUIT

Figure 15-6 shows a typical frequency converter for a small AM transistor radio. Note that L_2 at the bottom is the oscillator coil. It is tuned by C_{1C} and its trimmer capacitor C_{1D}.

The oscillator tuning of C_{1C} is ganged with the RF tuning of C_{1A} and its trimmer C_{1B} across one winding of the ferrite rod antenna. Transformer coupling from the antenna coil provides the RF signal to the base of $Q1$. Also coupled into the base of $Q1$ is the oscillator output from the tap on L_2.

The oscillator coil L_2 has energy inductively coupled by collector current through the top winding. This oscillator can be considered as having the tuned circuit at the base with a tickler feedback winding for the collector.

The RF signal input and oscillator injection voltage are heterodyned to produce an IF output in the collector circuit. T_1 is the IF transformer tuned to 455 kHz that couples the mixer output to the first IF amplifier.

DC Voltages Emitter self-bias is used on $Q1$ in the circuit of Fig. 15-6 for bias stabilization. The V_E of 1.8 V results from I_E through R_3, which is bypassed by C_3.

The base voltage V_B is 2.3 V. That value is derived from two sources. One is fixed bias from the R_1R_2 voltage divider. The other part of the bias is rectified voltage produced from the oscillator injection voltage at the base.

In operation, the net bias between base and emitter is $2.3 - 1.8 = 0.5$ V. That bias is actually at cutoff. However, the oscillator injection drives the base voltage into the active region for con-

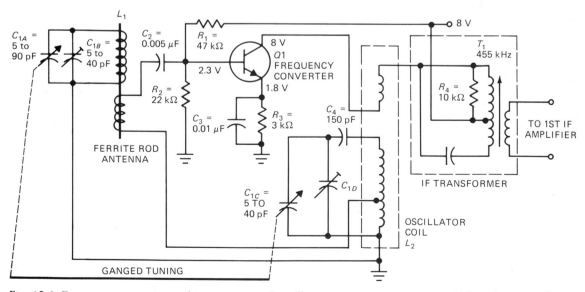

Fig. 15-6. Frequency converter combining mixer and oscillator in one transistor circuit. Values shown are for an AM radio with a 455-kHz IF output.

duction during the positive half-cycles of the oscillator voltage.

In the collector circuit, V_C is 8 V from the supply line at the tap on the primary of T_1, which is the first IF transformer. R_4 is a damping resistor across the top part of this winding.

Functions of the Components The entire oscillator-mixer circuit shown in Fig. 15-6 can be summarized by the following list of functions of the components.

$Q1$	Silicon NPN transistor used as an oscillator-mixer for frequency converter
L_1	Antenna and RF input transformer
C_{1A}	RF tuning capacitor
C_{1B}	Parallel-trimmer adjustment for C_{1A}
C_{1C}	Oscillator tuning capacitor
C_{1D}	Parallel-trimmer adjustment for C_{1C}
C_4	Oscillator series-padder capacitor
C_2	RF coupling capacitor and provides signal bias with R_2
R_1	Voltage divider with R_2 for base voltage
R_3	Emitter resistor
C_3	RF bypass capacitor for R_3
L_2	Oscillator coil
T_1	IF transformer tuned to 455 kHz
R_4	Damping resistor for T_1

Test Point Questions 15-4
(Answers on Page 371)

Which components of the frequency converter shown in Fig. 15-6 have the following functions?

a. Emitter self-bias resistor
b. First IF transformer
c. Oscillator transformer

15-5
EFFECT OF HETERODYNING ON THE MODULATED SIGNAL

The modulation of the RF carrier wave continues through the frequency converter to provide the same modulation in the IF signal as in the RF signal. The reason why can be illustrated by some

Table 15-1
IF Conversion for an RF Signal

RF Signal, kHz	Local Oscillator, kHz	Difference, kHz
Side frequency at 1005	1455	450
Carrier at 1000	1455	455
Side frequency at 955	1455	460

numerical values. Consider a modulated carrier of 1000 kHz in terms of its sideband frequencies. Assume the sideband frequencies extend ± 5 kHz from 995 to 1005 kHz. When that signal is coupled into the converter, the local oscillator beats with all the sideband frequencies to produce a new difference frequency for each. The oscillator frequency stays the same for any one station after it is tuned in. As a result, the sideband frequencies in the RF signal produce their own corresponding IF sidebands. Those values are listed in Table 15-1.

The carrier at 1000 kHz beats with the oscillator at 1455 kHz. The difference frequency in the IF output is $1455 - 1000 = 455$ kHz. The upper side frequency of 1005 kHz beats with the same oscillator frequency at 1455 kHz. That difference frequency is $1455 - 1005 = 450$ kHz. In addition, the lower side frequency at 995 kHz beats with the oscillator at 1455 kHz. That difference frequency in the IF output is $1455 - 995 = 460$ kHz.

Th preceding examples show that the modulation in the sidebands of the RF signal comes through the mixer with the same bandwidth of ± 5 kHz in the IF signal. The modulating information is still the same. The frequency conversion just shifts the RF carrier to a lower frequency for the IF section but with the original modulation.

Inversion of Upper and Lower Sidebands
The inversion of upper and lower sidebands is an interesting effect of heterodyning. Referring to the tabulated values, you can see that the upper sideband frequency at 1005 kHz in the original RF signal is converted to 450 kHz, which is lower than the IF carrier. The lower sideband frequency of 995 kHz in the RF signal becomes the higher sideband frequency at 460 kHz in the IF signal.

The inversion is only a result of operating the local oscillator above the incoming signal frequencies. Therefore, the upper sideband frequencies of the RF input signal are closer to the oscillator frequency and produce a lower difference frequency. The modulation is not changed by the inversion, however, because both sidebands have the same information. It should be noted, though, that when the oscillator beats below the RF signal frequencies, there is no inversion of the sidebands.

Test Point Questions 15-5
(Answers on Page 371)

A 1000-kHz carrier is modulated with 3-kHz audio. Give the following:

a. Upper-side frequency in the RF signal
b. The upper-side frequency converted in the IF signal

15-6
OSCILLATOR TUNING

It is the local oscillator frequency that determines which station is tuned in by the receiver. The RF circuits can amplify the RF signal of the station; but unless that carrier frequency is converted to an IF signal, the superheterodyne receiver will not produce any output. If a receiving station is located at the wrong frequency on the dial the problem is probably in the local oscillator.

Oscillator Tracking An example of tuning an AM radio is illustrated in Fig. 15-7. The receiver tuning for all stations in the AM frequency range is accomplished by making the oscillator track

455 kHz above the RF signal circuits. For any station in the bottom row of RF carrier frequencies, the oscillator frequency is 455 kHz higher at the values listed in the top row.

There is a constant difference of 455 kHz as the oscillator and RF circuits are tuned throughout their frequency ranges. The tuning dial on the receiver is marked with RF carrier frequencies, but the frequencies are received with an oscillator frequency 455 kHz higher.

Ganged Tuning The oscillator tracking is accomplished by varying the RF and oscillator circuits together. A two-section ganged tuning capacitor is shown in Fig. 15-8. Both the RF and oscillator sections are on the same shaft that moves the rotor plates. Minimum capacitance for the highest frequency results when the plates are out of mesh.

The higher range of frequencies for the oscillator is obtained by reducing the oscillator's L and C. In Fig. 15-8 the smaller section on the tuning capacitor is for the oscillator.

Note the two small trimmer capacitors at the side of the tuning gang. One is for the RF section, and the other is for the oscillator. Typical C values are 5 to 20 pF for each trimmer in parallel with its main section. They are used to make small adjustments for the minimum total C with the rotor out of mesh.

A separate trimmer capacitor is shown in Fig. 15-9. The mica is pressed by a screw to vary C. The schematic symbol uses a bar to indicate that C can be adjusted.

Oscillator Frequency Above or Below The oscillator can beat either above or below the RF

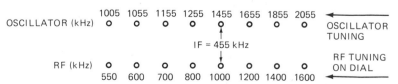

Fig. 15-7. How the local oscillator frequencies track 455 kHz above the RF signal frequencies.

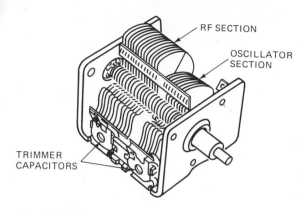

Fig. 15-8. Two-section ganged tuning capacitor. One section is for RF signal input to the mixer; the smaller section is for the oscillator.

Fig. 15-9. Compression-type trimmer (variable) capacitor with screwdriver adjustment. Tightening the mica sheets increases C. Size is ¾ in square.

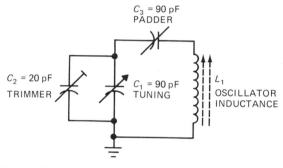

Fig. 15-10. Trimmer and padder adjustments in the tuning circuit of the local oscillator. Values shown are for an AM radio.

signal frequencies and still produce the same difference frequencies for the IF output. However, in practically all receivers, the oscillator frequencies are higher than the RF signal frequencies. That is necessary when the receiver must cover a wide frequency range.

The frequency range of the oscillator is much less when the oscillator beats above the RF signal. Some numerical examples will show why. Consider the ratio of highest to lowest oscillator frequency to cover the carrier frequencies of 540 to 1600 kHz. The lowest oscillator frequency is 540 + 455 = 995 kHz; the highest is 1600 + 455 = 2055 kHz. The ratio of the two is 2055/955 = 2.15, which means the oscillator must cover a frequency range a little more than 2:1 for all the stations in the band.

Consider the frequencies if the oscillator frequencies were below the RF signal. Then the lowest oscillator frequency would be 540 − 455 = 85 kHz. The highest frequency would be 1600 − 455 = 1145 kHz. The ratio is 1145/85 = 13.6. A tuning range of more than 13:1 would be required to cover all stations in the band. Actually, such a wide tuning range is just about impossible with one *LC* combination.

Trimmer and Padder Adjustments More details of the oscillator tuned circuit are shown in

Fig. 15-10. The series capacitor C_3 is called the *padder* capacitor because it adjusts the maximum C in the circuit. The padder is adjusted for oscillator tracking at the low end of the frequency range. Note that the series padder C_3 at 90 pF is much larger than the parallel trimmer C_2 at 20 pF.

Remember that connecting capacitors in series reduces the total C. The reciprocal formula applies. With equal values of 90 pF for C_3 and C_1, the total C is 90/2 = 45 pF. A padder is used when necessary to reduce the maximum value of the oscillator tuning capacitance.

The oscillator coil L_1 in the circuit of Fig. 15-10 has an adjustable slug. The adjustment is made for the required L at the low end of the frequency range. When the oscillator coil has the tuning slug, the padder capacitor may be omitted. A typical oscillator coil for an AM radio is shown in Fig. 15-11.

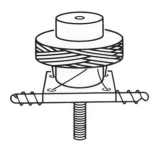

Fig. 15-11. Oscillator coil for an AM radio with L of 600 μH. Height is 1 in.

Test Point Questions 15-6
(Answers on Page 371)

a. Give the oscillator frequency for tuning in 710 kHz on an AM radio with a 455 kHz IF.
b. Give the oscillator frequency for tuning in 90 MHz on an FM radio with a 10.7 MHz IF.
c. Is the oscillator trimmer capacitor adjusted at the low or high end of the band?

15-7
RF AMPLIFIER CIRCUIT

Figure 15-12 shows a transistor RF amplifier in an FM receiver for the 88- to 108-MHz band. Because of weaker antenna signals in the VHF range, an RF amplifier stage is generally used before the mixer stage. The RF amplifier allows a better signal-to-noise ratio because the mixer then has more signal input.

The RF stage is also called a *preselector* or *tuned-radio-frequency* (TRF) amplifier. Before the super-heterodyne circuit was invented, TRF receivers were used with cascaded TRF stages to feed the audio detector.

In the amplifier of Fig. 15-12, Q1 is an N-channel JFET. It requires positive drain voltage and negative gate voltage for reverse bias. The input to Q1 at the gate is an RF signal from the antenna. The primary of the antenna transformer T_A is center-tapped with two terminals AA. The terminals connect to a twin-lead transmission line from a dipole antenna for the 88- to 108-MHz band. The secondary is tuned by C_A for maximum gate voltage at the desired RF signal frequency.

In the output circuit for the drain electrode, the primary of T_B is tuned for maximum gain. Tapping down on the coil reduces Z_L for the output circuit. The secondary couples the amplified RF signal to the mixer. The capacitors C_A and C_B are ganged with the local oscillator tuning, which is not shown in Fig. 15-12.

The JFET or an IGFET is often used for the RF amplifier circuit because of its ability to handle a wide range of input voltages at the gate. That feature results in less cross-modulation distortion from strong interfering signals. Also, the FET has less noise than bipolar transistors.

DC Voltages Drain voltage V_D is connected from the dc supply of 12 V through the primary of T_B. Since the coil has very little R, the V_D is 12 V without any appreciable IR drop.

The self-bias of 1 V at the source is produced by I_D through R_1. Since the gate has no current, it is at -1 V with respect to the source. That voltage serves as reverse gate bias for the JFET.

Functions of the Components The operation of the RF amplifier of Fig. 15-12 can be summarized as follows:

T_A	Antenna input RF coupling transformer
T_B	Mixer input RF coupling transformer
C_A	RF tuning capacitor for antenna input
C_B	RF tuning capacitor for mixer input
R_1	Source-bias resistor
C_1	RF bypass capacitor for R_1

RF Response Curve The RF response curve shows the relative RF gain from the antenna input to the mixer, where the RF signal is converted to an IF signal. The curve in Fig. 15-13a shows the RF gain in the FM radio receiver band of 88 to 108 MHz. In Fig. 15-13b the RF circuits have relatively uniform gain for all frequencies in the AM radio receiver band of 575 to 1605 kHz.

RF Alignment The alignment process consists of tuning the RF circuits for the required RF re-

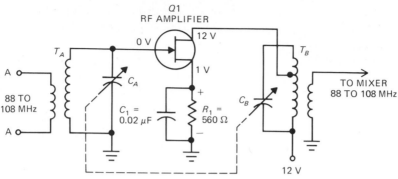

Fig. 15-12. RF amplifier circuit using a JFET in an FM receiver for 88 to 108 MHz. The trimmer capacitors are not shown. AA terminals indicate antenna input.

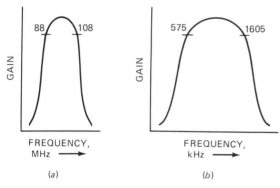

(a)   (b)

Fig. 15-13. RF response curves from the antenna input to the mixer. (a) FM broadcast band of 88 to 108 MHz. (b) AM broadcast band of 540 to 1600 kHz.

sponse curve. However, RF alignment usually is not as critical as oscillator alignment. Generally, the only adjustments for RF alignment of AM and FM receivers are to the RF trimmer capacitors on the tuning gang. They are adjusted at the high end of the band.

Test Point Questions 15-7
(Answers on Page 371)

Refer to Fig. 15-12.

a. To what frequency band is transformer T_B tuned?

b. What is the self-bias between the gate and source?

15-8
IMAGE FREQUENCIES

The main disadvantage of the superheterodyne circuit is that any signal at the intermediate frequency will be amplified in the IF section, regardless of whether the signal is the tuned station or an interfering signal. Any false signal is called a *spurious response*. The most important spurious response is the image of the desired signal. An image frequency differs from the oscillator frequency by the amount of the IF, just as the desired signal does, but the image frequency is higher instead of lower than the oscillator frequency.

As an example, suppose the receiver is tuned to a station at 600 kHz. The oscillator is then at $600 + 455 = 1055$ kHz. At the same time, any undesired signal at 1510 kHz that may be coupled into the mixer can also beat with the oscillator to produce the IF, because $1510 - 1055 = 455$ kHz.

Assuming the oscillator beats above the RF signal frequency, the image can be calculated from the following formula:

$$\text{Image frequency} = \text{RF} + (2 \times \text{IF}) \quad \textbf{(15-1)}$$

For the preceding example,

$$\begin{aligned}
\text{Image frequency} &= 600 + (2 \times 455) \\
&= 600 + 910 \\
&= 1510 \text{ kHz}
\end{aligned}$$

As another example, the image of 700 kHz is 700 + 910 = 1610 kHz. However, that image frequency is outside the band of the receiver. It is therefore less likely to be received because of the RF selectivity.

The problem of image-frequency interference is minimized by making the IF value as high as possible. That puts most of the image frequencies outside the RF band of the receiver.

Additional spurious responses that can be troublesome are interfering RF signals at or near the intermediate frequency of the receiver. It is the IF signal derived from the tuned RF signal frequency that is desired, not interfering radio signals at the intermediate frequency. Interference of that type can usually be eliminated, however, by an IF trap in the RF circuits.

Other spurious responses are caused by harmonics of the IF signal produced in the second detector and coupled back to the frequency converter. The local oscillator also can have harmonic frequencies.

Any combination of frequencies that beat with each other to produce the IF value can come through the receiver as interference. An RF amplifier as a preselector stage ahead of the mixer will reduce the relative amplitude of such spurious responses.

Test Point Questions 15-8
(Answers on Page 371)

Calculate the image frequency for a station at

a. 570 kHz
b. 1560 kHz

15-9
IF AMPLIFIER CIRCUITS

A typical 455-kHz section for an AM radio is shown in Fig. 15-14. There are two tuned amplifier stages using NPN transistors. Input to the base of $Q2$ is the 455-kHz signal from the mixer. The output from the collector of $Q3$ drives the diode detector CR1 to recover the audio modulation.

With two IF stages, three IF transformers are used. The reason is that the mixer output circuit, which produces the IF signal, needs an IF transformer. T_1 is the input IF transformer; T_3 is the output IF transformer; and T_2 is for interstage coupling. Each transformer has an adjustable slug that tunes the primary to 455 kHz. A typical IF transformer in its metal shield can is shown in Fig. 15-15. The overall voltage gain of the two IF stages is typically $100 \times 100 = 10,000$, which means that 0.2 mV of IF signal from the mixer can be amplified enough to feed a 2-V signal into the audio detector.

DC Voltages In the second IF amplifier $Q3$ (Fig. 15-14) the V_C of 8 V is supplied at the tap on the primary of T_3. There is practically no IR voltage drop across the small resistance of an IF coil. Self-bias of 1 V is used at the emitter of $Q3$; it is produced by emitter current through R_8. The bypass capacitor C_8 keeps the bias voltage steady. The base of $Q3$ is at 1.6 V, from the voltage divider from the supply line of 8 V. The divider consists of R_9, R_7, and R_6.

For forward bias on $Q3$, then, the net V_{BE} is 1.6 − 1.0 = 0.6 V. That is typical of class A operation with a silicon transistor. The IF and RF amplifier stages must operate class A for minimum distortion of the modulation on an AM signal.

The first IF amplifier is similar to $Q3$, but the base of $Q2$ returns to the automatic volume control (AVC) circuit for AVC bias. The purpose of AVC bias is to reduce the receiver gain automatically when signals are strong. To improve the AVC action, there is dc coupling for the bias between $Q2$ and $Q3$. That coupling is through R_7 between the $Q2$ emitter and $Q3$ base. As a result, the emitter voltage of $Q2$ affects the base voltage of $Q3$.

The bypass capacitors C_8, C_7, and C_6 are used to filter out any ac signal from the dc bias voltages. Only the IF transformers have an IF signal.

In the dc supply line, R_{13} drops the 9 V from the source to 8 V for the IF section. C_{12} bypasses R_{13} for ac signal. R_{13} also isolates the IF section from the 9 V supply to prevent mutual coupling of the signal through the power supply.

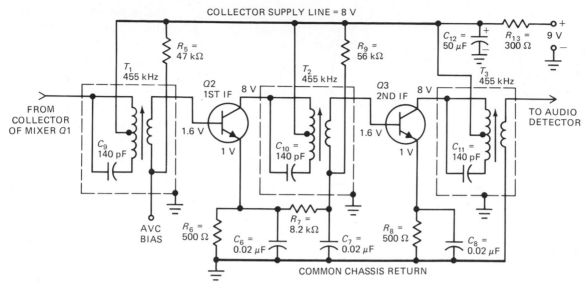

Fig. 15-14. Circuit of a two-stage transistor IF amplifier for 455 kHz. The input IF signal from the mixer is amplified enough to drive the audio detector.

Fig. 15-15. Double-tuned IF transformer in shield can. Height is 1½ in.

Functions of the Components

The functions of the components are as follows:

Q2	First IF amplifier
Q3	Second IF amplifier
T_1	IF input transformer from the mixer
T_2	Interstage IF coupling transformer
T_3	IF output transformer to the audio detector
R_5	Voltage-dropping resistor for the base bias of Q2
R_6	Emitter bias on Q2
C_6	Bypass capacitor for R_6
R_7	Dc coupling from the emitter of Q2 to the base of Q3
R_9	Voltage-dropping resistor for the base bias of Q3
C_6	Bypass capacitor for R_6
R_8	Emitter bias on Q3
C_8	Bypass capacitor for R_8
C_9, C_{10}, C_{11}	Tuning capacitors for the primary of the IF transformers
R_{13}	Voltage-dropping resistor; drops the supply voltage from 9 to 8 V
C_{12}	Bypass for R_{13}

IF Response Curve

A typical response curve of gain vs. frequency is shown in Fig. 15-16. It is actually a combined resonance curve of all the IF tuned circuits. The important requirements are (1) maximum gain at the IF but with uniform gain over a band wide enough for the sideband frequencies of the modulated signal and (2) sharp slope for the sides or skirts to reject frequencies not in the IF band.

The required IF bandwidth is typically ±5 kHz for 455 kHz, as in Fig. 15-16a, or ±0.1 MHz for

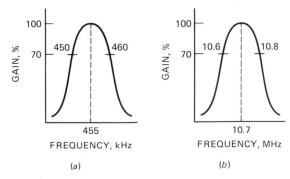

Fig. 15-16. IF response curves, from the mixer stage to the detector. (*a*) IF at 455 kHz for an AM radio. (*b*) IF at 10.7 MHz for an FM radio.

the 10.7-MHz response in Fig. 15-16*b*. Actually, the IF bandwidth as a percent of the center frequency is approximately the same: 2 percent in both cases.

Tubes, transistors, or linear IC units can be used in the IF section. With IC chips, though, the resonant circuits are not integrated. External coils and capacitors are still needed for the IF tuned circuits. However, a ceramic filter can be used instead of resonant *LC* circuits.

IF Selectivity   The IF response provides practically all the *adjacent-channel selectivity* of the receiver. That accounts for the capacity to reject frequencies close to the desired RF signal, such as a station in an adjacent channel. With sharp skirts on the IF response curve, the only frequencies amplified are those that can beat with the local oscillator to produce frequencies in the IF passband. There is very little gain for frequencies outside the IF response curve.

The excellent selectivity is the result of an IF value lower than the RF signal and the gain of cascaded IF stages. As an example of the advantage of a lower IF, suppose that the receiver is tuned to 1000 kHz and there is an interfering station at 1050 kHz. The 50-kHz separation between the stations is only 50/1000, or 5 percent, of the resonant frequency in a circuit tuned to 1000 kHz. At the IF of 455 kHz however, the same 50-kHz separation is 50/455, or about 10 percent.

As an example of the improvement with cascaded stages, suppose that an undesired frequency has a gain of 20 in one IF stage. That equals 20 percent response compared with 100 percent for 455 kHz. Remember that the overall gain is multiplied with cascaded stages. For two stages, the overall gain is 100 × 100 = 10,000 for 100 percent response at 455 kHz. At the undesired frequency, though, the overall gain is 20 × 20 = 400. Compared with 10,000, the gain of 400 is only 4 percent response for the two stages. The relative gain for the undesired frequency is reduced from 20 to 4 percent.

IF Alignment   The IF alignment of the receiver is carried out by tuning the IF circuits to the intermediate frequency. With the IF signal at the input of the mixer stage, the detector output is maximum when the IF transformers are tuned to the intermediate frequency. The adjustment begins with the last IF transformer and then continues stage by stage back to the mixer output circuit. More details of receiver alignment are given in Sec. 15-14.

Test Point Questions 15-9
(Answers on Page 371)

a. Which stage contains the primary of the first IF transformer?

b. Which stage contains the secondary of the last IF transformer?

c. In Fig. 15-14, what is the bias V_{BE} on Q3?

15-10
STANDARD INTERMEDIATE FREQUENCIES

The IF value is generally chosen to be as high as possible but below the RF signal frequencies for the receiver band. Advantages of the higher IF value are (1) less interference from image frequencies, (2) less oscillator radiation from the antenna through the RF signal circuits, and (3) reduced tuning range for the local oscillator. A lower IF has the advantages of more gain and selectivity with better stability. The main factor, though, is to reduce or eliminate image frequencies that could be in the desired RF band.

In any case, the exact IF value should not coincide with the frequency of any powerful radio service or with its harmonics. As a result, the following IF values have become standardized by the Electronic Industries Association (EIA) for receivers:

455 kHz	In AM receivers for the broadcast band of 535 to 1605 kHz
10.7 MHz	In FM receivers for the broadcast band of 88 to 108 MHz
41.25 MHz	Sound IF carrier in TV receivers with the lowest channel frequency of 54 MHz
41.75 MHz	Picture IF carrier in TV receivers
4.5 MHz	Intercarrier sound frequency in TV receivers

Actually, 455 kHz can also be used in CB receivers and VHF communications receivers in general. In this application, though, two frequency conversions are necessary.

Double Superheterodyne Circuit A 455-kHz IF signal in VHF receivers is used as a second IF value. In this system, the high RF carrier frequencies are beat down twice to obtain 455 kHz. There are two frequency conversions and two IF values, with the final output of 455 kHz as the second IF signal for the audio detector. This system is necessary for VHF receivers because the local oscillator frequency would otherwise be too close to the RF carrier frequency to produce a 455-kHz IF signal in one step.

Test Point Questions 15-10
(Answers on Page 371)

Give the standard IF values for each of the following:

a. AM radio receiver
b. FM radio receiver

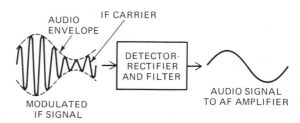

Fig. 15-17. The detector produces an audio output from the IF signal input.

15-11
AM DETECTORS

The function of the detector is illustrated in Fig. 15-17 for an IF signal that is amplitude-modulated with audio variations. In that form, the AM signal cannot provide the desired audio signal, even though it is in the envelope. The reason is that the envelope has equal and opposite variations for positive and negative polarities of the RF carrier wave. With respect to the AF variations, then, the average carrier voltage is zero.

To extract the audio envelope, the carrier wave must be rectified. Then either the peak or average amplitudes in the rectified output will vary at the audio rate. In addition, filtering out the IF carrier itself with a bypass capacitor results in detector output variations that correspond only to the desired audio signal. Either polarity of the IF signal input can be rectified in the detector circuit because the top and bottom envelopes have the same audio information. Remember that the amplitude variations of the envelope are not the sidebands, which are the frequency components.

Diode Detector Circuit In the circuit shown in Fig. 15-18a the last IF transformer T_3 supplies an input voltage to the cathode of diode detector $D1$. This is a semiconductor diode, either silicon or germanium. When an ac voltage input makes the cathode negative, diode current flows. No current can flow when the cathode is positive. Diode $D1$ is therefore a half-wave rectifier for the negative envelope of the IF signal input. Typical values of signal input are 1 to 3 V.

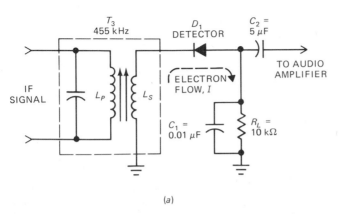
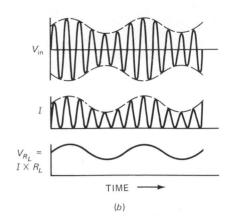

Fig. 15-18. Operation of a diode detector. (*a*) Circuit. (*b*) Waveforms.

Making the cathode negative corresponds to driving the anode positive, because the anode returns to chassis ground through R_L. Then the diode conducts forward current. Hole current is in the direction of the diode arrow, but electron flow in the circuit is shown in the opposite direction. This *I* flows from cathode to anode in the diode through R_L and returns to L_S through the common chassis-ground connections and to the diode cathode.

R_L is the diode load resistor that provides the rectified output. C_1 is the bypass capacitor in parallel with R_L to filter out the 455-kHz IF signal from the audio output. Actually, C_1 is made large enough to bypass variations down to the highest audio frequency of 20 kHz. Then it certainly can filter out 455 kHz.

Note that C_1 also serves as an ac return to chassis ground for the diode anode. As a result, all of the IF signal voltage input is applied across the diode. The signal waveforms are shown in Fig. 15-18*b*.

DC Level of the Detector Output No V^+ voltage is applied to the diode detector. The reason is that the rectified dc output from the detector should depend only on the amount of IF signal input. The detector is the only stage in the receiver that does not have a dc supply voltage.

The IF signal is an ac voltage that drives the diode into conduction each half-cycle. Since the detector rectifies the signal, the output circuit has a dc level that depends on the amount of signal input. For that reason, C_2 is needed to couple the audio signal output to the audio amplifier while blocking the dc level. In fact, using a voltmeter to measure the dc output voltage across the volume control is a good way to check on the amount of IF signal voltage input. Typical values are 1 to 3 V.

The polarity of the dc voltage output from the detector depends on whether R_L is in the anode or the cathode circuit of the diode. With R_L in the anode circuit the dc output is negative, as shown in Fig. 15-18*a*. When R_L is in the cathode circuit of the diode, the dc output is positive.

Types of Detector Circuits The diode detector is the circuit generally used in receivers because it has the least distortion when the signal input is above 1 V. Two other circuits that have been used in the past with tubes are the *grid-leak detector* and *plate detector*.

The grid-leak detector recovers the signal as audio variations in the amount of grid-leak bias, which is then amplified in the plate circuit. The detector has good sensitivity for very weak signals, but the disadvantage is high distortion.

In the plate detector the grid circuit is biased to cutoff so that only the input signal can produce plate current. The circuit is also called an *infinite-impedance detector*, because no current can flow in the grid input circuit.

Answer True or False.

a. The input to the detector is an IF signal.
b. The detector output is an ac signal on a dc axis.
c. A typical V^+ supply voltage for the detector is about $+5$ V.

15-12
MANUAL VOLUME CONTROLS

The amount of sound output from a loudspeaker depends on the power supplied by the audio amplifier. An increase in the audio signal to the loudspeaker produces louder sound.

In a receiver there are several possible ways to vary the amount of audio signal. One way is to vary the gain in the RF, IF, or audio amplifiers. However, practically all receivers have AVC to control the RF and IF gain. (See Chap. 16 on AVC.) The best way to control volume manually, therefore, is to adjust the amount of audio voltage.

Three circuits used for adjusting the audio signal are shown in Fig. 15-19. Two types of volume controls are shown in Fig. 15-20. Typical resistance values are 1 to 50 kΩ for the volume control used in transistor circuits.

Audio Level Control In Fig. 15-19a, R_L is a potentiometer used to tap off the amount of signal voltage into the first audio amplifier stage. The circuit is used in AM-FM receivers. The audio signal is coupled by C_C to terminal 3 at the top of R_L. It is the audio output from either the AM or FM detector in the receiver.

Assume 2 V of audio signal input. If the variable arm at lug 2 is set to the center of R_L, the voltage across 2 and 1 to ground will be 1 V. Then 1 V of audio signal is connected to the audio amplifier.

If the variable arm is moved to lug 3, all 2 V of the audio signal is connected to the amplifier. The control is wired so that rotating the shaft to the right increases R at lug 2 to increase the volume.

Turning the control completely to the left moves the variable arm to lug 1 at ground. Then there is zero audio voltage into the amplifier for no volume. Terminal 3 still has the 2 V of audio signal, but it is not connected to the audio amplifier.

Voltage Divider in Detector Output The volume control (Fig. 15-19b) functions in the same way as shown in Fig. 15-19a, but R_L is in the detector output circuit. That circuit can be used in an AM radio, which has just the one audio detector. C_C couples the ac signal to the audio amplifier but blocks the dc level of the detector output.

Current Divider in Detector Output In Fig. 15-19c the detector output is shown connected to

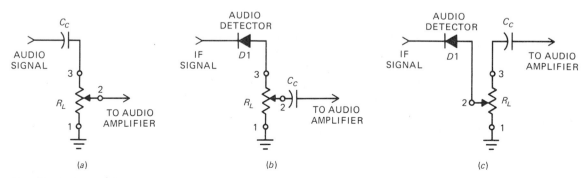

Fig. 15-19. Circuits for controlling volume manually. (a) Voltage divider for the signal input to the audio amplifier. (b) Voltage divider for the detector output. (c) Current divider for the detector output.

the variable arm of R_L at lug 2. Then R_L provides two resistance paths. The R between 1 and 2 is the detector load resistance, and the R between 2 and 3 is series resistance for the audio input circuit. C_C is the audio coupling capacitor.

Just as in the other two circuits, increasing R at lug 2 to ground increases the volume. Then the detector load between 2 and 1 has higher resistance for more voltage in the audio detector output. Also, the series resistance is reduced for the input to the audio amplifier.

Power On-Off Switch on Volume Control
The on-off switch for power is often mounted on the same shaft as the volume control, as shown in Fig. 15-20a. However, the control and the switch are in two separate circuits. The switch turns on the power supply but does not have any electrical connections to the signal circuits. The control varies the audio volume.

The power switch may be a push-button or rotary type. Pulling the shaft out or turning it to the right (clockwise) turns on the power. Further turning of the shaft changes the resistance of the control and varies the volume; the status of the on-off switch, however, does not change. Turning the shaft completely to the left (counterclockwise) reduces the resistance to zero and turns off the power.

Audio Taper   The variable resistor that is used for volume control is constructed with an element

that has nonuniform R values. The end used for low-volume settings has gradual changes in resistance. Finer changes in volume are necessary because the ear is more sensitive to increases or decreases in low levels. When the control is turned up to the right, the resistance changes are greater to provide greater changes in audio level.

A linear potentiometer is a control with uniform resistance along its length. If it is used for the volume control, it will be difficult to adjust for different settings at low volume. Note also that if the volume control is wired backwards, with the connections to terminals 3 and 1 reversed, the volume will increase when the shaft is turned to the left instead of the right.

Test Point Questions 15-12
(Answers on Page 371)

a. Does an increase in audio signal result in more or less sound from the loudspeaker?
b. In Fig. 15-19a to c, does volume increase with more or less R between lug 2 and ground?

15-13
COMPLETE CIRCUIT OF AN AM RADIO RECEIVER

The schematic diagram in Fig. 15-21 shows how the RF, IF, detector, and audio circuits fit together

(a)

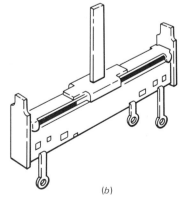

(b)

Fig. 15-20. Typical volume controls. (a) Variable resistor with an on-off switch. Diameter is 1 in. (b) Slide control without switch. Length is 2 in.

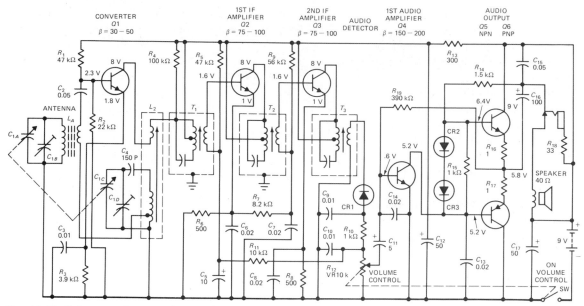

Fig. 15-21. Schematic diagram of a portable AM radio receiver. Values of C in microfarads. Values of R in ohms unless marked otherwise. Power supply is a 9-V battery. (*RCA Model RZG 343*)

in a receiver. Along the bottom, the heavier line is the common B⁻ return for all the amplifiers. The receiver is turned on by connecting the negative battery terminal to the common line with the on-off switch mounted on the volume control.

Along the top of the diagram, the beta (β) of each transistor is listed. The values correspond to the approximate current gain for the CE amplifiers.

RF Section The ferrite loopstick antenna L_A supplies RF signal to the base of $Q1$. An antenna coil can be seen in Fig. 15-22, which shows a typical circuit board for the entire receiver. Since $Q1$ is the frequency converter stage, it also has an input from the oscillator coil L_2. The mixer circuit is the same as that shown in Fig. 15-6. The IF signal at 455 kHz is produced in the collector circuit of $Q1$ with T_1 as the first IF transformer.

IF Section The two IF amplifiers $Q2$ and $Q3$ are the same as the amplifier of Fig. 15-14. They provide enough gain to drive the audio detector. The 455-kHz output at the collector of $Q3$ is coupled to the diode detector $CR1$ through the last IF transformer T_3.

Audio Detector Section The input IF signal is applied to the cathode of the detector $CR1$. R_{12} is variable as the volume control. The bypass C_{10} filters out the IF signal for audio voltage across R_{12}.

Audio Amplifier Section The volume control supplies an audio signal input through C_{11} to the base of $Q4$, the first audio amplifier. Its collector output drives the base circuits for $Q5$ and $Q6$ in the audio output stage. These are PNP and NPN transistors for complementary symmetry in push-pull operation.

In the audio output circuit, R_{16} and R_{17} are balanced emitter-bias resistors. The compensating diodes $CR2$ and $CR3$ keep constant bias voltage at the base of $Q5$ and $Q6$. Also, R_{15} balances the two base voltages. R_{14} provides negative feedback of the audio output signal to the base circuit to reduce distortion. C_{15} is a tone-control capacitor. It bypasses high audio frequencies to give more bass to the sound of the 3-in loudspeaker.

The audio output signal is taken from the junction of the emitter resistors R_{16} and R_{17} for the complementary pair. C_{16} couples the audio output to the loudspeaker. The jack switch at the top

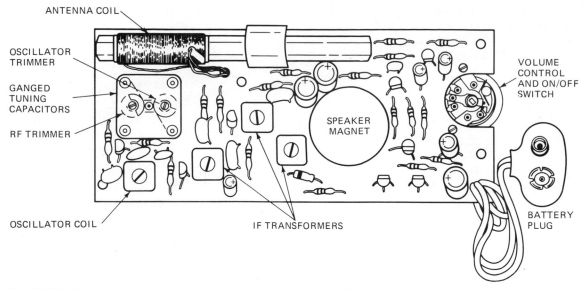

ANTENNA COIL

OSCILLATOR TRIMMER

GANGED TUNING CAPACITORS

RF TRIMMER

SPEAKER MAGNET

VOLUME CONTROL AND ON/OFF SWITCH

BATTERY PLUG

OSCILLATOR COIL

IF TRANSFORMERS

Fig. 15-22. Circuit board for a small, portable AM radio showing RF and IF adjustments for receiver alignment. Length is 5 in. (*Graymark International, Inc.*)

disconnects the speaker when the headphone is plugged in. R_{18} is a series-limiting resistor for the headphone.

Note that the collector of $Q5$ has the full 9 V of the battery for maximum power. Actually, $Q5$ and $Q6$ are in series with each other for dc voltage across the 9-V supply voltage.

Power Supply Except for $Q5$ and $Q6$, the transistor amplifiers have 8 V supplied through the 300-Ω R_{13} from the 9-V battery. C_{12} bypasses R_{13} to remove any ac signal voltage from the 8-V supply line.

The current drain on the 9-V battery is listed in Table 15-2 for different volume levels. Maximum undistorted audio output is 200 mW with a current drain of 38 mA.

Test Point Questions 15-13
(Answers on Page 371)

Refer to the receiver diagram in Fig. 15-21. Give the part number for:

a. Antenna
b. Oscillator coil
c. Detector
d. Volume control
e. Audio output coupling capacitor

15-14
RECEIVER ALIGNMENT

Receiver alignment is carried out by adjusting the tuned circuits to their correct frequencies. There

Table 15-2
Power Levels for the Portable AM Radio Shown in Fig. 15-21

Audio Output Power, mW	No Signal	50	100	180	200	Maximum
Battery Current, mA	7.5	21	30	36	38	44

are two parts to the alignment process: IF alignment and RF alignment. First, the IF transformers must be tuned to the intermediate frequency. The adjustment begins with the last IF transformer into the detector and works back to the mixer output circuit. The reason is that the IF signal is checked by observing the detector output. Correct alignment produces maximum output at either the detector or the loudspeaker in the audio output.

The main requirement of the RF alignment is setting the local oscillator to tune in the station frequencies on the dial. This procedure also has two parts, one at the low end and the other at the high end of the band. At the low end, either the oscillator coil or the padder capacitor is adjusted. The trimmer capacitor is used to adjust the frequencies at the high end.

After the oscillator frequency is set at the high end of the band, the RF trimmer is adjusted for maximum output at the same frequency. Sometimes the alignment can be improved by *rocking* the tuning capacitor to a slightly different setting and adjusting both trimmers again for maximum output. The idea is to obtain the best tracking of the oscillator and RF circuits.

Output Indicator Maximum output is usually checked by connecting an ac voltmeter across the voice coil and reading the audio signal. However, the ac audio voltage can also be checked in the detector output or any audio amplifier stage. A less exact method is just to listen for maximum sound. The best way is to observe audio output is with an oscilloscope to check for maximum output and minimum distortion.

Another possibility is to check for maximum dc voltage output across the detector load resistance using a dc voltmeter.

Input Test Signal The IF signal must be supplied by an RF signal generator, usually one that is modulated with a 400-Hz audio signal. The generator can also be used to provide the test signal in the RF alignment. However, a station signal can be used instead, and that signal is guaranteed to be at the exact frequency. At the low end of the band, a station of about 600 kHz can be used. A typical frequency at the high end is about 1400 kHz.

Signal Generator Connections A signal generator should not be connected to any circuit being aligned. The reason is that the generator connections cause detuning. For the IF signal input, connect the generator to the mixer input. The IF signal can even be connected into the antenna or RF circuits, because enough signal comes through to make the IF adjustments. For the RF signal input, connect the generator to a loop or wire to radiate signal into the antenna circuit.

In general, use as little input signal as possible. Set the output meter to its lowest range for maximum sensitivity to the adjustments. Also, have the volume control of the receiver at maximum.

Service Notes for Alignment The steps listed in Table 15-3 summarize the instructions in the manufacturer's service note for alignment of the receiver shown in Fig. 15-21. Location of the adjustments can be seen on the board shown in Fig. 15-22. Note that three frequencies are used for alignment here: 525 kHz for the oscillator low end RF, 1650 kHz for the oscillator trimmer, and 1400 kHz for the RF trimmer.

Table 15-3
Alignment Procedure for Receiver Shown in Fig. 15-21

Step	Generator Frequency	Set Dial	Adjust for Maximum
1	455 kHz	Gang open	IF transformers $T1$, $T2$, $T3$
2	1650 kHz	Gang open	Oscillator trimmer, C_{10}
3	525 kHz	Gang closed	Oscillator coil L_2
4	1400 kHz	1400 kHz	RF trimmer C_{1C}

In all the steps the generator is set for standard modulation of 400 Hz at 30 percent. Also, the generator is coupled to the antenna by radiation for all adjustments, including the IF alignment. For the output indicator, an ac voltmeter is connected across the loudspeaker voice coil.

Test Point Questions 15-14
(Answers on Page 371)

a. Should the RF trimmer capacitor be adjusted at 600 or 1400 kHz?

b. In IF alignment, is the last IF transformer adjusted first or last?

c. To check audio output, should a dc or an ac voltmeter be used?

15-15
SPECIALTY RADIOS

Besides the popular radios for AM and FM broadcasting and television, there are many types of receivers for special uses. For the most part, they operate in the VHF band of 30 to 300 MHz. Examples are shown in Figs. 15-23 and 15-24.

Multiband Receivers   The frequency ranges of multiband receivers are generally 30 to 50 MHz, 146 to 174 MHz, and possibly 450 to 512 MHz. A switch selects one of 3 to 8 separate bands in those frequency ranges. Typical services for the bands are listed in Table 15-4. Included in the public service bands are communications for police and fire departments, ambulance and taxi dispatching, air-traffic control, radio-paging services, marine weather, and marine communications.

Scanner Radio   The VHF receiver known as the scanner radio automatically checks for signals from 8 to 10 preselected stations. Each station is programmed by means of the required local oscillator frequency.

Radio-Paging   Radio-paging service operates on 27.255 MHz. A small portable receiver can pick up tone signals from a central station. See Fig. 15-24. When the receiver "beeps," it is being paged. As an example, a doctor can be paged with the beeper to call the office, normally by telephone.

Fig. 15-23. Specialty radio receiver for short-wave bands. (*Zenith Radio Corp.*)

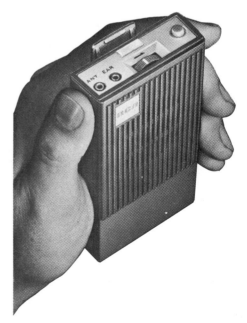

Fig. 15-24. Radio pager. (*RCA*)

Table 15-4
Typical Services for Multiband Radios

Bands	Frequencies, MHz	Notes
Short-wave band 1	5.95 to 6.2	Foreign broadcasts
Short-wave band 2	9.5 to 26	Foreign broadcasts
Citizens' band	26.96 to 27.41	Receive only
Public service	30 to 50	Low-band VHF
TV channels 2 to 6	54 to 88	Sound only
Weather broadcasts	162.4, 162.475, or 162.555	National weather service (NWS) station
Public service	146 to 174	High-band VHF
TV channels 7 to 13	174 to 216	Sound only
Public service	450 to 512	UHF

Radio-Phono-Tape Combinations As an example of radio-phono-tape combinations, an FM-AM radio can be combined with a phonograph for playing records. Also, a player for tape can be included by using either cassettes or eight-track cartridges. In all cases the phonograph or tape machine just supplies audio signal to the audio amplifier in the radio receiver, including its loudspeakers. For stereo, two audio inputs are used for left and right channels.

Test Point Questions 15-15
(Answers on Page 371)

Answer True or False.

a. Foreign broadcasts can be received on 54 to 60 MHz.

b. A scanner radio changes the local oscillator frequency automatically.

15-16
RECEIVER NOISE

Any undesired pulse disturbances in the signal that obscure the desired information are considered noise. There are two possible sources: noise generated in the receiver circuits and external noise.

Electrical equipment such as motors, ignition systems, and fluorescent lights generate noise fields. That type of external noise usually is picked up with the desired signal by the antenna. The internal noise produced by the receiver itself is *random noise*. It is just about impossible to filter out because it has no specific bandwidth; the frequency spectrum is flat. Random noise is also called *white noise* or *pink noise*.

Noise Limiter A receiver can have a noise limiter to reduce the effects of external noise pulses. The stage limits the amplitude swing of the signal to reduce the noise level. With AM or an audio signal, however, excessive amplitude limiting can distort the desired signal. With an FM signal, on the other hand, limiting can be used to eliminate all amplitude variations because the desired signal is in the frequency variations.

Random Noise In any device that conducts current, thermal agitation causes a random effect in the flow of charges. The effect is *thermal,* or *Johnson, noise.* Also, in tubes and transistors, *shot effect* in the emission of charges causes random variations in the current to the output electrode. Furthermore, in multigrid tubes, a partition effect also increases the random noise.

Because of those factors, the frequency converter or mixer stage produces the greatest amount of random noise in the receiver, especially since the conversion transconductance of the mixer is less than the mixer's g_m as a straight amplifier. The noise generated in the receiver is often called *IF noise,* then, because most of it is generated in the mixer producing the IF signal. A low-noise RF amplifier ahead of the mixer stage makes a big improvement.

Effects of Receiver Noise   In an audio output, noise produces a hissing or frying sound that can best be heard between stations with the manual volume control all the way up. It is very loud in high-gain receivers with high sensitivity. This assumes the receiver does not have a muting or squelch circuit to cut off the audio output between stations.

In a TV receiver with video signal, the internal noise produces white specks called *snow.* Snow is evidence that the signal is too weak compared with the amount of internal random noise. The solution is to supply more RF signal into the mixer stage. Usually the problem is caused by insufficient antenna signal.

Localizing Receiver Noise Troubles   To localize receiver noise troubles, it is helpful to distinguish between a weak signal with and without noise. In an AM radio the effect of noise is too much hiss in the sound; in television the effect is snow in the picture. In both cases the poor signal-to-noise ratio indicates not enough RF signal is being fed into the mixer stage. The trouble is in the antenna and RF circuits before the signal is converted to IF output.

Trouble in the IF or audio circuits also can cause weak or distorted signals. However, in that case there is no hiss or snow.

Test Point Questions 15-16
(Answers on Page 371)

a. Which contributes more to receiver noise, the mixer or the last IF stage?
b. A noise limiter does not reduce internal receiver noise. True or false?

15-17
TROUBLESHOOTING TECHNIQUES

The typical radio receiver has only a few trouble symptoms. The main symptoms are: no audio output, weak output with excessive background noise,

or weak, distorted audio resulting in garbled or hoarse voice reproduction. All of these symptoms are usually caused by a defective component. Examples of these defects are an open resistor, an open coil, a shorted or open capacitor, and a defective transistor or diode. Semiconductor diodes and transistors can be either open at the terminals or shorted at the junction.

Power Circuits   Receiver problems are most often found in the power circuits because of the heat generated by the relatively high currents. The power circuits involve the dc supply voltage for V^+ and the audio output stages.

If the receiver has no output at all, check all fuses first. Larger receivers generally have fuses in the dc power supply and in the loudspeaker circuits.

A low value of dc supply voltage can cause a weak, distorted audio output. If the problem is in a portable receiver, try a new battery. Remember battery voltage must always be checked under normal load conditions.

Audio Circuits   The audio section of a receiver includes all the circuits from the volume control to the loudspeaker. To determine whether the audio section is operating, touch the high side of the volume control. Normally, a loud 60-Hz hum will be heard.

IF Section   This section includes all the circuits from the mixer output to the detector input. The IF circuits are tuned to 455 kHz in AM radios and 10.7 MHz in FM radios. Test signals at these frequencies can be injected at the mixer input to check the IF section. (See Chap. 11, Sec. 11-13, for a description of how to use an RF signal generator for signal injection.)

RF Section   These circuits include the antenna, any RF amplification up to the mixer input, and the local oscillator stage. RF signal frequencies are 535 to 1605 kHz for AM radios or 88 to 108 MHz for FM radios. Test signals at these frequencies can be radiated into the antenna circuit to check the RF section. (See Chap. 11, Sec. 11-13.)

A typical RF problem is a weak antenna signal, possibly because of a broken antenna connection. The weak signal is usually accompanied by excessive background noise. If turning up the volume control results in increased noise, the mixer and IF amplifiers are operating normally.

If stations are not being tuned to their proper carrier frequencies, the problem is probably caused by improper oscillator tuning. It is the oscillator frequency that combines with the RF signal to produce the IF signal.

Indicators for V^+   The receiver must have the correct dc supply voltage in order to amplify the desired ac signal. This V^+ can be measured but several effects provide convenient indicators of its presence:

1. Hum in the output means the audio section has V^+ for amplification.
2. Receiver noise indicates operation of the mixer, IF section, and audio circuits.
3. A properly operating tuning indicator shows that the RF and IF sections are operating to produce a detector output.
4. If there is any output sound when the receiver is turned on and off, V^+ is present.

If the above indicators point to the absence of V^+, check all fuses in the receiver. However, a burned out fuse usually means the circuit was shorted or overloaded. Therefore merely replacing the fuse will not solve the problem. Further tests would be needed. If V^+ is lacking in a portable receiver, replace the battery with a new one.

DC Voltage Measurements   Probably the best thing to do when tests are necessary is to measure dc voltages around the circuits. The receiver is supposed to produce an ac signal output, but most troubles change the values of the dc voltages in the amplifier circuits.

1. Check for V^+ at the rectifier in the dc power supply. If there is no dc output, use an ac meter to check for proper input voltage. If an ac input voltage is present but there is no dc output voltage, the diode rectifier is open.

The dc output of the rectifier is the input to the filter choke or resistor in the power supply. If the rectifier has a dc output but there is no voltage at the output side of the filter, the series choke or resistor must be open.

2. Measure the collector voltage (V_C) at each transistor. Connect the dc voltmeter from the collector to ground. Start with the power amplifiers and work back toward the receiver input.
3. Measure the emitter voltage (V_E) at each transistor. The correct voltage shows that there is normal current in the collector and base circuits.
4. Measure the base voltage (V_B) and the bias voltage (V_{BE}).

Details about problems with the dc electrode voltages for transistors are explained in Chap. 2, Sec. 2-11. Remember that in dc coupled amplifiers, the dc voltages in each stage depend on the values in the preceding stage. Also the base and emitter voltages may be high but it is V_{BE} that determines the transistor bias.

Ohmmeter Measurements   Different methods can be used to localize receiver troubles to a particular stage but the ohmmeter resistance tests actually help locate the defective component. The ohmmeter must be used with power off in the circuit being checked. In making resistance measurements be sure to isolate or disconnect parallel paths. Use the setting for low-power ohms to prevent turning on any transistors when checking the external circuit. However, use the normal ohms range to test the internal junction of a diode or transistor.

Ohmmeter Click Test   When the ohmmeter is connected to an amplifier circuit, the meter's internal battery supplies a transient pulse signal that can produce a click in the audio output. Since power is on for this test, the highest ohms range must be used to minimize the possibility of damage to the ohmmeter. Just touch the ohmmeter test lead to the collector or base terminal for an instant. Work back from the last audio stage toward the antenna input to determine which stage does not produce a click in the output.

Bridging an Open Capacitor A coupling or bypass capacitor that is open will not affect the dc voltages though it will affect the audio output. The technique then is to connect a similar capacitor temporarily in parallel. If the output comes back to normal, the suspected capacitor must be open. The parallel capacitor does not have any effect across a shorted capacitor, but shorted capacitors will affect dc voltages.

Hum When the power-supply ripple at 60 or 120 Hz produces excessive hum in the audio output, the problem is generally insufficient filtering. The usual practice for curing excessive hum is just to replace the electrolytic filter capacitors. There is a possibility, though, that the hum is caused by an unbalanced push-pull audio output circuit and replacing the electrolytic capacitors will not solve the problem.

Test Point Questions 15-17
(Answers on Page 371)

Answer True or False.

a. Weak V^+ can cause a distorted audio output.
b. A defective power supply filter capacitor is common cause of hum.
c. A tuning indicator can operate without V^+.

SUMMARY

1. A radio receiver includes tuned amplifiers for the modulated carrier signal, an audio detector to recover the modulation, and an audio amplifier to drive the loudspeaker. A dc power supply is needed for the amplifier stages.
2. Receiver sensitivity is a gauge of the receiver's capacity to receive weak signals. Selectivity is a gauge of the receiver's capacity to reject interfering frequencies. Higher receiver gain provides increased sensitivity; more tuned circuits improve the selectivity.
3. The superheterodyne circuit uses a local oscillator to heterodyne, or beat, with the RF signal of the station and convert the carrier to the intermediate frequency. The oscillator frequency is varied to make all RF signals fit the IF band of the receiver.
4. It is the local oscillator that tunes the receiver to different stations, because the oscillator tracks above the RF signal frequencies by a constant amount equal to the intermediate frequency.
5. The frequency-converter or mixer section has the RF signal input and the oscillator input to produce the IF signal output.
6. Standard IF values are 455 kHz for AM radio receivers and 10.7 MHz for FM radios, in the commercial radio broadcast bands.
7. An IF stage is a class A amplifier to ensure minimum distortion of the modulation. The IF amplifier is tuned to the intermediate frequency of the receiver.
8. The image frequency of a station is above the oscillator frequency by the same amount the station frequency is below the oscillator frequency.
9. The audio detector rectifies the modulated IF signal and filters out the IF carrier to recover the audio modulation. A diode detector is the circuit generally used. No $V+$ voltage is used for the detector, because the diode must conduct only for the signal voltage.

10. Receiver alignment consists of tuning the IF circuits and adjusting for oscillator tracking with the RF circuits. The adjustments are for maximum signal output.
11. Receiver noise consists of random variations without any specific bandwidth. The effect is considered IF noise because most of it is produced in the frequency-converter stage generating the IF signal.
12. In the audio output the receiver noise produces a hissing or frying sound. In video the noise produces snow in the picture.
13. If there is no audio output or a weak output signal, check V^+ and measure all dc voltages on each transistor starting with the power amplifier.
14. An open or leaky filter capacitor in the power supply is a common cause of excessive hum in the output signal.

SELF-EXAMINATION
(Answers at back of book)

Match the numbers in the left column with the letters in the right column.

1. Standard IF value
2. Diode load resistor
3. IF noise
4. Zero beat
5. AM radio band
6. Collector or drain voltage
7. Station selector
8. Ferrite rod antenna
9. Receiver alignment
10. Oscillator trimmer adjustment
11. Image frequency
12. Receiver selectivity
13. Zero value of V_C
14. Excessive hum
15. Zero value of V_{BE}

(a) Dc power supply
(b) Ganged tuning
(c) RF amplifier input
(d) 455 kHz
(e) IF response
(f) Volume control
(g) Maximum output signal
(h) 535 to 1605 kHz
(i) Mixer stage
(j) Heterodyning
(k) Spurious response
(l) High end of band
(m) Shorted base-emitter junction
(n) Open R_L
(o) Leaky filter capacitor

ESSAY QUESTIONS

1. Name four main sections of a receiver.
2. Define sensitivity and selectivity of a receiver.
3. Give the function of each of the following stages in a superheterodyne receiver: RF amplifier, local oscillator, mixer, IF amplifier, audio detector, and audio output.
4. Make a drawing like Fig. 15-7 but for an FM receiver in the 88- to 108-MHz band with the IF at 10.7 MHz.
5. A receiver with IF of 455 kHz is tuned to different stations at 570, 660, 880, 1010, and 1560 kHz. For each of these, tabulate the resonant frequencies of the RF signal circuits, local oscillator, and IF amplifier.
6. Calculate the image frequencies for each of the stations in Question 5.

7. Give two types of spurious responses in a superheterodyne receiver.
8. Draw the schematic diagram of a diode detector with R_L in the cathode circuit. Indicate **a.** where the IF signal is applied, **b.** where the ac audio signal is taken out, **c.** the IF bypass capacitor, **d.** the polarity of the rectified dc voltage.
9. What is meant by audio taper in a volume control?
10. Why is receiver noise also called IF noise?
11. What is the effect of IF noise on the picture in a television receiver?
12. How are the oscillator padder and trimmer capacitors of the local oscillator in an AM radio adjusted?
13. How are the IF transformers in an AM radio adjusted?
14. Describe briefly how to align the AM radio receiver shown in Fig. 15-21.
15. Give the functions of L_1, L_2, T_1, C_{1A}, C_{1C}, R_3, and R_1 in the oscillator-mixer circuit shown in Fig. 15-6.
16. Give the functions of C_A, C_B, T_A, T_B, and R_1 in the RF amplifier circuit shown in Fig. 15-12.
17. What is the bandwidth of the RF response curve in Fig. 15-13*b?*
18. Give the functions of T_1, T_2, T_3, R_{13}, C_{12}, R_8, and R_9 in the IF amplifier shown in Fig. 15-14.
19. Give the functions of C_1 and C_2 in the audio detector circuit shown in Fig. 15-18.
20. Give the functions of L_A, L_2, T_1, T_2, T_3 CR1, Q4, R_{12}, C_{16}, and C_{15} in the complete AM radio receiver shown in Fig. 15-21.
21. Explain how an ohmmeter is used to find an open resistor, an open coil, a shorted capacitor, and a defective diode.
22. Give two indicators that show the presence of V^+ in a receiver.

PROBLEMS
(Answers to odd-numbered problems at back of book)

1. In the oscillator-mixer circuit shown in Fig. 15-6, calculate **a.** I_E through the 3-kΩ R_E to produce V_E of 1.8 V and **b.** reactance of C_3 at 540 kHz.
2. In the frequency converter shown in Fig. 15-6, calculate the reactance of the coupling capacitor C_2 at 540 kHz.
3. In the IF amplifier shown in Fig. 15-14, calculate **a.** total I through R_{13} for the 1-V drop and **b.** reactance of C_{12} at 455 kHz.
4. In the circuit shown in Fig. 15-14, calculate I_E through R_8.
5. In the AM radio receiver shown in Fig. 15-21, calculate the forward bias V_{BE} for the audio output transistors Q5 and Q6.
6. In the receiver diagram in Fig. 15-21, calculate the reactance of the 5-μF audio coupling capacitor C_{11} at 100 Hz.
7. For an IF transformer with C of 160 pF in the primary what value of parallel L is needed for resonance at 455 kHz?
8. If C is 80 pF in Problem 7, what value of L is needed?
9. Refer to the RF antenna input circuit shown in Fig. 15-6. **a.** The total C_T of C_{1A} and C_{1B} in parallel is 10 pF out of mesh at the high end of the band.

What value of L is needed for resonance at 1620 kHz? **b.** With that L, what value of C_T is needed for resonance at 540 kHz in mesh at the low end of the band?

10. **a.** What is the ratio of highest to lowest frequency in the 540- to 1600-kHz band? **b.** What is the ratio of highest to lowest frequency in the 88- to 108-MHz band?

11. **a.** Give the bandwidth, in percent, for the ratio of 10-kHz bandwidth compared with the IF resonant frequency of 455 kHz. **b.** Repeat part **a.** for 200 kHz compared with the IF resonant frequency of 10.7 MHz.

SPECIAL QUESTIONS

1. Describe briefly three types of specialty radios for services other than AM, FM, and TV broadcasting.
2. Give two examples of the double-superheterodyne circuit in receivers.
3. Give two methods of mounting the on-off switch of a radio receiver.
4. Give two applications of the audio input to the audio section of a radio receiver, besides the detected audio signal.
5. Describe briefly how an audio tone control can be used in a radio receiver.
6. Give two problems you have experienced with radio receivers and describe briefly how you would troubleshoot the set.

ANSWERS TO TEST POINT QUESTIONS

15-1	**a.**	Frequency converter or mixer	15-7	**a.**	88 to 108 MHz	15-13	**a.**	L_A
	b.	Detector		**b.**	-1 V		**b.**	L_2
	c.	C_3	15-8	**a.**	1480 kHz		**c.**	CR1
15-2	**a.**	455 kHz		**b.**	2470 kHz		**d.**	R_{12}
	b.	0 Hz	15-9	**a.**	Mixer		**e.**	C_{16}
15-3	**a.**	T		**b.**	Detector	15-14	**a.**	1400 kHz
	b.	T		**c.**	0.6 V		**b.**	First
	c.	F	15-10	**a.**	455 kHz		**c.**	ac
15-4	**a.**	R_3		**b.**	10.7 MHz	15-15	**a.**	F
	b.	T_1	15-11	**a.**	T		**b.**	T
	c.	L_2		**b.**	T	15-16	**a.**	Mixer
15-5	**a.**	1003 kHz		**c.**	F		**b.**	True
	b.	452 kHz	15-12	**a.**	More	15-17	**a.**	T
15-6	**a.**	1165 kHz		**b.**	More		**b.**	T
	b.	100.7 MHz					**c.**	F
	c.	High end						

Chapter 16
Automatic Volume Control

The automatic volume control (AVC) circuit is used in practically all receivers to prevent overload distortion on strong signals. The method is to control the gain of the RF and IF amplifiers automatically according to the amount of antenna signal. When the antenna signal is at a maximum, the AVC bias is high to reduce the receiver gain. After all, less gain is necessary for strong signals. The AVC bias is always in the polarity that reduces the amplifier gain. A weak signal produces less AVC bias. The receiver gain is then not reduced as much. As a result, the AVC circuit controls the receiver gain. The effect is to provide about the same volume for stations with different signal strengths.

A more general name for this process is *automatic gain control* (AGC). Control of the RF and IF gain is also used in television receivers for the picture signal. We can use the term AVC for volume control in a radio and AGC for control of the picture contrast in a television receiver. Automatic gain control can be used for audio amplifiers to limit the amplitude of an audio signal. The AVC and AGC circuits are essentially the same. More details are given in the following topics:

16-1
SIMPLE AVC

The circuit shown in Fig. 16-1 is not simple in operation, but it has only the basic needs for automatic gain control. They include an AVC rectifier, a filter, and the bias line to the stages controlled by the AVC bias. The AVC rectifier converts the IF signal to dc voltage for bias. The filter removes ac signal variations from the dc bias on the AVC line.

AVC Rectifier In Fig. 16-1 the AVC rectifier is a diode. It produces a dc output proportional to the amount of ac signal input from the last IF amplifier. R_L is the load resistance that provides the dc voltage output. Negative polarity of AVC bias is shown here, but the polarity can be positive for some transistor amplifiers.

AVC Filter C_1 and R_1 remove the ac ripple in the rectified output. The purpose is to make the AVC bias a steady dc voltage. The two ac components that must be filtered out are the IF signal and its audio variations in the rectified IF output.

The AVC filter has a relatively long RC time constant of 0.05 to 0.1 s. A shorter time constant will not filter out the audio signal. Too long a time constant cannot be used, however. The reason is that the AVC bias must change when the receiver is tuned to stations with different signal strengths.

AVC Bias Line The AVC bias line is the common line for dc bias to each RF or IF stage in which the gain is to be controlled. When tubes are used, the AVC bias line is the return circuit for the control grid. Then the AVC voltage varies the grid bias.

When transistors are used, the AVC bias line is usually the return circuit for the base electrode, although the AVC bias can be on the emitter instead. At the base the AVC bias makes small changes in the base voltage to control the base current.

AVC Decoupling Filters In the circuit of Fig. 16-1, R_2C_2 and R_3C_3 isolate each of the controlled amplifiers to prevent feedback of the signal between stages. The decoupling is needed when two or more stages are controlled by AVC bias.

The AVC decoupling is an important example of how isolation is necessary between amplifiers that have a common supply line. In this case the AVC bias line provides a common impedance. Feedback of the ac signal between the amplifiers is minimized, though, by the decoupling. Otherwise, feedback makes the amplifier unstable, with undesired oscillations.

How AVC Bias Reduces Gain When tubes are used, the g_m and mu are reduced as the control grid voltage is made more negative. The result is less gain as the negative AVC bias increases. Some tubes have a special construction for the control grid to accommodate a wide range of AVC bias voltages. They are the *variable-mu*, *remote-cutoff*, or *super control* pentodes.

In NPN and PNP junction transistors the beta (β) characteristic is less as the base current is reduced toward cutoff. Therefore, the current gain is reduced. In an FET the G_m is reduced by more negative gate bias for an N channel. Then the FET gain is less as the negative AVC bias increases.

The range of AVC bias voltages is about -2 to -20 V when tubes are used. When transistors are used, however, the AVC bias changes by less than 1 V. However, that small change in base voltage can control relatively large changes in base current.

Keep in mind that the AVC bias is only a dc axis for the ac signal in the controlled amplifier stage. In fact, the AVC bias line does not have any signal at all. The dc values can be increased, therefore, while the amount of amplified ac signal is decreased.

Stages Controlled by AVC Bias The AVC bias is usually applied to the RF amplifier and first IF stage in the receiver. It should be noted that the last IF amplifier generally is not controlled by AVC bias. The reason is that the stage has relatively large signal voltage, so that changing the bias can easily cause too much amplitude distortion. Also,

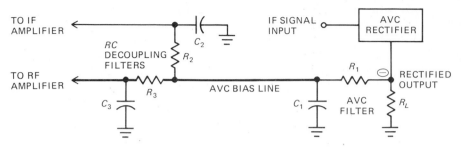

Fig. 16-1. Basic requirements of the AVC circuit shown for negative AVC bias voltage.

the mixer stage generally does not have AVC, because it is preferable not to change the dc bias for heterodyning.

Why AVC is Needed The advantage of automatic volume control in a receiver can be seen from the fact that it is easy to tune from station to station. In a radio without AVC, some stations would blast on too loud, with overload distortion, and others would be too weak. With AVC, all stations come on with about the same volume because the audio detector output is relatively constant. With AGC in a television receiver, all stations have about the same picture strength or contrast because the video signal from the detector has relatively constant amplitude.

Test Point Questions 16-1
(Answers on Page 385)

a. Is AVC bias greater or smaller when the antenna signal strength increases?
b. Is receiver gain smaller with more or with less AVC bias?
c. Is an AVC diode a rectifier or an amplifier?

16-2
REVERSE AND FORWARD AVC BIAS FOR TRANSISTORS

When NPN and PNP junction transistors are used, the gain can be reduced by two opposite methods. Either way, though, more AVC bias must decrease the beta characteristic to reduce the gain.

With reverse AVC bias the forward voltage between base and emitter is reduced toward zero. The result is less collector current I_C. Most important, though, the beta is reduced and the gain is less.

With forward AVC bias the forward voltage between base and emitter is increased toward saturation. For specific conditions, however, the beta can decrease although the average collector current increases. The reason is that beta depends on the changes in I_C rather than on the amount of current.

In general, reverse AVC is often used for IF amplifiers. Forward AVC is used with the antenna signal in the RF amplifier stage. The reason is that reverse AVC near cutoff on the first stage can cause cross modulation resulting in severe amplitude distortion by cross-modulation. The effect occurs when a strong antenna signal is rectified in an RF stage and is due to insufficient bias offset.

Polarities of AVC Bias The polarity of the AVC bias voltage depends on several factors, especially when junction transistors are used. These factors include the transistor type (NPN or PNP) and whether the AVC bias is connected to the base or the emitter. The choices can be simplified for the common case of reverse AVC at the base of an NPN transistor. Then AVC bias is negative to reduce the positive forward voltage at the base.

The negative polarity of reverse AVC happens to be the same as with tubes. Actually, negative grid voltage for tubes can be considered as reverse AVC. When the negative grid bias increases, it

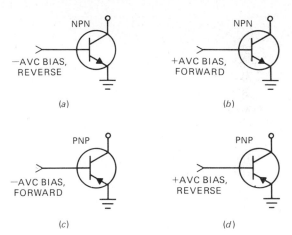

Fig. 16-2. Polarities of forward and reverse AVC bias on NPN and PNP transistors.

In all cases, however, more signal at the antenna produces more IF signal and higher AVC bias. Also, more AVC bias reduces the receiver gain. The AVC voltage is maximum for the strongest antenna signal, but the receiver gain is minimum.

Test Point Questions 16-2
(Answers on Page 385)

a. What is the polarity of the AVC bias for a pentode tube.
b. What is the polarity of reverse AVC bias at the base of an NPN transistor.
c. Give the polarity for reverse AVC at the gate of an N-channel FET.

---

reduces the amplifier gain as the operating point is shifted closer to cutoff.

Examples of AVC Polarities   Four possibilities of AVC polarities are shown in Fig. 16-2. With the NPN transistor (Fig. 16-2a) the negative AVC bias reduces positive V_{BE} at the base, toward cutoff, for reverse AVC. The positive AVC bias (Fig. 16-2b) increases V_{BE}, toward saturation, for forward AVC.

With the PNP transistors the negative AVC bias (Fig. 16-2c) increases the negative V_{BE}, toward saturation, for forward AVC. To have reverse AVC (Fig. 16-2d), the bias is positive to decrease the negative V_{BE}.

16-3
AVC BIAS FROM THE DETECTOR

In most AM radio receivers, the dc output of the audio detector is used to supply the AVC bias, as illustrated in Fig. 16-3. A separate AVC rectifier is not needed. Since the detected signal has dc and ac components, the detector output can have two separate dc and ac circuits with completely different functions.

Audio Signal Output   This signal is the ac component of the rectified output across the detector load R_1. The 0.01-μF C_1 bypasses R_1 for IF signal but not for audio frequencies. The audio signal is coupled by C_3 to the audio amplifier.

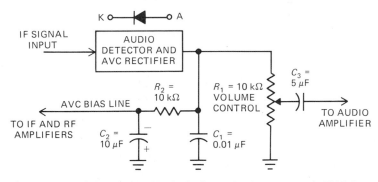

Fig. 16-3. Dual functions of a diode for audio detection and AVC bias. Values shown are typical for transistor AM radios.

DC Output   The average dc voltage across R_1 is proportional to signal strength. Therefore, this dc voltage can be used for AVC bias, after filtering out the audio components. In the AVC line, R_2 and C_2 form the AVC filter. The 10-kΩ R_2 isolates the AVC bias line from the detector output, while the 10 μF C_2 bypasses the audio signal.

The time constant for this RC filter is 0.01 MΩ $\times$ 10 μF = 0.1 s. That is long enough for filtering out the audio signal from the AVC bias line, but it is fast enough to allow the AVC bias voltage to change when different stations are tuned to.

It should be noted that the value of 10 μF for the AVC filter requires an electrolytic capacitor because of the large C. The positive side of C must be connected to chassis ground. The detector output is shown with negative polarity. That polarity results from the diode load R_1 in the anode circuit of the detector. Signal input is at the cathode.

It is important to realize that the AVC function does not interfere with the ability of the detector to provide an audio output signal. The audio signal and AVC bias line are in two separate paths. The audio circuit feeds the amplifier for the amplified signal to the loudspeaker. The AVC circuit just uses the dc component of the detector output. The dc voltage serves very well for AVC bias because it is proportional to the IF signal level.

Test Point Questions 16-3
(Answers on Page 385)

Refer to Fig. 16-3.

a. Which R is the detector load?
b. Which C is the AVC filter?

16-4
AMPLIFIER CIRCUIT WITH AVC BIAS

How the base electrode of a transistor IF amplifier returns to the AVC bias line for gain control is shown in Fig. 16-4. The amplifier Q2 has signal input to the base from L_S of the IF transformer T_1. The amplified IF signal output from the collector is coupled to the next IF amplifier. Specifically, Q2 is the first IF amplifier with an IF signal from the mixer stage.

To control the IF gain in Q2, the base voltage is varied by reverse AVC bias to control the base current. That function is provided for by connecting the low side of L_S to the AVC bias line instead of returning L_S to chassis ground.

AVC Variations   As the signal level at the antenna varies with the strength of signal from different stations, the AVC voltage will change to

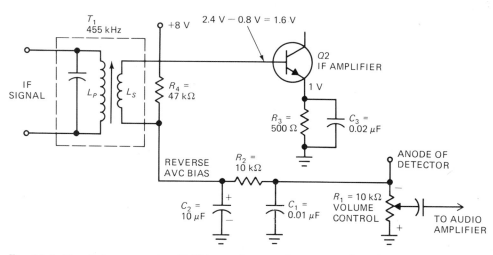

Fig. 16-4. Circuit for connecting AVC bias voltage to the transistor base for controlling IF gain.

control the bias at the base. At the signal level of the stronger stations, the AVC bias becomes more negative to reduce the forward bias which results in less gain. The changes in base voltage must be less than 0.1 V for a junction transistor, but it is the changes in base current that control the gain.

Return Paths for the Base   In an amplifier controlled by AVC bias (Fig. 16-4) it is important to remember that the AVC line is the return path for the base electrode. The path includes chassis ground and the return to the emitter. When there is no return path to the emitter, the base is in an open circuit.

The return path for dc voltage allows the forward bias to be applied between base and emitter. The return path of the ac signal allows all the input signal to be applied.

AVC Bias Control   The AVC voltage (Fig. 16-4) is obtained from the anode circuit of the audio detector; it results in negative AVC bias. However, positive voltage is also on this line through R_4. The connection is needed to have positive forward voltage at the base of the NPN transistor.

The voltage values at the base are 2.4 V from the supply of 8 V, with or without any signal, and -0.8 V from the AVC line with the signal input to the detector. The net result is $2.4 - 0.8 = 1.6$ V at the base with signal input for AVC bias.

The potential difference V_{BE} is $1.6 - 1.0 = 0.6$ V. That is the typical bias on a class A amplifier.

High and Low Sides of L_S   All the dc voltages must be applied to the same side of L_S, as shown in Fig. 16-4. That side then becomes the low side for the ac signal. The opposite side is the high side of L_S; it supplies the signal to the base of $Q2$.

The difference is that the high side has ac signal to ground but the low side does not. The low side is where the dc voltage is applied. There is no signal at that point because it has one or more ac bypass capacitors. If dc voltage with its bypass were applied to the high side, there would be no signal input to the amplifier.

In the circuit of Fig. 16-4 the dc return path for the base of $Q2$ includes R_2 and R_1 to the common chassis ground and R_3 back to the emitter. The ac return path includes the bypass capacitor C_2 in the AVC line and C_3 in the emitter circuit.

AVC Filter   In the AVC line, R_2 and C_2 form the filter to remove audio signal from the AVC bias. With R_2 of 10 kΩ and C_2 of 10 μF, the filter time constant is 0.01 MΩ $\times$ 10 μF = 0.1 s.

Note that the AVC side of the electrolytic capacitor C_2 is the positive terminal, although the AVC bias from the detector is negative. The reason is that the net voltage on the AVC line is positive from the $+8$ V to R_4.

Refer to Fig. 16-4.

a. Which two components form the filter on the AVC bias line?
b. With more antenna signal, will the base of $Q2$ go more or less positive?

16-5
SERIES AND PARALLEL FEED FOR THE AVC BIAS

The circuit shown in Fig. 16-4 has series feed because the AVC bias is in series with the IF input signal. In shunt feed the AVC bias would be connected at the base. However, that would require additional components to prevent shorting out the IF signal for the amplifier.

A comparison of feed methods is shown in Fig. 16-5. In Fig. 16-5a the series feed is used. An RF amplifier circuit is used to show a tuning capacitor with the rotor at chassis ground. Then C_1 is needed to prevent the AVC bias from being grounded. Also, R_1 is used to decouple the amplifier from other stages connected to the AVC bias line.

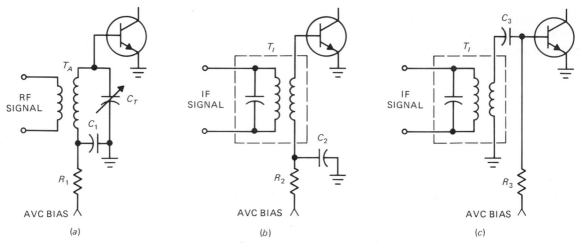

Fig. 16-5. Methods of connecting AVC bias voltage to the base of a transistor amplifier. (*a*) Series feed to the RF amplifier with a grounded tuning capacitor C_T. (*b*) Series feed to an IF amplifier. (*c*) Shunt feed to an IF amplifier.

In Fig. 16-5*b*, series feed is also shown for AVC bias but on the IF amplifier Q2 with a decoupling filter R_2C_2. When only one stage has the AVC bias, however, the decoupling filter is not used.

Parallel feed for the AVC bias is shown in Fig. 16-5*c*. In that circuit, R_3 connects the base to the AVC line for bias. C_3 prevents the dc bias voltage from being shorted through the grounded secondary coil in T_1. Also, R_3 is needed to prevent the IF signal from being shorted through the filter capacitor in the AVC bias line. The AVC filter is not shown.

In summary, there are two problems that need to be avoided in connecting the AVC bias to the controlled amplifier stage. The amplifier circuit should not short the dc bias to ground, and the AVC line should not short out the ac signal.

Test Point Questions 16-5
(Answers on Page 385)

Refer to the circuits shown in Fig. 16-5.

a. Is parallel feed for the AVC bias shown in Fig. 16-5*a*, *b*, or *c*?

b. Which capacitor is a bypass for RF signal in Fig. 16-5*a*?

16-6
TYPES OF AVC CIRCUITS

The circuit shown in Fig. 16-1 is considered simple AVC because it does not have any delay bias or any amplification of the AVC voltage. In addition, the AVC bias can be used for muting or squelch circuits that quiet the receiver between stations. Finally, the AVC voltage is also used for tuning indicators that show when a station is tuned in. The AVC bias is so useful because it has its maximum value when it is tuned to a station and its minimum value between stations.

Delayed AVC A disadvantage of AVC is that some bias is produced even with weak signals. Then the receiver gain is reduced when it would be better to have maximum gain. To overcome that problem, the AVC rectifier can have a delay bias to prevent any conduction for weak signals. As a result, there is no AVC bias at all until the signal is strong enough to require some reduction in receiver gain.

A separate AVC rectifier must be used with delayed AVC, as shown in Fig. 16-6. The AVC bias cannot be taken from the detector output because any delay bias voltage on the detector would prevent audio output. In Fig. 16-6 the delay bias voltage on the AVC diode $D2$ is -2 V at the anode. The reverse voltage used to prevent conduction is also called an *offset bias* on the diode.

The IF signal input is applied to $D1$ and $D2$. The audio detector $D1$ conducts for all signal levels, but the AVC diode $D2$ cannot conduct until the IF signal level is above 2 V to overcome the delay bias. As a result, no AVC bias is produced for weak signals.

For stations that can produce more than 2 V of IF signal, however, $D2$ conducts to provide AVC voltage proportional to the signal strength. The AVC adjustment $R1$ sets the amount of delay bias voltage for $D2$.

Amplified AVC The AVC circuit can be more effective in controlling the receiver gain when there are greater changes in AVC bias with changes in antenna signal level. Two methods used to obtain amplified AVC are illustrated in Figs. 16-7 and 16-8.

In Fig. 16-7 a dc amplifier is used to increase the level of AVC bias. The dc amplification is needed because the AVC bias is a steady dc voltage. Dc amplification means that no coupling capacitors can be used for the input control voltage and the amplified output.

With a CE stage the AVC amplifier inverts the polarity of the control voltage. If the input is negative, the amplified AVC bias is negative, or positive input will result in negative output.

In Fig. 16-8 a separate IF amplifier is employed to provide more signal input just for the AVC rectifier. The result is more voltage for the AVC bias. The purpose is to increase the amount of AVC bias without the need for a dc amplifier after the dc bias voltage is produced.

Overload Diode In small radios that have only one IF stage controlled by AVC bias, more control can be obtained by using an additional diode such as $D1$ in the circuit shown in Fig. 16-9. In that circuit, $Q1$ is the mixer stage producing IF output for T_1, which couples the signal to the first IF amplifier $Q2$. The base of $Q2$ is controlled by AVC

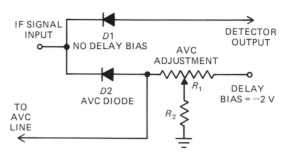

Fig. 16-6. Delay bias on AVC diode D2 but not on detector D1. The amount of bias delay voltage is adjusted by R_1.

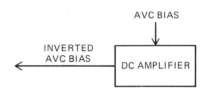

Fig. 16-7. DC amplifier used for AVC bias voltage. The AVC bias in the output is amplified and has inverted polarity.

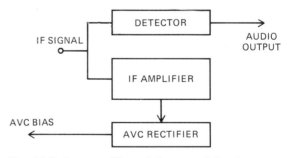

Fig. 16-8. Separate IF amplifier is used for the input signal to the AVC rectifier.

bias, but $Q1$ does not have AVC. However, its gain is controlled by the action of $D1$. The diode conducts on strong signal levels to reduce the gain of the mixer.

The gain is reduced by allowing conduction in $D1$ to damp the resonant circuit for the collector of $Q1$. Damping lowers the impedance and gain to reduce the IF output from the mixer. Whether $D1$ conducts is determined by the dc voltage between points L and H in Fig. 16-9. When the voltages at L and H are equal at 7 V, the potential

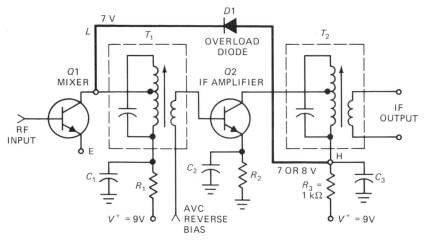

Fig. 16-9. Overload diode D1 used to reduce the Q of the collector-tuned circuit in the primary of T_1.

difference across $D1$ is 0 V. Then $D1$ cannot conduct and there is no effect on gain.

However, when the voltage at point H rises to 8 V, the $D1$ anode is 1 V more positive than the cathode. Then $D1$ conducts to lower the impedance of T_1. The diode is across the primary winding for the ac signal from the tap on the primary winding at point L to the bypass C_3 at point H. With $D1$ conducting, then, the IF output from the mixer is reduced.

The potential at point L is constant at 7 V because $Q1$ is not controlled by AVC. However, the potential at point H in the IF amplifier can change because $Q2$ is controlled by reverse AVC bias. On strong signals, more reverse bias reduces the collector current. Then the voltage drop across R_3 decreases and allows 8 V instead of 7 V at point H. As a result, $D1$ helps the action of the AVC circuit to reduce the receiver gain for strong signals. Note that $D1$ can decrease the amount of IF output from $Q1$ without changing the dc bias on the mixer. The overload diode provides about one-half the control of the receiver's gain.

AGC Level Adjustment In AVC circuits that have delay bias, or amplification, or both, a level adjustment is usually provided for local or distant reception. For distant stations with weak signals,

the receiver should have maximum gain with minimum AVC bias. Local stations need maximum AVC bias and minimum gain to prevent overload distortion on strong signals.

In general, the bias adjustment for the receiver's gain can be set as follows:

1. Tune to the strongest station.
2. Set the manual level, such as volume or contrast control, to maximum.
3. Adjust the AVC or AGC level for more receiver gain up to the point of overload distortion.
4. Finally, back off the control slightly to eliminate the distortion.

More receiver gain means louder sound for audio or stronger contrast for the television picture. Overload distortion in the picture is indicated by reversed black-and-white values and breakup into diagonal bars.

Test Point Questions 16-6
(Answers on Page 385)

Answer True or False.

a. In Fig. 16-6 the delay bias is -2 V for both $D1$ and $D2$.

b. In Fig. 16-8 the AVC rectifier has more IF signal input than the detector.

c. More AVC bias is needed for distant stations, than for local stations.

16-7
SQUELCH, OR MUTING, CIRCUITS

A receiver with AVC operates at maximum gain between stations because there is no input signal and no AVC bias. In high-gain receivers with more than two IF stages, amplification of the converter noise can produce a very loud hissing sound when the receiver is not tuned to a station. When a signal is received, the noise decreases as the AVC reduces the gain.

Still, it may be desirable to eliminate the receiver noise between stations. That feature is especially important in VHF receivers, which need high gain for weak antenna signals. The circuit used to reduce receiver noise between stations is called *squelch, muting,* or *quiet AVC.*

The squelching or muting, is done by cutting off the audio amplifier completely. Then there is no audio output at all. The receiver is dead as far as audio output is concerned. The muting must be done only when the receiver is between stations.

The AVC voltage indicates whether or not a signal is being received. When a station is tuned the AVC bias voltage is high; between stations the AVC is low. Therefore, the AVC bias can be used to operate a squelch circuit that cuts off the audio amplifier when there is no carrier signal.

The squelch operation is illustrated in Fig. 16-10. In short, when the squelch amplifier can conduct, it cuts off the audio amplifier.

It should be noted that the squelch operation is completely different from a noise-limiter circuit. The limiter can remove some interference pulses from the desired signal. However, the squelch operates to mute the receiver only when there is no carrier signal.

Squelch Circuit One type of squelch circuit, with two NPN transistors, is illustrated in Fig. 16-

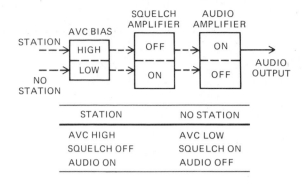

STATION	NO STATION
AVC HIGH	AVC LOW
SQUELCH OFF	SQUELCH ON
AUDIO ON	AUDIO OFF

Fig. 16-10. Squelch, or muting, circuit cuts off the audio output of the receiver between stations when there is no AVC bias.

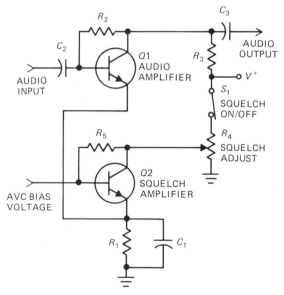

Fig. 16-11. Squelch circuit in which Q2 cuts off the audio amplifier Q1 when the AVC bias is low between stations.

11. Note that Q1 and Q2 have the common emitter resistor R_1 for self-bias on both stages.

First, assume that the squelch amplifier Q2 is off. Then Q1 has its normal forward bias to produce an audio output. Its V_{BE} is a combination of emitter self-bias from R_1C_1 and positive base voltage through R_2. The squelch amplifier is cut off by negative AVC voltage at the base when the receiver is tuned to a station.

Between stations, in the absence of a carrier signal, the receiver does not produce enough AVC voltage to cut off $Q2$. Now the squelch amplifier $Q2$ conducts. Positive base voltage is applied through R_5. This transistor is chosen to have more emitter current than $Q1$. As a result, the emitter bias voltage across R_1C_1 is enough to cut off $Q1$. Then there is no audio output.

Squelch Adjustments A problem with muting circuits is that any weak stations we may want to hear can be squelched. Two adjustments are provided. First, the squelch circuit can be disabled completely. In Fig. 16-11 the collector circuit for $Q2$ is opened by S_1. Then the squelch amplifier cannot conduct to cut off $Q1$.

With the squelch on, the operating point can be adjusted by varying R_4 (Fig. 16-11) to set the amount of collector voltage and emitter current for $Q2$. To adjust the circuit:

1. Tune in the weakest signal with the squelch off by opening S_1.
2. Close S_1 to turn on the squelch and adjust R_4 just to the point at which the station can be heard.

All stronger stations will also come through, because the AVC bias cuts off the squelch amplifier.

Test Point Questions 16-7
(Answers on Page 385)

Answer True or False.

a. In the squelch circuit of Fig. 16-10 the audio amplifier is on when the squelch amplifier is off.
b. In the circuit of Fig. 16-10 the squelch amplifier $Q2$ is off when the AVC bias is high.

16-8
TUNING INDICATORS

The AVC voltage increases when a station is tuned. Furthermore, the AVC voltage rises to maximum as the receiver is tuned more precisely to the carrier frequency. Therefore, the amount of AVC bias can be used as a tuning indicator. Three basic methods are used to show how sharply a station is being tuned:

1. A dc analog voltmeter is used to read the AVC bias. When the needle indicates maximum voltage, the station signal is the strongest and the tuning is the sharpest.
2. A dc analog milliameter is used to read average collector current in a stage controlled by AVC bias. With reverse AVC, the needle will indicate a minimum current at which point tuning of the station is sharpest.
3. An electron-ray indicator, or magic-eye tube, has a fluorescent screen that glows when excited by electrons. Where the screen is not luminated, the tube has a shadow. The amount of shadow can be controlled by AVC bias voltage applied to the tube.

Why I_C Decreases with More Reverse AVC In Fig. 16-12a the IF amplifier is shown with a dc milliameter reading an average I_C of 1.5 mA with a typical bias V_{BE} of 0.6 V. The conditions for the signal into the receiver are shown in Fig. 16-12b. Now the stage has more ac signal input but less dc collector current. The dc milliameter reads only average I_C.

Remember, this is a class A amplifier, in which current flows for the full cycle. In this case the average I_C depends only on the dc bias, not on the amount of ac signal. With a smaller forward bias V_{BE} of 0.58 V, the meter reads less I_C at 1.3 mA. The forward bias on the amplifier is reduced by more reverse AVC with a stronger antenna signal.

If the dc milliameter in Fig. 16-12 is used as a tuning indicator, it will dip to minimum for maximum signal. However, in order to make it read upscale, the meter is usually connected in a bridge circuit. Then the bridge can be unbalanced to make the tuning meter read upscale when the AVC bias increases. In that case, the maximum reading indicates the sharpest tuning.

S Meter The S meter is calibrated to indicate relative signal strength, it reads upscale as the sig-

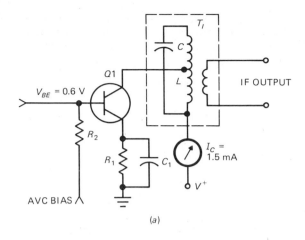

(a)

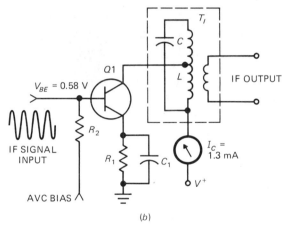

(b)

Fig. 16-12. Average dc collector current I_C decreases with an increase in the signal and more AVC bias. (a) No signal input, bias is 0.6 V with I_C = 1.5 mA. (b) With signal input, forward bias is reduced to 0.58 V by AVC voltage which reduces I_C to 1.3 mA.

nal increases because the dc milliameter is connected in a bridge circuit. It responds to changes in collector current with changes in AVC bias.

The S meter is generally used in communications receivers and CB radios.

Electron-Ray Indicators The AVC bias voltage is applied to an electron-ray indicator tube to control the amount of shadow on a green fluorescent screen. Some tubes have a built-in amplifier for the voltage to the electrode that controls the electron ray. Other tubes require a separate amplifier.

The following are typical tubes used as tuning indicators.

6BR8/EM80	Fan-type shadow; has built-in amplifier
6AF6G	Twin wedge-shaped shadows; needs separate amplifier
6AL7	Twin bar shadows; needs separate amplifier
6E5	Magic-eye tube with built-in amplifier; wedge-shaped shadow
6FG6/EM84	White screen; has shadow bar
DM70	Miniature tube

A simple type of indicator tube is the neon bulb, which glows when the gas is ionized. The bulb can be connected into a circuit that provides the required ionizing voltage to indicate maximum output for correct tuning.

Test Point Questions 16-8
(Answers on Page 385)

a. When the reverse AVC bias in the circuit of Fig. 16-12 increases, does I_C increase or decrease?
b. Will a dc voltmeter across the audio detector load resistor read maximum or minimum for maximum carrier signal?

SUMMARY

1. The AVC circuit controls the RF and IF gain of the receiver. The AVC bias is proportional to signal strength. More bias reduces the gain to prevent overload distortion. The result is relatively constant output from the audio detector.

2. The AVC rectifier produces dc voltage proportional to its IF signal input. Instead of having a separate AVC rectifier, however, the dc output voltage of the detector can be used for AVC bias.

3. The AVC filter removes the audio-signal variations from the dc bias voltage for AVC.

4. The AVC bias line is the return circuit for the amplifiers controlled by AVC bias. Usually, the AVC bias is applied to the control grid for tubes, the base of NPN and PNP transistors, or the gate of an FET.

5. Reverse AVC on transistors reduces the forward bias toward cutoff. Forward AVC increases the forward bias toward saturation. In both cases, more AVC bias reduces the gain.

6. In delayed AVC a separate AVC rectifier has a delay, or offset, voltage to prevent conduction on weak signals. There is no AVC bias until the ac signal is strong enough to overcome the dc delay bias at the AVC rectifier.

7. In amplified AVC, more AVC bias voltage is produced for more control of the receiver gain. A dc amplifier can be used for the AVC bias voltage itself, or an additional IF amplifier can be used for the signal input to a separate AVC rectifier.

8. In receivers with an AGC adjustment, vary the control for overload distortion on the strongest signal and then back off the setting.

9. In squelch or muting circuits the squelch amplifier cuts off the audio output of the receiver between stations when there is no AVC bias. When tuned to a station, the AVC bias cuts off the squelch amplifier to allow an audio output.

10. The AVC bias is used as a tuning indicator, because it rises to maximum for maximum signal input to the receiver.

SELF-EXAMINATION
(Answers at back of book)

1. Does more antenna signal produce more or less AVC bias?
2. Does more AVC bias produce more or less receiver gain?
3. Is a typical AVC filter capacitor 10 μF or 10 pF?
4. Is AVC bias at the control grid of a tube negative or positive?
5. Is reverse AVC voltage at the base of an NPN transistor positive or negative?
6. Does the circuit in Fig. 16-5a show series or parallel feed for AVC bias on the RF amplifier?
7. In Fig. 16-6, is the delay bias voltage on $D1$ or $D2$?
8. When the audio detector has R_L in the anode circuit, can it supply negative or positive AVC bias voltage?
9. Is the squelch amplifier on to cut off the audio output when the AVC bias is high or low?
10. With more reverse AVC bias, does the average collector current in an amplifier increase or decrease?

ESSAY QUESTIONS

1. Give the three main parts of a simple AVC circuit.
2. Give one advantage and one disadvantage of AVC.
3. What is the function of the AVC filter? Give a typical time constant.
4. Show a circuit for AVC bias from the audio detector output.
5. Compare reverse and forward AVC for transistors. Give polarities at the base of an NPN transistor.
6. Compare series and parallel feed for the AVC bias to the base of an IF amplifier.
7. What is the advantage of delayed AVC?
8. What is the purpose of a squelch or muting circuit?
9. What is the purpose of an S meter?
10. Explain briefly how to set the AGC level adjustment.
11. Explain briefly how to set the squelch level adjustment.
12. How would you use a dc voltmeter to check for best tuning?
13. How would you use a dc milliameter to check for best tuning?
14. Make a table listing the conditions on station and off station for AVC bias, the squelch amplifier, and audio amplifier.

PROBLEMS

(Answers to odd-numbered problems at back of book)

1. In Fig. 16-3, calculate the RC time constant for **a.** R_2C_2 and **b.** R_1C_1.
2. What C is needed for a time constant of 0.1 s with R of 1 MΩ?
3. In Fig. 16-4, calculate I_E for V_E of 1 V through the 500-Ω R_3.
4. What is I through a 1-kΩ R for an IR voltage drop of **a.** 1 V and **b.** 2 V?
5. In Fig. 16-9, calculate I_C through the 1-kΩ R_3 for V_C of **a.** 8 V and **b.** 7 V.

SPECIAL QUESTIONS

1. Give three types of receiver that can use the squelch or muting circuit.
2. Why might a communications receiver have a switch to short out the AVC bias circuit?
3. What do you consider the main advantage of automatic gain control in receivers?

ANSWERS TO TEST POINT QUESTIONS

16-1	**a.** Greater	16-3	**a.** R_1	16-6	**a.** F
	b. More		**b.** C_2		**b.** T
	c. Rectifier	16-4	**a.** R_2C_2		**c.** F
16-2	**a.** Negative		**b.** Less	16-7	**a.** T
	b. Negative	16-5	**a.** Fig. 16-5c.		**b.** T
	c. Negative		**b.** C_1	16-8	**a.** Decrease
					b. Maximum

Chapter 17
FM Receivers

An FM receiver, like most radio receivers, uses the superheterodyne circuit. The fact that the signal is frequency-modulated does not affect the heterodyning. (The details of frequency modulation are explained in Chap. 12.) As the local oscillator in the RF tuner beats with the FM signal, the original frequency swings of the RF carrier are still present in the IF signal, but around the lower intermediate frequency.

The intermediate frequency in FM receivers is 10.7 MHz. That frequency is much higher than the IF value of 455 kHz in AM receivers because the RF carrier frequencies for FM radio broadcasting are in the 88 to 108-MHz band.

An FM receiver can use amplitude limiting circuits to reject AM interference because the desired signal is in the frequency variations of the carrier. FM is much better than AM in its ability to eliminate the effects of noise and interference in the signal. More details are given in the following topics:

17-1   Circuits of an FM Receiver
17-2   RF Tuner Circuit
17-3   IF Amplifier Section
17-4   Limiters
17-5   The Discriminator
17-6   Quadrature Phase in Discriminator Transformer
17-7   Ratio Detector
17-8   Requirements of FM Detection
17-9   Discriminator Balance and IF Alignment
17-10   Ratio Detector Balance and IF Alignment
17-11   An FM-AM Receiver
17-12   Stereo FM Broadcasting
17-13   Stereo FM Receivers

17-1
CIRCUITS OF AN FM RECEIVER

As shown in the block diagram of Fig. 17-1 the antenna of an FM receiver is generally a half-wave dipole for the 88 to 108-MHz band. The length of the VHF dipole is approximately 3.5 ft.

The RF tuner includes an RF amplifier with the local oscillator and mixer stages. An RF amplifier is generally used ahead of the mixer to improve the signal-to-noise ratio for VHF receivers. The stage is tuned to the RF carrier frequencies in the 88- to 108-MHz band. The oscillator and mixer stages form the frequency-converter circuit. The circuit converts the RF signal to an IF signal at 10.7 MHz. The output of the mixer is the input to the IF section.

It is easy for the RF tuner to cover the entire FM radio band, with variable capacitive or inductive tuning, because 88 to 108 MHz is a relatively narrow range. The ratio of 108 MHz to 88 MHz, the highest and lowest frequencies, is only 1.23. The factor means that 108 MHz is only about 23 percent higher than 88 MHz.

The IF amplifier section is tuned to 10.7 MHz. Its bandwidth equals approximately 200 kHz, or 0.2 MHz.

Note that the bandwidth of 200 kHz at the IF of 10.7 MHz in an FM receiver is comparatively the same as 10 kHz at the IF of 455 kHz in an AM radio. The ratio is about 0.02 for either

$$\frac{0.2\ \text{MHz}}{10.7\ \text{MHz}} \approx \frac{10\ \text{kHz}}{455\ \text{kHz}} \approx 0.02$$

This means that the bandwidth of 200 kHz is only 2 percent of the resonant frequency at 10.7 MHz.

As a result, it is not difficult for the IF coupling transformers to have the required bandwidth.

Next is the limiter stage. This circuit is an IF amplifier tuned to 10.7 MHz, but the limiter provides a relatively constant output signal for different input levels. The stage is an overloaded amplifier that operates between saturation and cutoff. In addition, the limiter stage usually has signal bias that automatically adjusts itself to the amount of signal. Remember that the audio signal is in the frequency variations of the carrier, not the amplitude variations. Therefore, amplitude limiting can be used for an FM signal.

The amplified IF signal drives the input of the FM detector, which recovers the audio modulation. This circuit allows the frequency variations in the signal to provide equivalent amplitude variations that can be rectified by a diode. Generally, two diodes are used in a balanced detector circuit. Common examples are the discriminator circuit and ratio detector. The ratio detector also rejects undesired AM in the signal.

RF and IF Signal Levels   Note the values in Fig. 17-1. Assume an antenna input of 2 μV, which is a weak signal. That signal is amplified to 200 μV by the RF tuner. The overall gain then is 100, including the gains in the RF amplifier and frequency converter.

The IF gain is 10,000 to increase the input of 200 μV to 2 V of IF signal for the FM detector. That gain is provided by 2 to 4 IF stages, including the limiter. The diodes in the FM detector need at least 1 to 2 V of input signal for linear operation.

Mono and Stereo Receivers   The output of the FM detector includes the audio modulation with

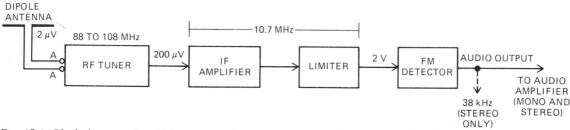

Fig. 17-1. Block diagram of an FM receiver. The stereo circuits are shown in Fig. 17-19.

a frequency range of 50 to 15,000 Hz. That audio signal is the input for the audio amplifier that drives the loudspeaker. In a monophonic, or mono, receiver just that signal is used for a single audio channel.

For stereophonic, or stereo, sound an additional audio signal is broadcast on a 38-kHz subcarrier. The 38-kHz signal is coupled to the stereo section of the receiver. Duplicate audio amplifiers are used with the stereo circuits that develop separate left and right audio signals. The stereo system is explained in Secs. 17-12 and 17-13.

Test Point Questions 17-1
(Answers on Page 412)

a. What are the frequencies of the FM radio band.
b. What is the standard IF value for FM radio receivers.
c. Is a limiter stage generally used in AM or FM receivers?
d. Is a 38 kHz subcarrier used in mono or stereo receivers?

17-2
RF TUNER CIRCUIT

In the tuner circuit of Fig. 17-2, $Q1$ is the RF amplifier and $Q3$ is the local oscillator for the mixer stage. The mixer $Q2$ beats the oscillator output with the RF signal input to produce the IF signal at 10.7 MHz. The oscillator and mixer stages form the frequency-converter section. Both the RF amplifier $Q1$ and the mixer $Q2$ use a dual-gate, N-channel MOSFET. The oscillator $Q3$ is an NPN bipolar transistor.

The field-effect transistor is used in the RF tuner because of its ability to handle a wide range of signal voltages. The dual-gate feature is convenient for $Q1$, because the RF input signal is applied to G_1 while G_2 is used for AVC bias. For the mixer $Q2$, also, G_1 is used for the RF input signal and the oscillator injection voltage is applied to G_2. The internal protective diodes at each gate are designed to prevent arcing from gate to channel.

RF Amplifier In the $Q1$ input circuit, L_1 indicates the antenna input transformer. It is standard practice to have the input impedance of 300 Ω for FM receivers, the same as for television receivers. Therefore, 300 Ω twin lead is generally used for the transmission line from a VHF dipole antenna.

The RF input signal is coupled by C_3 to G_1 of the RF amplifier $Q1$. In the tuned output circuit of the drain electrode, the tap on L_2 provides amplified RF signal through C_{12} to G_1 of the mixer.

The drain is connected to the dc supply of +15 V at the low side of L_2 through R_7 and its bypass capacitor C_8. Self-bias for the RF amplifier is provided by the source resistor R_4 and its bypass C_6. In addition to the source bias, the amplifier has AVC bias. The AVC bias line is connected directly to G_2 and through R_1 to G_1. R_3 is needed as a dc return path for the ac signal voltage and the dc bias.

Mixer Stage For the frequency conversion in $Q2$, the RF input signal is applied to G_1 and the other gate, G_2, has the oscillator injection voltage. It is coupled by C_{13} from the oscillator $Q3$. In that path the combination of C_{11} with L_3 is used as a 10.7-MHz trap to reject RF interference at the intermediate frequency.

The oscillator output at G_2 and the RF signal at G_1 together control the drain current. As a result, $Q2$ is the mixer that produces the IF signal for the drain output circuit. Here T_1 is the first IF transformer to supply the 10.7-MHz signal to the IF amplifier section.

The mixer stage has a combination of source and fixed bias. The stage does not use AVC bias because the amount of bias for the best heterodyning action is critical.

Local Oscillator $Q3$ uses the Hartley circuit with L_4 as the tapped oscillator coil. The tap is at the emitter connection. Feedback from one side of the tank circuit is coupled by C_{10} to the base electrode. The oscillator injection voltage for $Q2$ also is taken from that point to C_{13}.

The collector is effectively grounded for the ac signal by the bypass capacitor C_{22} in order to con-

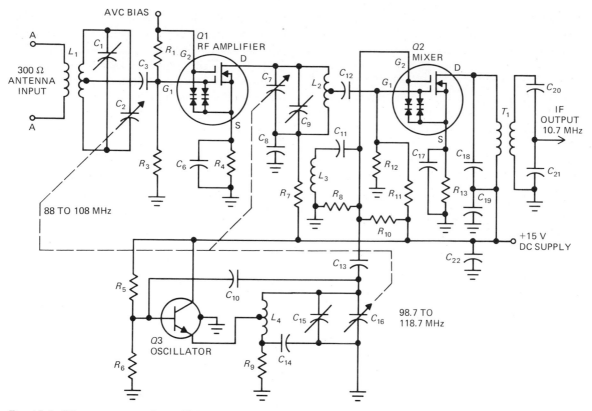

Fig. 17-2. RF tuner circuits for an FM receiver. MOSFETs are used in the RF amplifier and mixer stages; an NPN transistor is used for the local oscillator. (RCA)

nect the collector to the grounded side of the tuned circuit. With that circuit, one side of the tuning capacitor can be grounded.

Ganged Tuning   The dashed lines for C_2, C_7, and C_{16} in Fig. 17-2 indicate that the capacitors are on a common shaft for ganged tuning. C_2 tunes the RF amplifier input. Its trimmer capacitor is C_1. In the RF amplifier output, C_7 is the tuning capacitor with its trimmer C_9. Both the input and output circuits tune through the range of 88 to 108 MHz to provide RF input signal to the mixer.

The tuning capacitor for the local oscillator is C_{16} with its trimmer capacitor C_{15}. The oscillator tunes through the range of 98.7 to 118.7 MHz, beating 10.7 MHz above the RF signal frequencies. The calculations are: $88 + 10.7 = 98.7$ for the low end and $108 + 10.7 = 118.7$ for the high end.

RF Alignment   The alignment procedure for the circuit shown in Fig. 17-2 consists of adjusting the trimmer capacitors at the high end of the band, typically 106 MHz. First the oscillator trimmer C_{15} is set to receive 106 MHz at that point on the dial. Then the RF amplifier trimmers C_1 and C_9 are adjusted for maximum signal. We assume that the IF section is not out of alignment so that it can be used for checking the RF signal.

Test Point Questions 17-2
(Answers on Page 412)

Refer to Fig. 17-2.

a. Which transistor is the local oscillator stage?
b. Which capacitor couples the oscillator injection voltage?

17-3
IF AMPLIFIER SECTION

Two integrated-circuit (IC) units are used in the circuit of Fig. 17-3 to obtain enough IF amplification to drive the ratio detector. The function of the detector is to recover the audio modulation. The 10.7 MHz signal from the mixer in the RF tuner supplies the IF input signal to the first IF transformer T_1. The audio output of the ratio detector is taken from the R_1C_1 network at the right for the signal to the audio amplifier. No limiter is shown in the IF section, because the ratio detector rejects amplitude modulation.

These IF amplifiers are IC packages, but additional components are needed. The external parts include the 10.7-MHz transformers T_1, T_2, and T_3, each in its own shield can. The 9-V supply provides dc operating voltages for the amplifiers.

Also, two external diodes are used in the ratio detector circuit. This circuit is used often in FM receivers. Details of the ratio detector circuit are explained in Sec. 17-7.

When the input signal from the mixer is 200 μV, the IF output is about 2 V to drive the ratio detector. Each IC amplifier has a voltage gain of 1000, or 60 dB. However, the IF coupling transformers have an insertion loss. The overall IF voltage gain from input to output, therefore, is about 10,000, or 80 dB, for both IF amplifiers.

IF Response Curve The required selectivity is illustrated in Fig. 17-4a. The IF response curve is centered around 10.7 MHz and has a 3-dB bandwidth of ±100, or 200 kHz. An attenuation of 3 dB, or 3 dB down, corresponds to 70.7 percent of maximum response.

A bandwidth of 200 kHz is needed for the FM signal. The maximum frequency swing is ±75 kHz, but higher-order sidebands require slightly more bandwidth. In stereo receivers the IF bandwidth is generally 250 kHz because higher modulating frequencies are used for the multiplexed stereo signal.

The frequency response of the IF amplifier is provided by the transformers tuned to 10.7 MHz. In the circuit of Fig. 17-3, T_1 and T_2 are double-

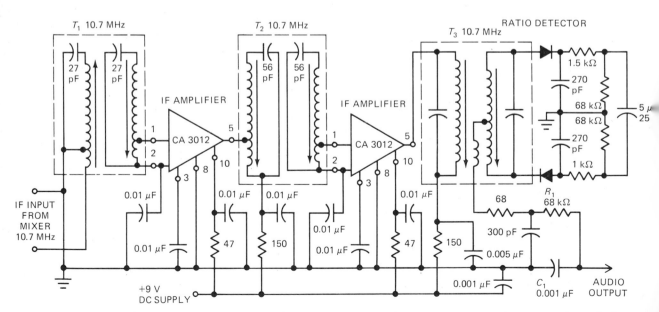

Fig. 17-3. Complete IF amplifier for 10.7 MHz. This circuit uses two CA 3012 integrated circuits and two diodes for the ratio detector. T_1, T_2, and T_3 are shielded IF transformers. R values are in ohms. (RCA)

tuned transformers with adjustable primary and secondary for IF alignment. T_3 is also tuned to 10.7 MHz, but a third winding is used for the ratio detector transformer. The IF transformers effectively provide a bandpass filter for 10.7 MHz ± 100 kHz, but they attenuate frequencies outside the IF passband. A typical transformer is shown in Fig. 17-4b.

Ceramic IF Filter In many FM receivers, a ceramic filter is used instead of a double-tuned transformer for the 10.7 MHz IF. Figure 17-5 illustrates the circuit. The advantage of a ceramic filter is sharper skirts on the IF response curve for better selectivity.

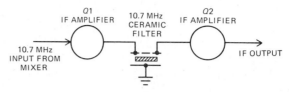

Fig. 17-5. Ceramic IF filter is used for IF coupling instead of a transformer.

17-4
LIMITERS

A limiter stage is an IF amplifier, but its main function is to remove AM interference from the FM signal. The undesired amplitude changes include atmospheric static and all types of electrical noise interference. Actually, any change in the peak amplitudes could produce interference in the audio output, because the FM signal is transmitted with a constant level.

An FM detector recovers the audio modulation in the frequency swings of the FM signal, but the detector also responds to amplitude changes. The discriminator circuit in particular is an FM detector that needs a limiter to provide a constant level of FM signal.

The idea of amplitude limiting is illustrated in Fig. 17-6. The input is an FM signal, but it has different amplitude levels because AM interference

---

Test Point Questions 17-3
(Answers on Page 412)

Refer to Fig. 17-3.

a. Which components are tuned to 10.7 MHz?
b. What is the dc supply voltage for the IF amplifiers?
c. Which pins of CA 3012 are used for the IF signal input?

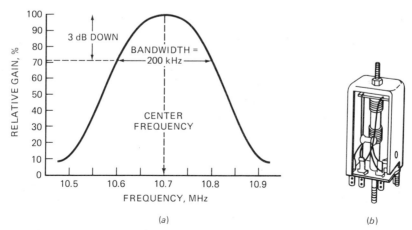

Fig. 17-4. (a) Typical IF response curve for 10.7 MHz. (b) Cutaway view of filter in a shielded can. Height is 1 in. (*Erie Technological Products, Inc.*)

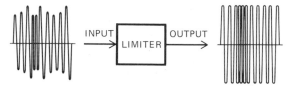

Fig. 17-6. How the limiter removes AM interference from an FM signal.

has been added. In the limiter circuit, though, the output level is constant for different input levels. Usually, the limiter can provide a constant output level for input signals that vary over a range of 10:1 in amplitude.

Limiter Circuit   In the circuit of Fig. 17-7, Q3 is the third IF stage serving as a limiter to drive a discriminator for the FM detector. The signal for the limiter is provided by the first and second IF amplifier stages. The limiter is the last IF stage because it needs enough signal for overload. Signals too weak to produce saturation cannot have any limiting effect.

The NPN transistor uses V^+ for collector voltage. V_C is dropped by R_2 so that saturation I_C can easily be produced by the input signal. C_2 is the RF bypass for R_2.

Signal Bias for the Limiter   The base bias for Q3 is a combination of signal bias with R_1C_1 and fixed bias from V^+ through R_3. The fixed bias is positive for forward polarity, but the signal bias is negative for a reverse offset.

Signal bias is an important requirement for the limiter stage so that the bias can automatically adjust itself to the signal level. When the signal amplitude increases, because of upward modulation by interference, the signal bias also increases. A reduced signal level with downward modulation means less bias. As a result, the varying bias at the base keeps the peak positive signal amplitudes clamped at the value of saturation collector current. The negative signal peaks are clipped by cutoff. Then the limiter output signal has a relatively constant amplitude. The frequency swings in the FM signal, however, are still the same for the input and output.

Limiter Time Constant   The R_1C_1 time constant determines the level of signal bias required to follow the changes in signal amplitude. The constant should be at least 1 μs to be more than 10 times the period of the 10.7-MHz signal. However, it cannot be too long or the bias will not change with the amplitudes of the AM interference. Typical values for the RC time constant are 1 to 40 μs.

Threshold Signal for Limiting   The threshold signal for limiting is the minimum voltage that will produce saturation. There must be enough signal amplitude to keep the limiter saturated, even with downward modulation of the AM interference.

For weaker signals below the threshold level, there is no limiting action. Then the limiter acts as an ordinary amplifier. That is why no limiting action is evident when the antenna signal input is very weak or when there is receiver noise between statons. A carrier signal is necessary to overload the limiter.

When transistors are used, a typical value for the threshold signal at the limiter is 1 V. Then the signal bias at the input can decrease or increase while still keeping the output signal clamped between saturation and cutoff.

Test Point Questions 17-4
(Answers on Page 412)

Answer True or False.

a. The limiter stage is usually the first IF amplifier.
b. The limiter output can drive a discriminator for the FM detector.
c. In Fig. 17-7 the RC time constant for the signal bias is 20 μs with C_1 of 0.02 μF and R_1 of 1 kΩ.

17-5
THE DISCRIMINATOR

The discriminator circuit is the basic method of detecting an FM signal. As shown in Fig. 17-8, its main characteristics are:

1. Two diode rectifiers with a balanced output circuit. The purpose of the balance is to provide

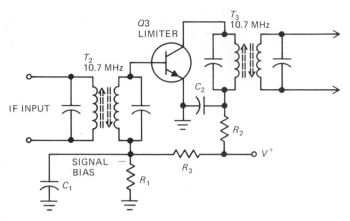

Fig. 17-7. Circuit of a limiter stage with signal bias in the base circuit.

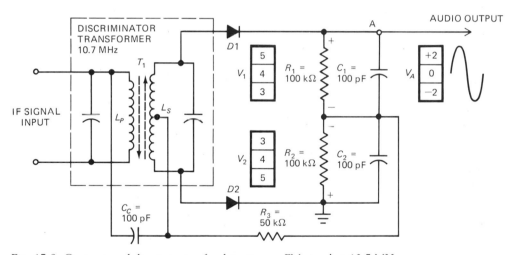

Fig. 17-8. Center-tuned discriminator for detecting an FM signal at 10.7 MHz.

zero output at the center frequency when both diodes have equal signals.

2. Discriminator input transformer for the two diodes. The transformer proportions the amount of IF signal for the diodes according to the frequency deviation above and below center.

3. The frequency response of the discriminator itself, separate from the IF amplifier, has the S-shaped curve shown in Fig. 17-9.

The discriminator circuit is an FM detector because it can distinguish between different frequencies in the input to provide different output voltages.

The discriminator circuit shown in Fig. 17-8 has several names. It is often called a *Foster-Seely discriminator* after the inventors. It is also called the *center-tuned discriminator* because both the primary and secondary of the input transformer are tuned to the center frequency. Probably the best name is *phase-shift discriminator*, because that describes how the input transformer proportions the IF signal voltage for the two diodes.

Balanced Diode Detection Each diode is a rectifier with its own load resistor. When positive signal voltage is applied to its anode, $D1$ conducts. Electron flow is cathode to anode through the top half of L_S to the center tap and returning to cath-

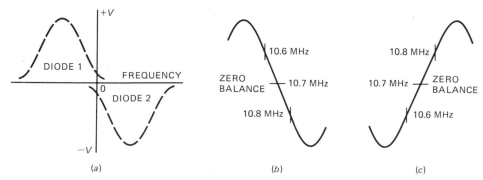

Fig. 17-9. S-response curve of a discriminator. (a) Individual responses with positive and negative outputs for D1 and D2 of Fig. 17-8. (b) Combined responses form the S curve. Balance is at the crossover frequency of 10.7 MHz. (c) Same curve with opposite polarity.

ode through R_3 and R_1. The bypass C_1 across R_1 is a filter to remove any IF signal from the audio output, as in the usual diode AM detector.

Note that the polarity of the rectified voltage across R_1 is positive at the cathode side of the diode. The amount of rectified output from $D1$ depends on the amount of IF signal provided by the $D1$ input circuit.

In the same way, the IF signal applied to $D2$ produces a rectified output across R_2. That voltage also is positive at the cathode side of $D2$. However, that point is at chassis ground.

The two voltages V_1 and V_2 are in series with each other because the output is taken from point A to ground across both R_1 and R_2. Those two voltage polarities, however, are opposing. The top of R_1 is positive to ground, and the top of R_2 is negative. When both V_1 and V_2 are 4 V, as an example, the net output at point A equals 0 V. Therefore, we can consider the two diodes of a discriminator as being in series-opposition for the rectified output. The zero-balance point occurs at the center frequency when both diodes have equal amounts of IF signal input.

Audio Output Voltage Assume now that the amount of IF signal voltage for the two diodes varies. When one increases, the other decreases. For example, the signal for $D1$ increases from 4 V to 5 V while the signal for $D2$ decreases from 4 V to 3 V. Then $D1$ will produce more output voltage while $D2$ produces less.

The output voltages are unbalanced now. V_1 rises to 5 V while V_2 drops to 3 V. The net output at point A is the difference between V_2 and V_1; then 5 V − 3 V = 2 V for the potential at point A to ground. That voltage is positive because the positive V_1 is greater than the negative V_2.

When the IF signals applied to the two diodes are reversed, $D2$ has more output than $D1$. For a symmetrical change, V_1 decreases from 4 V to 3 V while V_2 increases from 4 V to 5 V. The circuit is again unbalanced with output at A equal to 3 V − 5 V = −2 V. This time the net output is negative because V_2 is larger than V_1.

As the FM input continuously changes in frequency, the proportions of the IF signal for the two diodes follow the frequency variations. The resulting output at point A is the audio signal in the frequency-modulated IF input.

Discriminator Response Curve The S-shaped discriminator response curve for balanced detection is shown in Fig. 17-9. The output at the center frequency is zero. That is the balance point. It is also the crossover frequency, between positive and negative output voltages.

In Fig. 17-9a the separate response for the circuit with $D1$ is shown for a positive output and the response for $D2$ with a negative output. When the two responses are combined, the result is the S curve of Fig. 17-9b. The S curve of Fig. 17-9c is shown for opposite polarities. Either response is a typical S curve for the discriminator, depending on the polarity of the output for input frequencies above and below center.

The crossover is at 10.7 MHz for the center frequency of the IF signal. The bandwidth of the S curve is ± 0.1 MHz, or ± 100 kHz, on the linear slope of the response. The S curve is linear for about one-half to two-thirds of the distance between positive and negative peaks.

Audio Detection The S curve really shows how a discriminator detects the FM signal. At the center frequency the output is zero. When the signal swings above center, the discriminator produces one polarity of output voltage. An increase in frequency deviation results in greater output voltage. For one half-cycle of audio modulation, the discriminator produces one polarity of voltage proportional to the frequency deviation.

On the other half-cycle of audio, when the FM signal swings below the center frequency, the discriminator produces the opposite polarity of output voltage. That amplitude also follows the frequency deviations, but below the center frequency.

In summary, then, the changing dc voltages at the output of the discriminator correspond to the audio-modulating voltage that produces the frequency changes in the transmitted FM signal. The rate of the voltage changes is the same as the rate of the frequency swings, which is the audio-modulating frequency.

It does not matter which polarity of output voltage is produced by a frequency deviation above or below center. The only requirement is that the discriminator produce opposing polarities at the output and input.

Note that the discriminator diodes are balanced at the center frequency for any amount of input signal. If, for example, both diodes have 7 V, the output is still zero. At frequencies off center, though, the discriminator is not balanced. Then interfering amplitude changes in the IF signal are reproduced as changes in audio voltage. Therefore, a discriminator circuit is usually preceded by a limiter stage. The limiter has the function of eliminating any AM interference in the FM signal coupled to the discriminator.

Test Point Questions 17-5
(Answers on Page 412)

Refer to the discriminator circuit shown in Fig. 17-8.

a. What is the resonant frequency for the coupling transformer T_1?

b. Which diode produces a positive output voltage?

c. Which capacitor is the IF bypass for R_2?

d. If V_1 is 4.5 V and V_2 is 3.5 V, what is the output voltage at point A

17-6
QUADRATURE PHASE IN DISCRIMINATOR TRANSFORMER

The two diodes in a discriminator circuit are only half-wave rectifiers. They cannot discriminate between different frequencies. It is the input coupling circuit that proportions the amount of voltage for the diodes according to the instantaneous frequency of the FM signal. The phase angle of the tuned transformer is 90° at center frequency, but it varies at higher and lower frequencies.

Referring again to the discriminator shown in Fig. 17-8, note that the IF input signal is coupled two ways. One is by induction across the transformer from L_P to L_S. The other is through capacitive coupling by C_C from the primary to the center tap on the secondary. That path uses R_3 as a load resistor to prevent shorting the primary signal through the IF bypass C_2.

The induced voltage across L_S is applied to the diode anodes in push-pull because of the center tap. The capacitively coupled voltage is the primary signal, however. It is connected to the two diode anodes in parallel, not push-pull. Therefore, the IF signal voltage for each diode has two components:

1. Push-pull secondary voltages from opposite ends of L_S. Each diode has one-half the induced secondary voltage, of opposite polarity.
2. Parallel primary voltage from the center tap on L_S. Both diodes have the same primary voltage.

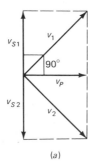 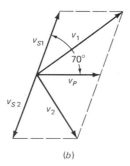 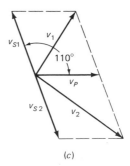

<table>
<tr><td>(a)</td><td>(b)</td><td>(c)</td></tr>
</table>

Fig. 17-10. Phase-angle diagrams for the center-tuned discriminator coupling transformer shown in Fig. 17-8. (a) Quadrature phase between v_S and v_P for resonance at the center frequency. V_1 and V_2 are equal for both diodes. (b) Above resonance, the phase angle is 70°; then V_1 is more than V_2. (c) Below resonance, the phase angle is 110°; then V_2 is more than V_1.

The signal input for the diodes is the resultant of those two voltages. The value of the resultant voltage is dependent upon the phase angle between them.

Quadrature Phase at Center Frequency The secondary voltage v_S is 90° out of phase with the primary voltage v_P at 10.7 MHz. That phase relation results because the transformer is resonant.

Phase-Angle Variations The phase angle varies around 90° as the FM signal goes above and below the center frequency. The change in phase angle between v_S and v_P changes the proportion of signal voltages for the diode. Either $D1$ or $D2$ then has more input than the other. The phasor diagrams are shown in Fig. 17-10. Note that the arrow for v_P, as the reference phasor, is horizontal.

Diode Voltages at Center Frequency. When diode voltages are at the center frequencys v_S and v_P are 90° out of phase; see Fig. 17-10a. Note that the center tap divides v_S into two equal voltages 180° out of phase. The v_{S1} for diode 1 is shown leading v_P by 90°, whereas v_{S2} for diode 2 lags v_P by 90°.

The voltage V_1 for $D1$ is the resultant of its two components v_{S1} and v_P. Similarly, the combined voltage V_2 for $D2$ is the resultant of v_{S2} and v_P. V_1 and V_2 are equal in this case because of the 90° phase angle between v_P and v_{S1} and between v_P and v_{S2}.

With equal amounts of signal voltage V_1 and V_2 applied to them, the two diodes have the same amount of rectified output. The net output from the balanced output circuit is zero then at the audio-takeoff point.

Diode Voltages Above Center Frequency In Fig. 17-10b the IF signal is above the center frequency. For example, v_P is shown 70°, out of phase with v_{S1}. The reason for the change in phase angle is that L_S is not resonant at a frequency significantly higher than 10.7 MHz (in this example, 40 kHz higher). The new phase angle of 70° shown in Fig. 17-10b makes v_P less out of phase with v_{S1}. Two phasors combine to produce a higher resultant value. At the same time, the phase angle between v_P and v_{S2} increases.

As a result, the combined voltage V_1 increases and V_2 decreases. More signal is applied to $D1$ and less to $D2$. In the discriminator circuit shown in Fig. 17-8, the net output voltage then is positive at point A.

Diode Voltages Below Center Frequency In Fig. 17-10c the phase angle swings in the opposite direction from that in Fig. 17-10b as the signal goes 40 kHz below center instead of above. The v_P is shown 110° out of phase with v_{S1} for diode 1; the v_P for diode 2 is 70° out of phase with v_{S2}. Either way, v_P is closer to being in phase with v_{S2} for diode 2. Then V_2 is greater than V_1. As a result, $D2$ produces more rectified output than $D1$. The net output voltage then is negative at point A in Fig. 17-8.

Test Point Questions 17-6
Answers on Page 412)

Answer True or False.

a. The phase angle between secondary and primary voltages is 90° at the center frequency because of resonance in the discriminator transformer.

b. In Fig. 17-8, R_3 is the diode load resistor for D2.

c. The primary signal v_P in Fig. 17-10 is coupled to the secondary by C_C in Fig. 17-8.

17-7
RATIO DETECTOR

The ratio detector circuit is probably the most common detector in FM receivers because it also provides a limiting effect. The ratio detector has a stabilized output voltage that is insensitive to AM in the FM signal.

Equivalent Circuit As shown in Fig. 17-11, the ratio detector circuit is almost like a discriminator, but the ratio detector has the two diodes in series. Note that D1 has the IF signal applied to its cathode and D2 has the signal at the anode. Further-

more, the audio output terminals A–B are at the junction of the two capacitors C_1 and C_2. However, the ratio detector and discriminator have the same type of S-response curve that is the characteristic of a balanced FM detector circuit.

In Fig. 17-11a the input transformer T_1 has the same function as in the discriminator. Both the primary and secondary tuned circuits are resonant at the IF center frequency. The secondary is center-tapped to produce equal voltages of opposite polarity for D1 and D2, and the primary signal voltage is applied in parallel to both diodes. T_1 is a phase-shift transformer to proportion the signal for the two diodes in accordance with the instantaneous frequency of the FM signal, just as in the discriminator. The transformer is shown in its equivalent form with L_M indicating the mutual induction between L_S and L_P.

In the ratio detector, one diode is reversed so that the two half-wave rectifiers are in series. The purpose is to have the diodes in series across the stabilizing voltage V_3 for the output side.

Another difference between the ratio detector and the discriminator is that the audio output from the detector is taken from point A at the junction of C_1 and C_2. The reason is that, because of the stabilizing voltage, no audio voltage exists across C_1 and C_2 in series. The potential difference between terminal A and point B at the center tap of the stabilizing voltage divider provides the audio output.

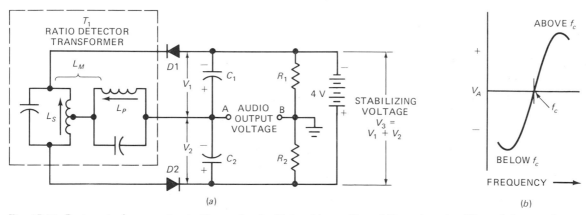

Fig. 17-11. Basic ratio detector circuit. T_1 couples the IF signal input. D_1 and D_2 are in series. The stabilizing voltage is across both diodes. The audio output is taken across points AB at the junction of C_1 and C_2.

The Stabilizing Voltage The stabilizing voltage V_3 is equal to the sum of the diode voltages V_1 and V_2. To make the ratio detector insensitive to AM interference, the total voltage V_3 must be stabilized so that it cannot vary at the audio-frequency rate. Then the audio output is produced at point A only when the ratio between V_1 and V_2 changes. That is why the circuit is called a ratio detector. The total voltage V_3 remains fixed by the stabilizing voltage source. A battery is shown here for V_3 only as the equivalent of the stabilized voltage for a ratio detector. Actually an RC circuit with a long time constant is used for the stabilizing voltage.

Audio Output Signal As each diode conducts, it produces the rectified output voltage V_1 or V_2 across C_1 or C_2. At the center frequency, the input transformer proportions the IF signal voltage equally for the two diodes, and the result is equal voltages across C_1 and C_2. The output voltage at the audio takeoff point A is zero, therefore, since V_1 and V_2 are equal. Note that those two voltages have opposite polarity at point A with respect to chassis ground.

When the FM signal input is above the center frequency, however, we can assume D1 has more IF signal input than D2. Then the rectified diode voltage V_1 is greater than V_2. That makes point A more positive and produces an audio output voltage of positive polarity.

Below the center frequency, D2 has more IF signal input and V_2 is greater than V_1. The result is an audio output voltage of negative polarity at point A. The response is illustrated by the ratio detector S curve in Fig. 17-11b, which is essentially the same as the discriminator response curve.

Signal Voltages An important characteristic of the ratio detector output circuit can be illustrated by numerical examples. Assume that the frequency deviation above the center frequency increases V_1 by 1 V. That makes point A more positive. At the same time, V_2 decreases by 1 V. That makes point A less negative by 1 V, which is the same as making it 1 V more positive. The two rectified diode volt-ages V_1 and V_2 thus produce the same voltage change of 1 V in the positive direction at point A. Since the audio signal is taken from point A, the amount of output is the same as though only one diode were supplying audio voltage corresponding to the frequency variations in the FM signal.

As a result, the audio voltage output of a ratio detector is one-half the output of a discriminator, where the audio signal voltages from the two diodes are combined in series with each other at the audio takeoff point. The output in the ratio detector must be taken from the junction of the two diode loads because there is no audio signal voltage across the stabilizing voltage source.

Typical Circuit If a battery were used for the stabilizing voltage, the diodes would operate only with a signal at least great enough to overcome the battery bias on each diode. A large capacitor such as C_3 in Fig. 17-12 is used instead. C_3 charges through the two diodes in series and automatically provides the desired amount of stabilizing voltage for the IF signal level. The 5-μF capacitance of C_3, generally called the *stabilizing capacitor*, is large enough to prevent the stabilizing voltage from varying at the audio-frequency rate. A discharge time constant of about 0.1 s is required; it is provided by R_1 in series with R_2 across C_3.

The tertiary winding L_t of the ratio detector transformer is used to couple the primary signal voltage in parallel with the two diodes. L_1 is wound directly over the primary winding for very close coupling, so that the phase of the primary signal voltage across L_p and L_t is practically the same. The construction of a ratio detector transformer with the tertiary winding is illustrated in Fig. 17-13. The arrangement with the tertiary winding is commonly used with the ratio detector circuit in order to match the high-impedance primary and the relatively low-impedance secondary, which is loaded by conduction in the diodes. The resistor R_3 in series with L_1 limits the peak diode current.

Because of the 90° phase difference between the voltages across the secondary and across L_1 at resonance, the circuit provides detection of the FM

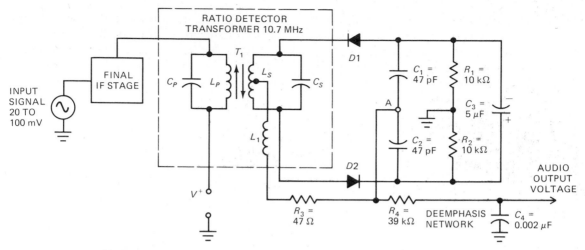

Fig. 17-12. Typical balanced ratio detector circuit. The 5-μF C_3 produces the required dc stabilizing voltage. The audio output is taken from point A at the junction of C_1 and C_2.

signal in the same manner as the phase-shift discriminator. Both the primary and secondary of the ratio detector transformer are tuned to the IF center frequency. The audio output voltage of the ratio detector is taken from the junction of the diode load capacitors C_1 and C_2, in series with the audio deemphasis network, to supply the desired audio signal for the first audio amplifier.

It should be noted that the amount of stabilizing voltage changes with the level of the input signal. Therefore, the stabilizing voltage can be used as a source of AVC bias in the receiver.

Single-ended Ratio Detector In Fig. 17-14 the ratio detector circuit is shown with an unbalanced output circuit. The input circuit, which is shown in its equivalent form, is the same as in the balanced circuit. However, only the one diode load capacitor C_A is used in the audio output circuit. Also, the stabilizing voltage across R_1 and C_S is not center-tapped. It is not necessary to ground the center point of the stabilizing voltage source.

The only effect of the ground connection in Fig. 17-14 is to change the dc level of the audio output signal at A with respect to chassis ground. The same audio signal variations are obtained, but around a dc voltage axis equal to one-half the stabilizing voltage. However, the audio output al-

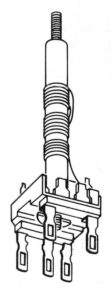

Fig. 17-13. Coil windings on ratio detector transformer. Height is 2 in.

ways has a zero dc level with respect to the midpoint of the stabilizing source.

Only the one diode load capacitor C_A is necessary in the output circuit to obtain the audio signal, because it serves as the load for both diodes. Each diode charges C_A in proportion to the IF signal input for each rectifier, but in opposite po-

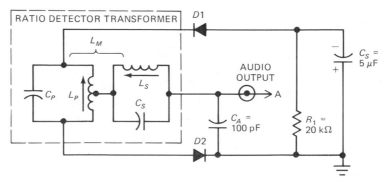

Fig. 17-14. Ratio detector with a single-ended or unbalanced output circuit.

larity. Therefore, the voltage across C_A is the same audio output signal as in the balanced ratio detector. The capacitance of C_A is doubled in the single-ended circuit because it replaces two capacitors that are effectively in parallel for the signal voltage in the balanced circuit.

Audio Deemphasis Network As explained in Chap. 12, the FM signal is transmitted with preemphasis for high audio frequencies in order to improve the signal-to-noise ratio. The time constant for the preemphasis network is 75 μs. At the receiver, equal deemphasis is necessary to restore the audio frequencies to their original relative amplitudes. The deemphasis network is usually in the output of the FM detector. In Fig. 17-12, R_4 and C_4 form the deemphasis network. Its time constant is 78 μs. This RC filter reduces the response for high audio frequencies because of less reactance for C_4.

Test Point Questions 17-7
(Answers on Page 412)

Refer to the ratio detector shown in Fig. 17-12.

a What is the value of the stabilizing capacitor?
b. Does terminal A have an audio output or a stabilizing voltage?
c. Which coil couples the primary signal to the two diodes?
d. Are $D1$ and $D2$ in series or in parallel to charge stabilizing capacitor C_3?

17-8
REQUIREMENTS OF FM DETECTION

The two main factors in the operation of the discriminator or ratio detector are balance at the center frequency and proper quadrature phase. Balance with zero output is useful because any amplitude changes at the input will have no effect on the output at the center frequency. Proper quadrature phase is important because the phase angle varies around 90° as the FM signal swings around the center frequency. This is the reason why a center-tuned detector can discriminate between different frequencies.

The discriminator balanced circuit is also used to indicate a change in the IF center frequency. Positive or negative dc control voltage is available when the center frequency is too high or too low. This is used for automatic frequency control (AFC) on the local oscillator for RF tuning. An AFC circuit with a varactor, or capacitive diode, is discussed in Chap. 11. The dc control voltage from the discriminator determines the capacitance of the varactor, which varies the oscillator frequency.

The principle of quadrature phase can also be applied in FM detector circuits by using an amplifier with either a transistor or vacuum tube. The *quadrature-grid detector* is an example. In this circuit, the last IF amplifier uses resonant circuits 90° out of phase.

It should be noted that the circuits of the IF amplifier and detector in an FM receiver also apply to the IF section of a television receiver for the associated FM sound signal. However, there are two differences. The IF center frequency is 4.5 MHz instead of 10.7 MHz. Also, the maximum frequency deviation is 25 instead of 75 kHz.

A special problem with the FM sound signal in television receivers is interference with the picture, even though the picture signal is AM. Any AM receiver can detect an FM signal by mistuning. That is why an interfering FM signal can be heard in an AM radio. The reason is *slope detection*, in which the signal frequency is on the sloping side instead of the center of the IF response curve. A sloping response means different amounts of gain at different frequencies. Then the frequency variations are converted to amplitude variations, which can be detected by a diode rectifier.

Test Point Questions 17-8
(Answers on Page 412)

Answer True or False.

a. A quadrature type of FM detector is tuned to the center frequency.
b. An FM signal can interfere with an AM radio signal because of slope detection.

17-9
DISCRIMINATOR BALANCE AND IF ALIGNMENT

The two requirements for balancing the discriminator and aligning the IF are:

1. The IF amplifier transformers must be tuned from the mixer output to the limiter input, for maximum signal at 10.7 MHz.
2. The discriminator transformer must be tuned for zero balance at 10.7 MHz and a symmetrical output above and below the center frequency.

Those circuits include all the transformers tuned to 10.7 MHz. They must be aligned at the same frequency for maximum gain and selectivity in the receiver.

Visual Response Curves In the sweep-frequency method of alignment, a visual response curve is produced on the oscilloscope screen to indicate correct tuning. The IF amplifier response should look like Fig. 17-4a with 10.7 MHz at the center and the required amount of bandwidth. The response of the discriminator is an S curve as shown in Fig. 17-9b or c. The crossover is at 10.7 MHz, and the curve has the required bandwidth for symmetrical peaks above and below the center frequency.

Steady-Frequency Method A conventional RF signal generator is used for an output at 10.7 MHz without any modulation. For a single frequency as the input signal, the rectified output is a steady dc voltage, not an audio signal. A dc voltmeter can be used for the output indicator, therefore, to show maximum or minimum signal.

The IF amplifiers are tuned for maximum output at 10.7 MHz. There is no indication of bandwidth on the voltmeter. However, the bandwidth will generally be normal when the transformer is tuned to the correct frequency.

The discriminator can be aligned with a dc voltmeter because the indications are so definite. The voltmeter used should have a sensitivity of at least 20,000 Ω/V to prevent detuning the discriminator transformer. The primary and secondary are tuned to the center frequency.

Tuning the Discriminator A signal generator provides a signal at the constant frequency of 10.7 MHz. The dc voltmeter is connected to the audio takeoff point and ground. Then, to tune the discriminator,

1. Tune the primary of the discriminator transformer for maximum output.
2. Tune the secondary for a *sharp* drop to zero.

The dc voltmeter is at the audio takeoff point for both adjustments.

It should be possible to produce either a positive or negative output voltage when tuning the secondary. Therefore, tune for a zero indication at the balance point, where the meter starts to swing from one polarity to the other. A gradual decrease to zero is the wrong indication. This reading means only that the circuit is being detuned away from resonance at the center frequency.

When the signal generator frequency is varied manually above and below 10.7 MHz, the dc output voltage should vary from zero to a maximum with opposite polarities at both sides of the center frequency. The polarity above and below center does not matter. However, the response should be symmetrical, with equal output voltages for the same amount of frequency change. The two points of maximum output voltage correspond to the two peaks on the discriminator S-response curve. They should have the required frequency separation.

When the discriminator response is not equal on both sides of the center frequency, try readjusting the transformer primary. This adjustment affects linearity and symmetry. The secondary adjustment determines the crossover frequency.

With an FM signal from a radio broadcast station as input, approximate adjustments can be made. Use the weakest possible signal. Adjust the secondary for minimum background noise and minimum distortion. Adjust the primary for maximum signal.

Test Point Questions 17-9
(Answers on Page 412)

In the steady-frequency method of tuning a discriminator to 10.7 MHz:

a. Is the primary or the secondary adjusted for zero balance?

b. Is the primary or the secondary adjusted for maximum output?

c. If the output is +4 V at 10.8 MHz, What should it be at 10.6 MHz?

17-10
RATIO DETECTOR BALANCE AND IF ALIGNMENT

The procedure for balancing the ratio detector differs from alignment with a discriminator in two ways:

1. The overall IF response, from the mixer output to the detector output, can be checked in one step. Measurements are made across the stabilizing capacitor in the output of the ratio detector.

2. For the zero balance adjustment of the ratio detector, the output indicator must be moved to the audio takeoff point. There is no detected audio signal voltage across the stabilizing capacitor.

This IF alignment is probably easier because all the IF circuits, including the primary of the ratio detector transformer, are tuned to 10.7 MHz for maximum stabilizing voltage.

For the sweep-frequency method, though, it is important to note that the IF response curve as viewed on the oscilloscope can be obtained only with the stabilizing capacitor disconnected temporarily. The capacitor is big enough to bypass the 60-Hz input to the oscilloscope needed for the response curve. However, that precaution does not apply to the balance adjustment for the ratio detector, because the S-response curve is taken from the audio takeoff point.

For the steady-frequency method with a dc voltmeter, the ratio detector transformer is tuned to 10.7 MHz as follows:

1. Tune the primary for maximum dc voltage across the stabilizing capacitor.

2. Move the dc voltmeter to the audio takeoff point.

3. Tune the secondary for balance, as with a discriminator.

When readjustments are necessary, the secondary determines the crossover frequency for balance. The primary is adjusted for linearity and symmetry.

In the case of a single-ended ratio detector circuit, as in Fig. 17-14, the balance point is not zero; it is equal to one-half the stabilizing voltage. Therefore, usual practice is to insert balancing resistors temporarily, just for the alignment. As shown in Fig. 17-15, R_2 and R_3 convert the output to a balanced output circuit. The voltmeter is connected from the audio takeoff point to the junction of the two resistors. Then the secondary of the ratio detector transformer can be aligned for zero balance at the center frequency.

(Answers on Page 412)

Test Point Questions 17-10
(Answers on Page 412)

In the steady-frequency method of aligning a ratio detector transformer:

a. Should L_P or L_S be adjusted with the dc voltmeter across the stabilizing capacitor?

b. Should L_P or L_S be adjusted with the dc voltmeter at the audio takeoff point?

c. Is balance at the center frequency set by tuning L_P or L_S?

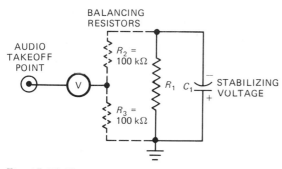

Fig. 17-15. Temporary resistor connections to convert a single-ended ratio detector to a balanced circuit for alignment.

17-11
AN FM-AM RECEIVER

Figure 17-16 shows the block diagram of the RF and IF circuits for a small FM-AM radio. The audio circuits are not shown, but they include first audio amplifier, driver, and push-pull power output stage. The receiver is for monophonic sound. The additional circuits needed for FM stereo reception are explained in Secs. 17-12 and 17-13.

Consider the operation for an FM signal. The antenna signal is coupled to the RF amplifier $Q1$. The output goes to $Q2$ as the frequency converter for the FM signal. These two stages operate only on the FM position of the selector switch S_{1A}. Also, S_{1B} connects the output of $Q2$ to the input of $Q3$.

The IF output at 10.7 MHz is amplified in three stages: $Q3$, $Q4$, and $Q5$. Then the ratio detector with $D1$ and $D2$ converts the FM signal to audio output that is connected to the volume control R_1 through S_{1C}.

For an AM signal, a loopstick antenna is connected to $Q3$ through S_{1B}. $Q1$ and $Q2$ are now disabled on AM operation. Also, $Q3$ operates as an AM frequency converter to produce IF output at 455 kHz. $Q4$ and $Q5$ are tuned to 455 kHz as IF amplifiers for the AM signal. The circuit of an IF amplifier for either 455 kHz or 10.7 MHz is

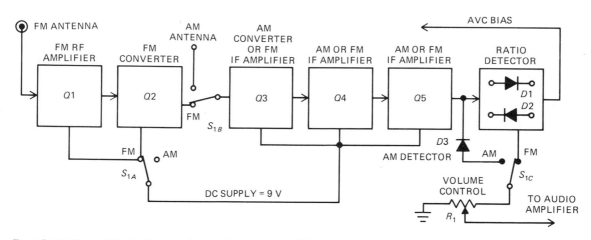

Fig. 17-16. Typical block diagram of a small monophonic FM-AM receiver. Q1 and Q2 operate on FM only. Audio amplifier circuits are not shown.

shown in Fig. 17-17. Finally, the AM signal at 455 kHz is rectified by the diode detector D1. It provides an audio signal for the volume control through S_{1C}.

Note that the ratio detector output is also shown supplying AVC bias to the RF and IF stages for automatic control. The dc bias comes from the stabilizing voltage, which indicates the signal level.

FM-AM Selector Switch The functions of the three sections of S_1 can be summarized as follows:

S_{1A} Connects the dc supply voltage to Q1 and Q2 only on FM. Those two stages can not operate on AM, because they lack a dc supply voltage when the switch is in the FM position.

S_{1B} Connects the FM signal output of Q2 to Q3 on FM when Q1 and Q2 are operating. However, that path is disconnected on AM and the AM antenna is connected to Q3 instead. Then Q3 serves as the converter for the AM signal.

S_{1C} Connects the audio output of the ratio detector to the volume control for an audio signal on FM operation. For AM operation the audio output of the AM detector is used. The volume control sets the amount of audio signal coupled to the audio amplifier.

Note that more amplifiers are used on FM than AM. The reason is that the 88- to 108-MHz FM broadcast range is in the VHF band of 30 to 300 MHz. At those frequencies the signal strength is usually much less than for the AM radio band of 535 to 1605 kHz. On FM operation the amplifiers are Q1, Q3, Q4, and Q5 and converter Q2 to feed the ratio detector. On AM converter Q3, and amplifiers Q4 and Q5 feed the diode detector D3.

Dual IF Amplifier for 10.7 MHz or 455 kHz
The two IF coupling transformers shown in Fig. 17-17 are used in the collector output circuit. They are in series with T_1 tuned to 10.7 MHz and T_2 for 455 kHz. Either IF signal can be amplified. The

collector circuit is returned to chassis ground, with a negative voltage at the emitter. The purpose is to keep the dc supply voltage off the tuned coils in the collector circuit.

When the input signal is 10.7 MHz, T_1 is resonant and serves as the IF interstage coupling transformer. The low side of the tuned circuit is effectively bypassed to chassis ground by C_2. That capacitor can tune L_2 to 455 kHz, but at the much higher frequency of 10.7 MHz, C_2 has very little reactance.

When the input signal is 455 kHz, T_2 is resonant to serve as the IF interstage coupling transformer. Then L_1 for 10.7 MHz has very little reactance at the much lower 455 kHz frequency.

Both T_1 and T_2 are step-down transformers from the collector output to the following base input circuit. At 10.7 MHz, L_3 has the output signal and L_4 serves effectively as an RF choke. At 455 kHz, L_4 has the output signal and L_3 has little reactance.

Test Point Questions 17-11
(Answers on Page 412)

a. In Fig. 17-16, which switch connects the FM and AM audio outputs to the volume control?
b. In Fig. 17-16, which two transistors are not used on AM?
c. In Fig. 17-17, which IF transformer is used for the FM signal?

17-12
STEREO FM BROADCASTING

The system of transmitting two separate audio signals for stereophonic sound was approved by the Federal Communications Commission (FCC) in 1961. Now practically all stations in the FM radio band of 88 to 108 MHz broadcast in stereo. The technique is based on multiplexing a 38-kHz subcarrier signal with the main RF carrier of the station. (The process of multiplexing multiple signals on one RF carrier is described in Chap. 12.) The modulation of the 38-kHz subcarrier is the additional audio signal needed for stereo.

Details of the stereo broadcast system are illustrated in Fig. 17-18 in terms of the modulating signals for the main RF carrier. All the signals produce frequency modulation of the transmitted RF carrier. Actually, the upper sidebands of the RF carrier are represented here, but the lower sidebands also are transmitted. The main point about this system is that the 38-kHz subcarrier signal is used for stereo sound.

For the sake of compatibility, the left and right audio signals are not transmitted in their original form. Instead, they are converted to left plus right (L + R) and left minus right (L − R) audio signals. The (L + R) means combining the original left and right audio signals when both are in phase. For (L − R), the left audio signal is combined with the right audio signal when both are out of phase by 180°. The circuits that combine the signals are called *matrix* networks.

The reason for the above conversion is to have the (L + R) audio signal, which can be used in monophonic receivers. The signal has all the audio information, but without any stereo effect. If the left and right signals were transmitted individually for stereo, a monophonic receiver would have only one without the other.

A stereo receiver uses the (L − R) signal in addition to (L + R). Those two signals can be matrixed, or combined, to provide the original left and right channels of audio signal. The combinations can be illustrated by using algebra, as follows:

$$(L + R) + (L − R) = 2L$$
$$(L + R) − (L − R) = 2R$$

Subtracting the (L − R) signal only means inverting the signal and adding. In one case the right signals cancel to leave only the left signal. In the other case the left signals cancel to leave only the right signal. The factor of 2 is only a relative amplitude. The left and right audio signals can then

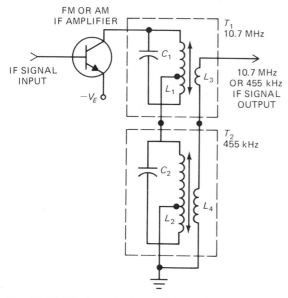

Fig. 17-17. Dual coupling transformers for the FM-AM receiver using a common amplifier for both IF signals. Base bias networks are not shown.

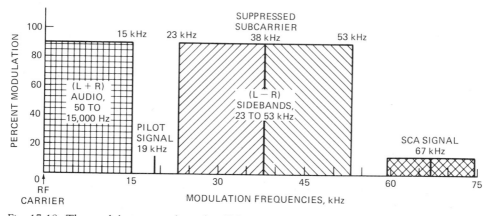

Fig. 17-18. The modulating signals used in FM stereo multiplexing. For 100 percent modulation, the frequency deviation of the RF carrier is 75 kHz.

be used to reproduce the stereophonic sound with two separate audio amplifier channels.

The (L − R) signal for stereo is transmitted on a 38-kHz subcarrier. That way it can be used in stereo receivers. The receivers separate the signal after detection of the main carrier signal, with resonant circuits tuned to 38 kHz.

The 38-kHz subcarrier is amplitude-modulated by the (L − R) signal. Double sidebands are produced, but the carrier is suppressed. The purpose is to minimize interference.

Since the subcarrier frequency is suppressed in transmission, the stereo receiver must reinsert 38 kHz for detection of the (L − R) modulation. For that reason the station broadcasts a 19-kHz pilot signal for the receiver circuits that detect the multiplexed stereo signal. The 19 kHz is doubled to provide 38 kHz as the reinserted subcarrier frequency.

It is not possible to use a 38-kHz pilot signal directly because it would be too difficult to separate the subcarrier frequency from the sideband frequencies in the modulated (L − R) signal. The 19 kHz can easily be doubled to 38 kHz in the receiver for the demodulator that detects the stereo signal.

It should be noted that 19 and 38 kHz are above the 50- to 15,000-Hz range of the audio circuits in the detector output of a mono receiver. Therefore, the stereo signal has no effect. In a stereo receiver, however, the multiplex circuits after the audio detector are tuned to the frequencies needed for the stereo signal.

At the right in Fig. 17-18 is the SCA signal; SCA is the abbreviation for Subsidiary Communications Authorization. The service, also called *storecasting*, allows broadcasting of background music for commercial use in addition to other special programming. To recover the SCA signal, multiplex circuits for 67 kHz are necessary. The bandwidth of the SCA signal is 59.5 to 74.5 kHz.

The 19-kHz pilot signal is also used in some receivers to energize an indicator lamp that shows the station is broadcasting in stereo. In monophonic broadcasting, there is no 19-kHz pilot signal and the indicator lamp does not light.

The functions of all the signals in FM stereo broadcasting can be summarized as follows:

(L + R)	Audio signal for monophonic and stereo receivers. AF range is 50 to 15,000 Hz
(L − R)	Audio signal for stereo receivers only. AF range is 50 to 15,000 Hz. Transmitted as AM sidebands of the 38-kHz subcarrier. These bands are 23 to 38 kHz and 38 to 53 kHz, or 38 ± 15 kHz.
38-kHz subcarrier	Has modulation sidebands of (L − R) signal. The 38 kHz is suppressed.
19-kHz pilot signal	Transmitted at one-half the 38-kHz subcarrier frequency to be doubled at the receiver.

Test Point Questions 17-12
(Answers on Page 412)

Give the frequencies of the following:

a. AF range of (L + R) signal
b. Stereo subcarrier
c. (L − R) sidebands of the stereo subcarrier
d. Pilot signal

17-13
STEREO FM RECEIVERS

The block diagram in Fig. 17-19 illustrates how the receiver recovers the left and right audio signals for stereo sound. Note that the receiver is divided into three main parts along the top of the diagram. The section at the left is the main part of the radio receiver for the RF carrier signal broadcast by the FM station. Included are the RF and IF circuits with an FM detector. The detector output has the multiplexed signals needed for stereo. They are amplified in the first audio stage, though, to provide a higher signal level for operation of the stereo

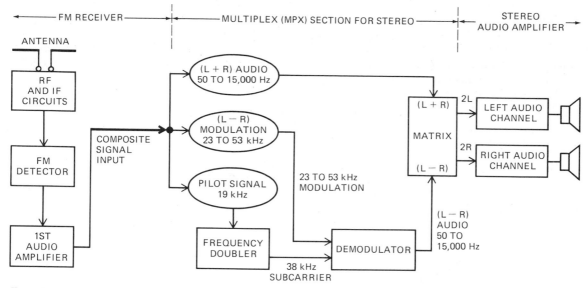

Fig. 17-19. Block diagram of the main circuits in a stereo FM receiver.

multiplex section. The input to the multiplex section is labeled *composite signal* because it includes the components needed for stereo.

In the multiplex (MPX) part of the receiver, consider first the matrix circuit in the output. It needs (L + R) and (L − R) audio signals for its input. They are combined to provide separate left and right audio signals for the separate audio amplifier channels. The (L + R) signal is present as an audio signal into the multiplex section. This signal is coupled directly to the matrix. However, the (L − R) signal is transmitted as the modulation sidebands of the 38-kHz stereo subcarrier. The sideband frequencies for the (L − R) signal have the frequency range of 23 to 53 kHz. Those frequencies must be converted to the 50 to 15,000 Hz range for (L − R) as an audio signal. The stereo demodulator has the function of providing the (L − R) audio signal to the matrix.

The stereo demodulator needs two inputs. One is the (L − R) modulation, and the other is the reinserted 38-kHz subcarrier. Remember that the subcarrier is suppressed in the transmitted signal. However, the 19-kHz pilot signal is a sample of the subcarrier at one-half the required frequency. Therefore, the pilot signal is doubled to 38 kHz and coupled to the stereo demodulator. Then the

modulation frequencies of 23 to 53 kHz beat with the subcarrier frequency of 38 kHz to recover the (L − R) audio signal of 50 to 15,000 Hz.

As a result, the matrix circuit has (L − R) audio input from the stereo demodulator and (L + R) audio input as a direct signal into the multiplex section. The matrix circuit must add those two signals for one output and subtract them for the other output.

When (L + R) and (L − R) are added, the R cancels, leaving just 2L. That output is the left audio signal; it has twice the amplitude of the input but without the right audio signal.

When (L + R) and (L − R) are subtracted, the L cancels but the output has 2R. That output is the right audio signal at twice the amplitude of the input but without the left audio signal.

Now the separate left and right audio signals can be coupled to two separate audio amplifiers. The loudspeakers reproduce audio signals for stereo sound.

To summarize the operation, the multiplex section has two functions: demodulation and matrixing. The demodulator recovers (L − R) audio from the modulation sidebands of the 38-kHz subcarrier. The matrix combines (L − R) and (L + R) audio signals to provide left and right audio signals.

Deemphasis In FM radio broadcasting, the audio signal is preemphasized to increase the signal-to-noise ratio for high audio frequencies. At the receiver the audio signal must be deemphasized. In stereo receivers, however, the deemphasis cannot be in the output of the FM detector. If it were, the 38-kHz stereo signal would be filtered out. Two deemphasis networks are used for the separate left and right audio channels in the output of the matrix circuit.

Filters With all the different frequencies in the multiplex section, it is necessary to separate the signals for each function without interference between them. The filters use LC tuned circuits.

Referring to the diagram in Fig. 17-19, each path for the three signals into the multiplex section needs a filter. At the top a low-pass filter is inserted for the (L + R) audio signal to cut off at 15 kHz. Then that path has the (L + R) audio signal of 50 to 15,000 Hz without any of the higher frequencies for the stereo signal.

In the middle path a bandpass filter for 23 to 53 kHz can be used for the (L − R) modulation

frequencies. Then there is no interference from the (L + R) audio signal.

In the bottom path a resonant circuit tuned to 19 kHz must be used to pick off the pilot signal. That input is used for the frequency doubler, which has an output circuit tuned to 38 kHz.

In addition, an SCA filter can be used to reject 59.5 to 74.5 kHz. The SCA signal is not used in FM radio receivers.

IC Package for Stereo Multiplex Circuits

The integrated circuit shown in Fig. 17-20 is a 14-pin DIP package. It includes the functions for subcarrier regeneration, stereo demodulator, and matrix circuits. Regeneration means producing the 38-kHz subcarrier for the demodulator. The composite signal that is needed for the input at pin 3 includes:

1. (L + R) audio at 50 to 15,000 Hz
2. (L − R) stereo modulation at 23 to 53 kHz
3. 19-kHz pilot signal

Note the tuned circuits at 19 and 38 kHz for producing the suppressed subcarrier needed for the

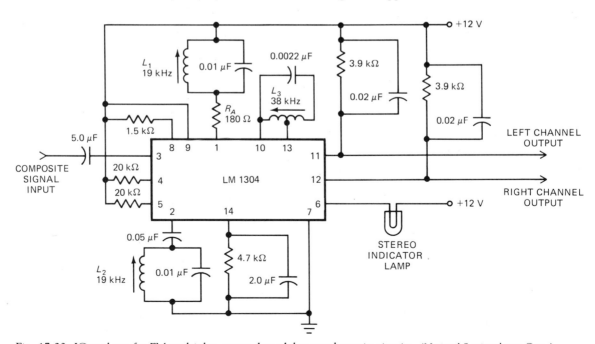

Fig. 17-20. IC package for FM multiplex stereo demodulator and matrix circuits. (*National Semiconductor Corp.*)

stereo demodulator. Those *LC* components are outside the IC package.

The stereo output includes:

1. The audio signal for the left audio channel at pin 11
2. The audio signal for the right audio channel at pin 12

Two *RC* networks are used for deemphasis of the audio outputs. The time constant for each is 78 μs, with 3.9 kΩ and 0.02 μF.

The dc supply of 12 V is connected to pin 6 through the stereo indicator. The lamp lights when the dc load current increases with a stereo signal. R_A at pin 1 determines the lamp sensitivity.

Test Point Questions 17-13
(Answers on Page 412)

Refer to Fig. 17-19.

a. Which block has the (L + R) and (L − R) audio signals for input?
b. To which stage is the (L − R) modulation coupled?
c. To which stage is the 19-kHz pilot signal coupled?
d. Which stage amplifies the composite stereo signal output from the FM detector?
e. What is the frequency of the output from the block labeled frequency doubler?

SUMMARY

1. Like an AM radio, an FM receiver is a superheterodyne; but the RF tuning is for 88 to 108 MHz, the IF value is 10.7 MHz with a bandwidth of 200 kHz for ± 75 kHz deviation, and an FM detector is used to recover the audio modulation.
2. Amplitude limiting in the IF section eliminates AM interference from the FM signal.
3. The discriminator is a basic FM detector circuit. Two diodes are used with a balanced output circuit. The input transformer is tuned to the center frequency.
4. The ratio detector is a balanced FM detector, like the discriminator, but the two diodes are in series.
5. The ratio detector is insensitive to AM interference because the stabilizing voltage across the output cannot change at the audio rate.
6. Both the discriminator and ratio detector have an S-response curve. It shows zero balance at the crossover for the center frequency with equal and opposite outputs above and below the center frequency.
7. For stereo FM broadcasting, the signals transmitted by multiplexing on one RF carrier are (L + R) audio, (L − R) modulation sidebands of a 38-kHz subcarrier, and a 19-kHz pilot signal. The encoding of the stereo signals is necessary for compatibility with monophonic receivers.
8. A stereo receiver has a multiplex section to decode the composite stereo signal out of the FM detector. The functions of the multiplex section are to (1) regenerate the 38-kHz subcarrier from the 19-kHz pilot signal, (2) demodulate the (L − R) sidebands of the 38-kHz subcarrier, and (3) matrix the (L + R) and (L − R) audio signals to produce left and right audio signals for the dual audio amplifiers.

SELF-EXAMINATION
(Answers at back of book)

Answer True or False.

1. The tuning range for the RF section in an FM radio is 88 to 108 MHz.
2. The IF value for the center frequency in FM radios is 4.5 MHz.
3. Maximum deviation is 75 kHz for 100 percent modulation of FM signals in the FM radio band.
4. A limiter rejects AM interference in the FM signal.
5. A discriminator is a balanced FM detector circuit.
6. Quadrature phase means 180°.
7. A ratio detector is like a discriminator, but it uses only one diode.
8. Stabilizing voltage is used for the ratio detector but not the discriminator.
9. The S-response curve is for the FM detector.
10. The part of the stereo signal used by monophonic receivers is (L −R) audio.
11. The stereo subcarrier frequency is 38 kHz.
12. The pilot signal for stereo is transmitted at 38 kHz.
13. The stereo demodulator output signal is (L + R) audio.
14. The matrix stereo output includes left and right audio signals.
15. An IF amplifier stage can be used for either 10.7 MHz or 455 kHz.

ESSAY QUESTIONS

1. Give two differences between AM and FM receivers.
2. Give the functions of the RF amplifier, mixer, and local oscillator stages in the RF tuner for an FM radio. Include typical frequencies for the input and output of each stage.
3. What is the function of the IF section in an FM radio?
4. Give the center frequency and maximum deviation of the IF signal in an FM radio.
5. What is the function of a limiter stage?
6. How does an FM detector differ from an AM detector?
7. Why is no dc supply voltage used for either the discriminator or ratio detector?
8. Why is the center-tuned discriminator a phase-shift detector?
9. Give two differences between the discriminator and the ratio detector circuits.
10. What is the function of the input transformer in the discriminator and ratio detector circuits?
11. What is the purpose of the stabilizing capacitor in a ratio detector? Give a typical value for this capacitor.
12. Compare response curves for the RF section, IF amplifier, and balanced FM detector. Mark the center frequency and edges of the passband.
13. What are two features of the S-response curve for the discriminator and ratio detector?

14. What are two features of the IF response curve?
15. Draw the diagram of a discriminator circuit and give the function of all components.
16. Draw the diagram of a balanced ratio detector circuit and give the function of all components.
17. Draw phasor diagrams for the two equal voltages with a phase angle of **a**. 90° **b**. 60° **c**. 120°. Show the resultant phasor for each example.
18. **a.** Why is deemphasis used in FM receivers? **b.** What is the required time constant for deemphasis?
19. What is meant by slope detection?
20. What is the function of a ceramic IF filter?
21. Describe briefly how to balance the discriminator and the ratio detector.
22. Refer to the block diagram of an FM-AM receiver in Fig. 17-16. Give the functions of all the stages.
23. How can one IF stage be used for both 10.7 MHz and 455 kHz?
24. Give the three components of the multiplexed stereo signal.
25. Why is the pilot signal 19 kHz and not 38 kHz?
26. Which audio signal is multiplexed on the 38-kHz subcarrier, $(L - R)$ or $(L + R)$? Why that one and not the other?
27. Name two signals that feed into the stereo demodulator and name the one output.
28. Give two signals that feed into the stereo matrix and name the two outputs.
29. What is meant by the SCA signal?
30. Refer to the RF tuner circuit in Fig. 17-2. Give the functions of R_4, R_7, C_{13}, and C_{16}.
31. Refer to the IF amplifier circuit in Fig. 17-3. Give the functions of T_1, T_2, T_3, $D1$ and $D2$.
32. Refer to the FM multiplex section in Fig. 17-20. Explain what makes the stereo indicator lamp go on.

PROBLEMS
(Answers to odd-numbered problems at back of book)

1. Calculate the range of local oscillator frequencies in an FM receiver for 88 to 108 MHz with the oscillator beating above the RF signal frequencies.
2. Calculate the ratio of highest to lowest frequency for the oscillator in Problem 1.
3. Calculate the RC time constant for deemphasis using the following values: **a.** R is 68 kΩ and C is 0.001 μF and **b.** R is 3.8 kΩ and C is 0.02 μF.
4. What R is needed with C of 0.01 μF for an RC time constant of 75 μs?
5. Calculate the RC time constant for stabilizing capacitor of 5 μF across 20 kΩ.
6. What value of C is needed for a time constant of 0.1 s with R of 20 kΩ?
7. What are the signal output frequencies for the stereo demodulator with 38 kHz beating against the $(L - R)$ modulation frequencies of 23 to 53 kHz?

SPECIAL QUESTIONS

1. In what ways do you think FM radio is better than AM?
2. In what ways do you think stereo FM is better than monophonic sound?
3. Why do you think FM is used for the sound signal in television, although AM is used for the picture signal?
4. Can you give any applications of multiplexing besides FM stereo broadcasting?
5. Why is amplitude limiting more important in FM receivers than in AM radio?
6. Show the diagram for a network of three resistors with two separate input signals and one added output signal.

ANSWERS TO TEST POINT QUESTIONS

17-1 a. 88 to 108 MHz
     b. 10.7 MHz
     c. FM
     d. Stereo

17-2 a. $Q3$
     b. C_{13}

17-3 a. T_1, T_2, and T_3
     b. 9 V
     c. 1 and 2

17-4 a. F
     b. T
     c. T

17-5 a. 10.7 MHz
     b. D1
     c. C_2
     d. 1 V

17-6 a. T
     b. F
     c. T

17-7 a. 5 μF
     b. Audio output
     c. L_1
     d. Series

17-8 a. T
     b. T

17-9 a. Secondary
     b. Primary
     c. -4 V

17-10 a. L_P
      b. L_S
      c. L_S

17-11 a. S_{1C}
      b. Q1 and Q2
      c. T_1

17-12 a. 50 to 15,000 Hz
      b. 38 kHz
      c. 23 to 53 kHz
      d. 19 kHz

17-13 a. Matrix
      b. Demodulator
      c. Doubler
      d. First audio
      e. 38 kHz.

Chapter 18
The Oscilloscope

An oscilloscope is one of the most important types of test equipment for checking electronic circuits because it can show the waveform of an applied voltage. The fluorescent screen of the cathode-ray tube (CRT), displays a graph of voltage amplitude variations with respect to time. The vertical axis represents voltage while the horizontal axis is a linear time base for the vertical signal.

With a sine-wave input, the oscilloscope shows sine waves on the screen. With an input of square waves or any other waveform, the screen pattern is a picture of the variations. The number of cycles displayed depends on the frequency of the input signal and a horizontal reference frequency. The oscilloscope not only can measure voltage, it can also be used to measure the frequency of its input signal.

Typical oscilloscopes have a frequency response from 0 Hz, for direct current, up to 40 MHz. RF and AF signals can be checked. Peak-to-peak amplitude measurements are possible. The oscilloscope also makes it possible to check for distortion by actually observing the test waveforms. More details about using the oscilloscope are explained in the following topics:

18-1
MAIN SECTIONS OF THE OSCILLOSCOPE

The oscilloscope is a complex but extremely versatile piece of electronic equipment. The front panel of a typical oscilloscope is shown in Fig. 18-1 which highlights three main sections. At the top, the CRT produces the illuminated display. Its electron beam produces a spot of light on the fluorescent screen. The brightness and focus controls are adjusted for a sharp, bright spot.

The spot can be positioned by voltage applied to the internal deflection plates of the CRT. The vertical- and horizontal-centering controls provide dc voltages to center the display.

The vertical (V) section amplifies the input signal for vertical deflection. Ac voltage deflects the beam up and down. At the same time, an internal sawtooth generator provides horizontal-deflection voltage to sweep the beam left and right. The result is that the vertical variations in the spot movements are spread out horizontally to form a trace pattern (Fig. 18-2).

Note that the signal to be displayed is applied to the vertical amplifier input terminal. An increase in the signal input increases the height or amplitude on the trace pattern. For that general use, no external signal is applied to the horizontal section. That section has its own internal sawtooth oscillator to produce horizontal-deflection voltage.

The vertical amplifier gain control can be used to adjust the height of the trace pattern on the CRT screen. The horizontal amplifier gain control can adjust the width of the trace.

The number of cycles of the vertical input signal shown on the screen can be set by the H sweep-frequency control. The reason is that this frequency determines the period of each horizontal sweep across the screen. When that time is longer than the period of the vertical input signal, more than one cycle is displayed. Usually, the horizontal sweep is adjusted for two cycles on the screen. The method is convenient because the transition from one cycle into the next can be seen and the size permits the details of the waveforms to be clear.

The H synchronizing or sync control is adjusted to lock the display in position on the screen instead of having it drift across the screen. Synchronization occurs when the H sweep frequency is an exact submultiple of the vertical signal frequency. In operation, the H frequency is adjusted for the

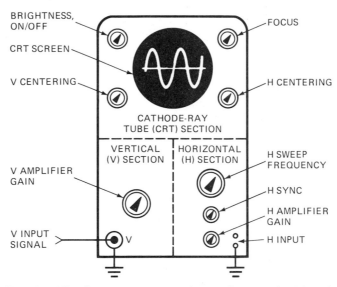

Fig. 18-1. The three main sections of an oscilloscope; the CRT, the vertical deflection section, and the horizontal deflection section.

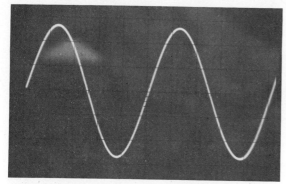

Fig. 18-2. Signal waveform on oscilloscope screen.

desired number of cycles and then the sync is turned up to lock in the pattern. As little sync as possible is used to minimize distortion of the display pattern.

The pattern is repeated so fast that it looks like a steady picture. Below about 30 Hz, though, the brightness of the trace flickers. At lower sweep frequencies, it is possible to see the spot moving.

Basically, the oscilloscope is a high-impedance voltmeter that shows the waveform of the ac signal. The vertical input terminals are connected across the voltage being checked. The input impedance is typically 1 MΩ shunted by 20 pF.

Test Point Questions 18-1
(Answers on Page 432)

In Fig. 18-1, which section is able to:

a. Display a point of light?
b. Determine the height of the display?
c. Determine how many cycles are displayed?

18-2
CATHODE-RAY TUBE (CRT)

Although vacuum-tube amplifiers are seldom used anymore, cathode-ray tubes are more common now than ever. Examples are shown in Fig. 18-3. The CRT is used to display the trace pattern in oscilloscopes, just as the picture tube is used in television receivers, and as a video monitor is used with computers. Its big advantage is that the electron beam can be controlled and deflected at high

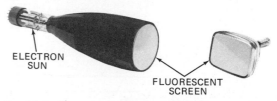

ELECTRON
SUN

FLUORESCENT
SCREEN

Fig. 18-3. Examples of the cathode-ray tube (CRT). Note electron gun in narrow glass neck. Screen at left has a 5-in. diameter. The inside of the glass faceplate is coated with a phosphorescent material. (*Dumont Electronics Corp.*)

frequencies. In oscilloscopes the CRT is usually mounted within a metal cover to shield the electron beam from interfering fields.

The internal structure of a CRT for oscilloscope use is illustrated in Fig. 18-4. Note that the main sections are the electron gun that produces a beam of electrons, the deflection plates to move the beam vertically and horizontally, and the phosphorescent screen to produce a point of light where the electron beam strikes. This CRT is suitable for oscilloscopes because it uses electrostatic deflection with varying voltage applied to the deflection plates. The oscilloscope screen is usually green, but blue is also used. The entire structure is in a vacuumed glass envelope.

Although the electrodes in the gun are constructed as cylinders, their functions are essentially the same as in vacuum-tube amplifiers. The cathode is heated to emit electrons. Grid G_1, next to the cathode, controls the space charge from the cathode. It has negative voltage, which can be varied to control the amount of beam current.

Electrons in the beam are attracted to the screen by an accelerating voltage that is positive with respect to the cathode. Grids G_2 and G_4 (Fig. 18-4) are the accelerating electrodes. The schematic base diagram for the CRT is shown in Fig. 18-5.

Each grid is a metal cylinder with a pinhole to form a narrow beam of electrons. In addition, the voltage on G_3 can be adjusted for focusing the beam. Grids G_2 and G_4 are tied together and are connected internally to the wall coating on the inside of the tube. The connection can be considered the final anode with the highest accelerating

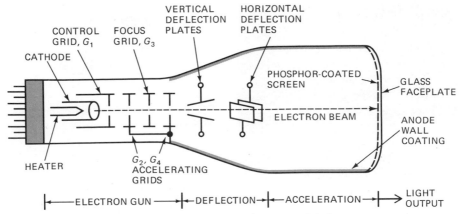

Fig. 18-4. Construction of CRT using electrostatic focus and deflection.

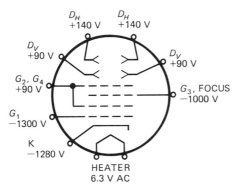

Fig. 18-5. Schematic symbol for CRT illustrated in Fig. 18-4. The typical dc voltages are shown for an inverted power supply with the highest negative voltage at the cathode.

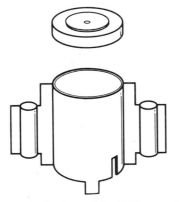

Fig. 18-6. Metal cylinder used for CRT control grid with aperture disk. The hole diameter is 0.04 in.

voltage. The internal coating is a film of conductive graphite called *aquadag*.

When the electrons strike the phosphor coating with high velocity, the electric energy is converted to light. Electrons are not attracted to the wall coating because the potential is equal at all points in any cross section of the beam path. In other words, the accelerating field is symmetrical around the electron beam. Furthermore, the phosphor-coated screen becomes charged to the anode potential by secondary emission. The final result is that the electron beam is accelerated to the screen to produce a point of light.

Electron Gun The electron gun is the entire assembly of metal cylinders. The electrodes are made of nickel or an alloy mounted on ceramic insulator supports. An example of the control grid is shown in Fig. 18-6. The small aperture of 0.04-in diameter restricts the electrons to a narrow beam. In addition, electronic focusing is used for a sharp spot of light.

Focusing The CRT in Fig. 18-4 uses *electrostatic focusing*. The voltage applied to G_3 in the electron gun can be adjusted for the smallest possible spot size. This focusing voltage is a little less than the voltage for G_2 and G_4. The general principle is to have the beam pass through a decelerating field, which makes the electrons converge toward the center.

Some types of CRT use *magnetic focusing*. An external coil around the neck of the tube produces a magnetic field which converges electrons in the beam. However, electrostatic focusing is generally used because it is more convenient.

Deflection   The method shown in Fig. 18-4 is *electrostatic deflection*. It uses two pairs of deflection plates. Deflection voltage applied to plates above and below the beam can move the electrons in the vertical direction. Also, the deflection voltage applied to the plates left and right of the beam produce horizontal motion.

The amount of voltage required is specified by the *deflection factor*. As a typical value, 30 V between the plates moves the beam 1 in [2.54 cm] with an accelerating voltage of 1 kV. That voltage is for the CRT itself. In an oscilloscope, much less voltage can be applied to the vertical amplifier, which provides the amount of vertical-deflection voltage required for the CRT.

In *magnetic deflection* a yoke containing two pairs of deflection coils is slipped over the neck of the CRT against the wide bell of the glass envelope. The magnetic field produced by the current in the external coils deflects the electron beam inside the glass neck.

Electrostatic deflection is generally used for the CRT in oscilloscopes. However, picture tubes for television receivers use magnetic deflection. The reason is that picture tubes have a larger screen area and much higher anode voltage. The combination would require very high deflection voltage for the deflection plates.

High Voltage for the Anode   For the oscilloscope CRT with a screen size of 3 or 5 in [7.6 to 12.7 cm] the high voltage is generally 1 to 5 kV, but 1400 V is a common value. The accelerating voltage is usually applied negative at the cathode of the oscilloscope. The cathode has the highest negative voltage, and the accelerating voltages progress to less negative values. The final anode is at, or close to, ground which is the highest positive potential with respect to the cathode. The purpose is to keep the anode and deflection plates close to the same potential.

For CRTs with a screen size of 10 to 25 in [25.4 to 63.5 cm] the high voltage at the anode is generally 10 to 30 kV. The anode connection is usually a recessed button in the wide glass bell.

CRT Type Numbers   The first digits give the screen size, in inches, for the diagonal length. For instance, the 5BP1 is a 5-in [12.7-cm] CRT for oscilloscopes. If the oscilloscope tube is rectangular, the screen size may be given in centimeters. The size of 8 × 10 cm is slightly larger than a 5-in [12.7-cm] diagonal.

The letters in the designation are for different tube types. However, the P number at the end specifies the phosphor coating for the screen. CRT type numbers do not specify the heater voltage, which is usually 6.3 V.

Screen Phosphors   The common types of screen phosphors are summarized in Table 18-1. P1 is a standard green phosphor used in oscilloscope tubes. P31 is a blue phosphor with a shorter persistence time that is used in high-frequency applications.

The *screen persistence* is the time taken by the emitted light to decay to 1 percent of its maximum value. Long persistence produces more light output from the screen, but a shorter decay time is needed when there are rapid changes in the visual display. A medium persistence time generally used is 5 to 50 ms.

To make the green P1 phosphor, a form of zinc silicate called *willemite* is generally used. A green

Table 18-1
Common Screen Phosphors

Phosphor Number	Color	Persistence	
P1	Yellowish green	Medium	Oscilloscopes
P4	White	Medium	Monochrome picture tube
P22	Green, blue, and red	Medium	Color picture tubes
P31	Green	Medium short	Oscilloscopes
P33	Orange	Very long	Radar

screen is popular for oscilloscopes because it is efficient and the eye is sensitive to green light.

Test Point Questions 18-2
(Answers on Page 432)

a. In Fig. 18-3, which is the focusing grid?
b. Does the oscilloscope CRT generally use magnetic or electrostatic deflection?
c. Does a green screen use phosphor P1, P4, or P22?
d. Is the typical high voltage for an oscilloscope CRT 1500 or 15,000 V?

18-3
Y AXIS FOR VOLTAGE AND X AXIS FOR TIME

With vertical deflection the spot of light moves up and down along the Y axis on the CRT screen. Horizontal deflection is along the X axis. Both axes are shown in Fig. 18-7 with reference to setting up the oscilloscope to show signal waveforms. A sine wave is connected to the vertical input terminals. The requirements are as follows:

1. Turn on the power. When the CRT warms up, a spot of light appears on the screen. There is no V or H deflection. Focus for a sharp spot and adjust the brightness. Center the spot as in Fig. 18-7a. If the spot is off the screen, adjust the centering controls. The internal horizontal sweep is off.
2. Turn on the internal horizontal sweep without any vertical deflection. The spot will move horizontally across the screen. A horizontal line is displayed, as in Fig. 18-7b, because the spot is repeating itself over the same area of the screen. This line illustrates the X axis.
3. Apply vertical deflection. If the internal horizontal deflection is off, the result will be the vertical line, shown in Fig. 18-7c, illustrating the Y axis.
4. With both vertical and horizontal deflection, you see the waveform of the vertical input signal, as in Fig. 18-7d.

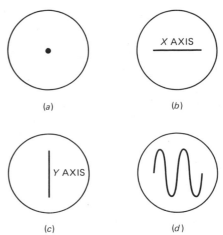

Fig. 18-7. Oscilloscope displays. (a) Light spot at the center of the screen without V and H deflection. (b) Horizontal line with H but without V deflection. (c) Vertical line with V but without H deflection. (d) Sinewave vertical input signal with V and H deflection.

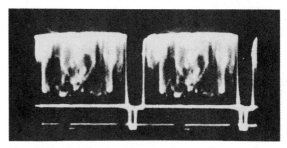

Fig. 18-8. Oscilloscope pattern for a video signal in a television receiver. Two cycles are shown.

Actually, it is better to turn on the oscilloscope with the internal sweep on. Then you start with a horizontal line. With just a spot, keep the brightness low to prevent burning the screen.

An example of oscilloscope waveforms of a video signal in a television receiver is shown in Fig. 18-8. This application is a common use of the oscilloscope.

Voltage Values on the Y Axis The vertical amplitude of the trace of any signal waveform indicates the peak-to-peak voltage. The screen usually has a plastic *graticule* overlay, marked in centimeter divisions as shown in Fig. 18-9. In Fig. 18-9a the vertical amplitude is shown without hor-

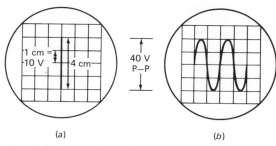

Fig. 18-9. A graticule used to measure voltage. (a) Voltage values are scaled on the Y axis. (b) Vertical amplitude of the waveform is 40 V p-p.

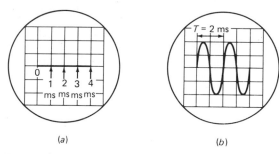

Fig. 18-10. Time and frequency values on the X axis. (a) The period T of the horizontal sweep time is 4 ms. (b) Vertical input signal has a period T of 2 ms.

izontal deflection. The vertical line is produced by turning off the horizontal sweep. The waveform in Fig. 18-9b is produced with horizontal deflection.

Either way, the amplitude shown is 40 V p-p. Each box is taken as 10 V in the vertical direction. The height of the vertical line or complete trace is four boxes. Therefore, this amplitude is 4 × 10 = 40 V. The measurement is p-p because it is between the two extreme amplitudes.

Time and Frequency Values on the X Axis

As an example of time and frequency values on the X axis, the internal sweep frequency in Fig. 18-10 is set for a horizontal trace time of 4 ms. Each box is then 1 ms. One cycle of the waveform shown takes two boxes, horizontally, equal to 2 ms.

This vertical signal frequency is 500 Hz. If it is increased to 1000 Hz, the pattern will show four cycles for the same horizontal sweep time.

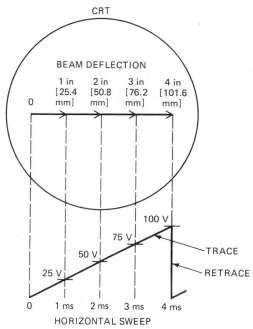

Fig. 18-11. How the horizontal sawtooth sweep voltage provides a linear time base on the X axis.

It should be noted that the oscilloscope graticule usually has ten boxes horizontally, in order to simplify calculations with respect to time.

Linear Time Base on the X Axis A sawtooth voltage is used for horizontal deflection to provide a linear time base. The idea is illustrated in Fig. 18-11. The linear rise is called a *ramp voltage*. It has a uniform change from 0 to 100 V. That part of the wave is the trace that is used to move the electron beam from the left edge of the screen to the right edge. The fast drop from 100 to 0 V is the retrace. Then the beam has a fast flyback to the left edge, where the next cycle starts. One cycle of sawtooth voltage includes the trace and retrace.

Actually, an ac deflection voltage is used to move the beam around the center. We can consider the changes starting from zero at the left edge, however, in order to visualize the entire horizontal sweep.

For the example of Fig. 18-11 the trace period is 4 ms. The ramp is divided into four parts with equal increases of 25 V. Each change of 25 V in the deflection voltage is shown moving the beam horizontally by 1 in [2.54 cm] on the screen of the CRT. Because of the linear change in deflection voltage, the beam moves at a uniform rate across the screen. Therefore, equal distances on the X axis correspond to equal intervals of time. The result is a linear time base for showing the waveform of a vertical input signal. Oscilloscope waveforms actually represent a graph of amplitude variations on the Y axis with respect to time on the X axis.

Test Point Questions 18-3
(Answers on Page 432)

a. In Fig. 18-9 the vertical deflection fills three boxes. What is the p-p voltage?
b. In Fig. 18-10 the trace shows four cycles. What is the frequency of the vertical input signal?
c. What time period is represented by the distance of 50.8 mm on the X axis in Fig. 18-11?

18-4
BLOCK DIAGRAM OF OSCILLOSCOPE

The block diagram of a typical oscilloscope is shown in Fig. 18-12. The CRT itself needs dc voltages for operation. The brightness or intensity control R_1 varies the negative bias at G_1, which is the control grid. For electrostatic focusing, R_2 varies the voltage at G_3.

R_3 and R_4 are V and H centering or positioning controls. The dc potential difference for each pair of deflection plates determines the average spot position. Then an ac voltage deflects the beam around the starting point. The electron beam is attracted toward the more positive plate.

Power Supplies All the dc voltages come from the power supply, shown at the bottom left in Fig. 18-12. It really has two sections. The low-voltage supply produces 140 V for the amplifiers and electron gun of the CRT. The high-voltage supply produces the accelerating potential needed for the CRT anode voltage. The output is -1300 V with an inverted power supply for negative voltage applied to the cathode of the electron gun.

The on-off switch is usually mounted on the shaft of the brightness control, but electrically the switch is in the ac line input as shown for S_1.

Vertical-Deflection Amplifier The signal waveform to be observed is connected to the V-input jack at the top left in Fig. 18-12. Shielded cable is generally used for the vertical input signal to minimize picking up any interference.

The vertical-deflection amplifiers are needed to provide enough signal voltage for the deflection plates. Typically, the oscilloscope has a sensitivity of 0.02 V/in for the vertical-input signal. For the 200 V needed by the vertical-deflection plates, the voltage gain is then $200/0.2 = 1000$. Two or more amplifier stages are used. The final amplifier has a balanced output to provide push-pull voltage for the pair of deflection plates.

The gain of the vertical-deflection amplifier is set to provide a trace that fills most of the screen from top to bottom. Without enough gain, the height of the trace would be too small. Too much signal, however, can deflect the peaks of the trace off the screen. In the diagram of Fig. 18-12 the vertical attenuator S_2 is a voltage divider that can tap off part of the vertical input signal. The division is usually in multiples of one-fifth or one-tenth. After the attenuator is set the vertical gain control R_9 is adjusted for the desired height of the trace pattern.

The total R of the attenuator is usually 1 MΩ, which provides the vertical input resistance of the oscilloscope. C_1 is a coupling capacitor to block dc voltage. Then the ac deflection in the trace pattern is above and below the center axis.

Although not shown here, the voltage divider for the vertical attenuator usually is frequency-compensated with shunt capacitors. The purpose is to provide the same voltage division for high and low frequencies.

Maximum vertical input signal is usually about 600 V, peak value, including dc and ac voltages. Higher voltage can cause arcing in the components of the vertical input circuit.

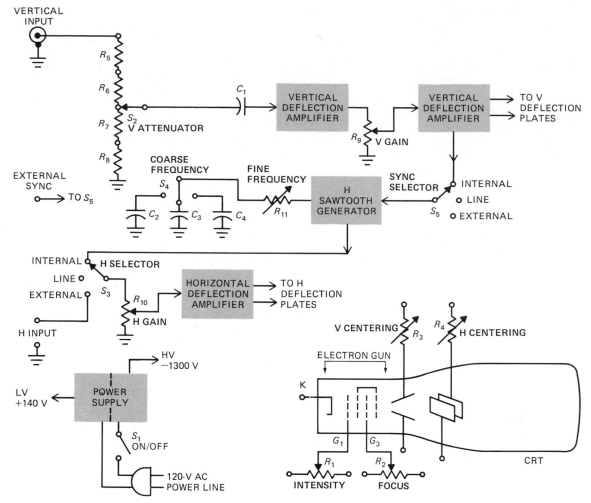

Fig. 18-12. Block diagram of oscilloscope circuits.

Horizontal Selector Switch   The horizontal section is more complicated than the vertical because the H-deflection amplifier can have any one of three different signals for the H-deflection plates. In Fig. 18-12, the H selector switch S_3 can be set for:

1. External H signal. Voltage must be applied to the H-input terminals. The connection is usually at the lower right corner on the front panel of the oscilloscope. That position is not being used in Fig. 18-12.

2. Line ac input. A sample of the ac power line voltage is used to obtain a sine wave at exactly 60 Hz. This connection is provided internally. That position also is not being used in Fig. 18-12.

3. Internal. This is the sawtooth generator that provides the internal horizontal sweep voltage. The selector switch S_3 is in this position because the oscilloscope is generally used this way. An external signal should not be connected in this case for horizontal deflection.

Horizontal-Deflection Amplifier For H-input signal from any of the three sources set by S_3, the amplifier provides enough output for the H-deflection plates. A balanced output circuit is used for push-pull voltage. The H-gain control varies the width of the trace pattern. Sensitivity of the H amplifier need not be as good as that of the V amplifier because there is generally no problem of weak signal for the H input.

Horizontal Sweep Frequency With a sawtooth generator to supply the internal horizontal sweep, its frequency depends on the time constant of the RC network that shapes the sawtooth voltage. In Fig. 18-12, S_4 is the coarse frequency control. It selects a different capacitor for each frequency range. Larger C values produce lower frequencies.

Within each range, R_{11} is variable for the fine frequency adjustment. It is sometimes called a *vernier control*. The switch S_1 is set for the frequency range and then R_{11} is varied for two cycles in the trace pattern. If two cycles cannot be obtained with the fine frequency control, change the coarse frequency adjustment and readjust the fine frequency control.

Sync Selector The sync selector switch, S_5 in Fig. 18-12, is for input to the H sweep generator, not the horizontal amplifier. The function of synchronization is to make the pattern remain stationary on the screen. The sync has no purpose, though, unless the sawtooth generator is on for internal horizontal sweep. Again, there are three possibilities:

1. External sync. This position of S_5 can provide an external synchronizing signal for the internal sweep generator. The sync voltage must be applied to the terminal on the front panel of the oscilloscope, which is connected to S_5. However, external sync is not being used in Fig. 18-12.
2. Line sync. This position of S_5 provides a sample of the 60-Hz ac line voltage for synchronizing the internal sweep generator. The line sync can be useful when the vertical input signal frequency is 60 Hz or an exact multiple. In Fig. 18-12, though, the line sync is not being used.
3. Internal sync. S_5 is in the position shown in Fig. 18-12 because the oscilloscope is generally used with internal synchronization. Note that a sample of the input signal is taken from the vertical amplifier to synchronize the horizontal sweep generator. In that way the sawtooth oscillator can be synchronized for any frequency of the vertical signal.

The positions for horizontal selector switch S_3 and sync selector switch S_5 have similar names, but the functions are entirely different. A horizontal deflection voltage of any type makes the CRT beam move across the screen. The type of deflection is set by S_3. However, sync voltage does not produce deflection. The sync only triggers the sawtooth oscillator to make its frequency exactly related to the vertical input signal. Different types of sync can be used for better triggering. The type of sync is set by S_5. However, the sync has no use unless the internal sawtooth generator is on. In other words, S_3 must be set for internal sweep to use different types of sync with S_5.

Test Point Questions 18-4
(Answers on Page 432)

Refer to Fig. 18-12.

a. Which control sets the height of the trace, R_9 or R_{10}?
b. Which switch turns on the internal sweep generator, S_3 or S_5?
c. Which control is varied for two cycles in the trace, R_{10} or R_{11}?

18-5
OSCILLOSCOPE PROBES

The oscilloscope probes are the leads used for connecting the vertical input signal to the oscilloscope. There are three possibilities: a direct lead, a low-capacitance probe (LCP), or a demodulator probe. Figure 18-13 shows a typical LCP, which is the probe generally used for oscilloscope measurements. Note the 9-MΩ isolating R_1 in the probe.

Direct Probe The direct probe is just a shielded lead without any isolating resistance. Shielding is needed to prevent pickup of interfering signals, especially with the high input resistance of 1 MΩ. Hum voltage at 60 Hz and stray RF signals are common examples of interference.

The shielded lead has relatively high capacitance. A typical value is 90 pF for 3 ft [0.9 m] of 50 Ω coaxial cable. Also, the oscilloscope has an input capacitance of about 20 pF. The total C is then 110 pF. Therefore, a direct probe can be used only when the added capacitance has little effect on the circuit being tested. It is usually used for 60-Hz power line or sine-wave audio signal measurements in a circuit with a relatively low resistance of several kilohms or less.

Low-Capacitance Probe (LCP) The low-capacitance probe (Fig. 18-13) has an internal 9-MΩ series resistance to isolate the capacitance of the cable and oscilloscope from the circuit connected to the probe tip. The input capacitance with the probe is about 10 pF.

The function of C_1 is to compensate the LCP for high frequencies. Its time constant with R_1 should equal the RC time constant of the input circuit for the vertical amplifier. C_1 is adjusted for minimum tilt on a square-wave signal.

The low-capacitance probe must be used for measurements when:

1. The signal frequency is high, that is, above audio frequencies.
2. The circuit resistance is high, more than about 100 kΩ.
3. The waveshape is nonsinusoidal, especially with sharp pulses.

Without the LCP the observed waveform can be distorted. The reason is that too much C changes the circuit while it is being tested.

However, it is important to remember that the series R of the low-capacitance probe reduces the signal amplitude being fed into the vertical amplifier of the oscilloscope. This voltage division is generally 1:10, as shown in Fig. 18-14. With 9 MΩ in series with 1 MΩ, the voltage across R_S

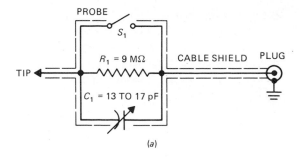

(a)

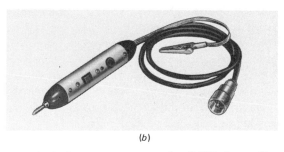

(b)

Fig. 18-13. Low-capacitance probe (LCP) for oscilloscopes. (a) Circuit. (b) Typical probe. Length of probe alone is 4 in. (RCA)

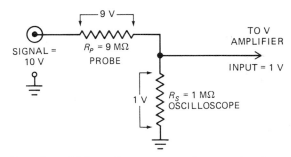

Fig. 18-14. Voltage division ratio of 1:10 with a low-capacitance probe.

applied to the oscilloscope is one-tenth of the applied signal voltage.

In summary, it is really preferable to use the low-capacitance probe for almost all oscilloscope measurements except those involving the 60-Hz ac power line. Remember, when using an LCP, to multiply by 10 for the actual signal amplitude. As an example, when the trace pattern on the screen indicates 2.4 V, the signal input at the probe is 24 V. The LCP is generally called the "10 times" probe.

Demodulator Probe The demodulator probe has an internal diode to detect the signal input to the probe. Its output is the envelope of an amplitude-modulated RF signal at the input. Polarity of the rectified dc output is usually negative.

The probe can be used for signal tracing in the RF and IF circuits of a receiver by looking at the output of each amplifier stage. There is usually a problem, though, in very low signal amplitudes.

Current Measurements Although the oscilloscope is an ac voltmeter, it can also be used for current measurements. The technique is to insert a low R in series with the current to be checked. The oscilloscope is used to measure the voltage across R. Then $I = V/R$. The inserted R must be much lower than the circuit resistance to prevent any appreciable change in the actual I.

Test Point Questions 18-5
(Answers on Page 432)

a. Is the value of C in a shielded direct probe about 10 pF or 100 pF?
b. The trace pattern measures 1.8 V with an LCP. What is the actual signal amplitude?
c. The demodulator probe has an internal diode detector. True or false?

18-6
SPECIAL OSCILLOSCOPE FEATURES

Although the block diagram in Fig. 18-12 illustrates the basic elements, most oscilloscopes have additional functions for convenience in special applications. However, a relatively simple oscilloscope is usually suitable for most radio and television servicing. A bandwidth of 10 MHz for the vertical-deflection amplifier is desirable for television. The video signal bandwidth is 4 MHz, which includes the 3.58-MHz color signal. Furthermore, additional bandwidth is needed for the sharp edges on pulse waveforms.

Two important frequencies for oscilloscopes are the AF values of 60 and 15,750 Hz. The 60 Hz is the ac power-line frequency. Incidentally, in most areas the power line voltage can be used as a 60 Hz reference frequency. Also, 60 Hz is the vertical-field scanning frequency for a television picture. The horizontal-line scanning frequency for television is 15,750 Hz.

TV Positions on Internal Sweep On the frequency-range switch for the internal sawtooth generator, oscilloscopes often have two positions marked V and H for television. At the V position the internal sweep is at 30 Hz, so that two cycles of 60-Hz signal can be observed. At the H position the internal sweep is at 7875 Hz, so that two cycles of the 15,750-Hz signal can be observed. An example is the pattern in Fig. 18-8, which shows a video signal for two horizontal scanning lines in a television receiver.

Do not confuse the abbreviation V for vertical deflection in television with V for the vertical input signal to an oscilloscope. The television V is 60 Hz. The oscilloscope vertical input signal can have any frequency. The V and H positions are used only in oscilloscopes for TV servicing.

Z Axis for Intensity Modulation The Y and X axes are for deflection of the electron beam in a CRT. In addition, though, the intensity of the beam can be varied by changing the control-grid voltage. That method of controlling the light output from the screen is intensity modulation of the electron beam. The intensity control is also considered as Z-axis modulation, only because the effect is not vertical or horizontal.

In oscilloscopes a separate Z-axis terminal may be provided for a connection to the control grid of the CRT. However, it is not used for the normal X-Y graph of the vertical input signal.

A p-p range of about 15 V at the Z terminal can vary the beam intensity between maximum and zero. No light output is considered as black or blanking level.

Polarity of Vertical Input Signal Oscilloscopes are generally made to show positive polarity in the upward direction on the CRT screen. Some oscilloscopes have a polarity-reversing switch, however. This inverts the display by 180°.

Direct Connections to Deflection Plates The direct connections to the deflection plates are used to bypass the frequency limitations of the vertical- and horizontal-deflection amplifiers. However, appreciable deflection voltage is needed with direct plate connections. Typically, 30 V produces 1 in of deflection.

Polarity of Internal Sync Positive and negative polarities are usually available with the sync selector switch. One may be better than the other to lock in the trace. The polarity that requires less sync voltage is preferred.

AC-DC Switch for V Input Signal The ac-dc switch for the V-input signal shorts out the input coupling capacitor for dc coupling if desired. Then the screen pattern is shifted from the center axis. The amount of offset is a measure of the dc level in the vertical input signal. A problem with dc coupling, though, is that the pattern is easily moved off the screen. Still, the dc position is useful for dc voltage measurements.

Trace Magnifier The trace magnifier circuit expands the horizontal deflection to make the trace bigger than the screen size. This permits observing more details in the trace that is on the screen.

Dual-Trace Oscilloscope The dual-trace oscilloscope can show two traces at the same time, one above the other, for two vertical input signals. There are two vertical amplifiers but just one electron beam. An internal electronic switch changes the signals to the vertical-deflection plates alternately from each amplifier. The switching rate is fast enough to make the changes invisible. The advantage of a dual trace is that it permits observing two simultaneous signals on the screen at the same time. Thus time and amplitude comparisons may be viewed directly.

Dual-Beam Oscilloscope The dual-beam oscilloscope also shows two traces but a special CRT

with two beams is used. This eliminates the need for electronic switching.

Triggered Sweep In a conventional oscilloscope the internal horizontal sweep is produced by a free-running sawtooth oscillator. The device oscillates to produce horizontal deflection with or without sync. That free-running method is called *recurrent sweep*. With triggered sweep the internal sweep oscillator does not produce horizontal deflection unless it is triggered into conduction by a sync voltage. The advantages are better synchronization and more exact control of the horizontal sweep time.

Oscilloscope Example Figure 18-15 shows a 5-in dual-trace oscilloscope with a triggered sweep. At the left are two vertical amplifiers for Y_1 and Y_2 input signals. Note the following characteristics:

V sensitivity	10 mV/cm to 20 V/cm
V attenuator	Eleven steps in 1-2-5 sequence
V frequency response	Dc and 2 Hz to 10 MHz
V modes	Y_1 or Y_2 for single trace Y_1 and Y_2 chopped at the rate of 200 kHz for dual trace

The internal horizontal sweep section on the right side of the oscilloscope in Fig. 18-15 provides a time base of 0.2 s/cm to 200 ns/cm. Those values mean 2 s to 2000 ns as the total period of time for the ramp voltage to sweep the 10 divisions across the screen. With internal sweep the horizontal amplitude is automatically set for 10 cm of deflection. The corresponding frequencies for the sawtooth voltage are 0.5 Hz at the low end and 500 kHz at the high end, approximately. Sweep time is varied in 1-2-5 steps with continuous variations for each position.

A useful value to keep in mind for time on the X axis is that three cycles of 60-Hz ac line voltage take exactly 50 ms.

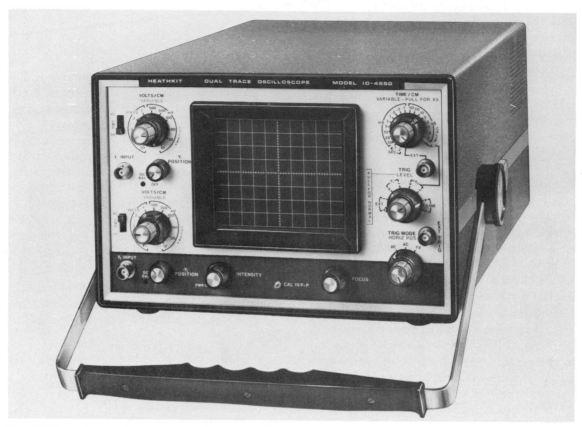

Fig. 18-15. A 5-in dual-trace, triggered-sweep oscilloscope. Length is 19.25 in [48.9 cm], and weight is 22 lb. [10 kg]. (*Heath Co.*)

The horizontal-deflection amplifier has a sensitivity of 0.1 V/cm and input impedance of 100 kΩ. Its frequency response is up to 1 MHz. Both the V and H inputs take a type BNC plug for the shielded coaxial cable.

Test Point Questions 18-6
(Answers on Page 432)

a. Which oscilloscope terminal is connected to the CRT control grid?

b. Does positive polarity of the vertical input signal move the CRT beam up or down?

c. Does a free-running sawtooth oscillator provide recurrent or triggered sweep?

d. With a sweep time of 1 ms across the screen, what is the approximate horizontal sawtooth frequency?

18-7
VERTICAL CALIBRATION

When the oscilloscope has a calibrated vertical attenuator, its values are usually in volts per centimeter. Each box on the screen graticule is 1 cm square, as shown in Fig. 18-15. As an example, with the attenuator on the 2-V position and a height of 3 cm for the trace, the oscilloscope input signal is 3 × 2 V = 6 V p-p. With the low-capacitance probe the vertical input signal is 10 × 6 V = 60 V. The vertical gain control must be at a preset position in order to use the calibrated values on the attenuator.

In addition, a calibrating voltage is often provided on the oscilloscope. It is generally a square wave 1 V p-p. To use it, connect it to the V input as a reference value to see the trace height that corresponds to the specified voltage.

If no calibration voltage is provided, a low voltage obtained from the ac power line can be used. The voltage, about 5 to 10 V, is then measured accurately with an ac voltmeter for its rms value. It can then be used the same way as the internal calibrating voltage. Its p-p value is 2.8 times the rms value for a sine wave.

Test Point Questions 18-7
(Answers on Page 432)

a. The calibrated vertical attenuator is on its 10-mV position and the trace height is 6 cm. What is the trace amplitude?

b. A calibrating input of 5 V rms produces a vertical deflection of 4 cm. What is the vertical voltage per centimeter?

18-8
LISSAJOUS PATTERNS FOR PHASE AND FREQUENCY COMPARISONS

Examples of Lissajous patterns are shown in Figs. 18-16 and 18-17. Those trace patterns on the oscilloscope screen are used to find the phase angle between two sine-wave voltages of the same frequency or for comparing sine waves of different frequencies. The purpose of the frequency comparison is to check an unknown frequency against a reference value.

The patterns are only for sine-wave signals. One is applied to the oscilloscope vertical input and the other to the horizontal input. The horizontal internal sweep is not used. Lissajous patterns are named after the man who first used them.

Both the vertical and horizontal deflection signals should have equal amplitude. This can be

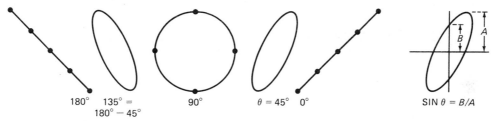

Fig. 18-16. Lissajous patterns on an oscilloscope screen. These patterns compare the phase relationships between two sine-wave voltages of the same frequency.

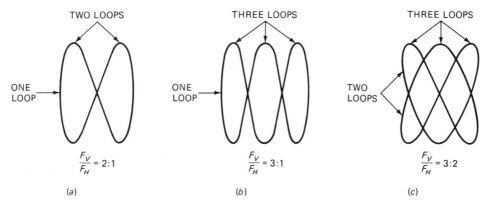

Fig. 18-17. Lissajous patterns on an oscilloscope screen. These patterns compare two sine waves of different frequencies. F_V is the frequency of the vertical signal; F_H is the frequency of the horizontal signal. The closed loops indicate the frequency ratio.

checked by adjusting the gain until the same height and width is obtained for each signal alone without the other.

Phase-Angle Comparisons Assume two sine waves have the same frequency. The combined trace looks like one of the patterns in Fig. 18-16. Consider the diagonal line for 0°. The two sine waves are in phase. At the start, the spot is at the center without any deflection. When the V signal increases in a positive direction to deflect the spot upward, the H signal also is positive and moves the spot the same amount to the right. Halfway to the peak voltage, the spot is halfway to the top, as shown by a dot in the figure. At the peak value for both the V and H signals, the spot is at its extreme top right position shown at the end of the diagonal line. That action occurs during the first quarter-cycle for both signals.

On the next quarter-cycle the spot repeats the same positions on the way back to the center. In the same way, during the negative half-cycle the spot moves diagonally down to the bottom left. The spot repeating over this path produces a diagonal line. When the two waves are 180° out of phase, the line slopes in the opposite direction.

Consider the circle pattern produced by two waves 90° out of phase. One signal is at maximum when the other is at zero. When the vertical signal forces the spot to the extreme top or bottom position, the spot is in the center horizontally. Also, when the horizontal signal is maximum for the extreme left and right positions, the spot is in the center vertically. Then the spot traces a circle pattern for all the V and H values 90° out of phase.

For the pattern of an ellipse, the phase angle θ can be determined by calculating the ratio of the two lengths B and A shown at the right in Fig. 18-16. As an example, when the B intercept is seven-tenths of A, the ratio is 0.7. Since sine θ equals 0.707 for 45°, the phase angle is 45°. Remember that those phase-angle patterns apply only to sine waves.

Frequency Comparisons In practical terms the patterns in Fig. 18-16 show that the frequency is the same for the vertical- and horizontal-input

signals. The pattern may drift between a diagonal line and the circle as the phase changes slowly. Even so, the pattern indicates a 1:1 frequency ratio.

When the vertical-input signal has a higher frequency than the horizontal input signal, the patterns in Fig. 18-17 are produced. To determine the frequency ratio, count the loops across either the top or bottom of the trace for F_V. Count only the closed loops; an open loop, such as a half-loop, is not counted at all. Similarly, count the closed loops at either side for F_H. The frequency ratio is then equal to F_V/F_H.

As an example, let the horizontal input be a 60-Hz ac voltage from the power line as a reference frequency. The vertical-input signal is from an audio signal generator. Its frequency calibration can be checked at the dial setting of 60 Hz. With a circle or line pattern, the frequency is exactly 60 Hz for the signal generator.

Next change the frequency dial to 120 Hz. Where the generator produces the pattern in Fig. 18-17a, the frequency is exactly $2 \times 60 = 120$ Hz. The 3:1 pattern in Fig. 18-17b shows the generator frequency is 180 Hz. Frequencies that are not exact multiples can also be compared, as in Fig. 18-17c. The 3:2 ratio shows the generator frequency is $3/2 \times 60 = 90$ Hz. In that way the generator frequency can be checked for patterns up to about 10 loops which would represent 600 Hz.

After the generator is calibrated, it can be used as the reference for checking an unknown frequency. Use the generator for the horizontal-input signal and connect the other to the oscilloscope vertical input. The unknown frequency can be determined by seeing how many loops it produces compared with the generator frequency.

Test Point Questions 18-8
(Answers on Page 432)

a. Does a circle pattern show the phase angle of 0°, 90°, or 180°?

b. Does a diagonal line pattern show a frequency ratio of 1:1 or 3:1?

c. The H reference frequency is 60 Hz. What is the vertical frequency that produces five loops?

18-9
HOW TO SEE A TELEVISION PICTURE ON THE OSCILLOSCOPE SCREEN

An oscilloscope CRT uses electrostatic deflection and a green screen, but its function is the same as that of a television picture tube in showing visual information. To use the oscilloscope to see the picture from a television receiver, these three requirements must be met (see Fig. 18-18).

1. Sawtooth deflection voltage at 60 Hz is connected to the oscilloscope V-input terminal. This voltage can be taken from the vertical-deflection oscillator in the TV receiver.
2. Sawtooth deflection voltage at 15,750 Hz is used for oscilloscope horizontal sweep. This frequency is the rate at which horizontal lines are produced in the television picture. Either the oscilloscope internal sweep can be used or the voltage can be taken from the horizontal-deflection oscillator in the TV receiver.

   The 60 Hz is the rate at which the horizontal scanning lines are moved downward in the television picture. With vertical and horizontal sawtooth deflection, the screen pattern is a rectangle filled with horizontal lines from top to bottom. The pattern is called the *scanning raster*. Adjust vertical and horizontal gain for the desired height and width.
3. The last requirement is a video signal from the receiver coupled to the Z-axis input of the TV CRT for intensity modulation. It is the video signal that has the voltage changes for the picture information.

   If the three conditions are satisfied, a green video picture will appear on the oscillosocope screen. The picture may look like a negative,

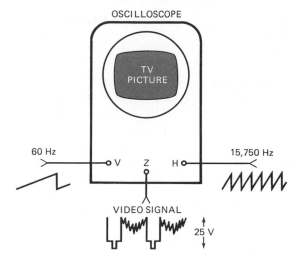

Fig. 18-18. Connections for reproducing a television picture on the oscilloscope screen.

depending on the polarity of the video signal. About 25-V p-p signal is needed.

All connections to the receiver should have a decoupling resistor of 5 to 50 kΩ to avoid detuning the circuits. When the picture rolls vertically, adjust the vertical scanning frequency. When the picture breaks into diagonal segments, adjust the horizontal scanning frequency.

Test Point Questions 18-9
(Answers on Page 432)

For a television picture on the oscilloscope screen:

a. Is video signal connected to the V or Z input?
b. What is the waveform for vertical deflection?
c. What is the frequency for horizontal deflection?

SUMMARY

1. The oscilloscope is a high-impedance ac voltmeter that shows the waveform of the signal voltage being measured.
2. A cathode-ray tube (CRT) is used to display the waveform. The CRT consists of an electron gun to produce a beam of electrons, vertical- and horizontal-deflection plates to move the beam, and a fluorescent screen to produce light output.

3. The main sections of an oscilloscope include a vertical amplifier for the input signal applied to the vertical-deflection plates, a horizontal amplifier for the horizontal-deflection plates, and an internal sawtooth generator to provide a linear time base for horizontal deflection.
4. To see the waveform of a signal voltage, the voltage is applied to the vertical-input terminal with internal sawtooth sweep for horizontal deflection. Set the frequency of the horizontal sweep for two cycles of vertical signal in the trace pattern. Then turn up just enough sync to make the pattern stay still.
5. Recurrent sweep means that the internal sawtooth generator operates with or without sync.
6. With triggered sweep, each cycle of the sawtooth generator is produced only with sync input.
7. A low-capacitance probe (LCP) has a series-isolating R that divides the input signal by one-tenth.
8. A Z-axis signal is input to the control grid of the CRT for intensity modulation of the electron beam.
9. The vertical calibration measures the amount of input signal voltage in terms of the height of the trace.
10. Lissajous patterns are used to compare the phase relationship of two sine waves. They are also used to determine the frequency of an unknown sine wave by comparing it to a known standard. Lissajous patterns are obtained by applying one sine wave to the vertical input and the other sine wave to the horizontal input. The internal sweep is not used.

SELF-EXAMINATION
(Answers at back of book)

Choose (a), (b), (c), or (d).

1. To which terminal is the signal to be viewed on the oscilloscope screen applied? (a) V; (b) H; (c) Z; (d) external sync?
2. What is typical operation of the horizontal section? (a) Internal sweep and sync off; (b) internal sweep with internal sync; (c) internal sync but external sweep; (d) external sync but internal sweep off.
3. Typical anode voltage for the oscilloscope CRT is (a) 1 to 5 kV, negative at the anode, (b) 2 kV negative at the cathode, (c) 30 kV positive at the anode, (d) 200 V negative at G3.
4. The screen graticule is calibrated for 5 V/cm. The height of the trace is 6 cm. The vertical trace then is (a) 5 V p-p, (b) 15 V p-p, (c) 30 V rms, (d) 30 V p-p.
5. The frequency of the internal horizontal sweep is 30 Hz with the vertical input at 60 Hz. How many cycles will appear on the screen? (a) One-half; (b) one; (c) two; (d) 60.
6. The frequency of the internal horizontal sweep is 1000 Hz. The trace has two

cycles. What is the frequency of the vertical input signal? (a) 500 Hz; (b) 1000 Hz; (c) 2000 Hz; (d) 5000 Hz.

7. The trace pattern measures 4 V with the low-capacitance probe used for vertical input signal. The actual signal at the probe is (a) 0.4 V; (b) 4 V; (c) 8 V; (d) 40 V.

8. Which terminal is connected to the control grid of the CRT? (a) Y; (b) X; (c) Z; (d) external sync.

9. For an oscilloscope with triggered sweep, the horizontal time base is set for 5 ms/cm. The entire width of 10 cm then represents the time of (a) 0.5 ms, (b) 1 ms, (c) 5 ms, (d) 50 ms.

10. The circle for a Lissajous pattern shows (a) 2:1 frequency ratio, (b) 1:1 frequency ratio, (d) 0° phase, (d) 180° phase.

ESSAY QUESTIONS

1. List three main sections of a CRT and give the function of each in producing a visual display.
2. Name the electrodes in an electron gun.
3. Give the functions for the following controls: intensity, focus, V and H positioning.
4. Compare two types of screen phosphors.
5. Show how to apply a sawtooth voltage to a pair of deflection plates.
6. Show how to apply high voltage to the anode with a negative supply.
7. Define deflection factor, Y axis, X axis, and Z axis.
8. Compare triggered sweep and recurrent sweep.
9. Compare internal and external sync. Why is internal sync generally used?
10. What is meant by line sync?
11. What is the function of the vertical attenuator?
12. What is the function of the internal horizontal sawtooth generator?
13. Explain briefly how to use the oscilloscope to see two cycles of an ac signal voltage.
14. a. When is the low-capacitance probe (LCP) used? b. Why does the LCP divide the signal input by one-tenth?
15. Compare a direct probe, an LCP, and a demodulator probe.
16. Name five special features of oscilloscopes included for special purposes and give the purpose of each.
17. Show the Lissajous pattern for two sine waves with a 1:1 frequency ratio and 90° out of phase.
18. Show the Lissajous patterns for sine waves with 2:1 and 1:2 frequency ratios.
19. a. Why is the horizontal internal sweep turned off for a Lissajous pattern?
    b. Show the trace pattern for a 2:1 frequency ratio with the internal sweep on.
20. Show the oscilloscope connections needed to reproduce a television picture on the oscilloscope screen.
21. Give two differences between a TV picture tube and the oscilloscope CRT.

PROBLEMS
(Answers to odd-numbered problems at back of book)

1. For a deflection of 25 V/in, what is the p-p sawtooth voltage needed for a 4 in deflection?
2. In a CRT the cathode is at -1280 V and G_1 is at -1300 V. What is the control-grid bias?
3. Calculate the approximate horizontal sweep frequencies for the following time periods: **a.** 0.0333 s, **b.** 0.0166 s, **c.** 0.1269 ms, **d.** 50 ms.
4. A sawtooth voltage has a linear ramp of 20 V/ms. **a.** What is the peak trace voltage after 10 ms? **b.** Calculate the approximate frequency of the sawtooth voltage, not counting retrace time.
5. Vertical deflection is calibrated for 1 V/cm on the screen graticule. **a.** What is the p-p signal for a trace height of 6 cm? **b.** What will be the height of the trace when the attenuator is changed to 2 V/cm?

SPECIAL QUESTIONS

1. Find the approximate cost of the following 5-in oscilloscopes:
   a. 4-MHz bandwidth, recurrent sweep, **b.** 10-MHz bandwidth, triggered sweep, calibrated vertical attenuator, **c.** 20-MHz bandwidth, dual-trace, triggered sweep and calibrated vertical attenuator. Which would you choose for your own work?
2. Why would an oscilloscope be considered more useful for checking a TV receiver than an FM radio?
3. Compare these two methods for localizing troubles to a defective stage: **a.** signal injection with a signal generator and **b.** signal tracing with an oscilloscope.
4. After a trouble has been localized to one stage, why is it generally necessary to use a volt-ohmmeter to find the defective component?

ANSWERS TO TEST POINT QUESTIONS

18-1	**a.** CRT	18-4	**a.** R_9	18-7	**a.** 60 mV
	b. Vertical		**b.** S_3		**b.** 3.5 V
	c. Horizontal		**c.** R_{11}	18-8	**a.** 90°
18-2	**a.** G_3	18-5	**a.** 100 pF		**b.** 1:1
	b. Electrostatic		**b.** 18 V		**c.** 300 Hz
	c. P1		**c.** True	18-9	**a.** Z
	d. 1500 V	18-6	**a.** Z axis		**b.** Sawtooth
18-3	**a.** 30 V		**b.** Up		**c.** 15,750 Hz
	b. 1000 Hz		**c.** Recurrent		
	c. 2 ms		**d.** 1000 Hz		

Bibliography

Bell, D.A.: *Fundamentals of Electronic Devices*, Reston Pub. Co., Reston, Va., 1975.

Chute, G.M. and R.D. Chute: *Electronics in Industry*, McGraw-Hill, New York, 1979.

Deboo, G.J. and C.N. Burrous: *Integrated Circuits and Semiconductor Devices*, McGraw-Hill, New York, 1971.

DeFrance, J.J.: *General Electronic Circuits*, Holt, Rinehart, and Winston, New York, 1972.

G.E. Transistor Manual, General Electric Company, Electronics Park, Syracuse, N.Y.

Grob, B.: *Basic Television*, 4th ed., McGraw-Hill, New York, 1975.

Ilardi, F.: *Computer Circuit Analysis*, Prentice-Hall, Englewood Cliffs, N.J., 1976.

Jung, W.G.: *IC Op-Amp Cookbook*, Howard W. Sams, Indianapolis, 1974.

Kaufman, M.: *Radiotelegraph Operator's License Q and A Manual*, Hayden, Rochelle Park, N.J., 1975.

Kiver, M.S.: *Transistors and Integrated Electronics*, McGraw-Hill, New York, 1972.

Lenk, J.: *Handbook of Basic Electronic Troubleshooting*, Prentice-Hall, Englewood Cliffs, N.J., 1979.

Malvino. A.P.: *Digital Computer Electronics*, McGraw-Hill, New York, 1977.

McGinty, G.F.: *Videocasette Recorders*, McGraw-Hill, New York, 1979.

Noll, E.M.: *Second-Class Radiotelephone License Handbook*, Howard W. Sams, Indianapolis, 1975.

Prensky, S.D.: *Electronic Instrumentation*, Prentice-Hall, Englewood Cliffs, N.J., 1971.

Radio Amateur's Handbook, American Radio Relay League, Newington, Conn., 1980.

RCA Solid-State Devices Manual, RCA, New York, 1975.

Schuler, C.A.: *Electronics: Principles and Applications*, McGraw-Hill, New York, 1979.

Schultz, L.E.: *How to Repair CB Radios*, McGraw-Hill, New York, 1980.

Shrader, R.L.: *Electronic Communication*, McGraw-Hill, New York, 1980.

Slurzberg, M. and Osterheld, D.: *Essentials of Communication Electronics*, McGraw-Hill, New York, 1973.

Stein, P.: *Graphical Analysis*, Hayden, Rochelle Park, N.J., 1964.

Stout, D. and Kaufman, M.: *Operational Amplifier Circuit Design*, McGraw-Hill, New York, 1976.

Temes, L.: *Communication Electronics for Technicians*, McGraw-Hill, New York, 1974.

Tocci, R.J.: *Fundamentals of Electronic Devices*, Merrill, Columbus, Ohio, 1970.

Tokheim, R.L.: *Digital Electronics*, McGraw-Hill, New York, 1979.

Wells, A.: *Audio Servicing: Theory and Practice*, McGraw-Hill, New York, 1980.

Appendix A
Electronic Frequency Spectrum

Table A-1
Frequency Ranges and Applications

Frequency or Wavelength*	Name	Applications
0 Hz	Steady direct current or voltage	DC motors, solenoids, relays, dc supply voltage for tubes, transistors, and IC Chips
16–16,000 Hz	Audio frequencies	60-Hz power, ac motors, audio amplifiers, microphones, loudspeakers, phonographs, tape recorders, high-fidelity equipment, public address systems, and intercoms
16–30 kHz	Ultrasonic frequencies or very low radio frequencies	Sound waves for ultrasonic cleaning, vibration testing, thickness gaging, flow detection, and sonar; electromagnetic waves for induction heating
30 kHz–30,000 MHz	Radio frequencies (see Appendix B)	Radio communications and broadcasting, including television, radio navigation, radio astronomy, satellites, and industrial, medical, scientific, and military radio
30,000–300,000 MHz or 1–0.1 cm	Extra-high frequencies	Experimental, weather radar, amateur, government
300,000–7600 Å	Infrared light rays	Heating, infrared photography
7600–3900 Å	Visible light rays	Color, illumination, photography
3900–320 Å	Ultraviolet rays	Sterilizing, deodorizing, medical
320–0.1 Å	x-rays	Thickness gages, inspection, medical
0.1–0.006 Å	Gamma rays	Radiation detection; more penetrating than hardest x-rays
Shortest of all electromagnetic waves	Cosmic rays	Exist in outer space; can penetrate 70 m of water or 1 m of lead

*Frequency and wavelength are inversely proportional to each other. The higher the frequency the shorter the wavelength, and vice versa.

AF and RF waves are generally considered in terms of frequency because the wavelength is so long. The exception is microwaves, which are often designated by wavelength because their frequencies are so high. Light waves, x-rays, and gamma rays are also generally considered by wavelength because their frequencies are so high. The units of wavelength are the micrometer, equal to 10^{-6} m; the nanometer, equal to 10^{-9} m; and the angstrom, Å, equal to 10^{-10} m.

The four main categories of electromagnetic radiation and their frequencies can be summarized as follows:

1. Radio frequency waves from 30 kHz to 300,000 MHz.
2. Heat waves or infrared rays from 1×10^{13} to 2.5×10^{14} Hz. *Infrared* means below the frequency of visible red light.
3. Visible light frequencies from about 2.5×10^{14} Hz for red up to 8×10^{14} Hz for blue and violet.
4. Ionizing radiation such as ultraviolet rays, x-rays, gamma rays, and cosmic rays from about 8×10^{14} Hz for ultraviolet light to above 5×10^{20} Hz for cosmic rays. *Ultraviolet* means above the frequencies of blue and violet visible light.

Note that radio waves have the lowest frequencies and therefore the longest wavelengths of any form of electromagnetic radiation.

The radio frequencies from 30 kHz to 300,000 MHz can be subdivided as follows:

1. Very low frequencies, or VLF band: below 30 kHz
2. Low frequencies, or LF band: 30 to 300 kHz
3. Medium frequencies, or MF band: 300 to 3000 kHz
4. High frequencies, or HF band: 3 to 30 MHz
5. Very high frequencies, or VHF band: 30 to 300 MHz
6. Ultra-high frequencies, or UHF band: 300 to 3000 MHz
7. Super-high frequencies, or SHF band: 3 to 30 GHz
8. Extra-high frequencies, or EHF band: 30 to 300 GHz

Note that each band has a 10:1 range in frequencies. In addition, each band is 10 times higher than the preceding lower band.

The SHF and EHF bands are listed in gigahertz (1 GHz = 1×10^9 Hz, or 1000 MHz). Microwaves are generally considered to be in the range of approximately 0.3 GHz and higher for wavelengths of 1 m or less.

Appendix B
FCC Frequency
Allocations

The main uses of the 30 kHz to 300,000 MHz frequency range are listed in Table B-1. AM radio broadcasting at 535 to 1605 kHz is in the MF band. FM radio broadcasting at 88 to 108 MHz is in the VHF band. Television broadcasting is in the VHF and UHF bands.

Television Channels See Table B-2. Each channel is 6 MHz wide for the FM audio signal and AM picture carrier signal, including the multiplexed 3.58-MHz color signal. Channels 2 to 6 and 7 to 13 are in the VHF band. Channels 14 to 83 are in the UHF band.

Citizens' Band (CB) Radio Channels In Table B-3 are listed the carrier frequencies for channels 1 through 40 for class D service.

Short-Wave Radio The bands listed in Table B-4 are used for international radio communications.

Amateur Radio Bands The amateur radio bands, together with permissible types of emission for the transmitter, can be found in the *ARRL Handbook*.

Satellites Frequency allocations for satellites include the following bands, in gigahertz: 3.7 to 42, 5.9 to 6.4, 10.7 to 13.2, 17.7 to 21.2, 30 to 31, and 36 to 41. Those bands are used for communications, meterology, and navigation.

Table B-1
Allocations from 30 kHz to 300,000 MHz

Band	Allocation	Remarks
30–535 kHz	Includes maritime communications and navigation, aeronautical radio navigation	Low and medium radio frequencies
535–1605 kHz	Standard radio broadcast band	AM broadcasting
1605 kHz– 30 MHz	Includes amateur radio, CB radio, loran, government radio, international shortwave broadcast, fixed and mobile communications, radio navigation, industrial, scientific, and medical radio	Amateur bands 3.5–4.0 MHz and 28–29.7 MHz; industrial, scientific, and medical band 26.95–27.54 MHz; citizen's band class D for voice is 26.965–27.045 MHz
30–50 MHz	Government and nongovernment, fixed and mobile	Includes police, fire, forestry, highway, and railroad services; VHF band starts at 30 MHz
50–54 MHz	Amateur	6-m band
54–72 MHz	Television broadcast channels 2 and 4	Also fixed and mobile services
72–76 MHz	Government and nongovernment services	Aeronautical marker beacon on 75 MHz
76–88 MHz	Television broadcast channels 5 and 6	Also fixed and mobile services
88–108 MHz	FM broadcast	Also available for facsimile broadcast; 88–92 MHz educational FM broadcast
108–122 MHz	Aeronautical navigation	Localizers, radio range, and airport control
122–174 MHz	Government and nongovernment, fixed and mobile, amateur broadcast	144–148 MHz amateur band
174–216 MHz	Television broadcast channels 7 to 13	Also fixed and mobile services
216–470 MHz	Amateur, government and nongovernment, fixed and mobile, aeronautical navigation, citizens' radio	Radio altimeter, glide path, and meteorological equipment; citizens' radio band 462.5–465 MHz; civil aviation 225–400 MHz; UHF band starts at 300 MHz
470–890 MHz	Television broadcasting	UHF television broadcast channels 14 to 83
890–3000 MHz	Aeronautical radio navigation, amateur broadcast, studio-transmitter relay	Radar bands 1300–1600 MHz
3000– 30,000 MHz	Government and nongovernment, amateur, radio navigation	Super-high frequencies (SHF); satellite communications
30,000– 300,000 MHZ	Experimental, government, amateur	Extra-high frequencies (EHF)

Table B-2
Television Channel Allocations

Channel Number	Frequency Band, MHz	Channel Number	Frequency Band, MHz
1*	—	42	638–644
2	54–60	43	644–650
3	60–66	44	650–656
4	66–72	45	656–662
5	76–82	46	662–668
6	82–88	47	668–674
		48	674–680
7	174–180	49	680–686
8	180–186	50	686–692
9	186–192	51	692–698
10	192–198	52	698–704
11	198–204	53	704–710
12	204–210	54	710–716
13	210–216	55	716–722
14	470–476	56	722–728
15	476–482	57	728–734
16	482–488	58	734–740
17	488–494	59	740–746
18	494–500	60	746–752
19	500–506	61	752–758
20	506–512	62	758–764
21	512–518	63	764–770
22	518–524	64	770–776
23	524–530	65	776–782
24	530–536	66	782–788
25	536–542	67	788–794
26	542–548	68	794–800
27	548–554	69	800–806
28	554–560	70	806–812
29	560–566	71	812–818
30	566–572	72	818–824
31	572–578	73	824–830
32	578–584	74	830–836
33	584–590	75	836–842
34	590–596	76	842–848
35	596–602	77	848–854
36	602–608	78	854–860
37	608–614	79	860–866
38	614–620	80	866–872
39	620–626	81	872–878
40	626–632	82	878–884
41	632–638	83	884–890

*The 44- to 50-MHz band was television channel 1 but is now assigned to other services.

Table B-3
Channels for Citizens' Band (CB) Radio‡

Channel	RF Carrier, MHz	Channel	RF Carrier, MHz
1	26.965	24	27.235
2	26.975	25	27.245
3	26.985	26	27.265
4	27.005	27	27.275
5	27.015	28	27.285
6	27.025	29	27.295
7	27.035	30	27.305
8	27.055	31	27.315
9*	27.065	32	27.325
10	27.075	33	27.335
11†	27.085	34	27.345
12	27.105	35	27.355
13	27.115	36	27.365
14	27.125	37	27.375
15	27.135	38	27.385
16	27.155	39	27.395
17	27.165	40	27.405
18	27.175		
19†	27.185		
20	27.205		
21	27.215		
22	27.225		
23	27.255		

*Channel 9 is for emergency communications only.

†Channels 11 and 19 are for auto traffic monitoring.

‡Channels 1 through 23 were the original CB channels. Channels 24 through 40 were added in 1976.

Table B-4
International Short-Wave Radio Allocations

Minor Bands, MHz	Major Bands, MHz
3.200–3.400	9.500–9.775
4.750–5.060	11.700–11.975
5.950–6.200	15.100–15.450
7.100–7.300	17.700–17.900
	21.450–21.750
	25.600–26.100

Appendix C
Standard Abbreviations

Abbreviations for Technical Terms

AF	Audio Frequency
AFC	Automatic Frequency Control
AFT	Automatic Fine Tuning
AGC	Automatic Gain Control
ALC	Automatic Limiting Control
AM	Amplitude Modulation
APC	Automatic Phase Control
AVC	Automatic Volume Control
BJT	Bipolar Junction Transistor
CB	Citizens' Band
D–A	Digital to Analog
DBS	Direct Broadcast Satellite
DIP	Dual In-line Package
DSB	Double Sidebands
DTL	Diode-Transistor Logic
EBS	Emergency Broadcast System
ECL	Emitter-Coupled Oscillator
ECO	Electron-Coupled Oscillator
FET	Field-Effect Transistor
FF	Flip-Flop
FM	Frequency-Shift Keying
FSK	Frequency-Shift Keying
IC	Integrated Circuit
IF	Intermediate Frequency
LCD	Liquid Crystal Display
LED	Light-Emitting Diode
LNA	Low-Noise Amplifier
LS	Loudspeaker
MOS	Metal-Oxide Semiconductor
MPX	Multiplex
MV	Multivibrator
OP AMP	Operational Amplifier
PA	Public Address (System), Power Amplifier
PAM	Pulse-Amplitude Modulation
PCM	Pulse-Code Modulation
PLL	Phase-Locked Loop
PM	Phase Modulation
PWM	Pulse-Width Modulation
RAM	Random Access Memory
RF	Radio Frequency
ROM	Read-Only Memory
SCR	Silicon Controlled Rectifier
S/N	Signal-to-Noise
SSB	Single Sideband
SWR	Standing-Wave Ratio
TE	Transverse Electric
TM	Transverse Magnetic
TTL or T²L	Transistor-Transistor Logic
TWT	Travelling Wave Tube
UHF	Ultra High Frequency
UJT	Unijunction Transistor
VCO	Voltage-Controlled Oscillator
VFO	Variable Frequency Oscillator
VHF	Very High Frequency
VR	Voltage Regulator

Abbreviations for Official Electronics Organizations

The abbreviations are given here in alphabetical order. The organizations include professional groups, government regulating agencies, and industry associations that promote standards.

ANSI   American National Standards Institute, 1430 Broadway, New York, NY 10018. Issues standards for graphic symbols in electricity and electronics.

ARRL   American Radio Relay League, Newington, CT 06111. For amateur radio operators.

ASEE   American Society for Engineering Education, One Dupont Circle, Washington, DC 20036. Professional group for engineering and technology educators.

EIA   Electronic Industries Association, 2001 Eye Street, NW, Washington, DC 20015. The main industry group for standards on receivers, semiconductor devices, and tubes. Formerly Radio, Electronics and Television Manufacturers Association (RETMA).

EIAJ   Electronic Industries Association of Japan. FCC Federal Communication Commission, Washington, DC 20554. The U.S. government agency that issues rules and regulations for transmitters, including radio and television broadcasting and citizens' band (CB) radio.

IEEE   Institute of Electrical and Electronics Engineers, 345 East 47 Street, New York, NY 10017. The professional group that provides standards for measurements, definitions, and graphic symbols diagrams in the electrical and electronics fields.

ISCET   International Society of Certified Electronics Technicians, 1715 Expo Lane, Indianapolis, IN 46224. Professional group for radio and television servicing.

JEDEC   Joint Electronic Devices Engineering Council. Professional group for standards on electron devices.

NAB   National Association of Broadcasters, 1771 N Street, NW, Washington, DC 20036.

NEMA   National Electrical Manufacturers Association, 155 East 44 Street, New York, NY 10017.

NESDA   National Electronic Service Dealers Association, 1715 Expo Lane, Indianapolis, IN 46224. Trade group.

NFPA   National Fire Protection Association, 470 Atlantic Avenue, Boston, MA 02210. Issues National Electrical Code (NEC).

SMPTE   Society of Motion Picture and Television Engineers, 862 Scarsdale Ave., Scarsdale, NY 10583. Professional group.

WARC   World Allocations Resources Committee, Geneva, Switzerland. International government group for frequency allocations in the radio spectrum.

Appendix D
Logarithms

A logarithm is an exponent. When a number is expressed as a power of 10, its exponent is the logarithm of the number to base 10. Logarithms to the base 10 are also called *common logarithms*.

Logarithms are useful as a short-cut method of multiplying and dividing, finding roots and powers, and simplifying graphical relationships. Some electronics formulas are expressed with logarithms. Graph paper, in which the ordinate (Y axis) is linear and the abscissa (X axis) is logarithmic, is used to show gain versus frequency. Since the frequency range used is often very broad, a linear frequency scale would be very long or the units would be very small. By using logarithmic values in the frequency scale, the entire range can be shown in very limited space.

Characteristic and Mantissa Since a logarithm is an exponent of 10, the logarithm of 10 is 1. That is, $10^1 = 10$. The logarithm of 100 is 2 ($10^2 = 100$). Other logarithms are shown below:

$$1000 = 10^3 \text{ or } 3 \text{ is the log of } 1000$$
$$10,000 = 10^4 \text{ or } 4 \text{ is the log of } 10,000$$
$$100,000 = 10^5 \text{ or } 5 \text{ is the log of } 100,000$$
$$1,000,000 = 10^6 \text{ or } 6 \text{ is the log of } 1,000,000$$

(Note that *log* is merely a short expression for logarithm.)

In each of the above cases, the log was a whole number since the exponents were whole numbers or exact powers of 10. However, all positive numbers have logarithms inasmuch as they can be expressed as some power of 10. The number 32, for example, is between 10^1 and 10^2, therefore its logarithm is somewhere between 1 and 2. In fact, the log of 32 is about 1.5 ($10^{1.5} \approx 32$). The logarithm of a number other than a whole number power of 10 is a decimal number. The number to the left of the decimal point is called the *characteristic* while the number to the right of the decimal point is called the *mantissa*. To find the logarithm of a number:

1. Find its characteristic. A decimal point follows the characteristic.
2. Find the mantissa and write it to the right of the decimal point.

The characteristic can be found by counting the number of digits to the left of the decimal point of the given number and subtracting 1. For example, the characteristic of 32 is 1; the characteristic of 320 is 2; the characteristic of 3.2 is 0.

A good way to visualize the characteristic is to write the number in standard scientific notation as a power of 10 with a coefficient between 1 and 10. Then the exponent of 10 is the characteristic. As examples, see table below.

The mantissa is usually found from printed tables called *log tables*. Only mantissas are given in such tables; the characteristic is found by the method given above. Log tables can often be found in math books and handbooks.

However, with the advent of the scientific electronic calculator the need for these tables has diminished.

An electronic calculator with scientific functions will display the entire logarithm, characteristic and mantissa, when the number is entered in the calculator and the log function used. Still, it is necessary to appreciate the meaning of logarithms, because logs are used in many electronic formulas.

Number	Scientific Notation	Characteristic of Logarithm
32	3.2×10^1	1
320	3.2×10^2	2
3290	3.2×10^3	3

Antilogarithms Sometimes it may be necessary to find the number whose logarithm is known. For example, to multiply two numbers, first find the logarithm (characteristic *and* mantissa) of each number. Then add the two logarithms. The number whose logarithm is equal to this sum, is the product of the two original numbers. The unknown number whose log is known, is called the *antilogarithm* or simply the *antilog*. For example, 100 is the antilog of 2; 1 is the antilog of 0 (remember $10^0 = 1$).

To find the antilog of a number:

1. Use a log table to determine the number associated with the mantissa.
2. Locate the decimal one more place than the characteristic of the log.

Suppose the antilog of 2.5172 is required. The characteristic is 2 and the mantissa is 5172. From a log table, the number whose mantissa is 5172 is 329 without a decimal point. Since the characteristic is 2, the antilog must have 3 places or the antilog is 329. If the original log were 3.5172, the antilog would be 3290. If the original log were 0.5172, the antilog would be 3.29.

Scientific notation can be useful for antilogs also. Write the digits corresponding to the mantissa, but put the number in scientific notation with a power of 10 equal to the characteristic. For the same example of the antilog of 2.5172, the number N is 3.29×10^2, which equals 329.

Again the scientific electronic calculator will produce the entire antilog. Thus by entering the entire logarithm in the calculator and pressing the antilog key (often denoted as 10^x) the entire antilog will be displayed.

Logarithms of Numbers Less Than 1 It is possible to find logarithms of numbers less than 1 by recalling the definition of a logarithm.

$$10^x = N$$
$$\text{or } \log N = x$$

If N is less than 1, the expression 10^x must be a fraction. This would be true if x were a negative number since $10^{-x} = 1/10^x$.

Again, scientific notation will make this clearer.

0.1	$= 1 \times 10^{-1}$	-1 is the log of 0.1
0.01	$= 1 \times 10^{-2}$	-2 is the log of 0.01
0.001	$= 1 \times 10^{-3}$	-3 is the log of 0.001

Similarly,

0.32	$= 3.2 \times 10^{-1}$	characteristic $= -1$
0.032	$= 3.2 \times 10^{-2}$	characteristic $= -2$
0.0032	$= 3.2 \times 10^{-3}$	characteristic $= -3$

However, when the mantissa is added to the characteristic, the result is a negative number that is the *difference* between the characteristic and the mantissa.

	Characteristic		Mantissa
Log 0.32 =	-1	$+$	0.5051
$= -0.4949$			

Very often the logs of numbers less than 1 are written in the form $A.M. - 10$ where $A - 10$ is the characteristic and M is the mantissa given in the log tables. Thus Log $0.32 = 9.5051 - 10$. This, of course, is the same as -0.4949 as given above.

Appendix E
Color Codes

Color codes are standardized by the Electronic Industries Association (EIA). Included are codes for chassis wiring connections and the values for carbon resistors and mica and ceramic capacitors.

Chassis Wiring The color of the wire indicates the function. Either a solid color or helical striping on white insulation can be used.

Red	V^+ from the dc power supply
Blue	Plate (amplifier tube), collector (transistor), or drain (FET)
Green	Control grid (tube), base (transistor), or gate (FET)
Yellow	Cathode (amplifier tube), emitter (transistor), or source (FET)
Orange	Screen grid (vacuum tube)
Brown	Heater (vacuum tube)
Black	Chassis ground
White	Return to AVC or AGC bias line

In addition, blue wire is used for the high side of antenna-input connections.

IF Transformers Used for interstage coupling in IF amplifiers. The terminals may be color-coded as follows:

Blue	High side of the primary to the amplifier plate, collector, or drain electrode
Red	Low side of the primary returning to V^+
Green	High side of the secondary for the signal output
White	Low side of the secondary
Violet	Can be used for an additional secondary output.

AF Transformers The AF transformers used for interstage coupling and power output in audio amplifiers often have color-coded leads as follows:

Blue	Primary to the amplifier plate, collector, or drain electrode. It indicates the end of a winding.
Red	Connection to V^+. Center tap of a push-pull winding.

Brown	Lead opposite to the blue lead for push-pull. Start of a primary winding.
Green	High side of the secondary for the signal output. The end of the secondary winding.
Black	Ground return.
Yellow	Center tap on the secondary.

Stereo Audio Channels The wiring of stereo audio channels is color-coded as follows:

Left Channel		Right Channel	
White:	High side	Red:	High side
Blue:	Low side	Green:	Low side

Power Transformers The leads of power transformers are color-coded as follows:

Start and end leads of the primary without a tap is black for both leads.
Tapped primary
  Common: black
  Tap: black and yellow
  End: black and red
High-voltage secondary is red for both start and end of the winding.
Center tap is red and yellow.
Low voltage secondary is green and brown.

Carbon Resistors For ratings of 2 W or less, carbon resistors are color-coded either with bands or the body-end-dot system; see Fig. E-1 and Table E-1. The colors are listed in Table E-2. They apply to resistor values in ohms and capacitors in picofarads.

The preferred values in Table E-3 also apply to resistors and capacitors. Only the basic value is listed, but multiples are available. For instance, resistors are 47, 470, 4700, and 47,000 Ω. There are more intermediate resistance values for lower tolerances. As an example, the next higher value above 47 can be 68, 56, or 51, depending on the tolerance.

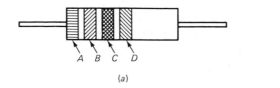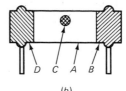

(a)

(b)

Fig. E-1. Resistor color codes. (a) Color stripes on *R* with axial leads. (b) Body-end-dot colors on *R* with radial leads.

Table E-1
Color codes for carbon resistors

Axial leads	Color	Radial leads
B and A	First significant figure	Body A
B and B	Second significant figure	End B
B and C	Decimal multiplier	Dot C
B and D	Tolerance	End D

Notes: Band A is double width for wire-wound resistors with axial leads. The body-end-dot system is a discontinued standard, but it may still be found on some old resistors. When resistors have color stripes and axial leads, body color is not used for color-coded value. Film resistors have five stripes; the fourth stripe is the multiplier and the fifth is tolerance.

Table E-2
Color Values for Resistor and Capacitor Codes

Color	Significant figure	Decimal multiplier	Tolerance,* %
Black	0	1	20
Brown	1	10	1
Red	2	10^2	2
Orange	3	10^3	3
Yellow	4	10^4	4
Green	5	10^5	5
Blue	6	10^6	6
Violet	7	10^7	7
Gray	8	10^8	8
White	9	10^9	9
Gold		0.1	5
Silver		0.01	10
No color			20

*Tolerance colors other than gold and silver for capacitors only.

Table E-3
Preferred Values for Resistors and Capacitors*

20% Tolerance	10% Tolerance	5% Tolerance
10	10	10
		11
	12	12
		13
15	15	15
		16
	18	18
		20
22	22	22
		24
	27	27
		30
33	33	33
		36
	39	39
		43
47	47	47
		51
	56	56
		62
68	68	68
		75
	82	82
		91
100	100	100

*Numbers and decimal multiples for ohms or pF.

Mica Capacitors See Figure E-2 for color-coding of mica capacitors. The first dot on a mica capacitor is white to indicate the EIA six-dot code. It may also be black for a military code. In either case, read the capacitance in picofarads from the next three color dots. Table E-2 lists tolerance colors for the fifth dot. The sixth dot indicates classes A to E of leakage and temperature coefficients. The maximum rating for dc working voltage is generally 500 V.

If a color is used in the first dot the capacitor is marked with the old EIA code. Use that dot and the next two dots for significant figures and the fourth dot as the decimal multiplier.

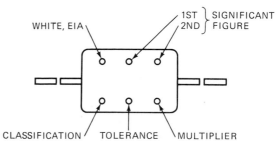

Fig. E-2. EIA six-dot code for mica capacitors.

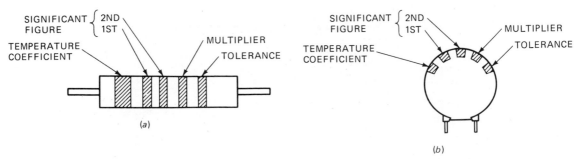

Fig. E-3. Color code for ceramic capacitors. (a) Tubular with axial leads. (b) Disk with radial leads.

Ceramic Capacitors Ceramic capacitors have stripes or dots with three or five colors. The construction may be tubular with axial leads or a disk with radial leads (Fig. E-3). When there are five colors, the first and last indicate the temperature coefficient and tolerance, as listed in Table E-4. The middle three colors give the picofarad value.

Ceramic disk capacitors often have the values printed on them. A value more than 1 is in picofarads, such as 47 pF. A value less than 1, say, 0.002, is in microfarads. The letter R may be used to indicate a decimal point; for example, $2R2 = 2.2$.

Table E-4
Color Code for Ceramic Capacitors

Color	Decimal multiplier	Tolerance		Temperature Coefficient ppm per °C
		Above 10 pF, %	Below 10pF, in pF	
Black	1	20	2.0	0
Brown	10	1		−30
Red	100	2		−80
Orange	1000			−150
Yellow				−220
Green		5	0.5	−330
Blue				−470
Violet				−750
Gray	0.01		0.25	30
White	0.1	10	1.0	500

Appendix F
Schematic Symbols

Schematic symbols are organized in separate tables generally according to device types.

Table F-1 General Symbols
    F-2 Vacuum Tubes
    F-3 Bipolar Junction Transistors
    F-4 Field-Effect Transistors (FET)

F-5 Semiconductor Diodes
F-6 Thyristors
F-7 Protective Devices for Semiconductors

Table F-1
General Symbols

Device	Symbol	Device	Symbol
AC voltage		Coil or inductance, Air-core	
		Iron-core	
Antenna, General		Variable	
Dipole		Powdered iron or ferrite slug	
Loop		Conductor General	
		Connection	
Battery, cell or dc voltage Long line positive		No connection	
		Current source	
Capacitor, General, fixed. Curved electrode is outside foil, negative or low-potential side		Cystal, piezoelectric	
		Fuse	
Variable		Ground, earth or metal frame Chassis or common return connected to one side of voltage source	
Ganged		Chassis or common return not connected to voltage source	
		Common return	

Device	Symbol	Device	Symbol
Jack		Resistor, Fixed	
Plug for jack	TIP SLEEVE	Tapped	
		Variable	
Key telegraph		Switch, (SPST)	
		(SPDT)	
		2-pole (DPDT)	
Loudspeaker, general			
Phones or headset		3-pole 3-circuit wafer	
Magnet, Permanent	PM	Shielding	
Electromagnet		Shielded conductor	
Microphone		Thermistor, general	T
Meters; letter or symbol to indicate range or function	A mA V	Thermocouple	
Motor	M	Transformer Air-core	
Neon bulb		iron-core	
Relay, coil		Autotransformer	
Contacts Normally open (NO) Normally closed (NC)	or	Link coupling	

Table F-2
Vacuum Tubes

Tube	Symbol	Electrodes	Tube	Symbol	Electrodes
Diode		P = Plate K = Cathode	Tetrode		P = Plate G_2 = Screen grid G_1 = Control grid K = Cathode
Triode		P = Plate G_1 = Control grid K = Cathode	Pentode		P = Plate G_3 = Suppressor grid G_2 = Screen grid G_1 = Control grid K = Cathode

Table F-3
Bipolar Junction Transistors*

Type	Symbol	Electrodes	Notes
NPN		C = Collector B = Base E = Emitter; hole current out from	Needs $+V_C$ reverse voltage for collector; forward bias V_{BE} of 0.6 V for silicon or 0.2 V for germanium
PNP		C = Collector B = Base E = Emitter; hole current into base	Needs $-V_C$ reverse voltage for collector; forward bias V_{BE} of 0.2 V for germanium or 0.6 V for silicon

*Type numbers for bipolar junction transistors start with 2 N, as in 2N1482, for two semiconductor junctions in the EIA system. Japanese transistors are marked 2SA for small-signal PNP, 2SB for power PNP, 2SC for small-signal NPN, and 2SD for power NPN types.

Table F-4
Field-Effect Transistors* (FET)

Type	Symbol	Electrodes	Notes
JFET		D = Drain G = Gate S = Source	Junction-type FET; arrow pointing to channel indicates P-gate and N-channel; reverse bias at junction
IGFET or MOSFET N-channel		D = Drain G = Gate S = Source	Insulated-gate; depletion or depletion-enchancement type; B is bulk or substrate connected internally to source
IGFET or MOSFET P-channel		D = Drain G = Gate S = Source	Arrow pointing away from channel indicates P-channel
IGFET or MOSFET, N-channel, enhancement		D = Drain G = Gate S = Source	Broken lines for channel show enhancement type
Dual-gate IGFET or MOSFET, N-channel		D = Drain G_2 = Gate 2 G_1 = Gate 1 S = Source	Either or both gates control amount of drain current I_D

*Classified by EIA as type A, B, or C. For an N-channel: depletion type A takes negative gate bias for a middle value of I_D; depletion-enhancement type B can operate with zero gate bias; enhancement type C needs positive gate bias.

Table F-5
Semiconductor Diodes*

Type	Symbol	Applications
Rectifier	A = anode K = cathode A ——▷|—— K	Conducts forward current with positive anode voltage
Zener	A ——▷|—— K	Voltage regulator; takes reverse voltage
Varactor	A ——▷|✦—— K	Capacitive diode; C varies with amount of reverse voltage; used for electronic tuning
Tunnel	A ——▷|—— K	Microwave diode; Esaki diode; same symbol for Schottky diode and Gunn diode
Photoconductive	A ——▷|—— K	Takes reverse voltage; high dark-resistance decreases with light input
Photovoltaic	A ——▷|—— K	No voltage applied; light input generates voltage output; solar cell
Photoemissive	A ——▷|—— K	Light emitting diode (LED) produces red, green, or yellow light output; takes forward voltage

*Type numbers for diodes start with 1N, for one semiconductor junction. An example is 1N2864.

Table F-6
Types of Thyristors

Name	Symbol	Electrodes	Notes
Silicon-controlled rectifier (SCR)	K, G, A	K = Cathode G = Gate A = Anode	Unidirectional with positive anode voltage; G can turn SCR on but not off
Triac	MT1, G, MT2	MT1 = Main terminal 1 G = Gate MT2 = Main terminal 2	Bidirectional with gate; G can turn triac on for either polarity, but not off
Diac	MT1, MT2	MT1 = Main terminal 1 MT2 = Main terminal 2	Bidirectional trigger without gate
Unijunction transistor (UJT)	B2, E, B1	B2 = Base 2 E = Emitter B1 = Base 1	Base has standoff voltage from dc dupply; V_E applied to start conduction

Type	Symbol	Notes
Thermistor		Resistance decreases with temperature; used as a series component
Voltage-dependent resistor (VDR)		Resistance decreases with V continuously; used as a series component
Varistor		Resistance decreases abruptly at specific voltage; used as a shunt component to shirt-circuit voltage transients

Appendix G
Digital
Logic Symbols

Table G-1

Name	Symbol	Notes	Name	Symbol	Notes
AND gate	A, B → Y	$AB = Y$	J-K flip-flop	J, CP, K → FF → Q, $\bar{Q}$	CP is clock pulse
OR gate	A, B → Y	$A + B = Y$	Clock pulse		Falling edge
Inverter	A → $\bar{A}$	$A = A$	Trigger		Rising edge
NAND gate	A, B → Y	$\overline{AB} = Y$			
NOR gate	A, B → Y	$\overline{A + B} = Y$	Operational amplifier, op amp	$-$, $+$ → V_o	For digital-analog circuits

Answers to Self-examinations

CHAPTER 1
1. (b) 2. (a) 3. (b) 4. (d) 5. (a) 6. (b) 7. (a) 8. (a) 9. (d) 10. (c)

CHAPTER 2
1. 6 V 2. 72 3. 180° 4. 160 μA 5. 8 mA 6. V_C = 8 V; I_C = 3 mA; I_B = 60 μA 7. v_C = 7.2 V p-p; i_C = 3.6 mA p-p; I_B = 80 μA p-p 8. 10 W 9. 43 W 10. 9 V

CHAPTER 3
1. (d) 2. (b) 3. (a) 4. (c) 5. (c) 6. (c) 7. (b) 8. (d) 9. (b) 10. (a) 11. (b) 12. (a)

CHAPTER 4
1. (d) 2. (a) 3. (b) 4. (a) 5. (b) 6. (a) 7. (b) 8. (b) 9. (c) 10. (c) 11. (c) 12. (c)

CHAPTER 5
1. 3 2. 10 3. 13 4. 6 5. 20 6. −3 7. −20 8. 12 9. 2 10. 2 11. 18 12. 26

CHAPTER 6
1. (c) 2. (f) 3. (j) 4. (i) 5. (b) 6. (a) 7. (d) 8. (h) 9. (g) 10. (e)

CHAPTER 7
1. T 2. T 3. F 4. F 5. T 6. F 7. F 8. T 9. T 10. F 11. F 12. T 13. T 14. T 15. F 16. T

CHAPTER 8
1. 0 2. 2 3. OR 4. All 5. $\overline{A + B} = Y$ 6. NAND 7. XOR 8. Clock 9. Flip-flop 10. HIGH 11. Astable 12. Four 13. RAM 14. Multiplexer 15. Converters 16. LED 17. 8 18. CMOS

CHAPTER 9
1. Linear 2. Transistor 3. More 4. Silicon dioxide 5. CMOS 6. TTL 7. Astable 8. T²L 9. 10 10. NPN 11. IGFET 12. HIGH 13. SSI 14. LOW 15. Bipolar transistors 16. Oscilloscope 17. Aluminum 18. CMOS 19. MOSFET 20. C

CHAPTER 10
1. (b) 2. (b) 3. (c) 4. (b) 5. (d) 6. (a) 7. (d) 8. (c) 9. (b) 10. (d)

CHAPTER 11
1. (b) 2. (d) 3. (a) 4. (c) 5. (a) 6. (c) 7. (d) 8. (a) 9. (c) 10. (d) 11. (b) 12. (c) 13. (d) 14. (a) 15. (c) 16. (d)

CHAPTER 12
1. (a) 2. (c) 3. (b) 4. (c) 5. (c) 6. (c) 7. (c) 8. (c) 9. (c) 10. (a) 11. (b) 12. (c)

CHAPTER 13
1. (k) 2. (h) 3. (j) 4. (p) 5. (o) 6. (r) 7. (c) 8. (a) 9. (b) 10. (m) 11. (d) 12. (q) 13. (f) 14. (s) 15. (i) 16. (l) 17. (e) 18. (n)

CHAPTER 14
1. (c) 2. (d) 3. (a) 4. (b) 5. (d) 6. (d) 7. (c) 8. (b) 9. (d) 10. (c) 11. (b) 12. (b)

CHAPTER 15
1. (d) 2. (f) 3. (i) 4. (j) 5. (h) 6. (a) 7. (b) 8. (c) 9. (g) 10. (l) 11. (k) 12. (e) 13. (n) 14. (o) 15. (m)

CHAPTER 16
1. More 2. Less 3. 10 μF 4. Negative 5. Negative 6. Series 7. $D2$ 8. Negative 9. Low 10. Decrease

CHAPTER 17

1. T 2. F 3. T 4. T 5. T 6. F 7. F 8. T
9. T 10. F 11. T 12. F 13. F 14. T 15. T

CHAPTER 18

1. (a) 2. (b) 3. (b) 4. (d) 5. (c) 6. (c)
7. (d) 8. (c) 9. (d) 10. (b)

Answers to Odd-numbered Problems

CHAPTER 1
1. a. 7 mA b. 31.8 Hz c. 3.98 MHz 3. a. 3.33 Ω b. 500 Ω c. 500 Ω 5. R_1 = 3.7 kΩ; R_2 = 0.8 kΩ

CHAPTER 2
1. 20 W 3. 4.375 W 5. 32.5 percent 7. 60 9. 66.7

CHAPTER 3
1. 12,000 3. 8 μF 5. 1 mV 7. a. 51 b. 19,608 Hz

CHAPTER 4
1. 14.125 and 0.035 ft 3. Eight 5. 1:6.5 7. I_{C_1} = 1 mA; I_{C_2} = 6 mA 9. 1592, 796, and 398 Ω

CHAPTER 5
1. a. 20 dB gain b. 40 dB gain c. -6 dB loss d. -3 dB loss 3. 42.9 dB 5. 9 dB 7. a. 20 mW b. ¼ mW c. 100 mV d. 0.3 mW 9. 70 dB

CHAPTER 6
1. 0.13 percent 3. 2.4 percent 5. 26.5 Ω and 13.25 Ω

CHAPTER 7
1. 10 A 3. 0.6 5. a. 12 V b. 12.7 V 7. 10.7 V 9. a. 62° b. 75°

CHAPTER 8
1. 59 3. 11001111_2 5. $C\bar{B}A + CB\bar{A} = Y$ 7. a. 3 b. 8 c. 111_2 or 7

CHAPTER 9
1. 8, 5, 4, and 1 3. Logic 1 5. 120 Ω 7. a. 10_2 b. LOW c. HIGH

CHAPTER 10
1. a. 0.3 MHz b. 3.67:1 3. a. 40 b. 100 5. a. 10 MHz b. 104 c. 33 d. 303 kHz 7. 0.1

CHAPTER 11
1. 765.7 kHz 3. 0.1 μH 5. 100 Ω

CHAPTER 12
1. 60 percent 3. 2.5 kW 5. ±30 Hz 7. 27 percent 9. a. 5 b. ±32 kHz 11. 91.902 MHz

CHAPTER 13
1. a. 300 m b. 20.7 m c. 3 m d. 0.375 m 3. a. 7.83 ft b. 4.7 ft c. 2.64 ft d. 0.78 ft 5. 15.5 7. 12 dB 9. 434 Ω

CHAPTER 14
1. 5, 2.5, and 0.5 cm 3. 1.875 cm 5. -196 dB

CHAPTER 15
1. a. 0.6 mA b. 29.5 Ω 3. a. 3.3 mA b. 0.007 Ω 5. $Q5$ = 0.6 V; $Q6$ = -0.6 V 7. 750 μH 9. a. 0.9 mH b. 90 pF 11. a. 2.2 percent b. 1.9 percent

CHAPTER 16
1. a. 0.1 s b. 0.0001 s 3. 2 mA 5. a. 1 mA b. 2 mA

CHAPTER 17
1. 98.7 to 118.7 MHz 3. a. 68 μs b. 76 μs 5. 0.075 s 7. 0 to 15,000 Hz

CHAPTER 18
1. 100 V 3. a. 30 Hz b. 60 Hz c. 7875 Hz d. 15,750 Hz 5. a. 6 V b. 3 cm

Index